MATHEMATICAL ANALYSIS (Ⅰ)

数学分析(Ⅰ)

Li Weimin

Shanghai Jiao Tong University

李为民　编

上海交通大學出版社

内 容 提 要

本书是为贯彻教育部教学改革精神，实施全英语授课需要而编写，此书书稿在实际教学中已获得广泛好评。其内容包括：实数系统和函数；序列极限；函数极限及连续性；导数和微分；中值定理和导数的应用；不定积分；定积分；定积分的应用；微分方程初步。

本书可作为大学数学系及要求较高的专业的本科生教材，也可作为大学教师教学用书或教学参考书。

图书在版编目(CIP)数据

数学分析.1/李为民编.—上海：上海交通大学出版社，2007
(2018重印)
ISBN 978-7-313-04888-2

Ⅰ.数...　Ⅱ.李...　Ⅲ.数学分析—研究生—教材　Ⅳ.O17

中国版本图书馆CIP数据核字(2007)第109180号

数学分析(1)

李为民　主编

上海交通大学出版社出版发行
(上海市番禺路951号　邮政编码：200030)
电话：64071208　出版人：谈　毅
当纳利(上海)信息技术有限公司 印刷　全国新华书店经销
开本：880mm×1230mm　1/32　印张：16.375　字数：416千字
2007年8月第1版　2018年9月第3次印刷
ISBN 978-7-313-04888-2/O·205　定价：48.00元

Preface

To cultivate the science and technique personnel of high quality for the 21th century, great efforts have been made to the teaching reform in undergraduate courses and graduate courses in such aspects as teaching contents, teaching methods, teaching means and the course features, etc. , This is of great immediate significance to carrying quality-education forward to the full. As one of the main reform orientations, English teaching or bilingual teaching in mathematics courses such as mathematical analysis has shown its promising prospect. The enforcement of teaching material construction is a current impending mission in implementation along this line. The primary textbooks of mathematical analysis in western countries are somewhat different in style, structure and layout as compared with that taught in our universities. The former tend to develop the analysis theory in the setting of general metric space as well as in Euclidean space. The primary goal of writing this book is to match the content with the level accessible to undergraduate students in China.

Mathematical analysis is a fundamental subject facing all branches of science which needs mathematics. It has its beginnings in the rigorous formulation of calculus. Though the preliminaries of mathematical analysis may date back hundred years ago, it remains a classic study and a thorough treatment of the fundamentals of calculus. As foundation of modern mathematics, mathematical analysis is endowed with features of rigorous logicality and precise description.

Mathematical analysis is the branch of mathematics most explicitly concerned with the notion of a limit, either the limit of a sequence or the limit of a function. This subject is usually studied in the context of real numbers. However, it can also be defined and studied in any space of mathematical objects that is equipped with a definition of "closeness"-a topological space, or more specifically "distance"-a metric space.

This book is intended to display the structure of analysis as a subject in its own right. The main objective of the text is to introduce students to fundamental concepts and standard theorems of analysis and to develop analytical techniques for attacking problems that arise in mathematical theory and applications of mathematics. Due to restriction of academic level and lack of experience, there may be mistakes and neglects in this book. All comments and suggestions are heartily welcome.

The publication of this book benefited from the financial support of Shanghai Jiao Tong University Office of Academic Affairs, which I appreciate greatly. It is also pleasure to record thanks to Professor Han Zhengzhi, Editors Chen Kejian and Dai Baicheng of Shanghai Jiao Tong University Press for their valuable comments and suggestions. Special thanks are due Editor Sun Qikun who carefully read the entire manuscript and made technical modifications which led to an improved layout of this book.

Contents

Chapter 1 Real number system and functions

Function is the most fundamental research object of mathematical analysis, and functions and other concepts studied in our subject are based on real numbers in some way, so we begin our study of analysis with a discussion of the real number system and functions.

§ 1.1 Real number system

Most applications of mathematics use real numbers. For purposes of such applications, it suffices to think of a real number as a decimal. A *rational* number is one that may be written as a finite or infinite repeating decimal, such as

$$2, -\frac{7}{4} = -1.75, 2.2689, \frac{20}{3} = 6.666\ldots.$$

An *irrational* number has an infinite decimal representation whose digits form no repeating pattern, such as

$$\sqrt{3} = 1.732\,050\,808\ldots, \pi = 3.141\,592\,653\,5\ldots.$$

The rational numbers and irrational numbers together constitutes the *real numbers* (*real number system*).

We have four infinite sets of familiar objects, in increasing order of complication:

$\mathbb{N}$: *the natural numbers* are defined as the set $\{1, 2, \ldots, n, \ldots\}$.

$\mathbb{Z}$: *the integers* are defined as the set $\{0, \pm 1, \pm 2, \ldots,$

$\pm n, \ldots \}$.

$\mathbb{Q}$: *the rational numbers* are defined as the set $\{p/q: p, q \in \mathbb{Z}, q \neq 0\}$.

$\mathbb{R}$: *the set of real numbers* (or *the reals*) is composed of the rational numbers and the irrational numbers.

Remark (1) We have natural conclusions $\mathbb{N} \subset \mathbb{Z} \subset \mathbb{Q} \subset \mathbb{R}$, where each inclusion is proper;

(2) The irrational number set is $\mathbb{R} \backslash \mathbb{Q}$.

The real number line are often presented geometrically as points on a line (called the *real line* or the *real axis*). A point is selected to represent 0 and another to represent 1, as shown in Figure 1 - 1. This choice determines the scale. Under an appropriate set of axioms for Euclidean geometry, each point on the real line corresponds to one and only one real number and, conversely, each real number is represented by one and only one point on the line. It is customary to refer to the *point* x rather than the point representing the real number x.

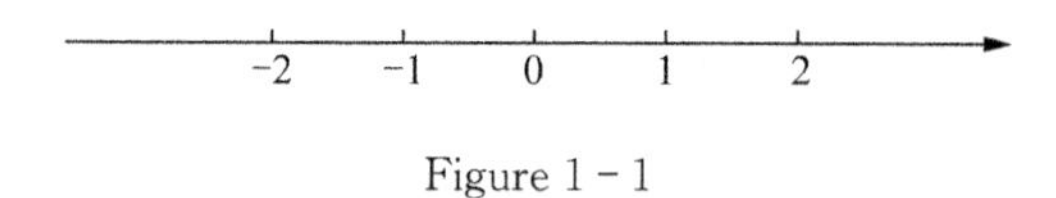

Figure 1 - 1

Geometrically, the inequality $x \leqslant b$ means that either x equals b or x lies to the left of b on the number line. The set of real numbers x that satisfy the double inequality $a \leqslant x \leqslant b$ corresponds to the line segment between a and b, including the endpoints. This set is sometimes denoted by $[a, b]$ and is called the closed interval from a to b. If a and b are removed from the set, the set is written as (a, b) and is called the open interval from a to b. The notation $(a, b]$ and $[a, b)$ etc. should be understood in a similar way.

Theorem 1.1.1 Given real number a and b such that $a \leqslant b + \varepsilon$ for every $\varepsilon > 0$. Then $a \leqslant b$.

Proof If $b < a$, take $\varepsilon = (a-b)/2$. Then

$$b+\varepsilon = b+\frac{a-b}{2} = \frac{a+b}{2} < \frac{a+a}{2} = a,$$

which yields a contradiction. □

Definition Let $x_0 \in \mathbb{R}$. If $x_0 \in (a, b)$, then (a, b) is called a *neighborhood* of x_0, denoted by $U(x_0)$, and $(a, b) \setminus \{x_0\}$ is called a *free-center neighborhood* of x_0, denoted by $U^o(x_0)$. In particular, if $\delta > 0$, then $(x_0 - \delta, x_0 + \delta)$ is called a δ-neighborhood of x_0, denoted by $U(x_0, \delta)$, and $(x_0 - \delta, x_0 + \delta) \setminus \{x_0\}$ is called a *free-center δ-neighborhood* of x_0, denoted by $U^o(x_0, \delta)$ (δ may be called the *radius of the neighborhood*), i. e.

$$U(x_0, \delta) = \{x \mid |x - x_0| < \delta\},$$
$$U^o(x_0, \delta) = \{x \mid 0 < |x - x_0| < \delta\}.$$

Properties of $\mathbb{R}$ We summarize the following properties of $\mathbb{R}$ that we work with.

Addition We can add and subtract real numbers exactly as we expect, and the usual rules of arithmetic hold-such results as $x+y = y+x$.

Multiplication In the same way, multiplication and division behave as we expect, and interact with addition and subtraction in the usual way. So we have rules such as $a(b+c) = ab+ac$. Note that we can divide by any number except 0. We make no attempt to make sense of $a/0$, even in the case when $a = 0$, so for us $0/0$ is meaningless. Formally these two properties say that $\mathbb{R}$ constructs a field algebraically, although it is not essential at this stage to know the terminology.

Order As well as the algebraic properties, $\mathbb{R}$ has an ordering on it, usually written as "$a > 0$" or "$\geqslant$". There are three parts to the property:

(1) **Trichotomy** For any $a \in \mathbb{R}$, exactly one of $a > 0$, $a = 0$ or $a < 0$ holds, where we write $a < 0$ instead of the formally correct $0 > a$; in words, we are simply saying that a number is either positive, negative or zero.

(2) **Addition** The order behaves as expected with respect to addition: if $a > 0$ and $b > 0$ then $a + b > 0$; i. e. the sum of positives is positive.

(3) **Multiplication** The order behaves as expected with respect to multiplication: if $a > 0$ and $b > 0$ then $ab > 0$; i. e. the product of positives is positive.

Now we extend the real number system by adjoining two "ideal points" $+\infty$ and $-\infty$.

The symbols $+\infty$ ("plus infinity") and $-\infty$ ("minus infinity") do not represent actual real numbers. Rather, they indicate that the corresponding line segment extends infinitely far to the right or left. The symbol ∞ ("infinity") usually designates either $+\infty$ or $-\infty$. An inequality that describes such an infinite interval may be written as $[a, +\infty)$, $(-\infty, a)$, etc.

Definition By the extended real number system $\mathbb{R}^*$ we shall mean the set of real numbers $\mathbb{R}$ with two symbols $+\infty$ and $-\infty$ which satisfy the following properties:

(1) If $x \in \mathbb{R}$, then we have $x + (+\infty) = +\infty$, $x + (-\infty) = -\infty$, $x - (+\infty) = -\infty$, $x - (-\infty) = +\infty$, $\dfrac{x}{+\infty} = 0$, $\dfrac{x}{-\infty} = 0$.

(2) If $x > 0$, then we have $x(+\infty) = +\infty$, $x(-\infty) = -\infty$.

(3) If $x < 0$, then we have $x(+\infty) = -\infty$, $x(-\infty) = +\infty$.

(4) $(+\infty) + (+\infty) = (+\infty)(+\infty) = (-\infty)(-\infty) = +\infty$, $(-\infty) + (-\infty) = (+\infty)(-\infty) = (-\infty)(+\infty) = -\infty$.

(5) If $x \in \mathbb{R}$, then we have $-\infty < x < +\infty$.

Note (1) As defined above, we denote $\mathbb{R} = (-\infty, +\infty)$, the set of real numbers, and $\mathbb{R}^* = [-\infty, +\infty]$, the set of extended real numbers. The points in $\mathbb{R}$ are said to be *finite* to distinguish them from the infinite points $-\infty$ and $+\infty$.

(2) For some of the later work concerned with limits, it is also convenient to introduce the terminology: Every open interval $(a, +\infty)$ is called a *neighborhood* of $+\infty$; every open interval $(-\infty, a)$ is called a *neighborhood* of $-\infty$.

Note that we write $a \geqslant 0$ if either $a > 0$ or $a = 0$. More generally, we write $a > b$ whenever $a - b > 0$.

Completion The set $\mathbb{R}$ has an additional property, which in contrast is much more mysterious-it is complete. It is this property that distinguishes it from $\mathbb{Q}$. Its effect is that there are always "enough" numbers to do what we want. Thus there are enough to solve any algebraic equation, even those like $x^2 = 2$ which can't be solved in $\mathbb{Q}$. In fact there are (uncountably many) more-all the numbers like π, certainly not rational, but in fact not even an algebraic number, are also in $\mathbb{R}$.

Definition Let $S \subseteq \mathbb{R}$ $(S \neq \varnothing)$.

(1) If there exists $t \in \mathbb{R}$ such that $x \leqslant t$ for any $x \in S$, then S is said to be *bounded above* and t is called an *upper bound* of S.

(2) Let t be an upper bound of S. If $t \leqslant d$ for any upper bound d of S, then t is called *the least upper bound* of S, which is denoted as $t = \text{l.u.b. } S$.

For example, let $S = \left\{ -\frac{1}{n} \middle| n = 1, 2, \ldots \right\}$. Then for any $d \in [0, +\infty)$, d is an upper bound of S, and l. u. b. $S = 0 \notin S$. Let $S = (0, 1]$. Then for any $d \in [1, +\infty)$, d is an upper bounds of S, and l.u.b. $S = 1 \in S$.

The completeness axiom Let $S \subseteq \mathbb{R}$ $(S \neq \varnothing)$. If S is bounded above, then there exists the least upper bound of S.

Definition The least upper bound of a number set S is also called *the supremum of* S, denoted as sup S.

By the definition of the supremum, it is easy to check the following:

Remark Let $S \subseteq \mathbb{R}$ $(S \neq \varnothing)$.

(1) If sup S exists, it is unique;

(2) The following two statements are equivalent:

ⓘ $t = \sup S$;

ⓘⓘ for any $x \in S$, $x \leqslant t$, and for any $a < t$ there exists $x \in S$ such that $x > a$.

Example Let $S = \left\{ \frac{n}{n+1} \middle| n = 1, 2, 3, \ldots \right\}$. Then $\sup S = 1$.

Proof Clearly, for any $x \in S$, $x \leqslant 1$.

Now, let $a < 1$. Take $x = \frac{n}{n+1}$, where $n = \left[\frac{a}{1-a}\right] + 1$. Then $x \in S$ with $x > a$. Thus $\sup S = 1$. □

Note If a number set S has no upper bound, denote $\sup S = +\infty$.

Definition Let $S \subseteq \mathbb{R}$ $(S \neq \varnothing)$.

(1) If there exists $b \in \mathbb{R}$ such that $x \geqslant b$ for any $x \in S$, then S is said to be *bounded below*, and b is called a *lower bound* of S.

(2) Let b be a lower bound of S. If for any lower bound d of S, $b \geqslant d$, then b is called *the greatest lower bound* of S, which is denoted as $b =$ g. l. b. S.

For example, let $S = \left\{ \frac{1}{n} \middle| n = 1, 2, \ldots \right\}$. Then for any $d \in (-\infty, 0]$, d is a lower bound of S, and g. l. b. $S = 0 \notin S$. Let $S = [1, 2)$. Then for any $d \in (-\infty, 1]$, d is a lower bound of S, and g.l.b. $S = 1 \in S$.

Theorem 1.1.2 Let $S \subseteq \mathbb{R}$ $(S \neq \varnothing)$. If S is bounded below, then there exists the greatest lower bound of S.

Proof Let $T=\{-x \mid x\in S\}$. Then T is bounded above. By the completeness axiom, there exists the least upper bound of T. Let $\beta=$l. u. b. T. It is easy to check that $-\beta=$g.l.b. S. □

Definition The greatest lower bound of a number set S is also called *the infimum of* S, denoted as inf S.

By the definition of the infimum, it is easy to check the following:

Remark Let $S\subseteq\mathbb{R}\ (S\neq\varnothing)$.

(1) If inf S exists, it is unique;

(2) The following two statements are equivalent:

ⓘ $b=\inf S$;

ⓘⓘ for any $x\in S$, $x\geqslant b$, and for any $a>b$ there exists $x\in S$ such that $x<a$.

Note If a number set S has no lower bound, denote $\inf S=-\infty$.

Definition Let $S\subseteq\mathbb{R}\ (S\neq\varnothing)$. If there exists $t\in\mathbb{R}$ such that $|x|\leqslant t$ for any $x\in S$, then S is said to be *bounded*, and t is called a *bound* of S; otherwise, S is said to be *unbounded*.

Clearly, S is bounded if and only if S is both bounded above and bounded below.

Definition Let $S\subseteq\mathbb{R}\ (S\neq\varnothing)$.

(1) If there exists $\alpha\in S$ such that $x\geqslant\alpha$ for any $x\in S$, then α is called the *minimum* of S, denoted as $\alpha=\min S$;

(2) If there exists $\beta\in S$ such that $x\leqslant\beta$ for any $x\in S$, then β is called the *maximum* of S, denoted as $\beta=\max S$.

By the definitions of the minimum and the maximum of a number set, it is routine to check the following:

Remark Let $S\subseteq\mathbb{R}\ (S\neq\varnothing)$.

(1) $\sup S=\min\{y \mid x\leqslant y,\ \forall x\in S\}$; $\inf S=\max\{y \mid x\geqslant y,$

$\forall x \in S\}$.

(2) $\min S \in S$ and $\max S \in S$, but $\inf S$ or $\sup S$ may be not an element of S.

(3) $\inf S \in S$ if and only if $\inf S = \min S$; $\sup S \in S$ if and only if $\sup S = \max S$.

Examples (1) Let $S = \left\{ \frac{n}{n+1} \middle| n = 1, 2, \dots \right\}$. Then $\inf S = \frac{1}{2} \in S$, $\sup S = 1 \notin S$, $\min S = \frac{1}{2}$, no $\max S$.

(2) Let $S = \left\{ (-1)^n + \frac{(-1)^{n+1}}{n} \middle| n = 1, 2, \dots \right\}$. Then $\inf S = -1 \notin S$, $\sup S = 1 \notin S$, no $\min S$, no $\max S$.

Theorem 1.1.3 Let $A, B \subseteq \mathbb{R}$ $(A, B \neq \varnothing)$. Then

(1) $\sup (A \cup B) = \max\{\sup A, \sup B\}$;

(2) $\inf (A \cup B) = \min\{\inf A, \inf B\}$.

Proof (1) If A or B has no upper bound, then $A \cup B$ has no upper bound, and in this case, $\sup (A \cup B) = +\infty = \max\{\sup A, \sup B\}$.

Now, we assume that both of A and B have upper bounds. For any $x \in A$, $x \in A \cup B$, and so $x \leqslant \sup (A \cup B)$. Thus, $\sup (A \cup B)$ is an upper bound of A. By the definition of the supremum, we have $\sup A \leqslant \sup (A \cup B)$; similarly, we have $\sup B \leqslant \sup (A \cup B)$. Hence, $\max\{\sup A, \sup B\} \leqslant \sup (A \cup B)$. On the other hand, let $x \in A \cup B$. Then $x \in A$ or $x \in B$, and so $x \leqslant \sup A$ or $x \leqslant \sup B$, i. e. $x \leqslant \max\{\sup A, \sup B\}$. Thus $\max\{\sup A, \sup B\}$ is an upper bound of $A \cup B$. So, we see that $\sup (A \cup B) \leqslant \max\{\sup A, \sup B\}$;

(2) can be proved by an analogous argument. □

Example Let $A \subseteq B$ $(\subseteq \mathbb{R})$ $(A, B \neq \varnothing)$. Then $\inf B \leqslant \inf A \leqslant \sup A \leqslant \sup B$.

Proof Clearly, $\inf A \leqslant \sup A$. Note $B = A \cup B$. We have

$\sup B=\sup(A\cup B)=\max\{\sup A, \sup B\}\geqslant\sup A$. Similarly, $\inf B\leqslant \inf A$, and then the result follows.

Theorem 1.1.4 (Dedekind gap theorem) Let $S, T\subseteq\mathbb{R}(S, T\neq\varnothing)$ such that $x\leqslant y$ for any $x\in S$ and any $y\in T$. Then

(1) $\sup S\leqslant\inf T$;

(2) Moreover, the following three assertions are equivalent:

ⓘ There exists uniquely $c\in\mathbb{R}$ such that $s\leqslant c\leqslant t$ for any $s\in S$ and any $t\in T$;

ⓘⓘ $\sup S=\inf T$;

ⓘⓘⓘ For any $\varepsilon>0$, there exist $x\in S$ and $y\in T$ such that $y-x<\varepsilon$.

Proof (1) By the condition, for any $y\in T$, y is an upper bound of S, and so $\sup S\leqslant y$ for any $y\in T$. Thus, $\sup S$ is a lower bound of T, which implies $\sup S\leqslant\inf T$.

(2) We first show ⓘ$\Leftrightarrow$ⓘⓘ.

ⓘ$\Rightarrow$ⓘⓘ: If $\sup S\neq\inf T$, by (1) we have $\sup S<\inf T$, and so there exist $x, y\in\mathbb{R}$ such that $\sup S<x<y<\inf T$. Thus, there exist two distinct numbers x and y such that $s\leqslant x\leqslant t$ and $s\leqslant y\leqslant t$ for any $s\in S$ and any $t\in T$, which yields a contradiction.

ⓘⓘ$\Rightarrow$ⓘ: Let $c:=\sup S=\inf T$. Clearly, $s\leqslant c\leqslant t$ for any $s\in S$ and any $t\in T$. We further show such number c is unique. Assume that there exists $d\in\mathbb{R}$ such that $s\leqslant d\leqslant t$ for any $s\in S$ and any $t\in T$. Then, $\sup S\leqslant d\leqslant\inf T$, and so $d=\sup S=\inf T$, i. e. $d=c$.

Now, we show ⓘⓘ$\Rightarrow$ⓘⓘⓘ.

ⓘⓘ$\Rightarrow$ⓘⓘⓘ: Note that for any $\varepsilon>0$, $\sup S-\frac{\varepsilon}{2}$ is not an upper bound of S and $\inf T+\frac{\varepsilon}{2}$ is not a lower bound of T. Thus, there exist $x \in S$ and $y \in T$ such that $x > \sup S - \frac{\varepsilon}{2}$ and $y < \inf T + \frac{\varepsilon}{2}$.

So, $y - x < \left(\inf T + \frac{\varepsilon}{2}\right) - \left(\sup S - \frac{\varepsilon}{2}\right) = \varepsilon$.

ⅲ $\Leftrightarrow$ ⅱ: Assume that for any $\varepsilon > 0$, there exist $x \in S$ and $y \in T$ such that $y - x < \varepsilon$. Since $\inf T \leqslant y$ and $\sup S \geqslant x$, by (1) $0 \leqslant \inf T - \sup S \leqslant y - x < \varepsilon$. Note ε is an arbitrary positive number. We conclude that $\sup S = \inf T$.

Now, we consider the density of rational numbers and irrational numbers in $\mathbb{R}$. First, some basic facts of natural numbers are listed below as propositions without proof.

Proposition

(1) Let $n \in \mathbb{N}$. Then $(n, n+1) \cap \mathbb{N} = \varnothing$.

(2) Let $n \in \mathbb{N}$ and $\varnothing \neq A \subseteq \{1, 2, \ldots, n\}$. Then A has the minimum and the maximum.

(3) For any $A \subseteq \mathbb{N}$ $(A \neq \varnothing)$, A has the minimum.

Proposition (Archimedean property) (1) For any $c \in \mathbb{R}$, there exists $n \in \mathbb{N}$ such that $n > c$.

(2) For any $\varepsilon > 0$, there exists $n \in \mathbb{N}$ such that $\frac{1}{n} < \varepsilon$.

Definition Let $S \subseteq \mathbb{R}$. If for any interval (a, b), $(a, b) \cap S \neq \varnothing$, then S is said to be *dense* in $\mathbb{R}$.

Theorem 1.1.5 The rational number set $\mathbb{Q}$ and the irrational number set $\mathbb{R} \backslash \mathbb{Q}$ are both dense in $\mathbb{R}$.

Proof Without loss of generality, let $0 \leqslant a < b$. We only need to show $(a, b) \cap \mathbb{Q} \neq \varnothing$ and $(a, b) \cap (\mathbb{R} \setminus \mathbb{Q}) \neq \varnothing$.

By Archimedean property, there exists $m \in \mathbb{N}$ with $\frac{1}{m} < b - a$ and $n \in \mathbb{N}$ with $n > mb$, i. e. $\frac{n}{m} > b$. Let $A = \left\{k \mid \frac{k}{m} < b, k \in \mathbb{N}\right\}$. Notice $1 \in A$. $\varnothing \neq A \subseteq \mathbb{N}$ and $k < n$ for any $k \in A$. Then, by the preceding proposition A has the maximum. Let $j = \max A$. Clearly, $j+1 \notin A$, and so $\frac{j+1}{m} \geqslant b$, also $\frac{j}{m} \geqslant b - \frac{1}{m} > b - (b-a) = a$. Since $j \in A$, $\frac{j}{m} < b$. Thus, $\frac{j}{m} \in (a, b) \cap \mathbb{Q}$, which implies $(a, b) \cap \mathbb{Q} \neq \varnothing$.

Since $b - \frac{j}{m} > 0$, by Archimedean property there exists $l \in \mathbb{N}$ such that $\frac{\sqrt{2}}{l} < b - \frac{j}{m}$, i. e. $\frac{j}{m} + \frac{\sqrt{2}}{l} < b$. Furthermore, since $\frac{j}{m} > a$, $\frac{j}{m} + \frac{\sqrt{2}}{l} > a$. Thus $\frac{j}{m} + \frac{\sqrt{2}}{l} \in (a, b)$. It is clear that $\frac{j}{m} + \frac{\sqrt{2}}{l} \in \mathbb{R} \setminus \mathbb{Q}$, and so $(a, b) \cap (\mathbb{R} \setminus \mathbb{Q}) \neq \varnothing$. □

§ 1.2 Inequalities

Theorem 1.2.1 (Bernoulli inequality) For any $n \in \mathbb{N}$ and any $a > -1$, $(1+a)^n \geqslant (1+na)$.

Proof If $n = 1$, clearly the inequality holds. Assume $(1+a)^k \geqslant (1+ka)$ for some $k \in \mathbb{N}$. Let $n = k+1$. $(1+a)^n = (1+a)^k \cdot (1+a) \geqslant (1+ka)(1+a) \geqslant 1+(k+1)a = 1+na$. □

Remark The following inequality is a generalization of Bernoulli inequality, which may be similarly proved by induction:

$(1+x_1)(1+x_2)\ldots(1+x_n) \geqslant 1+x_1+x_2+\ldots x_n$ where all $x_i > -1$ and have the same sign ($i = 1, 2, \ldots, n$).

Example Let $n \in \mathbb{N}$ ($n \geqslant 2$). Show that $\left(1+\frac{1}{n-1}\right)^n > \left(1+\frac{1}{n}\right)^{n+1}$.

Proof By Bernoulli inequality, $\left(1+\frac{1}{n^2-1}\right)^n > 1+\frac{n}{n^2-1} > 1+\frac{n}{n^2} = 1+\frac{1}{n}$. Thus

$$\left(\frac{1+\dfrac{1}{n-1}}{1+\dfrac{1}{n}}\right)^n > 1+\frac{1}{n},$$

$$\text{i. e. } \left(1+\frac{1}{n-1}\right)^n > \left(1+\frac{1}{n}\right)^{n+1}. \square$$

Absolute values are used throughout the text. We now present without proof inequalities about absolute values.

Proposition Let $a, b \in \mathbb{R}$.

(1) $-|a| \leqslant a \leqslant |a|$.

(2) if $|a| > b$ $(b \geqslant 0)$, $a > b$ *or* $a < -b$.

(3) $|a| < b \Leftrightarrow -b < a < b$.

(4) $|a+b| \leqslant |a| + |b|$.

(5) $|a-b| \geqslant |a| - |b|$

(6) $|ab| = |a||b|$.

(7) $|a/b| = |a| / |b|$.

(8) $|a+b| \geqslant ||a| - |b||$.

Example Prove that $|a| + |b| + |c| - |b+c| - |c+a| - |a+b| + |a+b+c| \geqslant 0$.

Proof By the above theorem, $(|a| + |b| + |c| - |b+c| - |c+a| - |a+b| + |a+b+c|)(|a| + |b| + |c| + |a+b+c|) = (|b| + |c| - |b+c|)(|a| - |b+c| + |a+b+c|) + (|c| + |a| - |c+a|)(|b| - |c+a| + |a+b+c|) + (|a| + |b| - |a+b|)(|c| - |a+b| + |a+b+c|) \geqslant 0$, from which the result follows directly. □

The following four inequalities are well-known.

Theorem 1.2.2 Let $a_i, b_i \in \mathbb{R}$ $(i = 1, 2, \dots, n)$.

(1) (Cauchy-Schwarz inequality)

$$\left(\sum_{i=1}^{n} a_i b_i\right)^2 \leqslant \left(\sum_{i=1}^{n} a_i^2\right)\left(\sum_{i=1}^{n} b_i^2\right)$$

(2) (Chebyshev inequality) Let $a_1 \geqslant a_2 \geqslant \cdots \geqslant a_n$ and $b_1 \geqslant$

$b_2 \geqslant \cdots \geqslant b_n$. Then

$$\left(\sum_{i=1}^{n} a_i\right)\left(\sum_{i=1}^{n} b_i\right) \leqslant n \sum_{i=1}^{n} a_i b_i$$

(3) (Minkowski inequality) Let a_i, $b_i \geqslant 0 (i = 1, 2, \ldots, n)$ and let $r > 1$. Then

$$\left[\sum_{i=1}^{n} (a_i + b_i)^r\right]^{\frac{1}{r}} \leqslant \left(\sum_{i=1}^{n} a_i^r\right)^{\frac{1}{r}} + \left(\sum_{i=1}^{n} b_i^r\right)^{\frac{1}{r}}.$$

(4) (Holder inequality) Let a_i, $b_i \geqslant 0$ $(i=1, 2, \ldots, n)$, and let p, $q \in \mathbb{R}$ such that $p > 1$ and $\frac{1}{p} + \frac{1}{q} = 1$. Then

$$\left(\sum_{i=1}^{n} a_i b_i\right)^2 \leqslant \left(\sum_{i=1}^{n} a_i^p\right)^{\frac{1}{p}} \left(\sum_{i=1}^{n} b_i^q\right)^{\frac{1}{q}}.$$

Proof We only prove (1). The proof of other three inequalities are left for readers.

Since

$$\left(\sum_{i=1}^{n} a_i^2\right)t^2 + 2\left(\sum_{i=1}^{n} a_i b_i\right)t + \sum_{i=1}^{n} b_i^2 = \sum_{i=1}^{n} (a_i t + b_i)^2 \geqslant 0,$$

we have

$$\Delta = \left(2 \sum_{i=1}^{n} a_i b_i\right)^2 - 4\left(\sum_{i=1}^{n} a_i^2\right)\left(\sum_{i=1}^{n} b_i^2\right) \leqslant 0,$$

i.e.

$$\left(\sum_{i=1}^{n} a_i b_i\right)^2 \leqslant \left(\sum_{i=1}^{n} a_i^2\right)\left(\sum_{i=1}^{n} b_i^2\right). \square$$

Example Let a_i, $b_i \in \mathbb{R}$ $(i=1, 2, \ldots, n)$ with $a_1 \geqslant a_2 \geqslant \cdots \geqslant a_n \geqslant 0$. Suppose for any $k \in \{1, 2, \ldots, n\}$, $a_1 + a_2 + \cdots + a_k \leqslant b_1 + b_2 + \cdots + b_k$. Show that $a_1^2 + a_2^2 + \cdots + a_n^2 \leqslant b_1^2 + b_2^2 + \cdots + b_n^2$.

Proof Let $a_{n+1} = 0$. Then by the condition, for any $k = 1, 2, \cdots, n$, $(a_k - a_k + 1)(a_1 + a_2 + \cdots + a_k) \leqslant (a_k - a_{k+1})(b_1 + b_2 + \cdots + b_k)$, and so $\sum_{k=1}^{n}(a_k - a_{k+1})(a_1 + a_2 + \cdots + a_k) \leqslant \sum_{k=1}^{n}(a_k - a_{k+1})(b_1 + b_2 + \cdots + b_k)$, which implies that

$$\sum_{k=1}^{n} a_k^2 \leqslant \sum_{k=1}^{n} a_k b_k.$$

Then, by Cauchy-Schwarz inequality we derive

$$\left(\sum_{k=1}^{n} a_k^2\right)^2 \leqslant \left(\sum_{k=1}^{n} a_k b_k\right)^2 \leqslant \left(\sum_{k=1}^{n} a_k^2\right)\left(\sum_{k=1}^{n} b_k^2\right).$$

i.e. $$a_1^2 + a_2^2 + \cdots + a_n^2 \leqslant b_1^2 + b_2^2 + \cdots + b_n^2. \square$$

Theorem 1.2.3 (The average number inequality) Let $a_i \in \mathbb{R}$ with $a_i > 0$ $(i = 1, 2, \cdots, n)$. Then

$$\sqrt[n]{a_1 a_2 \cdots a_n} \leqslant \frac{1}{n}(a_1 + a_2 + \cdots + a_n).$$

Proof If $n = 1$ or $n = 2$, the inequality holds clearly. Now we suppose it is true for $n - 1$ numbers. We show it remains to be true for n numbers.

Without loss of generality, let $a_1 \leqslant a_2 \leqslant \cdots \leqslant a_n$ and let $b = \frac{a_1 + a_2 + \cdots a_{n-1}}{n-1}$. Then, by the foregoing inequality, $a_n \geqslant b \geqslant \sqrt[n-1]{a_1 a_2 \cdots a_{n-1}}$, and so

$$\begin{aligned}\left(\frac{a_1 + a_2 + \cdots + a_n}{n}\right)^n &= \left[\frac{(n-1)b + a_n}{n}\right]^n = \left(b + \frac{a_n - b}{n}\right)^n \\ &= b^n + nb^{n-1}\left(\frac{a_n - b}{n}\right) + \cdots + \left(\frac{a_n - b}{n}\right)^n \\ &\geqslant b^n + nb^{n-1}\left(\frac{a_n - b}{n}\right) \\ &= b^{n-1}a_n \geqslant a_1 a_1 \cdots a_n. \square\end{aligned}$$

Examples (1) For any $x_i > 0 (i = 1, 2, \cdots, n)$, if $x_1 x_2 \cdots x_n =$

1, then $x_1+x_2+\cdots+x_n \geqslant n$.

Proof This conclusion follows directly from the average number inequality. □

(2) Prove $\left(1+\frac{1}{n}\right)^n \leqslant \left(1+\frac{1}{n+1}\right)^{n+1}$ for any $n=1, 2, \cdots, n$.

Proof By the average number inequality, $\sqrt[n+1]{x_1x_2\cdots x_{n+1}} \leqslant \frac{1}{n+1}(x_1+x_2+\cdots+x_{n+1})$ for any $x_i>0(i=1, 2, \cdots, n, n+1)$. Let $x_1=x_2=\cdots=x_n=1+\frac{1}{n}$ and $x_{n+1}=1$, we have $\sqrt[n+1]{\left(1+\frac{1}{n}\right)^n \cdot 1} \leqslant \frac{1}{n+1}\left[n\left(1+\frac{1}{n}\right)+1\right]=1+\frac{1}{n+1}$, i.e. $\left(1+\frac{1}{n}\right)^n \leqslant \left(1+\frac{1}{n+1}\right)^{n+1}$. □

(3) Prove

$$\sqrt[n]{x_1x_2\cdots x_n} \geqslant \frac{n}{\frac{1}{x_1}+\frac{1}{x_2}+\cdots+\frac{1}{x_n}}$$

for any $x_i>0(i=1, 2, \cdots, n)$.

Proof By the average number inequality, $\sqrt[n]{\frac{1}{x_1}\frac{1}{x_2}\cdots\frac{1}{x_n}} \leqslant \frac{1}{n}\left(\frac{1}{x_1}+\frac{1}{x_2}+\cdots+\frac{1}{x_n}\right)$, from which the result follows immediately. □

§ 1.3 Functions

In this section we will develop the concept of a function, which is the basic idea that underlines almost all mathematical and physical relationships, regardless of the form in which they are expressed. Basically, a function relates each element of a set with exactly one element of another set. These sets considered here are usually referred to the sets of real numbers.

Functions are truly fundamental to mathematics. In everyday

language we say, "The fuel needed to launch a rocket is a function of its payload" or "The patient blood pressure is a function of the drugs prescribed." In each case, the word *function* expressed the idea that knowledge of one fact tells us another. In mathematics, the most important functions are those in which knowledge of one number tells us another number. If we know the length of the side of a square, its area is determined. If the circumference of a circle is known, its radius is determined.

Calculus starts with the study of functions. This section lays the foundation for calculus by surveying the behavior of the most common functions, including powers, exponentials, logarithmic, and trigonometric functions. We also explore ways of presenting these functions.

§ 1. 3. 1 Concept of functions

Now we define and develop the concept of a function. Functions are used by mathematicians and scientists to describe the relationships between variable quantities and hence play a central role in calculus and its application.

Definition Let A and B be two (non-empty) number sets. A *function* of a variable x is a rule f that assigns to each value of $x \in A$ a unique number $f(x) \in B$, called *the function value at* x. The variable x is called the *independent variable of the function* and the variable $f(x)$ is called the *dependent variable of the function*. The number set A is called the *domain* of the function. If the domain is composed of an interval, it is also called the *domain interval* of the function. The domain of a function may be explicitly specified as part of the definition of a function or it may be understood from context. The *range* of a function is defined as the set of all function values, denoted as $f(A) := \{f(x) \mid x \in A\}$. Clearly, $f(A) \subseteq B$. The domain and the range of a function f are usually denoted by D_f and

R_f respectively.

A function $f(x)$ is usually denoted by $y = f(x)$. e. g. let $y = [x]$, where $[x]$ represents the largest integer which does not surpass x. For instance, $[4.7] = 4$, $[\pi] = 3$, $[-\pi] = -4$, $[-6.3] = -7$. Then, this is a function with $D_f = \mathbb{R}$ and $R_f = \mathbb{Z}$.

Remark

(1) When defining a function, besides the rule f it is also necessary to specify the domain of the function, which is the set of acceptable values of the independent variable x. If not specifically indicated, we will understand the domain of the function f to consist of all numbers $x \in \mathbb{R}$ for which $f(x)$ makes sense. For example, if $f(x) = x^2 - x + 1$, the domain is $\mathbb{R}$; if $f(x) = \dfrac{1}{x}$, the domain is $\mathbb{R}\setminus\{0\}$; if $f(x) = \sqrt{x}$, the domain is $D = \{x \in \mathbb{R} \mid x \geqslant 0\}$.

(2) The function f from A to B is also denoted intuitively as

$$f: A \to B \text{ or } A \xrightarrow{f} B.$$

(3) It is customary to write a function in the term of $y = f(x)$ or $y = y(x)$. The letters g, h, φ, ψ, F, G etc. can also be used for expressions of functions instead of f.

Examples Find the domain of the functions

(1) $f(x) = \sqrt{\dfrac{x^2(1-x^2)}{6-x-x^2}}$;

(2) $f(x) = \arcsin(x^2 - x - 1) + \sqrt{\ln x}$.

Solution.

(1) Let

$$\frac{x^2(1-x^2)}{6-x-x^2} = \frac{(1-x)(1+x)}{(3+x)(2-x)} \geqslant 0.$$

Solving this inequality, we have the domain $D = (-\infty, -3) \cup$

$[-1, 1] \cup (2, +\infty)$.

(2) Let

$$\begin{cases} -1 \leqslant x^2 - x - 1 \leqslant 1, \\ \ln x \geqslant 0. \end{cases}$$

Then

$$\begin{cases} x^2 - x \geqslant 0, \\ x^2 - x - 2 \leqslant 0, \\ x \geqslant 1. \end{cases} \quad \text{i.e.} \begin{cases} x(x-1) \geqslant 0, \\ (x-2)(x+1) \leqslant 0, \\ x \geqslant 1. \end{cases}$$

Hence, the domain of the function is $[1, 2]$.

Definition Often it is helpful to describe a function f geometrically, using a rectangular xy-coordinate system. Given any x in the domain of f, we can plot the point $(x, y) = (x, f(x))$. This is the point in the xy-plane whose y-coordinate is the value of the function at x. The set of all such points $(x, f(x))$ usually forms a curve in the xy-plane and is called the *graph* of the function $f(x)$.

It is possible to approximate the graph of $f(x)$ by plotting the points $(x, f(x))$ for a representative set of values of x and joining them by a smooth curve (cf. Figure 1-2). The more closely spaced the values of x, the closer the approximation.

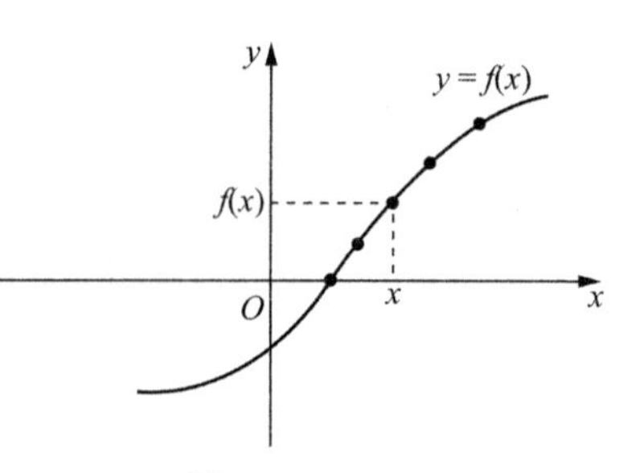

Figure 1-2

Note A function can also be presented differently in different part of its domain, for example:

$$f(x) = \begin{cases} x^2 - 1 & (x \in (-\infty, 0)), \\ 1 & (x = 0), \\ 3x + \sin x & (x \in (0, +\infty)). \end{cases}$$

Such a function is called a *piecewise function*.

§ 1. 3. 2 Some special kinds of functions

Now we will explore properties of functions in more detail and introduce some of the functions possessing special characteristics that will play a prominent role in our discussion of calculus.

Definition Suppose f is defined on D. If there is a constant K such that $f(x) \leqslant K$ for any $x \in D$, then f is said to be *bounded above* on D, and K is called an *upper bound* of f ; if there is a constant K such that $f(x) \geqslant K$ for any $x \in D$, then f is said to be *bounded below* on D, and K is called a *lower bound* of f ; if there is a positive constant K such that $|f(x)| \leqslant K$ for any $x \in D$, then f is said to be *bounded* on D, and K is called a *bound* of f ; if f is not bounded on D, then f is called an *unbounded* function on D.

Note The following statements are plain.

(1) f is bounded if and only if f is bounded both above and below.

(2) if K is an upper bound of f, then any $K_1 (\geqslant K)$ is also an upper bound of f; if K is a lower bound of f, then any $K_1 (\leqslant K)$ is also a lower bound of f; if K is a bound of f , then any $K_1 (\geqslant K)$ is also a bound of f . Thus, the (upper, lower) bound of a function is not unique if it exists.

(3) the characteristic of the graph of a bounded function with bound K is that it is located between two parallel straight lines $y = K$ and $y = -K$.

Examples (1) $f(x) = \sin x$ and $g(x) = \cos x$ are both bounded functions on $(-\infty, +\infty)$; $f(x) = x^2$ is only bounded below on $(-\infty, +\infty)$; $f(x) = x^3$ is neither bounded below nor bounded above on $(-\infty, +\infty)$.

(2) Prove the following functions are both bounded on $(-\infty, +\infty)$:

① $f(x) = \dfrac{1+x^2}{1+x^4}$;

② $f(x) = \dfrac{x}{1+x^2}$.

Proof ① If $|x|<1$, $0<f(x)\leqslant 1+x^2\leqslant 2$; if $|x|\geqslant 1$, $0<f(x)\leqslant(1+x^2)/(1+x^2)=1$. Thus $0<f(x)\leqslant 2$ for any $x\in(-\infty, +\infty)$.

Or we prove it as follows: $1+x^2\leqslant(1+x^2)^2\leqslant(1+x^2)^2+(1-x^2)^2=2(1+x^4)$, i. e. $0<f(x)\leqslant 2$.

② $|f(x)|=\left|\dfrac{x}{1+x^2}\right|\leqslant\left|\dfrac{x}{2\sqrt{x^2\cdot 1}}\right|=\dfrac{1}{2}$ for any $x\in(-\infty, +\infty)$. □

(3) Prove the function

$$f(x)=\frac{1}{x}\cos\frac{1}{x}$$

is unbounded on $U^\circ(0)$, any free-center neighborhood of 0.

Proof It is easy to see that for any $A>0$, there is $k\in\mathbb{N}$ such that $0<1/(2k[A]\pi)<\delta$, i. e. there is $x=1/(2k[A]\pi)\in U^\circ(0)$ such that

$$f(x)=2k[A]\pi\cos(2k[A]\pi)=2k[A]\pi>A. \ \square$$

(4) Prove by definition the function $f(x)=\tan x$ is unbounded on $(-\pi/2, \pi/2)$.

Proof We need to prove: for any $M>0$, there exists $x_0\in(-\pi/2, \pi/2)$ such that $\tan x_0>M$.

Let $x_1=\arctan M$. Then $-\pi/2<x_1<\pi/2$, and so we may take $x_0\in(x_1, \pi/2)$. Thus

$$\tan x_0-\tan x_1=\frac{\sin x_0}{\cos x_0}-\frac{\sin x_1}{\cos x_1}=\frac{\sin(x_0-x_1)}{\cos x_0\cos x_1}.$$

Note $\sin(x_0-x_1)>0$ as $0<x_0-x_1<\pi$, and $\cos x_0>0$, $\cos x_1>0$ as

$x_0, x_1 \in (-\pi/2, \pi/2)$. Then $\tan x_0 - \tan x_1 > 0$, which implies

$$\tan x_0 > \tan x_1 = \tan(\arctan M) = M. \square$$

Definition Suppose f is defined on D. If for any $x_1, x_2 \in D$, $x_1 < x_2$ implies $f(x_1) \leqslant f(x_2)$ ($f(x_1) < f(x_2)$), then f is said to be *increasing* (*strictly increasing*) on D; if for any $x_1, x_2 \in D$, $x_1 < x_2$ implies $f(x_1) \geqslant f(x_2)$ ($f(x_1) > f(x_2)$), then f is said to be *decreasing* (*strictly decreasing*) on D; if f is increasing or decreasing (strictly increasing or strictly decreasing) on D, then f is said to be *monotone* (*strictly monotone*) on D. Such a property is called *monotonicity*.

Examples (1) The function $f(x) = x^3$ is strictly increasing on $(-\infty, +\infty)$; $f(x) = x^2$ is strictly decreasing on $(-\infty, 0)$ and strictly increasing on $(0, +\infty)$; $f(x) = [x]$ is increasing but not strictly increasing on $(-\infty, +\infty)$.

(2) Prove the function $f(x) = 2x + \sin x$ is increasing on $(-\infty, +\infty)$.

Proof Let $x_2 > x_1$. Note $|\sin x| \leqslant |x|$. Then

$$\begin{aligned} f(x_2) - f(x_1) &= (2x_2 + \sin x_2) - (2x_1 + \sin x_1) \\ &= 2(x_2 - x_1) + 2\sin\frac{x_2 - x_1}{2}\cos\frac{x_2 + x_1}{2} \\ &\geqslant 2(x_2 - x_1) - 2\left|\sin\frac{x_2 - x_1}{2}\right| \geqslant 2(x_2 - x_1) - 2\left|\frac{x_2 - x_1}{2}\right| \\ &= x_2 - x_1 > 0. \end{aligned}$$

(3) Provided $f(x)$, $g(x)$ and $h(x)$ are all increasing functions with $g(x) \leqslant f(x) \leqslant h(x)$. Prove $g(g(x)) \leqslant f(f(x)) \leqslant h(h(x))$.

Proof Since $g(x) \leqslant f(x)$ and $f(x)$ is increasing, $f(g(x)) \leqslant f(f(x))$. Note also that $g(x) \leqslant f(x)$ implies $g(g(x)) \leqslant f(g(x))$. Hence $g(g(x)) \leqslant f(f(x))$. Similarly, we can prove $f(f(x)) \leqslant h(h(x))$. $\square$

Definition Suppose f is defined on $D=(-l, l)$ (or $[-l, l]$) where $l>0$. If for any $x \in D$, $f(-x)=-f(x)$, then f is called an *odd function*; if for any $x \in D$, $f(-x)=f(x)$, then f is called an *even function*. Also, f is said to be *one-to-one* if $f(x_1)=f(x_2)$ implies $x_1=x_2$.

Clearly, the graph of an odd function is always symmetric to the origin, and the graph of an even function is always symmetric to the y-axis.

Examples (1) $f(x)=\sin x$ is an odd function and $g(x)=\cos x$ is an even function; $f(x)=\sin x+\cos x$ is neither even nor odd; sign function

$$f(x)=\operatorname{sgn} x=\begin{cases} 1 & (x>0), \\ 0 & (x=0), \\ -1 & (x<0) \end{cases}$$

is an odd function.

(2) Consider the oddness and evenness of the following functions:

$$① \ f(x)=\ln \frac{1-x}{1+x};$$

$$② \ f(x)=\frac{(1+2^x)^2}{2^x}.$$

Solution ① The domain is $(-1, 1)$, and

$$f(-x)=\ln \frac{1+x}{1-x}=-\ln \frac{1-x}{1+x}=-f(x).$$

So, $f(x)$ is odd.

② The domain is $(-\infty, +\infty)$, and

$$f(-x)=\frac{(1+2^{-x})^2}{2^{-x}}=\frac{2^{-2x}(2^x+1)^2}{2^{-x}}$$

$$=\frac{(1+2^x)^2}{2^x}=f(x).$$

So, $f(x)$ is even.

(3) Show that any $f(x)$ defined on a symmetric interval $(-l, l)$ can be expressed as a sum of an odd function and an even function.

Proof Construct

$$g(x)=\frac{f(x)-f(-x)}{2} \text{ and } h(x)=\frac{f(x)+f(-x)}{2}$$

Then it is routine to check that $g(x)$ is an odd function and $h(x)$ is an even function such that $f(x)=g(x)+h(x)$. □

Definition Suppose f is defined on D. If there is a positive number T such that $f(x \pm T)=f(x)$ for any $x \in D$, then f is called a *periodical function*, and T is called a *period* of f.

If T is a period of f, then for any natural number n, nT is a period of f, i. e. a periodical function has a infinite number of periods. Thus, we may define the smallest period as the *primary period* of f. But notice that there does not always exist a primary period for a periodical function.

Examples (1) $f(x)=\sin x$ and $g(x)=\cos x$ are both periodical functions on $(-\infty, +\infty)$ with period 2π. $f(x)=x-[x]$ is a periodical function on $(-\infty, +\infty)$ with period 1.

(2) Discuss periodicity of the following functions defined on $(-\infty, +\infty)$:

① $f(x)=A\cos \lambda x+B \sin \lambda x \ (\lambda>0)$;

② $f(x)=2 \tan \frac{x}{2}-3\tan \frac{x}{3}$;

③ $f(x)=\sin x^2$;

④ $f(x)=c(\text{constant})$.

Solution

① Note

$$f(x+T)=f(x)$$
$$\Leftrightarrow A(\cos\lambda(x+T)-\cos\lambda x)+B(\sin\lambda(x+T)-\sin\lambda x)=0$$
$$\Leftrightarrow 2\sin\frac{\lambda T}{2}\left[\cos\lambda\left(x+\frac{T}{2}\right)-\sin\lambda\left(x+\frac{T}{2}\right)\right]=0$$

Hence, if $T=(2k\pi)/\lambda(k=\pm 1, \pm 2, \ldots)$, $\sin(\lambda T/2)=\sin k\pi\equiv 0$ and so $f(x+T)\equiv f(x)$. If $k=1$, $T=2\pi/\lambda$. Therefore, $f(x)$ is a periodical function with primary period $T=2\pi/\lambda$.

② Note

$$f(x+T)=f(x)\Leftrightarrow 2\tan\frac{T}{2}\left(1-\tan\frac{x+T}{2}\tan\frac{x}{2}\right)$$
$$-3\tan\frac{T}{3}\left(1-\tan\frac{x+T}{3}\tan\frac{x}{3}\right)=0$$

Thus, if $T=6k\pi$ $(k=\pm 1, \pm 2, \ldots)$, $\tan(T/2)\equiv\tan(T/3)\equiv 0$, i. e. $f(x+T)\equiv f(x)$. If $k=1$, $T=6\pi$. Therefore, $f(x)$ is a periodical function with primary period $T=6\pi$.

③ Note

$$f(x+T)=f(x)\Leftrightarrow 2\sin\left(xT+\frac{T^2}{2}\right)\cos\frac{(x+T)^2+x^2}{2}=0.$$

It is easy to see that for any $T\in\mathbb{R}$, this can not be an identity. So, $f(x)$ is not a periodical function.

④ Clearly, $f(x)$ is a periodical function with a period of any $T>0$. It has no primary period.

(3) Suppose $f(x)$ is defined on $(-\infty, +\infty)$ with $f(x+a)=\frac{1}{2}+\sqrt{f(x)-f^2(x)}$ $(a>0)$. Prove that $f(x)$ is a periodical function.

Proof $f(x+2a)=f((x+a)+a)$

$$=\frac{1}{2}+(f(x+a)-f^2(x+a))^{1/2}$$

$$
\begin{aligned}
&= \frac{1}{2} + \left[\left(\frac{1}{2} + \sqrt{f(x) - f^2(x)} \right) \right. \\
&\quad \left. - \left(\frac{1}{2} + \sqrt{f(x) - f^2(x)} \right)^2 \right]^{1/2} \\
&= \frac{1}{2} + \left[\left(f(x) - \frac{1}{2} \right)^2 \right]^{1/2} = f(x). \ \square
\end{aligned}
$$

§ 1. 3. 3 Operations of functions

Many functions we shall encounter later in the text can be viewed as combinations, or algebraic operations, of other functions, which will be the topic of this section.

Two functions f and g can be combined to form new functions $f+g$, $f-g$, fg and f/g in a manner similar to the way we add, subtract, multiply and divide real numbers.

Definition Let f_i be functions with domain D_i ($i = 1, 2$). Suppose $D = D_1 \cap D_2 \neq \varnothing$. Define the *sum*, *difference*, *product and quotient* of f_1 and f_2 respectively as follows:

(1) sum of f_1 and f_2: $g(x) = f_1(x) + f_2(x)$ ($x \in D$);

(2) difference of f_1 and f_2: $g(x) = f_1(x) - f_2(x)$ ($x \in D$);

(3) product of f_1 and f_2: $g(x) = f_1(x) f_2(x)$ ($x \in D$);

(4) quotient of f_1 and f_2: $g(x) = \dfrac{f_1(x)}{f_2(x)}$ ($x \in D^* = D_1 \cap \{x \mid f_2(x) \neq 0, x \in D_2\}$ if $D^* \neq \varnothing$).

These four operations concerning functions are called *fundamental operations* of functions.

Example Let

$$
f(x) = \begin{cases} 1 - x^2 & (x \leqslant 0), \\ x & (x > 0) \end{cases} \quad \text{and } g(x) = \begin{cases} -2x & (x < 1), \\ 1 - x & (x \geqslant 1). \end{cases}
$$

To find $f + g$, $f - g$ and fg, we must fit the pieces together, namely, we must break up the domain of both functions in the same manner:

$$f(x)=\begin{cases}1-x^2 & (x\leqslant 0),\\ x & (0<x<1),\\ x & (1\leqslant x),\end{cases}\quad g(x)=\begin{cases}-2x & (x\leqslant 0),\\ -2x & (0<x<1),\\ 1-x & (1\leqslant x).\end{cases}$$

The rest is straightforward:

$$(f+g)(x)=f(x)+g(x)=\begin{cases}1-2x-x^2 & (x\leqslant 0),\\ -x & (0<x<1),\\ 1 & (1\leqslant x),\end{cases}$$

$$(f-g)(x)=f(x)-g(x)=\begin{cases}1+2x-x^2 & (x\leqslant 0),\\ 3x & (0<x<1),\\ -1+2x & (1\leqslant x),\end{cases}$$

$$(fg)(x)=f(x)g(x)=\begin{cases}-2x+2x^3 & (x\leqslant 0),\\ -2x^2 & (0<x<1),\\ x-x^2 & (1\leqslant x).\end{cases}$$

Another important way of combining two functions $f(x)$ and $g(x)$ is to substitute the function $g(x)$ for every occurrence of the variable x in $f(x)$. The resulting function is called the composition (or composite) of f and g.

Definition Let $y=f(u)$ be a function with domain U and let $u=\varphi(x)$ be a function with domain X. Suppose $X_0=\{x\mid\varphi(x)\in U,\ x\in X\}\neq\varnothing$. Then for any $x\in X_0$, by the functions φ and f, there is a unique value of y corresponding to x. Thus a new function is determined, which is denoted as

$$y=f(\varphi(x)).$$

We call it the *composite function* of f and φ. The process to produce a composite function is called *composition* of functions. In a similar manner, composite of more than two functions can be formed.

Examples (1) Let

$$f(x)=\begin{cases}2x & (0\leqslant x\leqslant 1),\\ x^2 & (1<x\leqslant 2)\end{cases}, \text{ and } g(x)=\ln x.$$

Then

$$f(g(x))=\begin{cases}2g(x) & (0\leqslant g(x)\leqslant 1)\\ (g(x))^2 & (1<g(x)\leqslant 2)\end{cases}=\begin{cases}2\ln x & (1\leqslant x\leqslant \mathrm{e}),\\ \ln^2 x & (\mathrm{e}<x\leqslant \mathrm{e}^2),\end{cases}$$

$$g(f(x))=\ln f(x)=\begin{cases}\ln 2x & (0<x\leqslant 1),\\ \ln(x^2) & (1<x\leqslant 2).\end{cases}$$

(2) Let $f_n(x)=f(f(\cdots f(x)))$ and $f(x)=a+bx$. Show that

$$f_n(x)=a\cdot\frac{b^n-1}{b-1}+b^n x:$$

As $n=1$, $f_1(x)=a+bx$. Assume

$$f_k(x)=a\cdot\frac{b^k-1}{b-1}+b^k_x \quad \text{if} \quad n=k.$$

Now let $n=k+1$. Then

$$f_{k+1}(x)=f(f_k(x))=a+bf_k(x)=a+b\left(a\cdot\frac{b^k-1}{b-1}+b^k x\right)$$
$$=a\cdot\frac{b^{k+1}-1}{b-1}+b^{k+1}x. \ \square$$

The designation of one variable x as "independent" is really an arbitrary decision. Indeed it is possible, desirable, and often necessary to reverse the roles assigned to the variables, i. e. to determine x from a knowledge of y. This procedure is equivalent to expressing x as a function of y. This fact suggests us to introduce the concept of *inverse* function as follows:

Definition Let $y=f(x)$ be a function with domain D_f and range

R_f. If for any $y \in R_f$, there is a unique value $x \in D_f$ such that $f(x) = y$. In this way, a new function is defined on R_f, which is called the *inverse* function of $y = f(x)$, denoted by

$$x = f^{-1}(y)(y \in R_f).$$

In order to study an inverse function in detail, it is convenient to restore traditional roles to x and y. This means that the variables in $x = f^{-1}(y)$ are interchanged, so that the inverse function is written as

$$y = f^{-1}(x).$$

Definition A function f is said to be *one-to-one* if and only if there are no two points at which it takes on the same value, i. e. $f(x_1) = f(x_2)$ implies $x_1 = x_2$.

Note It is easy to check that any monotone function is a one-to-one function from its domain to its range.

By the definition of the one-to-one function and the inverse function, it is not difficult to derive the following:

Theorem 1.3.1 If f is a one-to-one function, then there is one and only one inverse function f^{-1} that is defined on the range of f and satisfies the equation $f(f^{-1}(x)) = x$ for all x in the range of f.

Remark (1) Trivially, if $f^{-1}(x)$ is an inverse function of $f(x)$, then $f(x)$ is an inverse function of $f^{-1}(x)$. The domain of f is exactly the range of f^{-1}, and the range of f is exactly the domain of f^{-1}.

(2) The combined operations $f^{-1}f$ are self-cancelling, in analogy with the law of exponents, i.e. $f^{-1}(f(x)) = f(f^{-1}(x)) = x$.

(3) Note carefully that $f^{-1}(x) \neq 1/f(x)$. Also note that if (a, b) is a point on the graph of $f(x)$, then the interchanged coordinates

(b, a) give a point on the curve of the inverse function $y=f^{-1}(x)$. This implies that the curves of a function and its inverse are just mirror images about the line $y=x$ (cf. Figure 1-3).

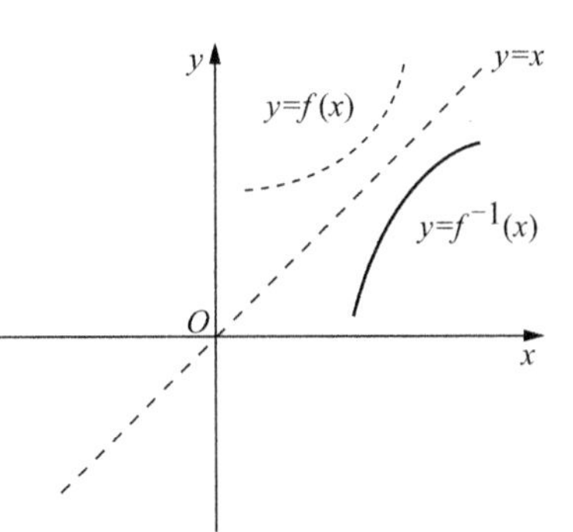

Figure 1-3

Examples (1) Let $y=f(x)=\sqrt[3]{x+\sqrt{1+x^2}}+\sqrt[3]{x-\sqrt{1+x^2}}$. Find its inverse function $f^{-1}(x)$.

Solution
$$\begin{aligned} y^3 &= \left(\sqrt[3]{x+\sqrt{1+x^2}}+\sqrt[3]{x-\sqrt{1+x^2}}\right)^3 \\ &= 2x-3\left[(x+\sqrt{1+x^2})^{1/3}+(x-\sqrt{1+x^2})^{1/3}\right] \\ &= 2x-3y, \end{aligned}$$

and so, $x=(y^3+3y)/2$. Thus, the inverse function $f^{-1}(x) = (x^3+3x)/2$.

(2) Let $y=f(x)=1+2\sin\frac{x-1}{x+1}$. Find its inverse function $f^{-1}(x)$.

Solution Since $\sin\frac{x-1}{x+1}=\frac{y-1}{2}$,

$$\arcsin\frac{y-1}{2}=\frac{x-1}{x+1}=1-\frac{2}{x+1},$$

which implies

$$x=\frac{1+\arcsin\frac{y-1}{2}}{1-\arcsin\frac{y-1}{2}} \quad \text{i.e. } f^{-1}(x)=\frac{1+\arcsin\frac{x-1}{2}}{1-\arcsin\frac{x-1}{2}}.$$

(3) Let $y=f(x)=(x^2+1)\operatorname{sgn} x$. Find its inverse function $f^{-1}(x)$.

Solution Note $\operatorname{sgn} x=\begin{cases}-1 & (\text{if } x<0), \\ 0 & (\text{if } x=0), \\ 1 & (\text{if } x>0).\end{cases}$

$$\text{So } y = \begin{cases} -(x^2+1) & (\text{if } x < 0) \quad (1) \\ 0 & (\text{if } x = 0) \quad (2) \\ x^2+1 & (\text{if } x > 0) \quad (3) \end{cases}$$

By (1), $x < 0$, $y < -1$ and $x = -\sqrt{-y-1}$;

by (2), $y = 0$, $x = 0$;

by (3), $x > 0$, $y > 1$ and $x = \sqrt{y-1}$.

$$\text{Hence, we have } x = \begin{cases} -\sqrt{-y-1} & (\text{if } y < -1), \\ 0 & (\text{if } y = 0), \\ \sqrt{y-1} & (\text{if } y > 1). \end{cases}$$

$$\text{Therefore, } f^{-1}(x) = \begin{cases} -\sqrt{-x-1} & (\text{if } x < -1), \\ 0 & (\text{if } x = 0), \\ \sqrt{x-1} & (\text{if } x > 1). \end{cases}.$$

(4) Let

$$y = \begin{cases} x & \big(x \in (-\infty, -1)\big), \\ -x^2 & \big(x \in [-1, 0]\big), \\ \ln(x+1) & \big(x \in (0, \mathrm{e}]\big). \end{cases}$$

Find its inverse function $f^{-1}(x)$.

Solution As $x \in (-\infty, -1)$, $x = y$ where $y \in (-\infty, -1)$; as $x \in [-1, 0]$, $x = -\sqrt{-y}$ where $y \in [-1, 0]$; as $x \in (0, \mathrm{e}]$, $x = \mathrm{e}^y - 1$ where $y \in (0, \ln(1+\mathrm{e})]$. So, its inverse function

$$y = f^{-1}(x) = \begin{cases} x & \big(x \in (-\infty, -1)\big), \\ -\sqrt{-x} & \big(x \in [-1, 0]\big), \\ \mathrm{e}^x - 1 & \big(x \in (0, \ln(1+\mathrm{e})\big). \end{cases}$$

§ 1.3.4 Primary functions

Functions are often grouped into families according to the form of their defining formulas or other common characteristics. We now discuss some of the most basic families of functions.

Definition The following functions are called *fundamental primary functions*:

(1) *constant function*: $f(x) = c$, $x \in (-\infty, +\infty)$ where c is a constant.

The constant function is a bounded, even and periodical function without the primary period.

(2) *power function*: $f(x) = x^\alpha$ where α is a real number.

ⓘ If α is a positive integer, its domain is $(-\infty, +\infty)$. Moreover, if α is odd, $f(x)$ is an odd function and strictly increasing; if α is even, $f(x)$ is an even function.

ⓘⓘ If α is a negative integer, its domain is $(-\infty, +\infty)\backslash\{0\}$; Also, if α is odd, $f(x)$ is an odd function; if α is even, $f(x)$ is an even function.

ⓘⓘⓘ If α is a rational number, the situation is relatively complicated. We only set two cases for discussion:

Example (1) $y = x^{\frac{1}{2}}$, $x \in [0, +\infty)$. It is the inverse function of the function $y = x^2$, and it is strictly increasing on $[0, +\infty)$.

Example (2) $y = x^{\frac{1}{3}}$, $x \in (-\infty, +\infty)$. It is the inverse function of the function $y = x^3$, and it is strictly increasing on its domain.

ⓘⓥ If α is an irrational number, its domain is specified as $(0, +\infty)$.

(3) *exponential function*: $f(x) = a^x$, $x \in (-\infty, +\infty)$ $(a > 0, a \neq 1)$.

If $a > 1$, it is strictly increasing; if $0 < a < 1$, it is strictly decreasing. The range of a^x is $(0, +\infty)$.

(4) *logarithmic function*: $f(x)=\log_a x$, $(a>0, a\neq 1)$.

The logarithmic function is the inverse function of the exponential function, and so its domain is $(0, +\infty)$ and its range $(-\infty, +\infty)$. The logarithmic function possesses the same monotonicity as the corresponding exponential function. Since $a^{\log_a x}=x$, if $a>0(a\neq 1)$ and $x>0$, the power function can be regarded as a composite function of the logarithmic function and the exponential function, i.e.

$$f(x)=x^a=(a^{\log_a x})^a=a^{a\log_a x}.$$

(5) *trigonometric function*. trigonometric functions are mainly composed of the following:

sine function: $y=\sin x$, $x\in(-\infty, +\infty)$,

cosine function: $y=\cos x$, $x\in(-\infty, +\infty)$,

tangent function: $y=\tan x$, $x\neq n\pi+(\pi/2)$, $n=0, \pm 1, \pm 2, \ldots$.

cotangent function: $y=\cot x$, $x\neq n\pi$, $n=0, \pm 1, \pm 2, \ldots$.

The sine function and the cosine function are both bounded periodical functions with a primary period 2π. The sine function is odd while the cosine function even. The tangent function and the cotangent function are both odd periodical functions with a primary period π. The tangent function is strictly increasing on intervals $\left(n\pi-(\pi/2), n\pi+(\pi/2)\right)$ $(n=0, \pm 1, \pm 2, \cdots)$ while the cotangent function strictly decreasing on intervals $\left(n\pi, (n+1)\pi\right)$ $(n=0, \pm 1, \pm 2, \cdots)$.

(6) *inverse trigonometric function*. Inverse trigonometric functions are mainly composed of the following:

inverse sine function: $y=\arcsin x$, $x\in[-1, 1]$, $y\in$

$[-\pi/2, \pi/2]$;

inverse cosine function: $y = \arccos x$, $x \in (-1, 1)$, $y \in [0, \pi]$;

inverse tangent function: $y = \arctan x$, $x \in (-\infty, +\infty)$, $y \in (-\pi/2, \pi/2)$;

inverse cotangent function: $y = \text{arccot}\, x$, $x \in (-\infty, +\infty)$, $y \in (0, \pi/2)$.

The range mentioned above is exactly called *range of principal value*. So, inverse trigonometric functions are bounded and strictly increasing.

Definition A function which is formed by finite times of the fundamental operations and the compositions of fundamental primary functions is called a *primary function*; a function which is not a primary function is called a *non-primary function*.

For example, the following functions are primary functions:

(1) *polynomial*: $y = a_0 x^n + a_1 x^{n-1} + \cdots + a_n$, where $a_i (i \in \{0, 1, \cdots, n\})$ are constants and $a_0 \neq 0$, n is a non-negative integer, called the *degree* of the polynomial.

(2) *rational function*: $y = P(x)/Q(x)$, where $P(x)$ and $Q(x)$ are both polynomials.

(3) *linear function*: $y = kx + b$, where the constant k is called the *slope* and the constant b is called *y-intercept*. The graph of this function is a straight line.

(4) *quadratic function*: $y = ax^2 + bx + c$, where a, b and c are constants with $a \neq 0$. The graph of this function is a parabola.

The following two well-known functions both defined on $[0, 1]$ are non-primary functions:

Dirichlet function:

$$f(x) = \begin{cases} 1 & \text{if } x \in [0, 1] \text{ and } x \text{ is rational,} \\ 0 & \text{if } x \in [0, 1] \text{ and } x \text{ is irrational.} \end{cases}$$

Riemann function:

$$f(x)=\begin{cases}\frac{1}{q} & \text{if } x=\frac{p}{q}\in[0,1] \text{ where } p, q\in\mathbb{N} \text{ and } p/q \text{ is irreducible fraction,}\\ 0 & \text{if } x\in[0,1] \text{is irrational, or } x=0, 1.\end{cases}$$

Definition A class of primary functions called *hyperbolic functions* is widely used in engineering. They are defined as follows:

(1) *Hyperbolic sine*: $y=\sinh x=(e^x-e^{-x})/2$;

(2) *Hyperbolic cosine*: $y=\cosh x=(e^x+^{e-x})/2$;

(3) *Hyperbolic tangent*: $y=\tanh x=\sinh x/\cosh x$;

(4) *Hyperbolic cotangent*: $y=\coth x=\cosh x/\sinh x$.

Hyperbolic sine and hyperbolic cosine are odd functions; hyperbolic tangent is an even function. The following proposition can be checked in a routine manner.

Proposition (1) $\sinh(u\pm v)=\sinh u\cdot\cosh v\pm\cosh u\cdot\sinh v$;

(2) $\cosh(u\pm v)=\cosh u\cdot\cosh v\pm\sinh u\cdot\sinh v$;

(3) $\cosh^2 u-\sinh^2 u=1$;

(4) $\sinh(2u)=2\sinh x\cosh u$;

(5) $\cosh(2u)=\cosh^2 u+\sinh^2 u=1+2\sinh^2 u=2\cosh^2 u-1$.

Remark Inverse hyperbolic functions are as follows:

(1) $\operatorname{arcsinh} x=\ln(x+\sqrt{x^2+1})$;

(2) $\operatorname{arccosh} x=\ln(x+\sqrt{x^2-1})\ (x\geqslant 1)$;

(3) $\operatorname{arctanh} x=(1/2)\ln[(1+x)/(1-x)]\ (x\in(-1,1))$.

Proof (1) Let $y=\sinh x=\frac{1}{2}(e^x-e^{-x})$. Then $2e^x y=2e^x(\frac{1}{2}(e^x-e^{-x}))=e^{2x}-1$, $e^{2x}-2ye^x-1=0$, $(e^x)^2-(2y)e^x-1=0$, $e^x=(2y\pm\sqrt{4y^2+4})/2=y\pm\sqrt{y^2+1}$.

Since $e^x>0$ for all x, $e^x=y+\sqrt{1+y^2}$. On taking natural logarithms of both sides, we get $x=\ln(y+\sqrt{1+y^2})$, i.e.

$\operatorname{arcsinh} x = \ln(x + \sqrt{x^2 + 1})$.

(2) As in part (1), we let $y = \cosh x$ and $2e^x y = 2e^x\left(\frac{1}{2}(e^x + e^{-x})\right) = e^{2x} + 1$, $e^{2x} - (2y)e^x + 1 = 0$, $e^x = (2y \pm \sqrt{4y^2 - 4})/2 = y \pm \sqrt{y^2 - 1}$. We observe that $\cosh x$ is an even function and hence it is not one-to-one. Since $\cosh(-x) = \cosh(x)$, we will solve for the larger x. On taking natural logarithms of both sides, we get $x_1 = \ln(y + \sqrt{y^2 - 1})$ or $x_2 = \ln(y - \sqrt{y^2 - 1})$. We observe that

$$x_2 = \ln(y - \sqrt{y^2 - 1}) = \ln\left[\frac{(y - \sqrt{y^2 - 1})(y + \sqrt{y^2 - 1})}{y + \sqrt{y^2 - 1}}\right]$$

$$= \ln\left(\frac{1}{y + \sqrt{y^2 - 1}}\right) = -\ln(y + \sqrt{y^2 - 1}) = -x_1.$$

Thus, $\operatorname{arccosh} x = \ln(x + \sqrt{x^2 - 1}) \quad (x \geqslant 1)$.

(3) We begin with $y = \tanh x$ and clear denominator to get

$$y = \frac{e^x - e^{-x}}{e^x + e^{-x}} \quad (|y| < 1),$$

$e^x[(e^x + e^{-x})y] = e^x[(e^x - e^{-x})]$, $(e^{2x} + 1)y = e^{2x} - 1$, $e^{2x}(y - 1) = -(1 + y)$, $e^{2x} = (1 + y)/(1 - y)$,

i.e. $$\operatorname{arctanh} x = \frac{1}{2}\ln\left(\frac{1+x}{1-x}\right) \quad (|x| < 1).$$

Exercises

1. Solve the following inequalities:

(1) $|x| > |x+1|$; (2) $|2x-1| < |x-1|$; (3) $|x+1| + |x-1| \leqslant 4$; (4) $|x-2| - |x| > 1$.

2. Determine whether $f(x)$ and $g(x)$ are a same function. Give your reason:

(1) $f(x)=x$, $g(x)=\sqrt{x^2}$;

(2) $f(x)=x$, $g(x)=(\sqrt{x})^2$;

(3) $f(x)=\lg x^2$, $g(x)=2\lg x$;

(4) $f(x)=\sqrt{\frac{1-x}{1+x}}$, $g(x)=\frac{\sqrt{1-x}}{\sqrt{1+x}}$.

3. Find the inverse functions of the following:

(1) $y=\lg(x+2)+1$; (2) $y=\arcsin\frac{x-1}{4}$;

(3) $y=\arccos\frac{1-x^2}{1+x^2}$.

4. Identify the primary functions by which the following composite function are composed:

(1) $y=\sin^3(1+2x)$; (2) $y=\sin\sqrt[3]{1-e^x}$;

(3) $y=10^{(2x-1)^2}$; (4) $y=\arctan[\tan(a^2+x^2)]^2$.

5. (1) Let $\varphi(x)=x^2$, $\psi(x)=2^x$. Find $\varphi(\psi(x))$, $\psi(\varphi(x))$, $\varphi(\varphi(x))$, $\psi(\psi(x))$.

(2) Let $f(x)=\frac{1}{2}(x+|x|)$ and

$$g(x)=\begin{cases} x & (\text{if } x<0), \\ x^2 & (\text{if } x\geqslant 0). \end{cases}$$

Find $f(g(x))$ and $g(f(x))$.

6. (1) Let $f(x)=\frac{1}{1-x}$. Find $f(f(x))$, $f(f(f(x)))$.

(2) Let $f(x)=x/\sqrt{1+x^2}$. Find $f(f(x))$, $f(f(f(x)))$, $\underbrace{f(\cdots f(x)\cdots)}_{nf's}$.

7. Identify the odd function or even function of the following:

(1) $y=(1-x)^{2/3}+(1+x)^{2/3}$; (2) $y=\frac{1}{2}(a^x-a^{-x})\ (a>0)$;

(3) $y = x \cdot \dfrac{a^x - 1}{a^x + 1}$ $(a > 0)$;　　　(4) $y = \log_a(x + \sqrt{x^2 + 1})$;

(5) $y = x + \sin x$.

8. Which of the following functions are periodical? Find the period and primary period of the periodical functions.

(1) $y = x \cos x$;　　　(2) $y = \sin^2 x$;

(3) $y = |\cos x|$;　　　(4) $y = \sin(\omega x + \varphi)$;

(5) $y = \cos \dfrac{x}{2} + 2 \sin \dfrac{x}{3}$;　　　(6) $y = \sin 2x + \tan \dfrac{x}{2}$.

9. Prove the following equalities:

(1) $\sinh x + \sinh y = 2\sinh \dfrac{x+y}{2} \cosh \dfrac{x-y}{2}$;

(2) $\cosh x - \cosh y = 2\sinh \dfrac{x+y}{2} \sinh \dfrac{x-y}{2}$;

(3) $(\cosh x \pm \sinh x)^n = \cosh nx \pm \sinh nx$;

(4) $\operatorname{arccosh} x = \pm \ln(x + \sqrt{x^2 - 1})$ $(1 \leqslant x < +\infty)$.

10. Prove the following functions are strictly monotone on the designated intervals:

(1) $y = 3x + 1$ on $(-\infty, \infty)$;

(2) $y = \sin x$ on $[-\pi/2, \pi/2]$;

(3) $y = \cos x$ on $[0, \pi]$.

11. Discuss the periodicity and boundedness of the Dirichlet function.

12. Prove the following statements:

(1) the sum of two odd functions is still odd;

(2) the sum of two even functions is still even;

(3) the product of an odd function and an even function is odd;

(4) the product of two odd functions is even;

(5) the product of two even functions is even.

13. Let $f(x)$ be a function defined on $[-l, l]$. Prove $f(x) + f(-x)$ is an even function and $f(x) - f(-x)$ is an odd function.

14. Suppose the graph of $f(x)$ is known. Sketch the graph of the following functions:

(1) $y=-f(x)$; (2) $y=f(-x)$;

(3) $y=-f(-x)$; (4) $y=|f(x)|$;

(5) $y=\operatorname{sgn}(f(x))$; (6) $y=\frac{1}{2}(|f(x)|+f(x))$;

(7) $y=\frac{1}{2}(|f(x)|-f(x))$.

15. Suppose the graphs of $y_1=f(x)$ and $y_2=g(x)$ are known. Sketch the graph of the following functions:

(1) $y=\max\{f(x), g(x)\}$; (2) $y=\min\{f(x), g(x)\}$;

(3) $y=\frac{1}{2}\{f(x)+g(x)+|f(x)-g(x)|\}$;

(4) $y=\frac{1}{2}\{f(x)+g(x)-|f(x)-g(x)|\}$.

16. Assume $f(x)$, $g(x)$ and $h(x)$ are increasing functions such that $f(x)\leqslant g(x)\leqslant h(x)$. Show that $f(f(x))\leqslant g(g(x))\leqslant h(h(x))$.

17. Assume $f(x)$ and $g(x)$ are increasing on (a, b). Show that $\varphi(x)=\max\{f(x), g(x)\}$ and $\psi(x)=\min\{f(x), g(x)\}$ are also increasing on (a, b).

18. Prove: $f(x)=2x+\sin x$ is a strictly increasing function on $(-\infty, +\infty)$.

19. Let a number set

$$S=\left\{\frac{a_1}{b_1}, \frac{a_2}{b_2}, \cdots, \frac{a_n}{b_n}\right\},$$

where $n\geqslant 2$, and a_i and b_i are all natural numbers ($i=1, 2, \cdots, n$). Prove that

$$\min S\leqslant\frac{a_1+a_2+\cdots+a_n}{b_1+b_2+\cdots+b_n}\leqslant\max S.$$

20. Let a, $b \in \mathbb{R}$. Show that $\max\{|a+b|, |a-b|, |b-1|\} \geqslant \frac{1}{2}$.

21. Let a, b, $c>0$. Prove that

$$a^a b^b c^c \geqslant (abc)^{\frac{a+b+c}{3}}.$$

22. Let m, $n \in \mathbb{N}$. Prove that $\min\{\sqrt[n]{m}, \sqrt[m]{n}\} \leqslant \sqrt[3]{3}$.

23. Find the supremum and the infimum of the following number sets and check you results:

(1) $S = \{x \mid x^2 < 2\}$; (2) $S = \{x \mid x = n!, n \in \mathbb{N}\}$;

(3) $S = \{x \mid x \in (0, 1), x \text{ is an irrational}\}$;

(4) $S = \{x \mid x = 1 - \frac{1}{2^n}, n \in \mathbb{N}\}$.

24. Find the supremum and the infimum of the ranges of following functions, and check whether they are also the maximum and the minimum:

(1) $f(x) = x^2$, $x \in (-1, 1)$;

(2) $f(x) = \sin x$, $x \in (0, \pi)$;

(3) $f(x) = \frac{1}{x}$, $x \in (0, 1)$.

Chapter 2 Sequence Limit

The concept of a limit is the fundamental building block on which all other calculus notions are based. Indeed, any "theoretical" development of calculus rests on an extensive use of the theory of limits. As we shall see later, the limit concept will allow us to define the notions of continuity, derivative and definite integral of a function independently of our geometric reasoning. In this chapter we discuss the limit of a sequence and we will discuss the limit of a function in the next chapter, where the sequence limit may be viewed as a discrete situation of the function limit.

§ 2.1 Concept of sequence limit

An *infinite sequence* is an ordered succession of numbers

$$a_1, a_2, \cdots, a_n, \ldots$$

where a_n is said to be the *general element* or *n-th term* of the sequence. The notation $\{a_n\}$ is used to describe the sequence. For example, if $a_n = 1/n$, the corresponding sequence is $\{a_n\}=\{1/n\}=1, 1/2, 1/3, 1/4, \ldots$; with $a_n = n/(1+n)$, the corresponding sequence is $\{n/(1+n)\} =1/2, 2/3, 3/4, \ldots$.

The elements of any sequence $\{a_n\}$ can be represented as points on the real line. In Figure 2 - 1, the elements of the sequence $\{a_n\}= \{1/n\}(n=1, 2, 3, \ldots)$ are shown and it is apparent that these points accumulate at the origin as n increases. This is equivalent to saying that the limit of this sequence is zero.

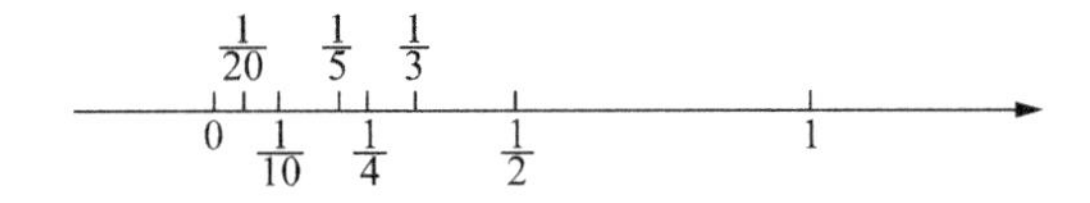

Figure 2 - 1

This observation can be translated into a precise definition of the limit of a sequence, which is also known as (ε, N) definition of a sequence.

Definition Let $\{a_n\}$ be a sequence and let A be a fixed number. If for any given $\varepsilon > 0$, there is a positive integer N such that the inequality

$$|a_n - A| < \varepsilon$$

is true for all $n > N$, then we say the sequence has the *limit* A, which is denoted as

$$\lim_{n \to +\infty} a_n = A.$$

The sequence is also said to *converge* to A or to be *convergent* to A.

Definition A sequence that does not converge is said to *diverge* or to be *divergent*. In particular, a sequence $\{a_n\}$ is said to diverge to $+\infty$, denoted by

$$\lim_{n \to +\infty} a_n = +\infty,$$

if for any positive number M, there exists some natural number N such that

$$a_n \geqslant M \text{ for any } n \geqslant N.$$

The sequence $\{a_n\}$ is said to diverge to $-\infty$, denoted by

$$\lim_{n \to +\infty} a_n = -\infty,$$

if for any positive number M, there exists some natural number N such that

$$a_n \leqslant -M \text{ for any } n \geqslant N.$$

The sequence $\{a_n\}$ is said to diverge to ∞, denoted by

$$\lim_{n \to +\infty} a_n = \infty,$$

if for any positive number M, there exists some natural number N such that

$$|a_n| \geqslant M \text{ for any } n \geqslant N.$$

Remarks (1) Since ε is an arbitrary positive number, 2ε, $(1/3)\varepsilon$ or ε^2, etc. are all arbitrary positive numbers. So, the ε in the inequality of the definition can be replaced by 2ε, $(1/3)\varepsilon$ or ε^2, etc.

(2) Also because of arbitrariness of ε, the symbol $<$ in the inequality of the definition can be replaced by $\leqslant$.

(3) In general, the positive integer N in the definition depends on ε, and the smaller the ε is, the greater the N ought to be. However, the N is not unique. In fact, if N satisfies the condition, $N+1$, $N+2$, ... also satisfy the condition.

(4) $\lim\limits_{n \to +\infty} a_n \neq A \Leftrightarrow$ there is a $\varepsilon > 0$ such that for any $N > 0$ there is $n > N$ with $|a_n - A| \geqslant \varepsilon$.

(5) The sequence $\{a_n\}$ is divergent $\Leftrightarrow$ for any $A \in \mathbb{R}$, $\lim\limits_{n \to +\infty} a_n \neq A$.

Examples (1) Let $a_n = (-1)^n$, $b_n = 2^{-n}$, $c_n = 2^n$, $d_n = (-1)^n/n$.

The sequence $\{a_n\}$ does not converge because its terms oscillate between -1 and 1. The sequence $\{b_n\}$ converges to 0. The sequence $\{c_n\}$ diverges to $+\infty$. The sequence $\{d_n\}$ converges to 0.

(2) Prove by definition that

$$\lim_{n \to +\infty} \left[\frac{1^2}{n^3} + \frac{2^2}{n^3} + \cdots + \frac{(n-1)^2}{n^3}\right] = \frac{1}{3}.$$

Analysis: Note

$$x_n = \frac{1^2}{n^3} + \frac{2^2}{n^3} + \cdots \frac{(n-1)^2}{n^3} = \frac{1}{n^3}[1^2 + 2^2 + \cdots + (n-1)^2]$$

$$= \frac{1}{n^3} \frac{(n-1)n(2n-1)}{6} = \frac{2n^2+3n+1}{6n^2}.$$

Then by

$$\left| x_n - \frac{1}{3} \right| = \frac{3n-1}{6n^2} < \frac{1}{2n} < \varepsilon,$$

we have $n > \frac{1}{2\varepsilon}$.

Proof For any $\varepsilon > 0$, Take $N = \left[\frac{1}{2\varepsilon}\right]$, then if $n > N$,

$$\left| x_n - \frac{1}{3} \right| < \varepsilon, \text{ i. e. } \lim_{n \to +\infty} x_n = \frac{1}{3}. \ \square$$

(3) Prove by definition that $\lim_{n \to +\infty} \sqrt[n]{n} = 1$.

Analysis: Note

$$|\sqrt[n]{n} - 1| = \sqrt[n]{n} - 1 < \varepsilon \Leftrightarrow n < (1+\varepsilon)^n = 1 + n\varepsilon + \frac{n(n-1)}{2}\varepsilon^2 + \cdots$$

$$\Rightarrow n < \frac{n(n-1)}{2}\varepsilon^2 \Leftrightarrow 1 < \frac{n-1}{2}\varepsilon^2 \Leftrightarrow n > \frac{2}{\varepsilon^2} + 1.$$

Proof For any $\varepsilon > 0$, take $N = [2/\varepsilon^2] + 1$, then $|\sqrt[n]{n} - 1| < \varepsilon$ whenever $n > N$. So, $\lim_{n \to +\infty} \sqrt[n]{n} = 1$. $\square$

(4) Suppose a sequence $\{x_n\}$ is convergent. Let $y_n = (x_1 + x_2 + \cdots + x_n)/n$. Prove by definition $\{y_n\}$ is also convergent and

$$\lim_{n \to +\infty} y_n = \lim_{n \to +\infty} x_n.$$

Proof Let $\lim_{n \to +\infty} x_n = a$. Then for any $\varepsilon > 0$, there is N_1 such that $| x_n - a | < \varepsilon/2$ whenever $n > N_1$. Let $d = \max\{| x_1 - a |,$

$|x_2 - a|, \ldots, |x_{N_1} - a|\}$. Clearly there is N_2 such that $d/n < \varepsilon/(2N_1)$ whenever $n > N_2$. Take $N = \max\{N_1, N_2\}$. Then if $n > N$,

$$\left|\frac{x_1 + x_2 + \cdots x_n}{n} - a\right| = \left|\frac{(x_1 - a) + (x_2 - a) + \cdots + (x_n - a)}{n}\right|$$

$$\leqslant \frac{|x_1 - a| + |x_2 - a| + \cdots + |x_{N_1} - a|}{n}$$

$$+ \frac{|x_{N_1} + 1 - a| + \cdots + |x_n - a|}{n}$$

$$< \frac{N_1 d}{n} + \frac{(n - N_1)\frac{\varepsilon}{2}}{n}$$

$$\leqslant N_1 \frac{\varepsilon}{2N_1} + \frac{\varepsilon}{2} = \varepsilon.$$

Thus

$$\lim_{n \to +\infty} y_n = a = \lim_{n \to +\infty} x_n. \ \square$$

§ 2.2 Properties of convergent sequences

Several of the most useful properties of limits are now set forth.

Theorem 2.2.1 (Uniqueness) The limit of a convergent sequence is unique.

Proof Suppose A and B are both limits of a convergent sequence $\{a_n\}$. Then for any given $\varepsilon > 0$, there is a positive integer N_1 such that $|a_n - A| < \varepsilon$ for any $n > N_1$; there is a positive integer N_2 such that $|a_n - B| < \varepsilon$ for any $n > N_2$. Thus for any $n > \max\{N_1, N_2\}$,

$$|A - B| \leqslant |a_n - A| + |a_n - B| < 2\varepsilon,$$

which implies $A = B$. $\square$

Theorem 2.2.2 (Boundedness) If sequence $\{a_n\}$ is convergent, then it is bounded, i. e. there is a number $M > 0$, such that $|a_n| \leqslant M$

for any $n \in \mathbb{N}$.

Proof Suppose $\{a_n\}$ converges to A. Then for $\varepsilon = 1$, there is a positive integer N such that for any $n > N$

$$|a_n - A| < \varepsilon = 1,$$

i. e. $A - 1 < a_n < A + 1$. Thus for any $n \geqslant 1$, we have

$$|a_n| \leqslant \max\{|a_1|, |a_2|, \cdots, |a_N|, |A-1|, |A+1|\}. \square$$

Theorem 2. 2. 3 (Sign-preserving) If sequence $\{a_n\}$ converges to $A \neq 0$, there is a positive integer N such that for any $n > N$, a_n preserves the same sign of A. Moreover, if $A > 0$, then $a_n > a > 0$ for some $a \in (0, A)$; if $A < 0$, then $a_n < a < 0$ for some $a \in (A, 0)$.

Proof Let $A > 0$. Take $a \in (0, A)$ and $\varepsilon = A - a$. Then there is a positive integer N such that for any $n > N$

$$|a_n - A| < \varepsilon = A - a,$$

which implies $a_n > a > 0$. For the case $A < 0$, it can be similarly proved. $\square$

Theorem 2. 2. 4 (Order-preserving) Suppose sequence $\{a_n\}$ converges to A and sequence $\{b_n\}$ converges to B. If there is a positive integer N such that $a_n \leqslant b_n$ whenever $n > N$, then $A \leqslant B$.

Proof Assuming $A > B$, we try to yield a contradiction. Take $\varepsilon = (A - B)/2$. Then there is a positive integer N_1 such that for any $n > N_1$, $a_n > A - \varepsilon = (A + B)/2$, and there is a positive integer N_2 such that for any $n > N_2$, $b_n < B + \varepsilon = (A + B)/2$. Thus as $n > \max\{N, N_1, N_2\}$,

$$a_n > (A + B)/2 > b_n,$$

which contradicts the given condition. $\square$

Theorem 2. 2. 5 (Squeezing law) Suppose sequences $\{a_n\}$ and $\{b_n\}$ both converge to A. If there is a positive integer N such that for any $n > N$, $a_n < c_n < b_n$, then sequence $\{c_n\}$ also converges to A.

Proof By the condition, for any $\varepsilon > 0$, there are positive integers N_1 and N_2 such that $A-\varepsilon < a_n$ whenever $n > N_1$, and $b_n < A+\varepsilon$ whenever $n > N_2$. Then for any $n > \max\{N, N_1, N_2\}$,

$$A-\varepsilon < a_n < c_n < b_n < A+\varepsilon,$$

which implies that $\{c_n\}$ also converges to A. □

Examples (1) Prove

$$\lim_{n\to+\infty}\left[\frac{1}{\sqrt{n^2+1}}+\frac{1}{\sqrt{n^2+2}}+\cdots+\frac{1}{\sqrt{n^2+n}}\right]=1.$$

Proof Let $y_n = \dfrac{n}{\sqrt{n^2+1}}$ and let $z_n = \dfrac{n}{\sqrt{n^2+n}}$.

$$\text{Note } z_n \leqslant \frac{1}{\sqrt{n^2+1}}+\frac{1}{\sqrt{n^2+2}}+\cdots+\frac{1}{\sqrt{n^2+n}} \leqslant y_n$$

and $\lim\limits_{n\to+\infty} y_n = \lim\limits_{n\to+\infty} z_n = 1$. Then the result follows from the squeezing law directly. □

(2) Suppose $a_1, a_2, \ldots, a_m$ are m positive numbers. Find

$$\lim_{n\to+\infty}\sqrt[n]{a_1^n+a_2^n+\cdots a_m^n}.$$

Solution Let $a=\max\{a_1, a_2, \ldots, a_m\}$. Since $a^n < a_1^n+a_2^n+\cdots+a_m^n < ma^n$, so $a < \sqrt[n]{a_1^n+a_2^n+\ldots a_m^n} < a\sqrt[n]{m}$. Note

$$\lim_{n\to+\infty} a = \lim_{n\to+\infty} a\sqrt[n]{m} = a.$$

Then by squeezing law we have

$$\lim_{n\to+\infty}\sqrt[n]{a_1^n+a_2^n+\cdots a_m^n} = \max\{a_1, a_2, \ldots, a_m\}. \square$$

Theorem 2. 2. 6 (Fundamental operations of sequence limit) Suppose sequences $\{a_n\}$ and $\{b_n\}$ are both convergent. Then

$\{a_n + b_n\}$, $\{a_n - b_n\}$, $\{a_n \cdot b_n\}$ are convergent and

(1) $\lim_{n\to+\infty}(a_n \pm b_n) = \lim_{n\to+\infty} a_n \pm \lim_{n\to+\infty} b_n$;

(2) $\lim_{n\to+\infty}(a_n \cdot b_n) = \lim_{n\to+\infty} a_n \cdot \lim_{n\to+\infty} b_n$;

(3) $\lim_{n\to+\infty}(ka_n) = k\lim_{n\to+\infty} a_n$, where k is any constant;

(4) if for any n, $b_n \neq 0$ and $\lim_{n\to\infty} b_n \neq 0$, then $\{a_n/b_n\}$ is convergent and $\lim_{n\to+\infty}\frac{a_n}{b_n} = \frac{\lim_{n\to+\infty} a_n}{\lim_{n\to+\infty} b_n}$.

Proof Suppose sequence $\{a_n\}$ converges to A and sequence $\{b_n\}$ converges to B. Then for any $\varepsilon > 0$, there are positive integers N_1 and N_2 such that for any $n > N_1$, $|a_n - A| < \varepsilon$ and for any $n > N_2$, $|b_n - B| < \varepsilon$.

(1), (2), (3): Let $N = \max\{N_1, N_2\}$. Then for any $n > N$,

$$|(a_n + b_n) - (A + B)| \leqslant |a_n - A| + |b_n - B| < 2\varepsilon,$$

which implies $\lim_{n\to+\infty}(a_n + b_n) = \lim_{n\to+\infty} a_n + \lim_{n\to+\infty} b_n$.

By boundedness of $\{b_n\}$, there is $M > 0$ such that $|b_n| < M$ for any n, and so for any $n > \max\{N_1, N_2\}$,

$$|a_n b_n - AB| \leqslant |a_n - A||b_n| + |A||b_n - B| < (M + |A|)\varepsilon,$$

which implies $\lim_{n\to+\infty}(a_n \cdot b_n) = \lim_{n\to+\infty} a_n \cdot \lim_{n\to+\infty} b_n$.

Let $b_n = k$ for any n. Then (3) follows from (2) directly.

Noting $a_n - b_n = a_n + (-1)b_n$, we see that $\lim_{n\to+\infty}(a_n - b_n) = \lim_{n\to+\infty} a_n + \lim_{n\to\infty}(-1)b_n = \lim_{n\to+\infty} a_n + (-1)\lim_{n\to+\infty} b_n = \lim_{n\to+\infty} a_n - \lim_{n\to+\infty} b_n$.

(4) Since $\lim_{n\to+\infty} b_n = B \neq 0$, by sign-preserving property, there is a positive integer N and $k \in (0, |B|)$ such that $|b_n| > k$ whenever $n > N$. Thus for any $n > \max\{N, N_2\}$,

$$\left|\frac{1}{b_n} - \frac{1}{B}\right| = \frac{|b_n - B|}{|b_n B|} < \frac{|b_n - B|}{k|B|} < \frac{\varepsilon}{k|B|},$$

which implies

$$\lim_{n\to+\infty}\frac{1}{b_n}=\frac{1}{B},$$

and so noting $a_n/b_n = a_n \cdot (1/b_n)$,

$$\lim_{n\to+\infty}\frac{a_n}{b_n}=\frac{\lim\limits_{n\to+\infty}a_n}{\lim\limits_{n\to+\infty}b_n}. \quad \square$$

Examples (1) Find

$$I=\lim_{n\to+\infty}\left(\frac{3n+1}{n}\cdot\frac{n+1}{n}\right).$$

Solution
$$\begin{aligned} I &= \lim_{n\to+\infty}\frac{3n+1}{n}\cdot\lim_{n\to+\infty}\frac{n+1}{n}\\ &= \lim_{n\to+\infty}\left(3+\frac{1}{n}\right)\cdot\lim_{n\to+\infty}\left(1+\frac{1}{n}\right)\\ &= \left(\lim_{n\to+\infty}3+\lim_{n\to+\infty}\frac{1}{n}\right)\left(\lim_{n\to+\infty}1+\lim_{n\to+\infty}\frac{1}{n}\right)=3\times1=3. \end{aligned}$$

(2) Find

$$I=\lim_{n\to+\infty}\frac{a^n}{a^n+1},\quad |a|\neq 1.$$

Solution If $|a|<1$, then by $\lim\limits_{n\to+\infty}a^n=0$,

$$I=\frac{\lim\limits_{n\to+\infty}a^n}{\lim\limits_{n\to+\infty}a^n+\lim\limits_{n\to+\infty}1}=\frac{0}{0+1}=0;$$

If $|a|>1$, then since $\lim\limits_{n\to+\infty}a^n=\infty$,

$$I=\lim_{n\to+\infty}\frac{1}{1+(1/a^n)}=\frac{1}{1+(1/\lim\limits_{n\to+\infty}a^n)}=\frac{1}{1+0}=1.$$

Definition Let $\{a_n\}$ be a sequence. If for any n, $a_n\leqslant a_{n+1}$, then $\{a_n\}$ is said to be *increasing*; if for any n, $a_n\geqslant a_{n+1}$, then $\{a_n\}$ is said to be *decreasing*; $\{a_n\}$ is said to be *monotone*, if $\{a_n\}$ is either increasing or decreasing.

If $a_n < a_{n+1}$ ($a_n > a_{n+1}$) for any n, then $\{a_n\}$ is said to be *strictly increasing* (*strictly decreasing*). If a sequence is strictly increasing or strictly decreasing, it is said to be *strictly monotone*.

Theorem (Stolz theorem I) Suppose sequence $\{y_n\}$ is strictly decreasing and $\lim\limits_{n\to\infty} x_n = \lim\limits_{n\to+\infty} y_n = 0$. If $\lim\limits_{n\to+\infty} \frac{x_n - x_{n+1}}{y_n - y_{n+1}}$ exists (i. e. $\in \mathbb{R}$) or$=\pm\infty$, then $\lim\limits_{n\to+\infty} \frac{x_n}{y_n}$ also exists and

$$\lim_{n\to+\infty} \frac{x_n}{y_n} = \lim_{n\to\infty} \frac{x_n - x_{n+1}}{y_n - y_{n+1}}.$$

Proof First, let $\lim\limits_{n\to\infty} \frac{x_n - x_{n+1}}{y_n - y_{n+1}} = a \in \mathbb{R}$. Then, for any $\varepsilon > 0$, there exists $N \in \mathbb{N}$ such that

$$\left|\frac{x_n - x_{n+1}}{y_n - y_{n+1}} - a\right| < \varepsilon, \text{ i. e. } a - \varepsilon < \frac{x_n - x_{n+1}}{y_n - y_{n+1}} < a + \varepsilon,$$

whenever $n \geqslant N$. Since $y_n > y_{n+1}$ for any n, if $n \geqslant N$,

$$(a-\varepsilon)(y_n - y_{n+1}) < x_n - x_{n+1} < (a+\varepsilon)(y_n - y_{n+1}),$$

$$(a-\varepsilon)(y_{n+1} - y_{n+2}) < x_{n+1} - x_{n+2} < (a+\varepsilon)(y_{n+1} - y_{n+2}),$$

$$\cdots\cdots\cdots\cdots\cdots\cdots\cdots\cdots\cdots\cdots\cdots\cdots\cdots\cdots\cdots\cdots\cdots\cdots$$

$$(a-\varepsilon)(y_{n+p-1} - y_{n+p}) < x_{n+p-1} - x_{n+p}$$

$< (a+\varepsilon)(y_{n+p-1} - y_{n+p})$, which sums into

$$(a-\varepsilon)(y_n - y_{n+p}) < x_n - x_{n+p} < (a+\varepsilon)(y_n - y_{n+p}).$$

Let $p\to\infty$. Noticing $\lim\limits_{p\to\infty} x_{n+p} = \lim\limits_{p\to\infty} y_{n+p} = 0$, we have $(a-\varepsilon)y_n \leqslant x_n \leqslant (a+\varepsilon)y_n$ for any $n \geqslant N$, which implies that $a - \varepsilon \leqslant \frac{x_n}{y_n} \leqslant a + \varepsilon$ whenever $n \geqslant N$, i. e. $\lim\limits_{n\to\infty} \frac{x_n}{y_n} = a$.

Now, let $\lim\limits_{n\to\infty}\dfrac{x_n - x_{n+1}}{y_n - y_{n+1}} = +\infty$. Then, for any $M > 0$, there is $N \in \mathbb{N}$ such that

$$\frac{x_n - x_{n+1}}{y_n - y_{n+1}} > M, \text{ i. e. } x_n - x_{n+1} > M(y_n - y_{n+1})$$

whenever $n \geqslant N$. In a way quite similar to what we do in the above case, we may derive $\lim\limits_{n\to\infty}\dfrac{x_n}{y_n} = +\infty$. Similar approach also applies to the case when $\lim\limits_{n\to\infty}\dfrac{x_n - x_{n+1}}{y_n - y_{n+1}} = -\infty$. □

Theorem (Stolz theorem Ⅱ) Suppose sequence $\{y_n\}$ is strictly increasing and $\lim\limits_{n\to+\infty} y_n = +\infty$. If $\lim\limits_{n\to+\infty}\dfrac{x_{n+1} - x_n}{y_{n+1} - y_n}$ exists (i. e. $\in \mathbb{R}$) or $= \pm\infty$, then $\lim\limits_{n\to+\infty}\dfrac{x_n}{y_n}$ also exists and

$$\lim_{n\to+\infty}\frac{x_n}{y_n} = \lim_{n\to\infty}\frac{x_{n+1} - x_n}{y_{n+1} - y_n}.$$

Proof Firstly, let $\lim\limits_{n\to\infty}\dfrac{x_{n+1} - x_n}{y_{n+1} - y_n} = 0$. Then, for any $\varepsilon > 0$, there exists $N_1 \in \mathbb{N}$ such that $|x_{n+1} - x_n| < \varepsilon(y_{n+1} - y_n)$ whenever $n \geqslant N_1$. Noticing $\lim\limits_{n\to+\infty} y_n = +\infty$, we may suppose $y_n > 0 (n \geqslant N_1)$. Thus

$$\begin{aligned} |x_{n+1} - x_{N_1}| &\leqslant |x_{n+1} - x_n| + |x_n - x_{n-1}| + \cdots + |x_{N_1+1} - x_{N_1}| \\ &< \varepsilon(y_{n+1} - y_n) + \varepsilon(y_n - y_{n-1}) + \cdots + \varepsilon(y_{N_1+1} - y_{N_1}) \\ &= \varepsilon(y_{n+1} - y_{N_1}), \end{aligned}$$

i. e.

$$\frac{|x_{n+1} - x_{N_1}|}{y_{n+1} - y_{N_1}} < \varepsilon (n \geqslant N_1).$$

Since $\lim\limits_{n\to+\infty} y_n = +\infty$, for any $\varepsilon > 0$, there exists $N_2 \in \mathbb{N}$ such that $\left|\dfrac{x_{N_1}}{y_{n+1}}\right| < \varepsilon$ whenever $n \geqslant N_2$. Take $N = \max\{N_1, N_2\}$. Then, for any

$n > N$,

$$\left|\frac{x_{n+1}}{y_{n+1}}\right| = \left|\frac{x_{n+1} - x_{N_1}}{y_{n+1} - y_{N_1}} \cdot \frac{y_{n+1} - y_{N_1}}{y_{n+1}} + \frac{x_{N_1}}{y_{n+1}}\right|$$

$$\leqslant \left|\frac{x_{n+1} - x_{N_1}}{y_{n+1} - y_{N_1}}\right| \cdot \left|1 - \frac{y_{N_1}}{y_{n+1}}\right| + \left|\frac{x_{N_1}}{y_{n+1}}\right| < \varepsilon + \varepsilon = 2\varepsilon,$$

which implies $\lim\limits_{n\to+\infty} \frac{x_n}{y_n} = 0$.

Secondly, let $\lim\limits_{n\to\infty} \frac{x_{n+1} - x_n}{y_{n+1} - y_n} = a < +\infty (a \neq 0)$. Let $z_n = x_n - a y_n$. Then, $\lim\limits_{n\to\infty} \frac{z_{n+1} - z_n}{y_{n+1} - y_n} = \lim\limits_{n\to\infty}\left(\frac{x_{n+1} - x_n}{y_{n+1} - y_n} - a\right) = a - a = 0$, and so $\lim\limits_{n\to+\infty} \frac{z_n}{y_n} = 0$. Thus, $\lim\limits_{n\to+\infty} \frac{x_n}{y_n} = \lim\limits_{n\to\infty}(\frac{z_n}{y_n} + a) = a$.

Thirdly, let $\lim\limits_{n\to\infty} \frac{x_{n+1} - x_n}{y_{n+1} - y_n} = +\infty$. Then, there exists $N \in \mathbb{N}$ such that $x_{n+1} - x_n > y_{n+1} - y_n > 0$ whenever $n \geqslant N$, which implies that $\{x_n\}$ is also strictly increasing. Then, noticing $x_n - x_N > y_n - y_N$ $(n > N)$, we have $\lim\limits_{n\to+\infty} x_n = +\infty$. Hence, $\lim\limits_{n\to+\infty} \frac{y_n}{x_n} = \lim\limits_{n\to\infty} \frac{y_n - y_{n-1}}{x_n - x_{n-1}} = 0$, i. e. $\lim\limits_{n\to+\infty} \frac{x_n}{y_n} = +\infty$.

Finally, we may similarly prove the result for the case when $\lim\limits_{n\to+\infty} \frac{x_{n+1} - x_n}{y_{n+1} - y_n} = -\infty$. □

Remark (1) In Stloz theorems, it is required that $\lim\limits_{n\to\infty} \frac{x_{n+1} - x_n}{y_{n+1} - y_n}$ exists (i. e. $\in \mathbb{R}$) or $= \pm\infty$. However, if $\lim\limits_{n\to\infty} \frac{x_{n+1} - x_n}{y_{n+1} - y_n} = \infty$, the result may not hold true. For example, let $x_n = (-1)^{n-1} n$, $y_n = n (n = 1, 2, \cdots)$. Then, $\lim\limits_{n\to\infty} \frac{x_{n+1} - x_n}{y_{n+1} - y_n} = \infty$, but the sequence $\{\frac{x_n}{y_n}\}$ is:

$1, -1, 1, -1, \cdots$, and so clearly, $\lim\limits_{n\to+\infty} \frac{x_n}{y_n}$ is non-existent.

(2) The condition in Stloz theorems is sufficient, but not necessary. i. e. the non-existence of $\lim\limits_{n\to\infty} \frac{x_{n+1} - x_n}{y_{n+1} - y_n}$ does not imply the non-existence of $\lim\limits_{n\to+\infty} \frac{x_n}{y_n}$. For example, let $x_n = 1 - 2 + 3 - 4 + \cdots + (-1)^{n-1} n$, $y_n = n^2 (n = 1, 2, \cdots)$. Then, it requires only routine verification to check that $\lim\limits_{n\to\infty} \frac{x_{n+1} - x_n}{y_{n+1} - y_n}$ does not exist, but $\lim\limits_{n\to+\infty} \frac{x_n}{y_n} = 0$.

Examples (1) Let $\{s_n\}$ be a sequence $(n = 0, 1, 2, \cdots)$ and $\sigma_n = \frac{1}{n+1}(s_0 + s_1 + \cdots + s_n)$. Let $a_n = s_n - s_{n-1} (n = 1, 2, \cdots)$. If $\lim\limits_{n\to\infty} na_n = 0$ and $\{\sigma_n\}$ is convergent, prove that $\{s_n\}$ is convergent and $\lim\limits_{n\to\infty} s_n = \lim\limits_{n\to+\infty} \sigma_n$.

Proof Note that $s_n - \sigma_n = s_n - \frac{s_0 + s_1 + \cdots + s_n}{n+1} = \frac{1}{n+1}[(s_n - s_0) + (s_n - s_1) + \cdots + (s_n - s_n)] = \frac{1}{n+1}\sum_{k=1}^{n} ka_k$. By Stolz theorem Ⅱ,

$$\lim_{n\to\infty} \frac{1}{n+1}\sum_{k=1}^{n} ka_k = \lim_{n\to\infty} \frac{\sum_{k=1}^{n+1} ka_k - \sum_{k=1}^{n} ka_k}{(n+2) - (n+1)} = \lim_{n\to\infty}(n+1)a_{n+1} = 0.$$

Hence, $\lim\limits_{n\to\infty} s_n = \lim\limits_{n\to+\infty} \sigma_n$. □

(2) Find $\lim\limits_{n\to+\infty} x_n$, where

$$x_n = \left(\frac{2}{2^2 - 1}\right)^{\frac{1}{2^{n-1}}} \cdot \left(\frac{2^2}{2^3 - 1}\right)^{\frac{1}{2^{n-2}}} \cdots \left(\frac{2^{n-1}}{2^n - 1}\right)^{\frac{1}{2}}.$$

Solution Note that

$$\ln x_n = \frac{1}{2^{n-1}}\ln\frac{2}{2^2-1} + \frac{1}{2^{n-2}}\ln\frac{2^2}{2^3-1} + \cdots + \frac{1}{2}\ln\frac{2^{n-1}}{2^n-1}$$
$$= \frac{1}{2^{n-1}}\left(\ln\frac{2}{2^2-1} + 2\ln\frac{2^2}{2^3-1} + \cdots + 2^{n-2}\ln\frac{2^{n-1}}{2^n-1}\right).$$

Then, by Stolz theorem Ⅱ, we have

$$\lim_{n\to+\infty}\ln x_n = \lim_{n\to+\infty}\frac{2^{n-2}\ln\frac{2^{n-1}}{2^n-1}}{2^{n-1}-2^{n-2}} = \lim_{n\to\infty}\frac{\ln\frac{2^{n-1}}{2^n-1}}{2-1}$$
$$= \lim_{n\to\infty}\ln\frac{2^{n-1}}{2^n-1} = \ln\frac{1}{2}.$$

Hence, $\lim\limits_{n\to+\infty} x_n = \frac{1}{2}$.

(3) Let $a_1 > 0$ and $a_{n+1} = a_n + \frac{1}{a_n}(n = 1, 2, \cdots)$. Prove that $\lim\limits_{n\to+\infty}\frac{a_n}{\sqrt{2n}} = 1$.

Proof We first assert that $\lim\limits_{n\to\infty} a_n = +\infty$. Because otherwise, since $\{a_n\}$ is (strictly) increasing, it is convergent and we may let $\lim\limits_{n\to+\infty} a_n = b \in \mathbb{R}$. Thus, $b = b + \frac{1}{b}$, which is a contradiction. Then, by Stolz theorem Ⅱ,

$$\lim_{n\to+\infty}\frac{a_n^2}{2n} = \lim_{n\to\infty}\frac{a_{n+1}^2 - a_n^2}{2(n+1)-2n} = \lim_{n\to\infty}\frac{1}{2}(a_{n+1}^2 - a_n^2)$$
$$= \lim_{n\to\infty}\frac{1}{2}\left[\left(a_n + \frac{1}{a_n}\right)^2 - a_n^2\right]$$
$$= \lim_{n\to\infty}\frac{1}{2}\left(2 + \frac{1}{a_n^2}\right) = 1.$$

Therefore, $\lim\limits_{n\to\infty}\frac{a_n}{\sqrt{2n}} = \lim\limits_{n\to\infty}\sqrt{\frac{a_n^2}{2n}} = 1$. □

(4) Let $p\in\mathbb{N}$. Find $\lim\limits_{n\to\infty}\frac{1^p + 2^p + \cdots + n^p}{n^{p+1}}$.

Solution Let $x_n = 1^p + 2^p + \cdots + n^p$ and let $y_n = n^{p+1}$. Then

$$\begin{aligned}
&\lim_{n\to\infty} \frac{1^p + 2^p + \cdots + n^p}{n^{p+1}} \\
&= \lim_{n\to\infty} \frac{x_{n+1} - x_n}{y_{n+1} - y_n} = \lim_{n\to\infty} \frac{(n+1)^p}{(n+1)^{p+1} - n^{p+1}} \\
&= \lim_{n\to\infty} \frac{(n+1)^p}{\left[n^{p+1} + (p+1)n^p + \frac{(p+1)p}{2!}n^{p-1} + \cdots + 1\right] - n^{p+1}} \\
&= \lim_{n\to\infty} \frac{\left(1+\frac{1}{n}\right)^p}{(p+1) + \frac{(p+1)p}{2!} \cdot \frac{1}{n} + \cdots + \frac{1}{n^p}} = \frac{1}{p+1}.
\end{aligned}$$

(5) Let $a_n = \dfrac{\sqrt{1} + \sqrt{2} + \cdots \sqrt{n}}{n\sqrt{n}}$. Find $\lim\limits_{n\to+\infty} a_n$.

Solution Let $x_n = \sqrt{1} + \sqrt{2} + \cdots + \sqrt{n}$ and let $y_n = n\sqrt{n}$. Then, $\{y_n\}$ is strictly increasing and $\lim\limits_{n\to+\infty} y_n = +\infty$. Thus

$$\begin{aligned}
\lim_{n\to\infty} a_n &= \lim_{n\to\infty} \frac{x_n}{y_n} = \lim_{n\to\infty} \frac{x_{n+1} - x_n}{y_{n+1} - y_n} \\
&= \lim_{n\to\infty} \frac{\sqrt{n+1}}{(n+1)\sqrt{n+1} - n\sqrt{n}} \\
&= \lim_{n\to\infty} \frac{\sqrt{n+1}\left[(n+1)\sqrt{n+1} + n\sqrt{n}\right]}{(n+1)^3 - n^3} = \frac{2}{3}.
\end{aligned}$$

§ 2.3 Fundamental theorems of sequence limit

Theorem 2.3.1 (The monotone convergence theorem) A monotone and bounded sequence must be convergent. Exactly,

(1) if a sequence $\{a_n\}$ is increasing and bounded above, then $\{a_n\}$ is convergent and $\lim\limits_{n\to+\infty} a_n = \sup\{a_n\}$;

(2) if a sequence $\{a_n\}$ is decreasing and bounded below, then

$\{a_n\}$ is convergent and $\lim_{n\to\infty} a_n = \inf\{a_n\}$.

Proof (1) Since $\{a_n\}$ is bounded above, by the completeness axiom, there exists a supremum. Denote $\sup\{a_n\} = L$. Then, for any $\varepsilon > 0$, $L - \varepsilon$ is not an upper bound, i. e. there is some $N \in \mathbb{N}$ such that $a_N > L - \varepsilon$. As $\{a_n\}$ is increasing, $a_n > L - \varepsilon$ for any $n > N$, i. e. $0 \leqslant L - a_n < \varepsilon$. Thus, for any $\varepsilon > 0$, there is $N \in \mathbb{N}$ such that $|L - a_n| < \varepsilon$ whenever $n > N$. So, $\{a_n\}$ is convergent and $\lim_{n\to+\infty} a_n = L = \sup\{a_n\}$.

(2) can be proved in a similar manner. □

Examples (1) Show that $\lim_{n\to+\infty}\left(\frac{10}{1} \cdot \frac{11}{3} \cdots \frac{n+9}{2n-1}\right)$ exists.

Proof Let $x_n = \frac{10}{1} \cdot \frac{11}{3} \cdots \frac{n+9}{2n-1}$. Then it is easy to check that $\frac{x_{n+1}}{x_n} = \frac{n+10}{2n+1} < \frac{2n+1}{2n+1} = 1(\text{if } n > 9)$, and so $\{x_n\}$ is decreasing if $n > 9$. Clearly, $x_n > 0$, i. e. $\{x_n\}$ is bounded below. So by Theorem 2. 3. 1, $\{x_n\}$ is convergent. □

(2) Let $x_1 > 0$ and $x_{n+1} = \frac{3(1+x_n)}{3+x_n}(n = 1, 2, \cdots)$. Prove the sequence $\{x_n\}$ is convergent.

Proof Since $0 < x_n < \frac{3(1+x_{n-1})}{3+x_{n-1}} \leqslant \frac{3(1+x_{n-1})}{1+x_{n-1}} = 3$, $\{x_n\}$ is bounded. Note that $x_{n+1} - x_n = \frac{6(x_n - x_{n-1})}{(3+x_n)(3+x_{n-1})}$, which implies that $x_{n+1} - x_n$ and $x_n - x_{n-1}$ have a same sign. Thus $\{x_n\}$ is monotone. Then by Theorem 2. 3. 1 we have the result. □

(3) Let $a > 0$ and let $\{x_n\}$ be defined as follows: $x_0 > 0$; $x_{n+1} = \frac{1}{2}\left(x_n + \frac{a}{x_n}\right)(n = 0, 1, 2, \cdots)$. Prove that $\{x_n\}$ is convergent and $\lim_{n\to+\infty} x_n = \sqrt{a}$.

Proof Since $x_{n+1} = \frac{1}{2}\left(x_n + \frac{a}{x_n}\right) \geqslant \sqrt{x_n \cdot \frac{a}{x_n}} = \sqrt{a}$, $\{x_n\}$ is bounded below. Note that $x_{n+1} - x_n = \frac{1}{2}\left(x_n + \frac{a}{x_n}\right) - x_n = \frac{a}{2x_n} - \frac{1}{2}x_n \leqslant \frac{x_n^2}{2x_n} - \frac{1}{2}x_n = 0$. So, $\{x_n\}$ is decreasing. Thus this sequence is convergent.

Let $\lim_{n\to+\infty} x_n = A$, and so by $x_{n+1} = \frac{1}{2}\left(x_n + \frac{a}{x_n}\right)$, we have $A = \frac{1}{2}\left(A + \frac{a}{A}\right)$, which implies $A = \sqrt{a}$. □

(4) Suppose $x_0 = a > 0$, $y_0 = b > 0$, $x_{n+1} = \sqrt{x_n y_n}$, $y_{n+1} = \frac{x_n + y_n}{2}$ $(n = 0, 1, 2, \cdots)$. Prove $\{x_n\}$ and $\{y_n\}$ are both convergent, and $\lim_{n\to+\infty} x_n = \lim_{n\to+\infty} y_n$.

Proof Clearly for any $n \in \{1, 2, \cdots\}$, $x_{n+1} \leqslant y_{n+1}$ and also $x_{n+1} = \sqrt{x_n y_n} \geqslant \sqrt{x_n x_n} = x_n$, $y_{n+1} = \frac{x_n + y_n}{2} \leqslant \frac{y_n + y_n}{2} = y_n$, which implies that $\{x_n\}$ $(n = 1, 2, \cdots)$ is increasing and bounded above, whereas $\{y_n\}$ $(n = 1, 2, \cdots)$ is decreasing and bounded below. So, these two sequences are both convergent.

Now, let $\lim_{n\to+\infty} x_n = A$ and $\lim_{n\to+\infty} y_n = B$. Then from $y_{n+1} = (x_n + y_n)/2$ it follows that $B = (A+B)/2$, i. e. $A = B$. □

(5) Suppose $\{x_n\}$ is a sequence such that $0 < x_n < 1$ and $(1 - x_n)x_{n+1} > \frac{1}{4}$ for any $n = 1, 2, \cdots$. Prove that $\{x_n\}$ is convergent and find $\lim_{n\to+\infty} x_n$.

Proof and solution Let $f(x) = (1-x)x$. Clearly, $\max_{x\in(0,1)} f(x) = x(1-x)\big|_{x=\frac{1}{2}} = \frac{1}{4}$ and so $(1-x_n)x_n \leqslant \frac{1}{4} < (1-x_n)x_{n+1}$. Thus, $x_{n+1} > x_n$, i. e. $\{x_n\}$ is increasing and bounded above. By the above theorem, $\{x_n\}$ is convergent.

Let $\lim_{n \to +\infty} x_n = a$. Then, by taking limit on both sides of $(1-x_n)x_n \leqslant \frac{1}{4} < (1-x_n)x_{n+1}$, we have $(1-a)a \leqslant \frac{1}{4} \leqslant (1-a)a$, i. e. $(1-a)a = \frac{1}{4}$. So, $\lim_{n \to +\infty} x_n = a = \frac{1}{2}$. □

(6) Let the sequence $\{a_n\}$ be: $\sqrt{2}$, $\sqrt{2+\sqrt{2}}$, $\sqrt{2+\sqrt{2+\sqrt{2}}}$, $\cdots$. Show that $\{a_n\}$ is convergent and find its limit.

Proof and solution Obviously, $\{a_n\}$ is increasing. Now we show it is bounded by induction. $a_1 = \sqrt{2} < 2$. Assume $a_n < 2$. Then $a_{n+1} = \sqrt{2+a_n} < \sqrt{2+2} = 2$, which implies $0 < a_n < 2$ for any n. So, by the preceding theorem, $\{x_n\}$ is convergent.

Let $\lim_{n \to +\infty} a_n = a$. Then, by taking limit on both sides of $a_{n+1} = \sqrt{2+a_n}$, we have $a = \sqrt{2+a}$. As $a = -1$ is impossible, $\lim_{n \to +\infty} a_n = a = 2$.

(7) Let $x_0 = 1$, $x_{n+1} = \frac{x_n}{1+x_n} + 1$. Prove the sequence $\{x_n\}$ is convergent and find its limit.

Proof and solution Clearly, $0 < x_n < 2$, and so $\{x_n\}$ is bounded. Now we show the sequence is monotone by induction. $x_1 = \frac{x_0}{1+x_0} + 1 = \frac{3}{2} > x_0$. Provided $x_k > x_{k-1}$, $x_k(1+x_{k-1}) > x_{k-1}(1+x_k)$, i. e. $\frac{x_k}{1+x_k} > \frac{x_{k-1}}{1+x_{k-1}}$, and so $x_{k+1} > x_k$. Thus, $\{x_n\}$ is monotone. Hence, the sequence $\{x_n\}$ is convergent.

Let $\lim_{n \to +\infty} x_n = x$. Then $x = \frac{x}{1+x} + 1$, and so $\lim_{n \to +\infty} x_n = x = \frac{1+\sqrt{5}}{2}$. □

(8) Let $x_1 \in (0, 1)$ and $x_{n+1} = x_n(1-x_n)$ for any $n = 1, 2, \cdots$. Prove $\lim_{n \to +\infty} nx_n = 1$.

Proof Since $x_1 \in (0, 1)$, by induction it is easy to check that

$x_n \in (0, 1)$ for any $n=1, 2, \cdots$. Thus, $0<\frac{x_{n+1}}{x_n}=1-x_n<1$, which implies that $\{x_n\}$ is decreasing and bounded below. Then, $\lim\limits_{n\to\infty} x_n = a$ for some $a \in \mathbb{R}$, and so $a = a(1-a)$. Hence, $\lim\limits_{n\to+\infty} x_n = a = 0$. Now let $b_n = \frac{1}{x_n}$. Then, $\{b_n\}$ is increasing and $\lim\limits_{n\to\infty} b_n = +\infty$. By Stolz theorem,

$$\begin{aligned}\lim_{n\to\infty} nx_n &= \lim_{n\to\infty}\frac{n}{bn} = \lim_{n\to\infty}\frac{(n+1)-1}{b_{n+1}-b_n} = \lim_{n\to\infty}\frac{1}{b_{n+1}-b_n}\\ &= \lim_{n\to\infty}\frac{x_n x_{n+1}}{x_n - x_{n+1}} = \lim_{n\to\infty}\frac{x_n[x_n(1-x_n)]}{x_n - x_n(1-x_n)}\\ &= \lim_{n\to\infty}(1-x_n) = 1. \ \square\end{aligned}$$

Proposition The sequence $\left\{\left(1+\frac{1}{n}\right)^n\right\}$ is convergent.

Proof Let $a_n = \left\{\left(1+\frac{1}{n}\right)^n\right\}$. Then $a_n = 1+n\cdot\frac{1}{n}+\frac{n(n-1)}{2!}\cdot\frac{1}{n^2}+\frac{n(n-1)(n-2)}{3!}\cdot\frac{1}{n^3}+\cdots+\frac{1}{n^n}\leqslant 1+1+\frac{1}{2!}+\frac{1}{3!}+\cdots+\frac{1}{n!}\leqslant 1+1+\frac{1}{2}+\frac{1}{2^2}+\cdots+\frac{1}{2^{n-1}}=1+\frac{1\cdot\left(1-\frac{1}{2^n}\right)}{1-\frac{1}{2}}=3-\frac{1}{2^{n-1}}<3$, which implies $\{a_n\}$ is bounded above. Recall that $\left(1+\frac{1}{n}\right)^n < \left(1+\frac{1}{n+1}\right)^{n+1}$ (an example of Theorem 1. 2. 3, the average number inequality), i. e. the sequence $\{a_n\}$ is (strictly) increasing. Then the result follows directly from the monotone convergence theorem. $\square$

Remark

(1) Denote $\lim\limits_{n\to+\infty}\left(1+\frac{1}{n}\right)^n = \mathrm{e}$, and denote $\log_{\mathrm{e}} x = \ln x$, called the *natural logrithmic function*. It can be proved that e is an

irrational number with $e \approx 2.171\,828\,182\,845\,9$. We will see later this e is an important constant in mathematical analysis.

(2) It is easy to check: $\lim\limits_{n\to+\infty}\left(1+\dfrac{1}{n+1}\right)^n = \lim\limits_{n\to+\infty}\left(1+\dfrac{1}{n}\right)^{n+1} = e$; $\lim\limits_{n\to+\infty}\left(1+\dfrac{a}{n}\right)^n = e^a$ for any $a \in \mathbb{R}$; in particular, $\lim\limits_{n\to+\infty}\left(1-\dfrac{1}{n}\right)^n = \dfrac{1}{e}$.

(3) For any $n = 1, 2, 3, \cdots$, $\left(1+\dfrac{1}{n}\right)^n < e < \left(1+\dfrac{1}{n}\right)^{n+1}$.

Proof Since $\left\{\left(1+\dfrac{1}{n}\right)^n\right\}$ is strictly increasing and $\lim\limits_{n\to+\infty}\left(1+\dfrac{1}{n}\right)^n = e$, $\left(1+\dfrac{1}{n}\right)^n < e$; Recall also that $\left\{\left(1+\dfrac{1}{n}\right)^{n+1}\right\}$ is strictly decreasing (an example of Theorem 1.2.1, Bernoulli inequality) and $\lim\limits_{n\to+\infty}\left(1+\dfrac{1}{n}\right)^{n+1} = \lim\limits_{n\to+\infty}\left[\left(1+\dfrac{1}{n}\right)^n\left(1+\dfrac{1}{n}\right)\right] = e$. So, $\left(1+\dfrac{1}{n}\right)^{n+1} > e$.

Example Let $a_n = 1+\dfrac{1}{2}+\cdots+\dfrac{1}{n}-\ln n$. Show $\{a_n\}$ is convergent.

Proof By the foregoing remark, $\left(1+\dfrac{1}{n}\right)^n < e < \left(1+\dfrac{1}{n}\right)^{n+1}$, which implies that $n\ln\left(1+\dfrac{1}{n}\right) < 1 < (n+1)\ln\left(1+\dfrac{1}{n}\right)$. Thus

$$n < \frac{1}{\ln\left(1+\frac{1}{n}\right)} < n+1, \text{ or } \frac{1}{n} > \ln\left(1+\frac{1}{n}\right) > \frac{1}{n+1}.$$

Then, $a_{n+1} - a_n = \dfrac{1}{n+1} - \ln(n+1) + \ln n = \dfrac{1}{n+1} - \ln\left(1+\dfrac{1}{n}\right) < 0$, i.e. $\{a_n\}$ is (strictly) decreasing. Furthermore, $a_n = 1+\dfrac{1}{2}+\cdots+\dfrac{1}{n}-\ln n > \ln\left(1+\dfrac{1}{1}\right)+\ln\left(1+\dfrac{1}{2}\right)+\cdots+\ln\left(1+\dfrac{1}{n}\right)-\ln n = \ln\left(\dfrac{2}{1}\,\dfrac{3}{2}\cdots\dfrac{n+1}{n}\right)-\ln n = \ln(n+1)-\ln n > 0$, i.e. $\{a_n\}$ is bounded below. Hence, $\{a_n\}$ is convergent. □

Note Let $\lim\limits_{n\to+\infty}\left(1+\frac{1}{2}+\cdots+\frac{1}{n}-\ln n\right)=r$, which is called *Euler number*. It can be proved that the Euler number $r\approx 0.577\,215$ (proof omitted).

Example Find $\lim\limits_{n\to+\infty}\left(\frac{1}{n+1}+\frac{1}{n+2}+\cdots+\frac{1}{2n}\right)$.

Solution Let $b_n=\frac{1}{n+1}+\frac{1}{n+2}+\cdots+\frac{1}{2n}$. Then, $b_n=\left(1+\frac{1}{2}+\cdots+\frac{1}{2n}\right)-\left(1+\frac{1}{2}+\cdots+\frac{1}{n}\right)=\left(1+\frac{1}{2}+\cdots+\frac{1}{2n}-\ln 2n\right)-\left(1+\frac{1}{2}+\cdots+\frac{1}{n}-\ln n\right)+\ln 2n-\ln n$. Thus, $\lim\limits_{n\to+\infty}b_n=\lim\limits_{n\to+\infty}\left(1+\frac{1}{2}+\cdots+\frac{1}{2n}-\ln 2n\right)-\lim\limits_{n\to+\infty}\left(1+\frac{1}{2}+\cdots+\frac{1}{n}-\ln n\right)+\lim\limits_{n\to+\infty}\ln\frac{2n}{n}=r-r+\ln 2=\ln 2$.

Note Using Euler number, we can also find the following:

(1) $\lim\limits_{n\to+\infty}\left(1-\frac{1}{2}+\frac{1}{3}-\cdots+(-1)^{n+1}\frac{1}{n}\right)=\ln 2$.

(2) $\lim\limits_{n\to+\infty}\frac{1}{\ln n}\left(1+\frac{1}{2}+\cdots+\frac{1}{n}\right)=1$.

Theorem 2.3.2 (The nested interval theorem) Let I_n be a series of intervals with $I_n=[a_n, b_n]\,(n=1, 2, \cdots)$ such that $[a_1, b_1]\supseteq[a_2, b_2]\supseteq\cdots\supseteq[a_n, b_n]\supseteq\cdots$ (called *a nest of closed intervals*). If $\lim\limits_{n\to+\infty}(b_n-a_n)=0$, then

(1) there exists $c\in\mathbb{R}$ such that $\lim\limits_{n\to+\infty}a_n=\lim\limits_{n\to+\infty}b_n=c$;

(2) $\bigcap_{n=1}^{+\infty}[a_n, b_n]=\{c\}$.

Proof (1) Since $a_n\leqslant a_{n+1}\leqslant b_{n+1}\leqslant b_n$, $a_1\leqslant a_n\leqslant b_n\leqslant b_1$ for any n. So, $\{a_n\}$ is increasing and bounded while $\{b_n\}$ is decreasing and bounded. Then, by the monotone convergence theorem, $\lim\limits_{n\to+\infty}a_n$ and

$\lim\limits_{n\to+\infty} b_n$ are both existent. Thus, $\lim\limits_{n\to+\infty} b_n - \lim\limits_{n\to+\infty} a_n = \lim\limits_{n\to+\infty} (b_n - a_n) = 0$, i. e. there exists $c \in \mathbb{R}$ such that $\lim\limits_{n\to+\infty} a_n = \lim\limits_{n\to+\infty} b_n = c$.

(2) Noticing that $c = \sup\{a_n\} = \inf\{b_n\}$, we have $a_n \leqslant c \leqslant b_n$ for any n, i. e. $c \in \bigcap_{n=1}^{+\infty}[a_n, b_n]$. Suppose there exists $d \in \bigcap_{n=1}^{+\infty}[a_n, b_n]$. Then, $|d-c| \leqslant b_n - a_n$ for any n. Since $\lim\limits_{n\to+\infty}(b_n - a_n) = 0$, it follows that $d = c$. This verifies the second statement. □

Remark (1) If $I_n(n = 1, 2, \cdots)$ are nested intervals but not all are closed, then the result is not necessarily true.

For example, $\bigcap_{n=1}^{+\infty}\left(0, \dfrac{1}{n}\right) = \varnothing$, because otherwise, if there exists $c \in \bigcap_{n=1}^{+\infty}\left(0, \dfrac{1}{n}\right)$, then $0 < c < \dfrac{1}{n}$ for any n, and so $c \leqslant \lim\limits_{n\to+\infty} \dfrac{1}{n} = 0$, a contradiction.

(2) Suppose $\{(a_n, b_n)\}$ is a sequences of intervals such that $a_1 < a_2 < \cdots < a_n < \cdots < b_n < \cdots < b_2 < b_1$ and $\lim\limits_{n\to+\infty}(b_n - a_n) = 0$. Then there exists uniquely $c \in \mathbb{R}$ such that $a_n < c < b_n$ for any n.

Proof Clearly, $[a_1, b_1] \supseteq [a_2, b_2] \supseteq \cdots \supseteq [a_n, b_n] \supseteq \cdots$. By the nested interval theorem, there exists uniquely a number $c \in \bigcap_{n=1}^{+\infty}[a_n, b_n]$. It suffices to show $c \neq a_n$ and $c \neq b_n$ for any n: if $c = a_N$ for some N, as $a_N < a_{N+1}$, $c \notin [a_{N+1}, b_{N+1}]$, which yields a contradiction; if $c = b_N$ for some N, as $b_N > b_{N+1}$, $c \notin [a_{N+1}, b_{N+1}]$, which also yields a contradiction. □

As an application of the nested interval theorem, we introduce the sequencability of a number set as follows:

Definition Let $S \subseteq \mathbb{R}$. If all elements in S can be arranged as a sequence: $a_1, a_2, \cdots, a_n, \cdots$, i. e. $S = \bigcup_{i=1}^{+\infty}\{a_i\}$, S is said to be *sequencable* (or *countable*), otherwise S is said to be *unsequencable* (or *uncountable*).

Proposition For any $a, b \in \mathbb{R}\ (a < b)$, $[a, b]$ is unsequencable.

Proof Suppose $[a, b]$ is sequencable, i. e. there is a sequence $\{x_i\}$ such that $[a, b] = \bigcup_{i=1}^{+\infty}\{x_i\}$. By trisecting $[a, b]$, we may get one subinterval of $[a, b]$, say, $[a_1, b_1]$, such that $x_1 \notin [a_1, b_1]$; by trisecting $[a_1, b_1]$, we may get one subinterval of $[a_1, b_1]$, say, $[a_2, b_2]$, such that $x_2 \notin [a_2, b_2]$. By continuing the manipulation in this manner, we may obtain a sequence of closed intervals $\{[a_n, b_n]\}$ satisfying the following three properties:

(1) $[a_{n+1}, b_{n+1}] \subseteq [a_n, b_n]$ for any n;

(2) $\lim\limits_{n\to+\infty}(b_n - a_n) = \lim\limits_{n\to+\infty}\frac{1}{3^n}(b-a) = 0$;

(3) $x_1, x_2, \cdots, x_n \notin [a_n, b_n]$ for any n.

Then, on one hand, by the nested interval theorem there is uniquely a number $r \in \bigcap_{n=1}^{+\infty}[a_n, b_n] \subseteq [a, b]$, while on the other hand, $r \neq x_n$ for any n, and so $r \notin \bigcup_{i=1}^{+\infty}\{x_i\} = [a, b]$. This is a contradiction. □

The next corollary follows directly from the foregoing proposition.

Corollary For any $a, b \in \mathbb{R}\ (a<b)$, each of $(a, b]$, $(a, b]$ and (a, b) is unsequencable.

Definition Let $E \subseteq \mathbb{R}$ and let $H = \{(\alpha, \beta) \mid \alpha, \beta \in \mathbb{R}, \alpha<\beta\}$ be a set of open intervals. If for any $x \in E$, there exists an open interval $(\alpha, \beta) \in H$ such that $x \in (\alpha, \beta)$, then H is called an *open (interval) covering* of E, and we also say E is *covered* by H. Moreover, if $|H| < +\infty$, H is called an *finite open (interval) covering* of E; otherwise, H is called an *infinite open (interval) covering* of E.

Examples (1) Let $E = \{x \mid x \in [0, 1] \cap \mathbb{Q}\}$. Then $H = \left\{\left(-\frac{1}{2}, \frac{1}{2}\right), \left(\frac{1}{3}, \frac{3}{2}\right)\right\}$ is a finite open covering of E.

(2) Let $E_1 = (0, 1)$, $E_2 = [0.1, 0.9]$, $E_3 = [0, 0.5]$ and $H = \left\{\left(\frac{1}{n}, 1\right) \mid n = 2, 3, \cdots\right\}$. Then, H is an (infinite) open covering of

E_1 and E_2; however, as $0 \notin \left(\frac{1}{n}, 1\right)$ for any $n=2, 3, \cdots$, H is not an open covering of E_3.

Theorem 2.3.3 (Heine-Borel finite open covering theorem) Let H be an open covering of $[a, b]$. Then there exists $H_0 \subseteq H$ with $|H_0| < +\infty$ such that $[a, b]$ is covered by H_0. (i. e. For any open covering H of $[a, b]$, there must exist a finite sub-covering H_0 ($\subseteq H$) of $[a, b]$).

Proof Suppose that the result is not true. Divide $[a, b]$ into $\left[a, \frac{a+b}{2}\right]$ and $\left[\frac{a+b}{2}, b\right]$. Then at least one of them, denoted by $[a_1, b_1]$, can not be covered by finite intervals in H; Divide $[a_1, b_1]$ into $\left[a_1, \frac{a_1+b_1}{2}\right]$ and $\left[\frac{a_1+b_1}{2}, b_1\right]$. Then at least one of them, denoted by $[a_2, b_2]$, can not be covered by finite intervals in H. By continuing this manipulation in a similar manner, we have a nest of closed intervals: $[a, b] \supseteq [a_1, b_1] \supseteq [a_2, b_2] \supseteq \cdots \supseteq [a_n, b_n] \supseteq \cdots$ with $\lim_{n\to+\infty}(b_n - a_n) = \lim_{n\to+\infty}\frac{b-a}{2^n} = 0$. Then, by the nested interval theorem, there is uniquely a number $x_0 \in [a_n, b_n] \subseteq [a, b]$ for any n and $\lim_{n\to+\infty} a_n = \lim_{n\to+\infty} b_n = x_0$. Since $[a, b]$ is covered by H, there is $(\alpha, \beta) \in H$ such that $x_0 \in (\alpha, \beta)$. Thus, $x_0 \in [a_N, b_N] \subseteq (\alpha, \beta)$ for some N, which implies that $[a_N, b_N]$ is covered by finite intervals (exactly one interval here) of H, a contradiction of the construction of $[a_n, b_n]$. □

Remark

(1) For an open covering of an interval which is not closed, the result may not hold true.

For example, $H = \left\{\left(0, 1-\frac{1}{n}\right) \mid n = 2, 3, \cdots\right\}$ is clearly an open covering of the interval $I = (0, 1)$, but there is no finite sub-covering H_0 ($\subseteq H$) of I.

(2) If the covering of a closed interval is not open (i. e. the intervals in the covering of a closed interval are not all open), the result may not hold true.

For example, $H = \left\{\left(\frac{1}{n}, 1\right) \mid n = 2, 3, \cdots\right\} \cup \{[-1, 0]\}$ is clearly a covering of the interval $I = \left[0, \frac{1}{2}\right]$. However, there is no finite sub-covering $H_0 (\subseteq H)$ of I. Otherwise, we may suppose that I is covered by finite intervals of H, say, $H_0 = \left\{\left(\frac{1}{n_1}, 1\right), \left(\frac{1}{n_2}, 1\right), \cdots, \left(\frac{1}{n_t}, 1\right)\right\} \cup \{[-1, 0]\}$. Let $m = \max\{n_1, n_2, \cdots, n_t\}$. Clearly, for any $x \in \left(0, \frac{1}{m}\right)$, x does not belong to any interval in H_0, which yields a contradiction.

Definition Let $S \subseteq \mathbb{R}$ and let $a \in \mathbb{R}$. If for any neighborhood of a, denoted by $U(a)$, $|U(a) \cap S| = +\infty$, then the point a is called a *cluster point* (or *accumulation point*) of S, i. e. if any neighborhood of a contains infinitely many points of S, then the point a is called a cluster point of S. Denote $\mathrm{CP}(S) = \{a \mid a \text{ is a cluster point of } S\}$, called the *cluster point set* of S.

Example Let $S_1 = \left\{\frac{1}{n} \mid n = 1, 2, \cdots\right\}$, $S_2 = [0, 1]$, $S_3 = (0, 1)$, $S_4 = \{x \mid x \in [0, 1] \cap \mathbb{Q}\}$, $S_5 = \left\{(-1)^n + \frac{1}{n} \mid n = 1, 2, \cdots\right\}$, $S_6 = \mathbb{N}$. Then, $\mathrm{CP}(S_1) = \{0\}$, $\mathrm{CP}(S_2) = S_2 = [0, 1]$, $\mathrm{CP}(S_3) = [0, 1]$, $\mathrm{CP}(S_4) = [0, 1]$, $\mathrm{CP}(S_5) = \{-1, 1\}$, $\mathrm{CP}(S_6) = \varnothing$.

Some simple facts concerning cluster points are listed below.

Remark (1) Let $S \subseteq \mathbb{R}$ and let $a \in \mathbb{R}$. Then the following three statements are equivalent:

ⓘ $a \in \mathrm{CP}(S)$;

ⓘⓘ for any $\delta > 0$, there exists $x \in S (x \neq a)$ such that $x \in U(a, \delta)$;

(iii) For any free-center neighborhood of a, denoted by $U^0(a)$, $U^0(a) \cap S \neq \varnothing$.

(2) If $S \subseteq \mathbb{R}$ with $|S| < +\infty$, then $CP(S) = \varnothing$, where $|S|$ is the cardinality of the set S (i. e. the number of the elements in S).

Theorem 2.3.4 (The cluster point theorem) Let $S \subseteq \mathbb{R}$. If S is bounded and $|S| = +\infty$, then there exists at least one cluster point of S.

Proof Since S is bounded, there is an interval $[a, b]$ such that $S \subseteq [a, b]$, Divide $[a, b]$ into $\left[a, \frac{a+b}{2}\right]$ and $\left[\frac{a+b}{2}, b\right]$. Then at least one of them, denoted by $[a_1, b_1]$, contains infinitely many points of S; Divide $[a_1, b_1]$ into $\left[a_1, \frac{a_1+b_1}{2}\right]$ and $\left[\frac{a_1+b_1}{2}, b_1\right]$. Then at least one of them, denoted by $[a_2, b_2]$, contains infinitely many points of S. By continuing this manipulation in a similar manner, we have a nest of closed intervals: $[a, b] \supseteq [a_1, b_1] \supseteq [a_2, b_2] \supseteq \cdots \supseteq [a_n, b_n] \supseteq \cdots$ with $\lim\limits_{n \to +\infty}(b_n - a_n) = \lim\limits_{n \to +\infty} \frac{b-a}{2^n} = 0$. Then, by the nested interval theorem, there is uniquely a number $\xi \in [a_n, b_n]$ for any n with $\lim\limits_{n \to +\infty} a_n = \lim\limits_{n \to +\infty} b_n = \xi$. Thus, for any $\varepsilon > 0$, there is $N \in \mathbb{N}$ such that $[a_N, b_N] \subseteq U(\xi, \varepsilon)$ and so $|U(\xi, \varepsilon) \cap S| = +\infty$, which implies that ξ is a cluster point of S. □

A cluster point of a number set can also be characterized in terms of sequence, which will be presented as the following:

Theorem 2.3.5 Let $S \subseteq \mathbb{R}$. Then ξ is a cluster point of $S \Leftrightarrow S$ contains a sequence $\{x_n\}$ such that $\lim\limits_{n \to +\infty} x_n = \xi$ and $x_i \neq x_j$ whenever $i \neq j$.

Proof $\Rightarrow$: Since ξ is a cluster point of S, $S \cap U^0(\xi, 1) \neq \varnothing$, and so there exists a point, say, $x_1 \in S \cap U^0(\xi, 1)$. Let $\delta_1 = \min\left\{\frac{1}{2}, |\xi - x_1|\right\} (> 0)$. Then, there exists $x_2 \in S \cap U^0(\xi, \delta_1)$ such that

$x_2 \neq x_1$ and $|\xi - x_2| < \frac{1}{2}$. Similarly, let $\delta_2 = \min\left\{\frac{1}{2^2}, |\xi - x_2|\right\}(>0)$. Then, there exists $x_3 \in S \cap U^0(\xi, \delta_2)$ such that $x_3 \neq x_1$, $x_3 \neq x_2$ and $|\xi - x_3| < \frac{1}{2^2}$; let $\delta_3 = \min\left\{\frac{1}{2^3}, |\xi - x_3|\right\}(>0)$. Then, there exists $x_4 \in S \cap U^o(\xi, \delta_3)$ such that $x_4 \neq x_1$, $x_4 \neq x_2$, $x_4 \neq x_3$ and $|\xi - x_4| < \frac{1}{2^3}$; by repeating this manipulation in a similar manner, we obtain a sequence $\{x_n\} \subseteq S$ such that $|\xi - x_n| < \delta_{n-1} \leqslant \frac{1}{2^{n-1}}$ and $x_i \neq x_j$ whenever $i \neq j$. Clearly, $\lim\limits_{n \to +\infty} x_n = \xi$.

$\Leftarrow$: Since $\lim\limits_{n \to +\infty} x_n = \xi$, for any $\varepsilon > 0$, there exists $N \in \mathbb{N}$ such that $|x_n - \xi| < \varepsilon$ whenever $n > N$. Noticing that $\{x_n\} \subseteq S$ and $x_i \neq x_j (i \neq j)$, we have $|U(\xi, \varepsilon) \cap S| = +\infty$. Therefore, ξ is a cluster point of S. □

The next theorem describes another version of the foregoing theorem, the verification of which is similar and so omitted.

Theorem 2.3.6 Let $S \subseteq \mathbb{R}$. Then ξ is a cluster point of $S \Leftrightarrow$ there exists a sequence $\{x_n\} \subseteq S \backslash \{\xi\}$ such that $\lim\limits_{n \to +\infty} x_n = \xi$.

Examples (1) Let $S = [a, b]$. Prove that $x \in S$ if and only if $x \in CP(S)$.

Proof Let $x \in S$. Then, clearly for any $\delta > 0$, $|U(x, \delta) \cap S| = +\infty$, and so $x \in$ CP(S).

If $x \notin S$, without loss of generality, let $x < a$ and let $\delta = \frac{a - x}{2}$ (> 0), then $U(x, \delta) \cap S = \varnothing$, which implies that $x \notin$ CP(S). □

(2) Let $S \subseteq \mathbb{R}$. Then CP (CP(S)) $\subseteq$ CP(S).

Proof If CP(CP(S)) $= \varnothing$, the result is trivial. Now, let CP(CP(S)) $\neq \varnothing$, and let $b \in$ CP(CP(S)). Then, for any $\varepsilon > 0$, there exists $h \in$ CP(S) $(h \neq b)$ such that $h \in U(b, \varepsilon)$. Let $\delta = \min\{b + \varepsilon -$

$h, h-b+\varepsilon\}$. Then, $U(h, \delta) \subseteq U(b, \varepsilon)$. Since $h \in \mathrm{CP}(S)$, $|U(h, \delta) \cap S| = +\infty$, and so $|U(b, \varepsilon) \cap S| = +\infty$, which implies that $b \in \mathrm{CP}(S)$. □

Definition Let $S \subseteq \mathbb{R}$ and let $x_0 \in \mathbb{R}$. If there exists $\delta > 0$ such that $U(x_0, \delta) \cap S = \{x_0\}$, then x_0 is called an *isolated point* of S. Denote $\mathrm{IP}(S) = \{x \mid x \text{ is an isolated point of } S\}$.

For example, let $S = \left\{0, 1, \frac{1}{2}, \frac{1}{3}, \frac{1}{4}, \cdots\right\}$, then $\mathrm{IP}(S) = S\backslash\{0\}$; let $S = \left\{1 - \frac{1}{n} \mid n = 1, 2, \cdots\right\}$, then $\mathrm{IP}(S) = S$.

By the definition of a cluster point and an isolated point of a number set, it is conventional to check the following:

Remark (1) For any $S \subseteq \mathbb{R}$, there exist $C, I \subseteq S$ such that $C \cup I = S$, $C \cap I = \varnothing$, $C \subseteq \mathrm{CP}(S)$ and $I \subseteq \mathrm{IP}(S)$, i.e. for any point x of S, either x is a cluster point of S or x is an isolated point of S.

(2) For any $S \subseteq \mathbb{R}$, $S \subseteq \mathrm{CP}(S) \cup \mathrm{IP}(S)$, but, in general $S \neq \mathrm{CP}(S) \cup \mathrm{IP}(S)$.

(3) For any $S \subseteq \mathbb{R}$, $\mathrm{CP}(S) \cup \mathrm{IP}(S) = \varnothing$.

Definition Let $\{a_n\}$ be a sequence. If $\{i_1, i_2, i_3, \cdots\} \subseteq \{1, 2, 3, \cdots\}$ with $i_1 < i_2 < i_3 < \cdots$, then $\{a_{i_n}\}$ is called a *subsequence* of $\{a_n\}$.

Note: Let $\{a_{i_n}\}$ be a subsequence of $\{a_n\}$. Then, obviously $i_n \geqslant n$ for any n.

Proposition Let $\{a_n\}$ be a sequence. Then, $\lim_{n\to+\infty} a_n = a$ if and only if $\lim_{n\to+\infty} a_{i_n} = a$ for any subsequence $\{a_{i_n}\}$ of $\{a_n\}$.

Proof Suppose $\lim_{n\to+\infty} a_n = a$. Then, for any $\varepsilon > 0$, there is $N \in \mathbb{N}$ such that $|a_n - a| < \varepsilon$ whenever $n \geqslant N$. Take $N_1 = i_N$, then for any $n > N_1$, $i_n \geqslant n > i_N \geqslant N$, and so $|a_{i_n} - a| < \varepsilon$, i.e. $\lim_{n\to+\infty} a_{i_n} = a$.

Now, Suppose for any subsequence $\{a_{i_n}\}$ of $\{a_n\}$, $\lim_{n\to+\infty} a_{i_n} = a$.

As $\{a_n\}$ is also a subsequence of itself, trivially $\lim_{n\to+\infty} a_n = a$. □

Examples (1) Let $\{a_n\}$ be a sequence. If $\lim_{n\to+\infty} a_{2n} = a$ and $\lim_{n\to+\infty} a_{2n+1} = a$, then $\lim_{n\to+\infty} a_n = a$.

Proof By the condition, for any $\varepsilon > 0$, there exists $K_1 \in \mathbb{N}$ such that $|a_{2k} - a| < \varepsilon$ whenever $k > K_1$, also there exists $K_2 \in \mathbb{N}$ such that $|a_{2k+1} - a| < \varepsilon$ whenever $k > K_2$. Thus, for any $\varepsilon > 0$, there exists $N = \max\{K_1, K_2\}$ such that $|a_n - a| < \varepsilon$ whenever $n > N$, i. e. $\lim_{n\to+\infty} a_n = a$.

(2) Prove that $\lim_{n\to+\infty} \sin n$ is non-existent.

Proof Assume there exists $a \in \mathbb{R}$ such that $\lim_{n\to+\infty} \sin n = a$. Then, $\lim_{n\to+\infty} \sin(n+2) = a$, and so $\lim_{n\to+\infty} [2\sin 1 \cos(n+1)] = \lim_{n\to+\infty} [\sin(n+2) - \sin n] = a - a = 0$. Thus, $\lim_{n\to+\infty} \cos(n+1) = 0$, i. e. $\lim_{n\to+\infty} \cos n = 0$. Considering that $\{\sin(2n)\}$ is a subsequence of $\{\sin n\}$ and $\lim_{n\to+\infty} \sin n$ exists by assumption, by the above proposition we have $\lim_{n\to+\infty} \sin n = \lim_{n\to+\infty} \sin(2n) = \lim_{n\to+\infty} (2\sin n \cos n) = 2a \cdot 0 = 0$. Furthermore, $\lim_{n\to+\infty} (\sin^2 n + \cos^2 n) = 0 + 0 = 0$, which contradicts $\sin^2 n + \cos^2 n = 1$. □

It is easy to see that a sequence is not necessarily convergent if one of its subsequences is convergent. Nevertheless, we have the following:

Proposition Let $\{a_n\}$ be a monotone sequence. Then, $\lim_{n\to+\infty} a_n = a$ if and only if there exists a subsequence $\{a_{i_n}\}$ of $\{a_n\}$ with $\lim_{n\to+\infty} a_{i_n} = a$.

Proof Necessity is evidently true by the above proposition. We now assume that there exists a subsequence $\{a_{i_n}\}$ of $\{a_n\}$ with $\lim_{n\to+\infty} a_{i_n} = a$. Without loss of generality, suppose $\{a_n\}$ is an increasing sequence. Then, for any $\varepsilon > 0$ there is $N \in \mathbb{N}$ such that $|a_{i_n} - a| < \varepsilon$ whenever $n \geqslant N$, i. e. $0 < a - a_{i_n} < \varepsilon$ whenever $n \geqslant N$ since $\{a_n\}$ is increasing. Take $N_1 = i_N$. If $n \geqslant N_1$, $|a_n - a| = a - a_n \leqslant a - a_{N_1} = a - a_{iN} < \varepsilon$. Therefore, $\lim\limits_{n \to +\infty} a_n = a$. □

As we know, a convergent sequence must be bounded, but the converse may not be true, i. e. a bounded sequence may not be convergent. However, a bounded sequence must contain a convergent subsequence. To prove this assertion, we need a lemma as follows:

Lemma Any sequence $\{a_n\}$ has a monotone subsequence $\{a_{i_n}\}$.

Proof First let us suppose the number set $\{a_n \mid n = 1, 2, \cdots\}$ does not possess a maximum. Take $i_1 = 1$. Then there is $i_2 > i_1$ with $a_{i_2} > a_{i_1}$. As $\{a_n \mid n > i_2\}$ does not possess a maximum too, there is $i_3 > i_2$ with $a_{i_3} > a_{i_2}$. Evidently, by repeating a similar manipulation we can construct an increasing subsequence $\{a_{i_n}\}$ of $\{a_n\}$.

Now, we suppose $\{a_n \mid n = 1, 2, \cdots\}$ possesses a maximum, say, a_{i_1}, and we further consider $\{a_n \mid n > i_1\}$. If it does not possess a maximum, as we have just proved, the sequence $\{a_{i_1+1}, a_{i_1+2}, a_{i_1+3}, \cdots\}$ has an increasing subsequence, which is obviously also an increasing subsequence of the sequence $\{a_n\}$. If $\{a_n \mid n > i_1\}$ possesses a maximum, say, a_{i_2}, then $a_{i_2} \leqslant a_{i_1}$, and we further consider $\{a_n \mid n > i_2\}$. By carrying on this process repeatedly, either we have an increasing subsequence $\{a_{j_n}\}$ of $\{a_n\}$ or we may construct a decreasing subsequence $\{a_{i_n}\}$ of $\{a_n\}$. □

Theorem 2.3.7 (Weierstrass theorem) Let $\{a_n\}$ be a sequence. If $\{a_n\}$ is bounded, then it must have a convergent subsequence $\{a_{i_n}\}$.

Proof By lemma $\{a_n\}$ has a monotone subsequence $\{a_{i_n}\}$, which is also bounded since $\{a_n\}$ is bounded. Then, the result follows directly from the monotone convergence theorem. □

Theorem 2.3.8 (Bolzano-Weierstrass theorem) For any sequence $\{x_n\}$ in $[a, b]$, there exists a subsequence $\{x_{i_n}\} \subseteq \{x_n\}$ such that $\lim_{n\to+\infty} x_{i_n} \in [a, b]$.

Proof Clearly, $\{x_n\}$ is bounded, and so by the foregoing theorem there exists a subsequence $\{x_{i_n}\} \subseteq \{x_n\}$ such that $\lim_{n\to+\infty} x_{i_n} = \alpha$ for some $\alpha \in \mathbb{R}$. Since $a \leqslant x_{i_n} \leqslant b$ for any n, $a \leqslant \lim_{n\to+\infty} x_{i_n} \leqslant b$, i. e. $\lim_{n\to+\infty} x_{i_n} = \alpha \in [a, b]$. □

Definition Let $A \subseteq \mathbb{R}$. If for any sequence $\{x_n\} \subseteq A$, there exists a subsequence $\{x_{i_n}\} \subseteq \{x_n\}$ such that $\lim_{n\to+\infty} x_{i_n} \in A$, then the number set A is said to be *compact*.

In terms of compactness, Bolzano-Weierstrass theorem can be presented in another version as the next:

Theorem 2.3.8* (Bolzano-Weierstrass theorem) For any $a, b \in \mathbb{R}$ with $a \leqslant b$, $[a, b]$ is compact.

Note It is not hard to check that $(a, b]$ or $[a, b)$ or (a, b) is not compact; the number set $\left\{0, 1, \frac{1}{2}, \frac{1}{3}, \cdots\right\}$ is compact, while $\left\{1, \frac{1}{2}, \frac{1}{3}, \cdots\right\}$ is not compact.

If a sequence $\{a_n\}$ converges to a number p, its terms must ultimately become close to p and hence close to each other. This property is stated more formally in the next theorem.

Theorem 2.3.9 (Cauchy criterion for convergence) A sequence $\{a_n\}$ is convergent if and only if for any $\varepsilon > 0$, there exists a positive integer N such that $|a_m - a_n| < \varepsilon$ whenever $m, n > N$.

Proof Necessity. Let $p = \lim_{n\to+\infty} a_n$. Then for any $\varepsilon > 0$, there is $N \in \mathbb{N}$ such that $|a_n - p| < \varepsilon/2$ whenever $n > N$. Then $|a_m - p| < \varepsilon/2$ if $m > N$. So for any $m, n > N$, $|a_n - a_m| \leqslant |a_n - p| + |a_m - p| < \varepsilon/2 + \varepsilon/2 = \varepsilon$.

Sufficiency. We first show $\{a_n\}$ is bounded. Take $\varepsilon = 1$. Then, there exists $N \in \mathbb{N}$ such that $|a_m - a_n| < 1$ whenever $m, n > N$. So, for any $m > N$, $|a_m - a_{N+1}| < 1$, and so $|a_m| \leqslant |a_{N+1}| + 1$. Thus,

$|x_n| \leqslant \max\{|a_1|, |a_2|, \cdots, |a_N|, |a_{N+1}|+1\}$ for any $n = 1, 2, \cdots$, i. e. $\{a_n\}$ is bounded. By Weierstrass theorem, there is a subsequence $\{a_{i_n}\} \subseteq \{a_n\}$ such that $\{a_{i_n}\}$ is convergent. Let $\lim\limits_{n\to+\infty} a_{i_n} = a$. Then, for any $\varepsilon > 0$, there exists $N_1 \in \mathbb{N}$ such that $|a_{i_n} - a| < \frac{\varepsilon}{2}$ whenever $n > N_1$; by condition, there exists $N_2 \in \mathbb{N}$ such that $|a_m - a_n| < \frac{\varepsilon}{2}$ whenever $m, n > N_2$. By taking $N = \max\{N_1, N_2\}$, we see that $|a_n - a| < |a_n - a_{i_n}| + |a_{i_n} - a| < \frac{\varepsilon}{2} + \frac{\varepsilon}{2} = \varepsilon$ whenever $n > N$, i. e. $\lim\limits_{n\to+\infty} a_n = a$, and so $\{a_n\}$ is convergent. □

Remark Let $\{a_n\}$ be a sequence. By Cauchy criterion for convergence, we have the following statements:

(1) $\{a_n\}$ is convergent if and only if there exists a positive integer N such that $|a_{n+p} - a_n| < \varepsilon$ whenever $n > N$ and $p \in \mathbb{N}$;

(2) $\{a_n\}$ is divergent if and only if there is $\varepsilon > 0$ such that for any $N \in \mathbb{N}$ there exist $m, n > N$ with $|a_m - a_n| \geqslant \varepsilon$;

(3) $\{a_n\}$ is divergent if and only if there is $\varepsilon > 0$ such that for any $N \in \mathbb{N}$ there exist $n > N$ and $p \in \mathbb{N}$ with $|a_{n+p} - a_n| \geqslant \varepsilon$.

Definition A sequence $\{x_n\}$ is called a *Cauchy sequence* if it satisfies the following condition (called the *Cauchy condition*):

For any $\varepsilon > 0$ there is a positive integer N such that $|x_m - x_n| < \varepsilon$ whenever m, n $>$ N.

Theorem 2. 3. 9 states that a sequence is convergent if and only if it is a Cauchy sequence.

Examples (1) Let

$$x_n = \frac{\sin 1}{2} + \frac{\sin 2}{2^2} + \cdots + \frac{\sin n}{2^n}.$$

Prove the sequence $\{x_n\}$ is convergent.

Proof For any $\varepsilon > 0$, since $1/2^n \to 0$, we can find a positive *number* N such that $1/2^n < \varepsilon$ whenever $n > N$. Then, if $m > n > N$, we see

$$|x_n - x_m| = \left|\frac{\sin(n+1)}{2^{n+1}} + \frac{\sin(n+2)}{2^{n+2}} + \cdots + \frac{\sin m}{2^m}\right|$$

$$\leqslant \left|\frac{\sin(n+1)}{2^{n+1}}\right| + \left|\frac{\sin(n+2)}{2^{n+2}}\right| + \cdots + \left|\frac{\sin m}{2^m}\right|$$

$$\leqslant \frac{1}{2^{n+1}} + \frac{1}{2^{n+2}} + \cdots + \frac{1}{2^m} = \frac{1}{2^n}\left(\frac{1}{2} + \frac{1}{2^2} + \cdots + \frac{1}{2^{m-n}}\right)$$

$$< \frac{1}{2^n} < \varepsilon.$$

Hence, from Cauchy criterion it follows that the sequence $\{x_n\}$ is convergent. □

(2) Let

$$a_n = 1 + \frac{1}{2^2} + \frac{1}{3^2} + \cdots + \frac{1}{n^2}.$$

Prove the sequence $\{a_n\}$ is convergent.

Proof Without loss of generality, let $m > n$. Then

$$|a_m - a_n| = \frac{1}{(n+1)^2} + \frac{1}{(n+2)^2} + \cdots + \frac{1}{m^2}$$

$$\leqslant \frac{1}{n(n+1)} + \frac{1}{(n+1)(n+2)} + \cdots + \frac{1}{(m-1)m}$$

$$= \left(\frac{1}{n} - \frac{1}{n+1}\right) + \left(\frac{1}{n+1} - \frac{1}{n+2}\right) + \cdots + \left(\frac{1}{m-1} - \frac{1}{m}\right)$$

$$= \frac{1}{n} - \frac{1}{m} < \frac{1}{n}.$$

Thus, for any $\varepsilon > 0$, by taking $N = \left[\frac{1}{\varepsilon}\right]$, we see that $|a_m - a_n| < \varepsilon$ whenever $m > n > N$. Therefore, $\{a_n\}$ is convergent. □

(3) Let

$$a_n = 1 + \frac{1}{2} + \frac{1}{3} + \cdots + \frac{1}{n}.$$

Prove the sequence $\{a_n\}$ is divergent.

Proof Let $\varepsilon = \frac{1}{2}$. For any $N \in \mathbb{N}$, take $n > N$ and $p = n$. Then

$$\begin{aligned} |a_{n+p} - a_n| &= \frac{1}{n+1} + \frac{1}{n+2} + \cdots + \frac{1}{2n} \\ &\geqslant \frac{1}{2n} + \frac{1}{2n} + \cdots + \frac{1}{2n} = \frac{n}{2n} = \frac{1}{2} = \varepsilon. \end{aligned}$$

So, the sequence $\{a_n\}$ is divergent.

(4) Let

$$x_n = a_0 + a_1 q + a_2 q^2 + \cdots + a_n q^n,$$

where constants $|q| < 1$ and $|a_k| < M$ (M is a positive constant, $k = 0, 1, 2, \cdots$). Prove the sequence $\{x_n\}$ is convergent.

Proof Analysis: Let $p \in \mathbb{N}$.

$$\begin{aligned} |x_{n+p} - x_n| &= |a_{n+1} q^{n+1} + a_{n+2} q^{n+2} + \cdots + a_{n+p} q^{n+p}| \\ &< M(|q|^{n+1} + |q|^{n+2} + \cdots + |q|^{n+p}) \\ &< M \frac{|q|^{n+1}}{1 - |q|} = M_1 |q|^{n+1}, \end{aligned}$$

where $M_1 = \frac{M}{1 - |q|}$. Requiring $M_1 |q|^{n+1} < \varepsilon$, i. e. $(n+1) \ln |q| < \ln \frac{\varepsilon}{M_1}$, we have $n > \frac{\ln \frac{\varepsilon}{M_1}}{\ln |q|} - 1$.

So, for any $\varepsilon > 0$, there exists

$$N = \left[\frac{\ln \frac{\varepsilon}{M_1}}{\ln |q|} - 1 \right],$$

where $M_1 = \dfrac{M}{1-|q|}$, such that $|x_{n+p} - x_n| < \varepsilon$ whenever $n > N$ and $p \in \mathbb{N}$. Hence, the sequence $\{x_n\}$ is convergent by Cauchy criterion for convergence. □

Remark In this section, we prove the main theorems along the following path with the completeness axiom as a starting point:

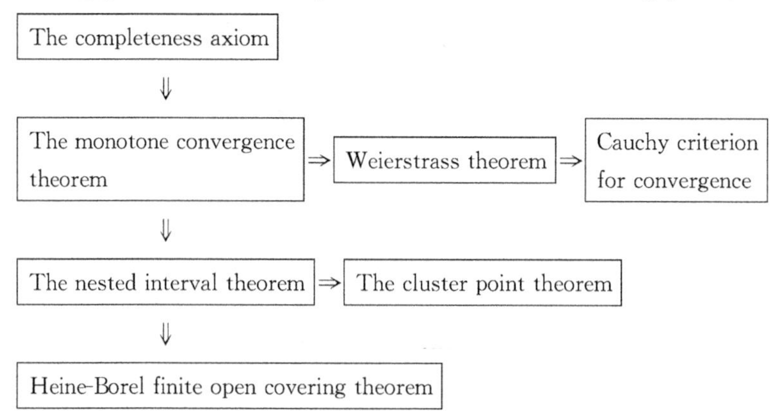

In fact, all these theorems are equivalent to each other, i. e. we can start from any theorem to verify the other theorems. We give four presentations as follows, and the others are left for readers.

(1) Heine-Borel finite open covering theorem ⇒ Weierstrass theorem

Proof Since $\{a_n\}$ is bounded, there exist m, $M \in \mathbb{R}$ with $m \leqslant a_n \leqslant M (n = 1, 2, \cdots)$. We first show that there is $c \in [m, M]$ such that for any neighborhood of c, $U(c)$, there exist infinitely many terms of $\{a_n\}$ in $U(c)$. Otherwise, assume for any $x \in [m, M]$, there is a neighborhood $U(x, \delta_x)$ containing at most a finite number of terms of $\{a_n\}$. Thus, $H = \{U(x, \delta_x) \mid x \in [m, M]\}$ constructs an open covering of $[m, M]$. By Heine-Borel finite open covering theorem, there is a finite sub-covering, i. e. there are $x_j^* (j = 1, 2, \cdots, k)$ such that

$$[m, M] \subseteq \bigcup_{j=1}^{k} U(x_j^*, \delta^*_{x_j}),$$

which implies that $[m, M]$ contains at most a finite number of terms of $\{a_n\}$, contradicting $m \leqslant a_n \leqslant M$ $(n = 1, 2, \cdots)$.

Let $x_{i_1} \in U(c, 1)$. Then, there exists $x_{i_2} \in U(c, \frac{1}{2})$ with $i_2 > i_1$; there exists $x_{i_3} \in U\left(c, \frac{1}{3}\right)$ with $i_3 > i_2 > i_1$; in general, there exists $x_{i_n} \in U(c, \frac{1}{n})$ with $i_n > i_{n-1} > \cdots > i_2 > i_1$. Clearly, $\{a_{i_n}\}$ is a convergent subsequence of $\{a_n\}$. □

(2) The cluster point theorem $\Rightarrow$ Weierstrass theorem

Proof If $|\{a_n \mid n = 1, 2, \cdots\}| < +\infty$, i. e. the sequence $\{a_n\}$ is composed of only finite distinct numbers, then at least one of them appears infinitely many times in the sequence and this is clearly a convergent subsequence of $\{a_n\}$. Now, we suppose the sequence $\{a_n\}$ is composed of infinitely many distinct numbers, denoted by the number set S. Then, S is bounded and $|S| = +\infty$. By the cluster point theorem, there exists at least one cluster point of S, say ξ. Just as in the proof of Theorem 2. 3. 5, we can construct a subsequence $\{a_{i_n}\}$ of $\{a_n\}$ such that $\lim_{n \to +\infty} a_{i_n} = \xi$, i. e. $\{a_{i_n}\}$ is a convergent subsequence of $\{a_n\}$. □

(3) The Borel finite open covering theorem $\Rightarrow$

The nested interval theorem

Proof First we show by contradiction that there exists $c \in \mathbb{R}$ such that $c \in [a_n, b_n]$ for any n. If for any $x \in \mathbb{R}$, there exists $[a_{k(x)}, b_{k(x)}]$ such that $x \notin [a_{k(x)}, b_{k(x)}]$ where $k(x) \in \{1, 2, \cdots\}$. Then, there exists $\delta_x > 0$ such that $U(x, \delta_x) \cap [a_{k(x)}, b_{k(x)}] = \varnothing$. In particular, for any $x \in [a_1, b_1]$, there exists $\delta_x > 0$ such that

$U(x, \delta_x) \cap [a_{k(x)}, b_{k(x)}] = \varnothing$. On the other hand, since

$$[a_1, b_1] \subseteq \bigcup_{x \in [a_1, b_1]} U(x, \delta_x),$$

(i. e. $[a_1, b_1]$ is covered by $\{U(x, \delta_x) \mid x \in [a_1, b_1]\}$). So, by the Borel finite open covering theorem, there exists a finite sub-covering, say,

$$\{U(x_i, \delta_{x_i}) \mid i = 1, 2, \cdots, n\} \subseteq \{U(x, \delta_x) \mid x \in [a_1, b_1]\},$$

such that

$$[a_1, b_1] \subseteq \bigcup_{i=1}^{n} U(x_i, \delta_{x_i}).$$

Let $j = \max\{k(x_1), k(x_2), \cdots, k(x_n)\}$. As $[a_1, b_1] \supseteq [a_2, b_2] \supseteq \cdots \supseteq [a_n, b_n] \supseteq \cdots$, $[a_j, b_j] \subseteq [a_{k(x_i)}, b_{k(x_i)}]$ for any $i = 1, 2, \cdots, n$, and so $U(x_i, \delta_{x_i}) \cap [a_j, b_j] = \varnothing$ since $U(x_i, \delta_{x_i}) \cap [a_{k(x_i)}, b_{k(x_i)}] = \varnothing (i = 1, 2, \cdots, n)$. Therefore,

$$[a_j, b_j] \cap \bigcup_{i=1}^{n} U(x_i, \delta_{x_i}) = \varnothing,$$

which contradicts

$$[a_j, b_j] \subseteq [a_1, b_1] \subseteq \bigcup_{i=1}^{n} U(x_i, \delta_{x_i}).$$

Hence, there exists $c \in \mathbb{R}$ such that $c \in [a_n, b_n]$ for any n.

Furthermore, since $a_n \leqslant c \leqslant b_n$ for any n and $\lim_{n \to +\infty} (b_n - a_n) = 0$, it follows immediately that $\lim_{n \to +\infty} a_n = \lim_{n \to +\infty} b_n = c$. □

(4) Cauchy criterion for convergence ⇒ The completeness axiom

Proof Suppose $S \subseteq \mathbb{R}$ $(S \neq \varnothing)$ such that S is bounded above. We show there exists the least upper bound of S. Let M be an upper bound of S. Since $S \neq \varnothing$, there is $x_0 \in S$. Now we determine y_1 from

x_0 and M following the next procedure: If x_0 is an upper bound of S, let $y_1 = x_0$; otherwise, consider $\frac{x_0 + M}{2}$. If $\frac{x_0 + M}{2}$ is an upper bound of S, let $y_1 = \frac{x_0 + M}{2}$; if $\frac{x_0 + M}{2}$ is not an upper bound of S, let $y_1 = M$. Then, y_1 satisfies the following:

ⓘ y_1 is an upper bound of S;

ⓘⓘ $[y_1 - \frac{M - x_0}{2}, y_1] \cap S \neq \varnothing$.

Then, for $y_1 - \frac{M - x_0}{2}$ and y_1 we repeat the last manipulation to determine y_2, which satisfies the following:

ⓘ y_2 is an upper bound of S;

ⓘⓘ $[y_2 - \frac{M - x_0}{2^2}, y_2] \cap S \neq \varnothing$;

ⓘⓘⓘ $|y_2 - y_1| \leqslant \frac{M - x_0}{2}$.

We repeat the foregoing process to find y_3, y_4, …, y_n, …, where $y_n (n = 1, 2, 3, \cdots)$ satisfies the following:

ⓘ y_n is an upper bound of S;

ⓘⓘ $[y_n - \frac{M - x_0}{2^n}, y_n] \cap S \neq \varnothing$;

ⓘⓘⓘ $|y_n - y_{n-1}| \leqslant \frac{M - x_0}{2^{n-1}}$.

As $m > n$,

$$|y_m - y_n| = |y_m - y_{m-1}| + \cdots + |y_{n+1} - y_n|$$
$$\leqslant \frac{M - x_0}{2^n} + \cdots + \frac{M - x_0}{2^{m-1}} \leqslant \frac{M - x_0}{2^{n-1}}.$$

Thus, $\{y_n\}$ is a Cauchy sequence, and we may let $\lim_{n \to +\infty} y_n = a$. We proceed to show that a is the least upper bound of S. First, we assert that a is an upper bound of S. Because otherwise, there is $x_1 \in S$ with $x_1 > a$. Since $\lim_{n \to +\infty} y_n = a$, there exists $N \in \mathbb{N}$ such that $y_{N+1} < x_1$, which contradicts that y_n is an upper bound of S for any n.

Assume that a is not the least upper bound of S. Then, there is $c < a$ such that c is an upper bound of S. However, this contradicts the statement $\left[y_n - \frac{M - x_0}{2^n}, y_n\right] \cap S \neq \varnothing$ (Note: $y_n \geqslant a$). Hence, a is the least upper bound of S. □

§ 2.4 Upper limit and lower limit of a sequence

In this section, we will discuss a generalization of a sequence limit, namely, upper limit and lower limit. First, we introduce the concept of a cluster point of a sequence.

Definition Let $\{x_n\}$ be a sequence. If there exists a subsequence $\{x_{i_n}\} \subseteq \{x_n\}$ such that $\lim_{n \to +\infty} x_{i_n} = b (\in \mathbb{R})$, then b is called a *cluster point* of the sequence $\{x_n\}$.

Remark (1) Let $\{x_n\}$ be a sequence and let $a \in \mathbb{R}$. it is easy to check that the following two statements are equivalent to each other:

ⓘ The point a is a cluster point of the sequence $\{x_n\}$;

ⓘⓘ Any neighborhood of the point a contains infinitely many terms of $\{x_n\}$.

(2) The definition of a cluster point of a sequence is different from that of a cluster point of a number set. For example, the sequence 0, 1, 0, 1, 0, 1, ⋯ has two cluster points 0 and 1. But, this sequence is composed of only two points 0 and 1, and there is no cluster points of the number set $S = \{0, 1\}$.

Example The sequence $\left\{(-1)^n \frac{n}{n+1}\right\}$ has cluster points -1

and 1; the sequence $\left\{\sin\frac{n\pi}{4}\right\}$ has cluster points 0, -1, 1, $-\frac{\sqrt{2}}{2}$ and $\frac{\sqrt{2}}{2}$.

Theorem 2.4.1 A bounded sequence $\{x_n\}$ has at least one cluster point, and it has the maximal cluster point and the minimal cluster point.

Proof We use the nested interval theorem to find a cluster point.

Since the sequence $\{x_n\}$ is bounded, there is $M > 0$ such that $|x_n| < M$ for any n. Dividing $[-M, M]$ into two subintervals with equal length. Clearly, at least one of these two subintervals contains infinitely many terms of $\{x_n\}$, which is denoted by $[a_1, b_1]$. Similarly, dividing $[a_1, b_1]$ into two subintervals with equal length, and denote the one that contains infinitely many terms of $\{x_n\}$ as $[a_2, b_2]$. By repeating this process, we obtain a nest of closed intervals $[a_1, b_1] \supseteq [a_2, b_2] \supseteq \cdots \supseteq [a_n, b_n] \supseteq \cdots$ such that each $[a_n, b_n]$ contains infinitely many terms of $\{x_n\}$ and $b_n - a_n = \frac{M}{2^{n-1}} \to 0$ $(n \to +\infty)$. Then, by the nested interval theorem, there exists one point $\xi \in [a_n, b_n] (n = 1, 2, \cdots)$ with $\lim\limits_{n\to+\infty} a_n = \lim\limits_{n\to+\infty} b_n = \xi$. Thus, for any $\varepsilon > 0$, there is $N \in \mathbb{N}$ such that $[a_N, b_N] \subseteq (\xi - \varepsilon, \xi + \varepsilon)$. By the property of $[a_N, b_N]$ and the arbitrariness of ε, we see that any neighborhood of the point ξ contains infinitely many terms of $\{x_n\}$. So, ξ is a cluster point of $\{x_n\}$.

As for the proof of the existence the maximal cluster point of $\{x_n\}$, we merely need to choose each $[a_n, b_n]$ in the nested intervals in the following way: Divide $[a_{n-1}, b_{n-1}]$ into $I_1 = \left[a_{n-1}, \frac{a_{n-1} + b_{n-1}}{2}\right]$ and $I_2 = \left[\frac{a_{n-1} + b_{n-1}}{2}, b_{n-1}\right]$. If I_2 contains infinitely many terms of

$\{x_n\}$, take $[a_n, b_n] = I_2$; otherwise, take $[a_n, b_n] = I_1$. This choice guarantees that each $[a_n, b_n]$ contains infinitely many terms of $\{x_n\}$ and also that there are only a finite number of terms of $\{x_n\}$ located on the right hand side of $[a_n, b_n]$. Then, the above-determined ξ must be the maximal cluster point of $\{x_n\}$. Because otherwise, we may let ζ be another cluster point of $\{x_n\}$ with $\zeta > \varepsilon$. Then, in $(\zeta - \frac{\zeta-\xi}{3}, \zeta + \frac{\zeta-\xi}{3})$ there are infinitely many terms of $\{x_n\}$. However, $(\zeta - \frac{\zeta-\xi}{3}, \zeta + \frac{\zeta-\xi}{3})$ is located completely on the right hand side of $[a_n, b_n]$ as n is large enough, which yields a contradiction to the choice of $[a_n, b_n]$ $(n = 1, 2, \cdots)$.

In an analogous way, i. e. in the foregoing process if I_1 contains infinitely many terms of $\{x_n\}$, take $[a_n, b_n] = I_1$; otherwise, take $[a_n, b_n] = I_2$, then we can obtain the minimal cluster point of $\{x_n\}$. □

Definition Let $\{x_n\}$ be a bounded sequence.

(1) the maximal cluster point of $\{x_n\}$ is called *the upper limit* of $\{x_n\}$, denoted by $\overline{\lim\limits_{n\to+\infty}} x_n$.

(2) the minimal cluster point of $\{x_n\}$ is called *the lower limit* of $\{x_n\}$, denoted by $\underline{\lim\limits_{n\to+\infty}} x_n$.

Remark By Theorem 2. 4. 1, a bounded sequence $\{x_n\}$ must have the upper limit and the lower limit.

Examples (1) $\overline{\lim\limits_{n\to+\infty}} (-1)^n \frac{n}{n+1} = 1$, $\underline{\lim\limits_{n\to+\infty}} (-1)^n \frac{n}{n+1} = -1$.

(2) $\overline{\lim\limits_{n\to+\infty}} \sin\frac{n\pi}{4} = 1$, $\underline{\lim\limits_{n\to+\infty}} \sin\frac{n\pi}{4} = -1$.

(3) Let a sequence $\{x_n\}$ be: $\frac{1}{2}, \frac{1}{2}, \frac{1}{4}, \frac{3}{4}, \frac{1}{8}, \frac{7}{8}, \cdots, \frac{1}{2^n}, \frac{2^n-1}{2^n}, \cdots$. Find all cluster points of $\{x_n\}$, $\overline{\lim\limits_{n\to+\infty}} x_n$ and $\underline{\lim\limits_{n\to+\infty}} x_n$.

Solution Since $\lim\limits_{n\to+\infty}\frac{1}{2^n}=0$ and $\lim\limits_{n\to+\infty}\frac{2^n-1}{2^n}=1$, cluster points of $\{x_n\}$ are 0 and 1, and so $\overline{\lim\limits_{n\to+\infty}}x_n=1$, $\underline{\lim\limits_{n\to+\infty}}x_n=0$.

(4) Let a sequence $\{x_n\}$ be: $1, \frac{1}{2}, 1+\frac{1}{2}, \frac{1}{3}, 1+\frac{1}{3}, \frac{1}{2}+\frac{1}{3}, \frac{1}{4}, 1+\frac{1}{4}, \frac{1}{2}+\frac{1}{4}, \frac{1}{3}+\frac{1}{4}, \frac{1}{5}, \cdots, \frac{1}{n}, 1+\frac{1}{n}, \frac{1}{2}+\frac{1}{n}, \cdots, \frac{1}{n-1}+\frac{1}{n}, \frac{1}{n+1}, \cdots$. Find all cluster points of $\{x_n\}$, $\overline{\lim\limits_{n\to+\infty}}x_n$ and $\underline{\lim\limits_{n\to+\infty}}x_n$.

Solution Notice that the convergent subsequences of $\{x_n\}$ are: $\left\{\frac{1}{n}\right\}$, $\left\{1+\frac{1}{n}\right\}$, $\left\{\frac{1}{2}+\frac{1}{n}\right\}$, $\left\{\frac{1}{3}+\frac{1}{n}\right\}$, $\cdots$ and $\lim\limits_{n\to+\infty}\frac{1}{n}=0$, $\lim\limits_{n\to+\infty}\left(1+\frac{1}{n}\right)=1$, $\lim\limits_{n\to+\infty}\left(\frac{1}{2}+\frac{1}{n}\right)=\frac{1}{2}$, $\lim\limits_{n\to+\infty}\left(\frac{1}{3}+\frac{1}{n}\right)=\frac{1}{3}$, $\cdots$.

Then, cluster points of $\{x_n\}$ are: $0, 1, \frac{1}{2}, \frac{1}{3}, \cdots$, and so $\overline{\lim\limits_{n\to+\infty}}x_n=1$ and $\underline{\lim\limits_{n\to+\infty}}x_n=0$.

(5) Let $x_n=3\left(1-\frac{1}{n}\right)+2(-1)^n(n=1, 2, \cdots)$. Find all cluster points of $\{x_n\}$, $\overline{\lim\limits_{n\to+\infty}}x_n$ and $\underline{\lim\limits_{n\to+\infty}}x_n$.

Solution Consider the sequence $\{y_n\}=\{2(-1)^n\}$. Then, cluster points of $\{y_n\}$ are 2 and -2, and so $\overline{\lim\limits_{n\to+\infty}}y_n=2$ and $\underline{\lim\limits_{n\to+\infty}}y_n=-2$. Notice that $\lim\limits_{n\to+\infty}3\left(1-\frac{1}{n}\right)=3$. Thus, cluster points of $\{x_n\}$ are 5 and 1, and so $\overline{\lim\limits_{n\to+\infty}}x_n=5$ and $\underline{\lim\limits_{n\to+\infty}}x_n=1$.

The following theorem can be proved by the definition of upper (lower) limit and Theorem 2.4.1. The proofs are left for readers.

Theorem 2.4.2 Let $\{x_n\}$ be a bounded sequence. Then

(1) $\underline{\lim\limits_{n\to+\infty}}x_n\leqslant\overline{\lim\limits_{n\to+\infty}}x_n$;

(2) $\underline{\lim}_{n\to+\infty}(-x_n) = -\overline{\lim}_{n\to+\infty} x_n$, $\overline{\lim}_{n\to+\infty}(-x_n) = -\underline{\lim}_{n\to+\infty} x_n$;

(3) for any subsequence $\{x_{i_n}\} \subseteq \{x_n\}$, $\underline{\lim}_{n\to+\infty} x_{i_n} \geqslant \underline{\lim}_{n\to+\infty} x_n$, $\overline{\lim}_{n\to+\infty} x_{i_n} \leqslant \overline{\lim}_{n\to+\infty} x_n$;

(4) $\lim_{n\to+\infty} x_n = a$ if and only if $\overline{\lim}_{n\to+\infty} x_n = \underline{\lim}_{n\to+\infty} x_n = a$.

Theorem 2.4.3 Let $\{x_n\}$ be a bounded sequence and let a, $b \in \mathbb{R}$. Then

(1) $\overline{\lim}_{n\to+\infty} x_n = a$ if and only if: for any $\varepsilon > 0$, there exists $N \in \mathbb{N}$ such that $x_n < a + \varepsilon$ whenever $n > N$ and there is a subsequence $\{x_{i_n}\} \subseteq \{x_n\}$ such that $x_{i_n} > a - \varepsilon$ for any n;

(2) $\underline{\lim}_{n\to+\infty} x_n = b$ if and only if: for any $\varepsilon > 0$, there exists $N \in \mathbb{N}$ such that $x_n > a - \varepsilon$ whenever $n > N$ and there is a subsequence $\{x_{i_n}\} \subseteq \{x_n\}$ such that $x_{i_n} < a + \varepsilon$ for any n.

Proof We only prove the necessity of (1). The proof of other assertions will be left for readers.

(1) "$\Rightarrow$": (by contradiction) Suppose that there is $\varepsilon > 0$, for any $N \in \mathbb{N}$, there exists $n > N$ such that $x_n \geqslant a + \varepsilon$. Then

for $N = 1$, there exists $i_1 > 1$ with $x_{i_1} \geqslant a + \varepsilon$;

for $N = i_1$, there exists $i_2 > i_1$ with $x_{i_2} \geqslant a + \varepsilon$;

for $N = i_2$, there exists $i_3 > i_2$ with $x_{i_3} \geqslant a + \varepsilon$;

… … … … … … … … … … … … … … … … … … … …

Thus, we have a subsequence $\{x_{i_n}\} \subseteq \{x_n\}$ such that $x_{i_n} \geqslant a + \varepsilon (n = 1, 2, \cdots)$.

Since $\{x_n\}$ is bounded, $\{x_{i_n}\}$ is bounded. Then, by Weierstrass theorem, there exists a subsequence $\{x_{j_m}\} \subseteq \{x_{i_n}\}$ $(m = 1, 2, \cdots)$ such that $\{x_{j_m}\}$ is convergent. Let $b = \lim_{m\to\infty} x_{j_m}$. Thus, b is a cluster point of $\{x_n\}$ with $b \geqslant a + \varepsilon$. But, on the other hand, $a = \overline{\lim}_{n\to+\infty} x_n$ is a maximal cluster point of $\{x_n\}$, which yields a contradiction.

Therefore, for any $\varepsilon > 0$, there exists $N \in \mathbb{N}$ such that $x_n < a + \varepsilon$ whenever $n > N$.

Since $\overline{\lim_{n\to+\infty}} x_n = a$, there is a subsequence $\{x_{j_m}\} \subseteq \{x_n\}$ $(m = 1, 2, \cdots)$ with $\lim_{m\to\infty} x_{j_m} = a$, and so for any $\varepsilon > 0$ there is $M \in \mathbb{N}$ such that $|x_{j_m} - a| < \varepsilon$ whenever $m > M$. Then, $x_{j_m} > a - \varepsilon$ whenever $m > M$. Let $i_1 = j_{M+1}$, $i_2 = j_{M+2}$, $i_3 = j_{M+3}$, $\cdots$. Hence, $\{x_{i_n}\} \subseteq \{x_n\}$ such that $x_{i_n} > a - \varepsilon$ for any n. □

Theorem 2.4.3 can be expressed equivalently in another version as follows:

Theorem 2.4.3* Let $\{x_n\}$ be a bounded sequence and let $a, b \in \mathbb{R}$. Then

(1) $\overline{\lim_{n\to+\infty}} x_n = a$ if and only if: for any $\alpha \in \mathbb{R}$, if $\alpha > a$, $\{x_n\}$ has at most a finite number of terms which are greater than α; if $\alpha < a$, $\{x_n\}$ has infinitely many terms which are greater than α.

(2) $\underline{\lim_{n\to+\infty}} x_n = b$ if and only if: for any $\beta \in \mathbb{R}$, if $\beta < b$, $\{x_n\}$ has at most a finite number of terms which are less than β; if $\beta > b$, $\{x_n\}$ has infinitely many terms which are less than β.

Remark Let $\{a_n\}$ be a bounded sequence and let $b \in \mathbb{R}$.

(1) If $\overline{\lim_{n\to+\infty}} a_n < b$, then there exists $N \in \mathbb{N}$ such that $a_n < b$ whenever $n \geqslant N$;

(2) If $\underline{\lim_{n\to+\infty}} a_n > b$, then there exists $N \in \mathbb{N}$ such that $a_n > b$ whenever $n \geqslant N$.

Proof (1) Let $\overline{\lim_{n\to+\infty}} a_n = a$. By the above theorem, for any $\varepsilon > 0$ there exists $N \in \mathbb{N}$ such that $a_n < a + \varepsilon$ whenever $n \geqslant N$. Thus, noticing the arbitrariness of ε we have $a_n \leqslant a < b$ whenever $n \geqslant N$.

(2) Let $\underline{\lim_{n\to+\infty}} a_n = a$. Then, for any $\varepsilon > 0$, there exists $N \in \mathbb{N}$ such that $a_n > a - \varepsilon$ whenever $n \geqslant N$. Thus, $a_n \geqslant a > b$ whenever $n \geqslant N$. □

Theorem 2.4.4 Let $\{x_n\}$ and $\{y_n\}$ be both bounded sequences.

(1) If there exists $N \in \mathbb{N}$ such that $x_n \leqslant y_n$ whenever $n > N$, then $\underline{\lim}_{n\to+\infty} x_n \leqslant \underline{\lim}_{n\to+\infty} y_n$ and $\overline{\lim}_{n\to+\infty} x_n \leqslant \overline{\lim}_{n\to+\infty} y_n$;

(2) $\underline{\lim}_{n\to+\infty} x_n + \underline{\lim}_{n\to+\infty} y_n \leqslant \underline{\lim}_{n\to+\infty}(x_n+y_n) \leqslant \underline{\lim}_{n\to+\infty} x_n + \overline{\lim}_{n\to+\infty} y_n$ (or $\overline{\lim}_{n\to+\infty} x_n + \underline{\lim}_{n\to+\infty} y_n$) $\leqslant \overline{\lim}_{n\to+\infty}(x_n+y_n) \leqslant \overline{\lim}_{n\to+\infty} x_n + \overline{\lim}_{n\to+\infty} y_n$;

(3) If $x_n, y_n \geqslant 0$ for any n, then $\underline{\lim}_{n\to+\infty} x_n \cdot \underline{\lim}_{n\to+\infty} y_n \leqslant \underline{\lim}_{n\to+\infty}(x_ny_n) \leqslant \underline{\lim}_{n\to+\infty} x_n \cdot \overline{\lim}_{n\to+\infty} y_n$ ($\overline{\lim}_{n\to+\infty} x_n \cdot \underline{\lim}_{n\to+\infty} y_n$) $\leqslant \overline{\lim}_{n\to+\infty}(x_ny_n) \leqslant \overline{\lim}_{n\to+\infty} x_n \cdot \overline{\lim}_{n\to+\infty} y_n$;

(4) If $\underline{\lim}_{n\to+\infty} x_n > 0$, then $\overline{\lim}_{n\to+\infty} \dfrac{1}{x_n} = \dfrac{1}{\underline{\lim}_{n\to+\infty} x_n}$.

Proof We just prove part of (2): $\underline{\lim}_{n\to+\infty} x_n + \underline{\lim}_{n\to+\infty} y_n \leqslant \underline{\lim}_{n\to+\infty}(x_n + y_n) \leqslant \underline{\lim}_{n\to+\infty} x_n + \overline{\lim}_{n\to+\infty} y_n$. Other statements may be similarly proved.

Let $\underline{\lim}_{n\to+\infty} x_n = a$ and let $\underline{\lim}_{n\to+\infty} y_n = b$. By Theorem 2.4.3, for any $\varepsilon > 0$, there exists $N_1 \in \mathbb{N}$ such that $x_n > a - \varepsilon$ whenever $n > N_1$, and there exists $N_2 \in \mathbb{N}$ such that $y_n > a - \varepsilon$ whenever $n > N_2$. Take $N = \max\{N_1, N_2\}$. Then, $x_n + y_n > (a-\varepsilon) + (b-\varepsilon) = a+b-2\varepsilon$ whenever $n > N$. Thus, $\underline{\lim}_{n\to+\infty}(x_n + y_n) \geqslant a + b - 2\varepsilon$, which, because of the arbitrariness of ε, implies that $\underline{\lim}_{n\to+\infty}(x_n+y_n) \geqslant a+b = \underline{\lim}_{n\to+\infty} x_n + \underline{\lim}_{n\to+\infty} y_n$.

Notice that $\underline{\lim}_{n\to+\infty}(x_n + y_n) - \overline{\lim}_{n\to+\infty} y_n = \underline{\lim}_{n\to+\infty}(x_n + y_n) + \underline{\lim}_{n\to+\infty}(-y_n) \leqslant \underline{\lim}_{n\to+\infty}[(x_n+y_n)+(-y_n)] = \underline{\lim}_{n\to+\infty} x_n$. Thus, $\underline{\lim}_{n\to+\infty}(x_n + y_n) \leqslant \underline{\lim}_{n\to+\infty} x_n + \overline{\lim}_{n\to+\infty} y_n$. □

Example Suppose $\{x_n\}$ is a bounded sequence and the sequence $\{x_{2n}+2x_n\}$ is convergent. Show that $\lim_{n\to+\infty} x_n$ exists.

Proof Since $\{x_{2n} + 2x_n\}$ is convergent, $\lim_{n\to+\infty}(x_{2n} + 2x_n) = \overline{\lim}_{n\to+\infty}(x_{2n} + 2x_n) = \underline{\lim}_{n\to+\infty}(x_{2n} + 2x_n)$. Thus, $\overline{\lim}_{n\to+\infty} x_n + 2\underline{\lim}_{n\to+\infty} x_n \geqslant$

$\overline{\lim\limits_{n\to+\infty}} x_{2n} + 2 \underline{\lim\limits_{n\to+\infty}} x_n \geqslant \underline{\lim\limits_{n\to+\infty}} (x_{2n} + 2x_n) = \overline{\lim\limits_{n\to+\infty}} (x_{2n} + 2x_n) \geqslant \underline{\lim\limits_{n\to+\infty}} x_{2n} + 2 \overline{\lim\limits_{n\to+\infty}} x_n \geqslant \underline{\lim\limits_{n\to+\infty}} x_n + 2 \overline{\lim\limits_{n\to+\infty}} x_n$. Then, $\underline{\lim\limits_{n\to+\infty}} x_n \geqslant \overline{\lim\limits_{n\to+\infty}} x_n$, and so $\underline{\lim\limits_{n\to+\infty}} x_n = \overline{\lim\limits_{n\to+\infty}} x_n$. Hence, $\lim\limits_{n\to+\infty} x_n$ exists. □

Remark Let $\{x_n\}$ and $\{y_n\}$ be both sequences. Assume that $\{y_n\}$ is convergent. Then

(1) $\overline{\lim\limits_{n\to+\infty}} (x_n + y_n) = \overline{\lim\limits_{n\to+\infty}} x_n + \lim\limits_{n\to+\infty} y_n$;

(2) $\underline{\lim\limits_{n\to+\infty}} (x_n + y_n) = \underline{\lim\limits_{n\to+\infty}} x_n + \lim\limits_{n\to+\infty} y_n$.

Proof We just prove (1), and (2) can be similarly proved.

Since $\overline{\lim\limits_{n\to+\infty}} x_n + \lim\limits_{n\to+\infty} y_n = \overline{\lim\limits_{n\to+\infty}} x_n + \underline{\lim\limits_{n\to+\infty}} y_n \leqslant \overline{\lim\limits_{n\to+\infty}} (x_n + y_n) \leqslant \overline{\lim\limits_{n\to+\infty}} x_n + \overline{\lim\limits_{n\to+\infty}} y_n = \overline{\lim\limits_{n\to+\infty}} x_n + \lim\limits_{n\to+\infty} y_n$, all inequalities there can be changed into equalities. □

Theorem 2.4.5 Let $\{a_n\}$ be a sequence. Then

(1) $\overline{\lim\limits_{n\to+\infty}} a_n = \lim\limits_{n\to+\infty} \sup\limits_{k\geqslant n}\{a_k\} = \inf\limits_{n}\{\sup\limits_{k\geqslant n}\{a_k\}\}$;

(2) $\underline{\lim\limits_{n\to+\infty}} a_n = \lim\limits_{n\to+\infty} \sup\limits_{k\geqslant n}\{a_k\} = \sup\limits_{n}\{\inf\limits_{k\geqslant n}\{a_k\}\}$.

Proof We only prove (1), and (2) can be similarly proved. We first show $\overline{\lim\limits_{n\to+\infty}} a_n = \lim\limits_{n\to+\infty} \sup\limits_{k\geqslant n}\{a_k\}$.

If $\{a_n\}$ is unbounded above, then $\overline{\lim\limits_{n\to+\infty}} a_n = +\infty$; in this case, we also see that $\{a_n, a_{n+1}, a_{n+2}, \cdots\}$ is unbounded above for any $n \in \mathbb{N}$, which implies $\sup\{a_n, a_{n+1}, a_{n+2}, \cdots\} = +\infty$. So, $\lim\limits_{n\to+\infty} \sup\{a_n, a_{n+1}, a_{n+2}, \cdots\} = +\infty$. Thus

$$\overline{\lim_{n\to+\infty}} a_n = \lim_{n\to+\infty} \sup_{k\geqslant n}\{a_k\}.$$

Now, we suppose $\{a_n\}$ is bounded above. Then, for any $n \in \mathbb{N}$, $\{a_n, a_{n+1}, a_{n+2}, \cdots\}$ is bounded above. By the completeness axiom, there exists $\lambda_n := \sup\{a_n, a_{n+1}, a_{n+2}, \cdots\}$ for any $n \in \mathbb{N}$. Clearly, the sequence $\{\lambda_n\}$ is decreasing. We now consider the following two cases:

Case 1. $\{\lambda_n\}$ is bounded below. In this case, $\{\lambda_n\}$ is convergent. Let $\lim\limits_{n\to+\infty}\lambda_n = A \in \mathbb{R}$. We first show A is a cluster point of $\{a_n\}$. Note that for any $\varepsilon > 0$, there exists $N \in \mathbb{N}$ such that $\lambda_n \in (A-\varepsilon, A+\varepsilon)$ whenever $n > N$. Take $N_1 > N$. Then, $\lambda_{N_1} \in (A-\varepsilon, A+\varepsilon)$. Since $\lambda_{N_1} = \sup\{a_{N_1}, a_{N_1+1}, \cdots\}$, for any $\varepsilon_1 > 0$, there exists $n_1 \geqslant N_1$ such that $a_{n_1} > \lambda_{N_1} - \varepsilon_1$. So, specifically for $\varepsilon_1 = \lambda_{N_1} - (A-\varepsilon)$ (>0), there exists $n_1 \geqslant N_1$ such that $a_{n_1} > \lambda_{N_1} - [\lambda_{N_1} - (A-\varepsilon)] = A-\varepsilon$. Thus, $a_{n_1} \in (A-\varepsilon, \lambda_{N_1}] \subseteq (A-\varepsilon, A+\varepsilon)$. Similarly, we may take $N_2 > n_1 (> N)$ such that $\lambda_{N_2} \in (A-\varepsilon, A+\varepsilon)$. Then, there exists $n_2 \geqslant N_2 (> n_1)$ such that $a_{n_2} \in (A-\varepsilon, \lambda_{N_2}] \subseteq (A-\varepsilon, A+\varepsilon)$. Proceeding in this manner, we obtain a sub-sequence $\{a_{n_k}\} \subseteq \{a_n\}$ such that for any $\varepsilon > 0$, $A-\varepsilon < a_{n_k} < A+\varepsilon$, and so A is a cluster point of $\{a_n\}$. We now show that A is the greatest cluster point of $\{a_n\}$. Let B be a cluster point of $\{a_n\}$. Then, there is a subsequence $\{a_{n_i}\} \subseteq \{a_n\}$ with $\lim\limits_{i\to\infty} a_{n_i} = B$. Since $\lambda_{n_i} = \sup\{a_{n_i}, a_{n_i+1}, \cdots\}$, $a_{n_i} \leqslant \lambda_{n_i}$ for any $i \in \mathbb{N}$, and so $\lim\limits_{i\to\infty} a_{n_i} \leqslant \lim\limits_{i\to\infty}\lambda_{n_i} = A$. Hence, $B \leqslant A$, which implies that A is the greatest cluster point of $\{a_n\}$, i. e. $\overline{\lim\limits_{n\to+\infty}} a_n = A = \lim\limits_{n\to+\infty}\lambda_n$.

Case 2 $\{\lambda_n\}$ is unbounded below. Then, for any $M > 0$, there exists $N \in \mathbb{N}$ such that $\lambda_N < -M$. Since $\{\lambda_n\}$ is decreasing, $\lambda_n \leqslant \lambda_N < -M$ whenever $n > N$, i. e. $\lim\limits_{n\to+\infty}\lambda_n = -\infty$. On the other hand, since $\lambda_n = \sup\{a_n, a_{n+1}, \cdots\}$, $a_n \leqslant \lambda_n$ for any $n \in \mathbb{N}$. Thus, $\overline{\lim\limits_{n\to+\infty}} a_n \leqslant \overline{\lim\limits_{n\to+\infty}}\lambda_n$, and so $-\infty \leqslant \underline{\lim\limits_{n\to+\infty}} a_n \leqslant \overline{\lim\limits_{n\to+\infty}} a_n \leqslant \overline{\lim\limits_{n\to+\infty}}\lambda_n = \lim\limits_{n\to+\infty}\lambda_n = -\infty$. Therefore $\overline{\lim\limits_{n\to+\infty}} a_n = \lim\limits_{n\to+\infty}\lambda_n (= -\infty)$.

Now we show the second equality in (1). We still consider two cases:

Case 1 $\{\lambda_n\}$ is bounded below. Since $\{\lambda_n\}$ is decreasing, by Theorem 2. 3. 1, $\lim\limits_{n\to+\infty}\lambda_n = \inf\{\lambda_n \mid n = 1, 2, \cdots\}$.

Case 2 $\{\lambda_n\}$ is unbounded below. Then, $\inf\{\lambda_n \mid n=1, 2, \cdots\} = -\infty$ and so $\lim\limits_{n\to+\infty} \lambda_n = \inf\{\lambda_n \mid n=1, 2, \cdots\}$. □

Exercise

1. Prove by (ε, N) definition the following sequence limits:

(1) $\lim\limits_{n\to+\infty} (-1)^n \dfrac{1}{n^2} = 0$; (2) $\lim\limits_{n\to+\infty} \dfrac{3n+1}{2n-1} = \dfrac{3}{2}$;

(3) $\lim\limits_{n\to+\infty} \dfrac{1}{n}\sin\dfrac{n\pi}{2} = 0$; (4) $\lim\limits_{n\to+\infty} \dfrac{n!}{n^n} = 0$;

(5) $\lim\limits_{n\to+\infty} \sin\dfrac{\pi}{n} = 0$; (6) $\lim\limits_{n\to+\infty} \dfrac{n}{a^n} = 0 (a>1)$;

(7) $\lim\limits_{n\to+\infty} (\sqrt{n+1} - \sqrt{n}) = 0$; (8) $\lim\limits_{n\to+\infty} \dfrac{1+2+\cdots+n}{n^3} = 0$;

(9) $\lim\limits_{n\to+\infty} a_n = 1$, where $a_n = \begin{cases} \dfrac{n-1}{n} & (\text{if } n \text{ is even}), \\ \dfrac{\sqrt{n^2+n}}{n} & (\text{if } n \text{ is odd}). \end{cases}$

2. First estimate the following limits, and then prove by definition your conclusion:

(1) $\lim\limits_{n\to+\infty} \dfrac{4n^2+1}{3n^2-5}$; (2) $\lim\limits_{n\to+\infty} \dfrac{\sqrt{n^2+a^2}}{n}$.

3. Suppose $\lim\limits_{n\to+\infty} u_n = A > 0$. Show that there exists a positive integer N such that $u_n > 0$ for any $n > N$.

4. Prove the existence of the limits of the following sequences:

(1) $u_n = \dfrac{1}{3+1} + \dfrac{1}{3^2+1} + \cdots \dfrac{1}{3^n+1}$;

(2) $u_n = \dfrac{1}{2} \times \dfrac{3}{4} \times \cdots \times \dfrac{2n-1}{2n}$;

(3) $x_1 = \sqrt{c}, \cdots, x_{n+1} = \sqrt{c + x_n}, \cdots (c>0)$;

(4) $u_1 = 4, \cdots, u_{n+1} = \dfrac{3u_n+2}{u_n+4}, \cdots$

5. Prove by (ε, N) definition:

(1) the limit of the sequence $\left\{\frac{1}{n}\right\}$ is not 1;

(2) the limit of the sequence $\left\{\frac{n}{n+1}\right\}$ is not 0.

6. Prove: if $\lim\limits_{n\to+\infty} a_n=a$, then $\lim\limits_{n\to+\infty} |a_n|=|a|$. What about the reverse statement?

7. Suppose $\lim\limits_{n\to+\infty} a_n=a$, $\lim\limits_{n\to+\infty} b_n=b$ and $a\neq b$. Show that the sequence $a_1, b_1, a_2, b_2, \cdots, a_n, b_n, \cdots$ is divergent.

8. Find the following sequence limits:

(1) $\lim\limits_{n\to+\infty}\dfrac{(-2)^n+3^n}{(-2)^{n+1}+3^{n+1}}$; (2) $\lim\limits_{n\to+\infty}(\sqrt{n^2+n}-n)$;

(3) $\lim\limits_{n\to+\infty}(\sqrt[n]{1}+\sqrt[n]{2}+\cdots+\sqrt[n]{10})$; (4) $\lim\limits_{n\to+\infty}\dfrac{\frac{1}{2}+\frac{1}{2^2}+\cdots+\frac{1}{2^n}}{\frac{1}{3}+\frac{1}{3^2}+\cdots+\frac{1}{3^n}}$;

(5) $\lim\limits_{n\to+\infty}\frac{1}{2}\times\frac{3}{4}\times\cdots\times\frac{2n-1}{2n}$; (6) $\lim\limits_{n\to+\infty}\dfrac{\sum_{p=1}^{n}p!}{n!}$.

9. Let $\lim\limits_{n\to+\infty} a_n=a$, $\lim\limits_{n\to+\infty} b_n=b$ and $a<b$. Show that there exists $N\in\mathbb{N}$ such that $a_n<b_n$ whenever $n>N$.

10. Let $\{a_n\}$ be a bounded sequence and $\lim\limits_{n\to+\infty} b_n=0$. Prove $\lim\limits_{n\to+\infty}(a_nb_n)=0$.

11. Using Cauchy criterion to prove the convergence or the divergence of the following sequences:

(1) $a_n=\dfrac{1}{1\times 2}+\dfrac{1}{2\times 3}+\cdots+\dfrac{1}{n(n+1)}$ is convergent;

(2) $a_n=(-1)^n n$ is divergent;

(3) $a_n=\sin\dfrac{n\pi}{2}$ is divergent.

12. Find the following limits:

(1) $\lim\limits_{n\to+\infty}\left(1-\dfrac{1}{n}\right)^n$; (2) $\lim\limits_{n\to+\infty}\left(\dfrac{n+1}{n-1}\right)^n$;

(3) $\lim\limits_{n\to+\infty}\left(1+\frac{k}{n}\right)^{mn}$;　　(4) $\lim\limits_{n\to+\infty}\left(1+\frac{1}{n}+\frac{1}{n^2}\right)^{n}$.

13. Prove: if $a_n>0$ and $\lim\limits_{n\to+\infty}\frac{a_n}{a_{n+1}}=b>1$, then $\lim\limits_{n\to+\infty}a_n=0$.

14. Prove that the limit of an increasing sequence is not less than any term of this sequence, and that the limit of an decreasing sequence is not greater than any term of this sequence.

15. Find the following limits:

(1) $\lim\limits_{n\to+\infty}\sqrt[n]{n^3+3^n}$;　　(2) $\lim\limits_{n\to+\infty}\frac{\sqrt[3]{n^2}\sin n^2}{n+1}$;

(3) $\lim\limits_{n\to+\infty}\frac{n^5}{e^n}$;　　(4) $\lim\limits_{n\to+\infty}(\sqrt{n+2}-2\sqrt{n+1}+\sqrt{n})$.

(5) $\lim\limits_{n\to+\infty}\sum_{k=1}^{n}\frac{1}{\sqrt{n^2+k}}$;　　(6) $\lim\limits_{n\to+\infty}\frac{\ln n}{n^2}$.

16. Prove the following limits:

(1) $\lim\limits_{n\to+\infty}\frac{n}{2^n}=0$;　　(2) $\lim\limits_{n\to+\infty}n^2q^n=0\ (|q|<1)$;

(3) $\lim\limits_{n\to+\infty}\frac{\lg n}{n^a}=0(a>0)$;　　(4) $\lim\limits_{n\to+\infty}\frac{1}{\sqrt[n]{n!}}=0$.

17. Let $\lim\limits_{n\to+\infty}a_n=a$. Prove $\lim\limits_{n\to+\infty}\frac{a_1+a_2+\cdots+a_n}{n}=a$, and using this to find $\lim\limits_{n\to+\infty}\frac{1+\frac{1}{2}+\frac{1}{3}+\cdots+\frac{1}{n}}{n}$.

18. Prove: if there exists $N\in\mathbb{N}$ and $k\in(0,1)$ such that $a_{n+1}\leqslant ka_n$ whenever $n>N$, then $\lim\limits_{n\to+\infty}a_n=0$.

19. Suppose $\{a_n\}$ is an increasing sequence and $\{b_n\}$ is a decreasing sequence. If $\lim\limits_{n\to+\infty}(a_n-b_n)=0$, prove that $\lim\limits_{n\to+\infty}a_n$ and $\lim\limits_{n\to+\infty}b_n$ are both existent and equal to each other.

20. Let $\{a_n\}$ be a sequence. Suppose there is a number M such that for any n, $A_n=|a_2-a_1|+|a_3-a_2|+\cdots+|a_n-a_{n-1}|\leqslant M$.

Prove that $\{A_n\}$ and $\{a_n\}$ are both convergent.

21. Prove: (1) Let $\{a_n\}$ be an increasing sequence. If it has no upper bound, then $\lim\limits_{n\to+\infty} a_n = +\infty$.

(2) Let $\{a_n\}$ be a decreasing sequence. If it has no lower bound, then $\lim\limits_{n\to+\infty} a_n = -\infty$.

22. Give examples to explain: The nested interval theorem, the monotone convergence theorem and the completeness axiom do not hold true in $\mathbb{Q}$, the rational number set.

23. Let M be an upper bound of a number set S and $M \notin S$. Prove that there exists a sequence $\{x_n\} \subseteq S$ such that $x_i \neq x_j$ ($i \neq j$) and $\lim\limits_{n\to+\infty} x_n = M$.

24. Suppose a sequence $\{x_n\}$ is unbounded. Show that there is a subsequence $\{x_{i_n}\}$ such that $\lim\limits_{n\to+\infty} |x_{i_n}| = +\infty$.

25. Prove that a monotone sequence must be convergent if it contains a convergent subsequence.

26. Let $H = \left\{ \left(\frac{1}{n+2}, \frac{1}{n} \right) \middle| \, n = 1, 2, \cdots \right\}$ be a set of open intervals. Answer the following questions:

(1) Is H an open covering of $(0, 1)$?

(2) Is H a finite open covering of $\left(0, \frac{1}{2}\right)$?

(3) Is H a finite open covering of $\left(\frac{1}{100}, 1\right)$?

27. Find the upper limit and the lower limit of the following sequences:

(1) $\{1 + (-1)^n\}$; (2) $\left\{(-1)^n \frac{n}{2n+1}\right\}$; (3) $\{2n+1\}$;

(4) $\left\{\frac{2n}{n+1} \sin \frac{n\pi}{4}\right\}$; (5) $\left\{\frac{n^2+1}{n} \sin \frac{\pi}{n}\right\}$; (6) $\left\{\sqrt[n]{\left|\cos \frac{n\pi}{3}\right|}\right\}$.

Chapter 3 Function limits and continuity

In this chapter, we first study the concept of the limit of a function, which conveys the intuitive idea that a $f(x)$ can be made arbitrarily close to a number A by taking x suffciently large or, in another case, by taking x suffciently close to a point x_0. In the later part of this chapter, we focus on the topic of continuity of a function, which is intimately linked with the concept of the limit. The concept of the limiting value for a function at a point x_0 is particularly important when x_0 is a point at which $f(x)$ has not been defined, or where $f(x)$ is not continuous.

§ 3. 1 Concept of function limits

We will discuss the concept of function limits in two cases, namely, as x approaches infinity and as x approaches a number.

§ 3. 1. 1 Function limit as $x \to \infty$

Definition (1) Let function f be defined on $(-\infty, +\infty)$ and let A be a number. If for any $\varepsilon > 0$, there is a number $M > 0$ such that for any x with $|x| > M$, $|f(x) - A| < \varepsilon$, then we say f has a *limit* A *as* x approaches ∞, and denote

$$\lim_{x \to \infty} f(x) = A$$

or

$$f(x) \to A(x \to \infty).$$

(2) Let function f be defined on $[a, +\infty)$ and let A be a number. If for any $\varepsilon>0$, there is a number $M>0$ such that for any $x>M$, $|f(x)-A|<\varepsilon$, then we say f has a *limit* A as x approaches $+\infty$, and denote

$$\lim_{x\to+\infty} f(x) = A$$

or

$$f(x) \to A(x \to +\infty).$$

(3) Let function f be defined on $(-\infty, a]$ and let A be a number. If for any $\varepsilon>0$, there is a number $M>0$ such that for any $x<-M$, $|f(x)-A|<\varepsilon$, then we say f has a *limit* A *as* x approaches $-\infty$, and denote

$$\lim_{x\to-\infty} f(x) = A$$

or

$$f(x) \to A(x \to -\infty).$$

Remark (1) It is plain to see that

$$\lim_{x\to\infty} f(x) = A \Leftrightarrow \lim_{x\to+\infty} f(x) = A \text{ and } \lim_{x\to-\infty} f(x) = A;$$

(2) $\lim\limits_{x\to\infty} f(x) \neq A \Leftrightarrow$ there is a $\varepsilon>0$ such that for any $M>0$ there is some x with $|x|>M$ but $|f(x)-A| \geqslant \varepsilon$;

(3) $f(x)$ has no limit as $x\to\infty\Leftrightarrow$for any $A\in\mathbb{R}$, $\lim\limits_{x\to\infty} f(x) \neq A$.

Example Prove by definition the following limits:

① $\lim\limits_{x\to-\infty} \arctan x = -\frac{\pi}{2}$; ② $\lim\limits_{n\to\infty} \frac{3x^2-1}{x^2+3} = 3$;

③ $\lim\limits_{x\to+\infty} (\sqrt{x+1}-\sqrt{x}) = 0$.

Proof ① Analysis: Let $\varepsilon > 0$. Then $\left|\arctan x - \left(-\frac{\pi}{2}\right)\right| =$

$$\left|\arctan x+\left(\frac{\pi}{2}\right)\right|<\varepsilon \Leftrightarrow -\varepsilon-\left(\frac{\pi}{2}\right)<\arctan\ x<\varepsilon-\left(\frac{\pi}{2}\right) \Rightarrow x< \tan\left(\varepsilon-\left(\frac{\pi}{2}\right)\right)=-\tan\left(\frac{\pi}{2}-\varepsilon\right).$$

Thus, for any $\varepsilon>0\left(\varepsilon<\frac{\pi}{2}\right)$, take $M=\tan\left(\frac{\pi}{2}-\varepsilon\right)$, then

$$\left|\arctan x-\left(-\frac{\pi}{2}\right)\right|<\varepsilon, \text{ whenever } x<-M.$$

② Since

$$\left|\frac{3x^2-1}{x^2+3}-3\right|=\frac{10}{x^2+3}<\frac{10}{x^2},$$

so, for any $\varepsilon>0$, take $M=\sqrt{10/\varepsilon}$, then

$$\left|\frac{3x^2-1}{x^2+3}-3\right|<\varepsilon \text{ whenever } |x|>M.$$

③ Since

$$|(\sqrt{x+1}-\sqrt{x})-0|=\frac{1}{\sqrt{x+1}+\sqrt{x}}<\frac{1}{\sqrt{x}}.$$

Hence, for any $\varepsilon>0$, take $M=\frac{1}{\varepsilon^2}$, then

$$|(\sqrt{x+1}-\sqrt{x})-0|<\varepsilon \text{ whenever } x>M. \ \square$$

§3.1.2 Function limit as $x\to x_0$

Now we turn to the notion of a function limit as x approaches a fixed number, say x_0, instead of infinity, that is denoted by

$$\lim_{x\to x_0} f(x)=A.$$

The interpretation of the foregoing expression is that $f(x)$ will approach A as x approaches the value x_0. This is an entirely correct qualitative description, but a quantitative criterion is required for analytical work. Since the only relevant quantities that can be measured are the differences $f(x)-A$ and $x-x_0$, the limit must be described in terms of these numbers.

The language of a precise statement, one without flaws or loopholes, seems strange initially but some thought and a little practice will make it familiar and useful. Frequent reference will be made to "arbitrarily small numbers," and it is convenient to have special names for this purpose. Following established usage, let ε denote a small number, which we are free to select at will, and δ another which is then determined. In other words, ε is given and δ is found. (For example, if $\varepsilon=0.0001$, it might turn out that $\delta=0.003$.) These ideas could also be exerted as x tends to x_0 from the left side or the right side. So armed, we can give the precise meaning of the definition as follows.

Definition Let A be a number.

(1) If for each $\varepsilon>0$, there exists some $\delta>0$ such that

$$|f(x)-A|<\varepsilon, \text{ whenever } x_0<x<x_0+\delta,$$

then $f(x)$ is said to have the *right-hand limit* A as x approaches x_0^+ (i. e. as x approaches x_0 from the right side of x_0), which is denoted as

$$\lim_{x\to x_0^+} f(x)=A.$$

(2) If for each $\varepsilon>0$, there exists $\delta>0$ such that

$$|f(x)-A|<\varepsilon, \text{ whenever } x_0-\delta<x<x_0,$$

then $f(x)$ is said to have the *left-hand limit* A *as* x approaches x_0^- (i. e. as x approaches x_0 from the left side of x_0), which is denoted as

$$\lim_{x \to x_0^-} f(x) = A.$$

(3) If for each $\varepsilon > 0$, there exists some $\delta > 0$ such that

$$|f(x) - A| < \varepsilon, \text{ whenever } 0 < |x - x_0| < \delta,$$

then the function $f(x)$ is said to have the *(two-sided) limit* A *as* x approaches x_0, which is written as

$$\lim_{x \to x_0} f(x) = A.$$

The right-hand or the left-hand limit of f is also called *one-sided limit* of f at x_0.

In other words, $\lim\limits_{x \to x_0^+} f(x) = A$ means: no matter how small ε is chosen, the quantity $|f(x) - A|$ can be made less than this ε, provided x is suffciently near to x_0 (but not equal to x_0). The number ε is a measure of how close we wish $f(x)$ to be to A, while the number δ is then a measure of the closeness of x to point x_0, as illustrated in Figure 3 - 1.

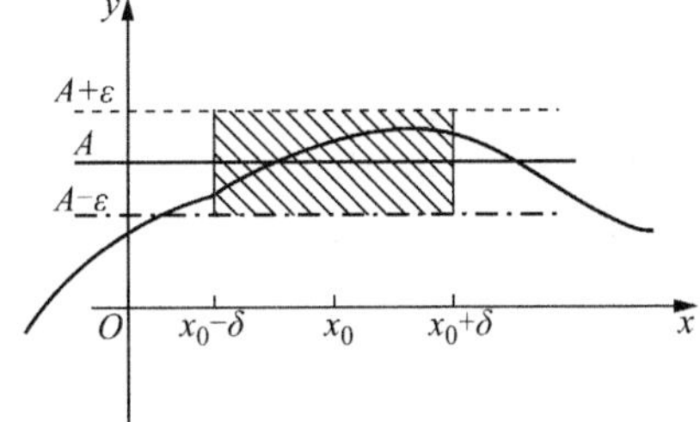

Figure 3 - 1

Remark (1) The above definition of the limit of $f(x)$ at x_0 is also known as ε-δ definition, in which the δ is corresponding to the N in the ε - N definition of a sequence limit. In general, the number δ depends on ε, but is not uniquely determined by ε. Usually, the smaller the ε is, the smaller the δ ought to be. Also because of arbitrariness of ε, the symbol $<$ in the inequality of the definition can be replaced by $\leqslant$.

(2) Note that by limit we study the behavior of the function as x approaches x_0 (not equals x_0). So the choice of $x = x_0$ is ruled out in the definition of the limit of $f(x)$ at x_0, i. e. whether or how $f(x)$ is

defined at x_0 itself turns out to be irrelevant to the discussion of limits.

(3) It is easy to see that

$$\lim_{x\to x_0} f(x) = A \Leftrightarrow \lim_{x\to x_0^+} f(x) = A \text{ and } \lim_{x\to x_0^-} f(x) = A.$$

(4) $\lim\limits_{x\to x_0} f(x) \neq A \Leftrightarrow$ there is an $\varepsilon>0$ such that for any $\delta>0$ there is some x with $0<|x-x_0|<\delta$ but $|f(x)-A| \geqslant \varepsilon$.

(5) $f(x)$ has no limit as $x\to x_0 \Leftrightarrow$ for any $A\in(-\infty, \infty)$, $\lim\limits_{x\to x_0} f(x) \neq A$.

The procedure for proof of the existence of a limit begins with the quantity $|f(x)-A|$. By manipulating this, the magnitude of δ is then determined which guarantees that the difference $|f(x)-A|$ is less than the preassigned arbitrarily small ε. The existence of a limit requires only that a number δ be found; it does not have to be an optimum value, such as the largest δ corresponding to a given ε.

Examples (1) Prove by definition that

$$\lim_{x\to -1}(5x+2)=-3.$$

Proof Note that $|(5x+2)-(-3)|=5|x-(-1)|$. So in order to make $|(5x+2)-(-3)|<\varepsilon$ for any $\varepsilon>0$, we merely need take $\delta=\varepsilon/5$. That is, for any $\varepsilon>0$, there is $\delta=\varepsilon/5$ such that $|(5x+2)-(-3)|<\varepsilon$ whenever $0<|x-(-1)|<\delta$. Therefore, by the definition we see that

$$\lim_{x\to -1}(5x+2)=-3. \square$$

(2) Prove by definition that

$$\lim_{x\to 0}\cos x=1.$$

Proof Note $|\cos x-1|=2\left|\sin^2\frac{x}{2}\right|<2\cdot\left(\frac{x}{2}\right)^2=\frac{x^2}{2}$. Thus, for any $\varepsilon>0$, take $\delta=\sqrt{2\varepsilon}$. Then for any x with $0<|x|<\delta$, $|\cos x-1|<\varepsilon$. □

(3) Prove by definition that

$$\lim_{x\to 2}\frac{x^2}{x+1}=\frac{4}{3}.$$

Analysis: Note

$$\left|\frac{x^2}{x+1}-\frac{4}{3}\right|=\left|\frac{(3x+2)(x-2)}{3(x+1)}\right|.$$

By taking $\delta_1=1$, if $|x-2|<\delta_1$, $|3x+2|=|3(x-2)+8|\leqslant 3|x-2|+8<11$ and $|3(x+1)|=|3(x-2)+9|>9-3|x-2|>9-3=6$, and so

$$\left|\frac{3x+2}{3(x+1)}\right|<\frac{11}{6}<2.$$

Thus

$$\left|\frac{(3x+2)(x-2)}{3(x+1)}\right|<2|x-2|.$$

Now by letting $2|x-2|<\varepsilon$, we have $|x-2|<\varepsilon/2$.

Proof For any $\varepsilon>0$, take $\delta=\min\{1, \varepsilon/2\}$, then

$$\left|\frac{x^2}{x+1}-\frac{4}{3}\right|<\varepsilon$$

whenever $0<|x-2|<\delta$. □

(4) Prove by definition that

$$\lim_{x\to 1}\frac{2x}{\sqrt{x^2+1}}=\sqrt{2}.$$

Analysis: Note

$$\left|\frac{2x}{\sqrt{x^2+1}}-\sqrt{2}\right|=\left|\frac{2x-\sqrt{2}\sqrt{x^2+1}}{\sqrt{x^2+1}}\right|$$

$$= \frac{2|x^2-1|}{\sqrt{x^2+1}\cdot|2x+\sqrt{2(x^2+1)}|}.$$

First restrict $|x-1|<\frac{1}{2}$, i. e. $\frac{1}{2}<x<\frac{3}{2}$, and then $|2x+\sqrt{2(x^2+1)}|>1$ and $|x+1|<\frac{5}{2}$. Thus

$$\frac{2|x^2-1|}{\sqrt{x^2+1}\cdot|2x+\sqrt{2(x^2+1)}|}$$
$$<2|x^2-1|=2|x+1|\cdot|x-1|$$
$$<2\times\frac{5}{2}|x-1|=5|x-1|.$$

Proof For any $\varepsilon>0$, take

$$\delta=\min\left\{\frac{1}{2}, \frac{\varepsilon}{5}\right\}.$$

Then for any x with $0<|x-1|<\delta$,

$$\left|\frac{2x}{\sqrt{x^2+1}}-\sqrt{2}\right|<\varepsilon. \square$$

(5) Let

$$f(x)=\begin{cases} x+2 & (x\leqslant 2), \\ \dfrac{1}{x-2} & (x>2). \end{cases}$$

Prove that $\lim\limits_{x\to 2} f(x)$ does not exist.

Proof As $x<2$, $|f(x)-4|=|x-2|=2-x$. For any $\varepsilon>0$, take $\delta=\varepsilon$, and so $|f(x)-4|<\varepsilon$ if $0<2-x<\delta$, i. e. $\lim\limits_{x\to 2^-} f(x)=4$.

Now let $x>2$. For any $M>0$, take $\delta=1/M$, then

$$f(x)=\frac{1}{x-2}>M,$$

if $0<x-2<\delta$, i. e. $\lim_{x\to 2^+} f(x)=+\infty$.

Therefore, $\lim_{x\to 2^-} f(x)\neq \lim_{x\to 2^+} f(x)$, and so $\lim_{x\to 2} f(x)$ does not exist. □

(6) Let

$$f(x)=\begin{cases}0 & (\text{if } x \text{ is rational}),\\ 1 & (\text{if } x \text{ is irrational}).\end{cases}$$

Prove by definition that $\lim_{x\to 0} f(x)$ does not exist.

Proof That is to show: for any real number A, $\lim_{x\to 0} f(x)\neq A$.

First let $A=0$. Take $\varepsilon=0.1>0$. Then for any $\delta>0$, there exists irrational $x\in U^\circ(0,\delta)$ with $|f(x)-A|=|f(x)|=1>0.1=\varepsilon$. Now let $A\neq 0$. Take $\varepsilon=|A|/2>0$. Then for any $\delta>0$, there exists rational $x\in U^\circ(0,\delta)$ with $|f(x)-A|=|0-A|=|A|>|A|/2=\varepsilon$. □

§ 3.2 Properties of function limits

In this section, we only consider the properties of the limit $\lim_{x\to x_0} f(x)$. For other types of the limits such as $x\to x_0^\pm$ and $x\to(\pm)\infty$, there are analogous results, and the proofs merely need to be modified correspondingly.

Theorem 3.2.1(Uniqueness) If the limit $\lim_{x\to x_0} f(x)$ exists, then it is unique.

Proof Suppose A and B are both limits of f as $x\to x_0$. Then for any given $\varepsilon>0$, there are δ_1 and δ_2 such that

for any x with $0<|x-x_0|<\delta_1$, $|f(x)-A|<\varepsilon$,

and for any x with $0<|x-x_0|<\delta_2$, $|f(x)-B|<\varepsilon$,

which implies

$$|A-B| \leqslant |f(x)-A| + |f(x)-B| < 2\varepsilon.$$

Thus we have $A = B$ and the result follows.

Theorem 3.2.2 (Local boundedness) If the limit $\lim\limits_{x \to x_0} f(x)$ exists, then there is a free-center neighborhood of point x_0, $U^o(x_0)$, such that f is bounded on $U^o(x_0)$.

Proof Suppose A is the limit of f as $x \to x_0$. Then for $\varepsilon = 1$, there is a number $\delta > 0$ such that for any $x \in U^o(x_0, \delta)$,

$$|f(x)-A| < \varepsilon = 1,$$

which implies f is bounded on $U^o(x_0, \delta)$. □

Theorem 3.2.3 (Locally sign-preserving) If there exists the limit $\lim\limits_{x \to x_0} f(x) = A > 0 (< 0)$, then for any $r \in (0, |A|)$, there is $U^o(x_0)$, a free-center neighborhood of point x_0, such that $f(x) > r > 0$ ($f(x) < -r < 0$) for any $x \in U^o(x_0)$.

Proof Let $A > 0$. Then for $\varepsilon = A - r$, there is $\delta > 0$ such that

$$|f(x)-A| < \varepsilon = A - r$$

for any $x \in U^o(x_0, \delta)$, which implies $f(x) > r > 0$ on $U^o(x_0)$. For the case $A < 0$, the result can be similarly proved. □

Theorem 3.2.4 (Locally order-preserving) Suppose the limits of f and g both exist as x approaches x_0. If there is $U^o(x_0)$, such that $f(x) \leqslant g(x)$ for any $x \in U^o(x_0)$, then $\lim\limits_{x \to x_0} f(x) \leqslant \lim\limits_{x \to x_0} g(x)$.

Proof Let $\lim\limits_{x \to x_0} f(x) = A$ and $\lim\limits_{x \to x_0} g(x) = B$. Assuming $A > B$. Take $\varepsilon = (A-B)/2$. Then there are $\delta_1 > 0$ and $\delta_2 > 0$ such that for any x with $0 < |x - x_0| < \delta_1$,

$$|f(x)-A| < \varepsilon,$$

and for any x with $0 < |x - x_0| < \delta_2$,

$$|f(x)-B| < \varepsilon.$$

Thus for any x with $0 < |x - x_0| < \min\{\delta_1, \delta_2\}$,

$$g(x) < B + \varepsilon = \frac{A+B}{2} = A - \varepsilon < f(x),$$

which clearly contradicts the given condition. □

Theorem 3. 2. 5 (Squeezing law) Suppose there is a free-center neighborhood of point x_0 with radius δ, $U^o(x_0, \delta)$, such that $f(x) \leqslant h(x) \leqslant g(x)$ for any $x \in U^o(x_0, \delta)$. If

$$\lim_{x \to x_0} f(x) = \lim_{x \to x_0} g(x) = A,$$

then

$$\lim_{x \to x_0} h(x) = A.$$

Proof By the condition, for any $\varepsilon > 0$ there exist $\delta_1, \delta_2 > 0$ such that

$|f(x) - A| < \varepsilon$, whenever $0 < |x - x_0| < \delta_1$,

and $|g(x) - A| < \varepsilon$, whenever $0 < |x - x_0| < \delta_2$.

Therefore for any x with $0 < |x - x_0| < \min\{\delta, \delta_1, \delta_2\}$,

$$A - \varepsilon < f(x) \leqslant h(x) \leqslant g(x) < A + \varepsilon,$$

i. e.

$$|h(x) - A| < \varepsilon. \ \square$$

Examples (1) Show $\lim\limits_{x \to x_0} x\left[\frac{1}{x}\right] = 1$.

Proof First we consider the case where $x \to 0^+$. Let $0 < x < 1$. Then $[1/x] \leqslant 1/x < [1/x] + 1$, and

$$\frac{1}{\left[\frac{1}{x}\right]+1} < x \leqslant \frac{1}{\left[\frac{1}{x}\right]}, \quad \frac{1}{1+\left(1/\left[\frac{1}{x}\right]\right)} = \frac{\left[\frac{1}{x}\right]}{\left[\frac{1}{x}\right]+1} < x\left[\frac{1}{x}\right] \leqslant 1.$$

Note $\lim_{x\to0^+}\frac{1}{1+\left(1/\left[\frac{1}{x}\right]\right)}=1$. Thus, $\lim_{x\to0^+}x\left[\frac{1}{x}\right]=1$. Similarly, we can prove that $\lim_{x\to0^-}x\left[\frac{1}{x}\right]=1$. □

(2) Evaluate the following limits:

① $\lim_{x\to0}\frac{\sin[x^2\sin(1/x)]}{x}$; ② $\lim_{x\to+\infty}\left(\frac{a^x+b^x}{2}\right)^{1/x}$ $(a, b>0)$.

Solution ① As

$$0\leqslant\left|\frac{\sin[x^2\sin(1/x)]}{x}\right|\leqslant\left|\frac{x^2\sin(1/x)}{x}\right|\leqslant\left|\frac{x^2}{x}\right|=|x|\to0,$$

$$\lim_{x\to0}\frac{\sin[x^2\sin(1/x)]}{x}=0.$$

② Note as $x>0$,

$$\frac{\max\{a, b\}}{2^{1/x}}=\left[\frac{(\max\{a, b\})^x}{2}\right]^{1/x}\leqslant\left(\frac{a^x+b^x}{2}\right)^{1/x}$$
$$\leqslant\left[\frac{2(\max\{a, b\})^x}{2}\right]^{1/x}=\max\{a, b\},$$

and

$$\lim_{x\to+\infty}\frac{\max\{a, b\}}{2^{1/x}}=\max\{a, b\}.$$

Hence,

$$\lim_{x\to+\infty}\left(\frac{a^x+b^x}{2}\right)^{1/x}=\max\{a, b\}.$$

The following theorem allows us to reduce the computation of limits for combinations of functions to computation of limits involving the constituent functions.

Theorem 3.2.6 (Fundamental operations of function limits)

Suppose $\lim_{x\to x_0}f(x)$ and $\lim_{x\to x_0}g(x)$ are both existent. Then

(1) $\lim\limits_{x\to x_0}(f(x)+g(x))$ is existent and

$$\lim_{x\to x_0}(f(x)+g(x))=\lim_{x\to x_0}f(x)+\lim_{x\to x_0}g(x);$$

(2) $\lim\limits_{x\to x_0}(f(x)\cdot g(x))$ is existent and

$$\lim_{x\to x_0}(f(x)\cdot g(x))=\lim_{x\to x_0}f(x)\cdot\lim_{x\to x_0}g(x);$$

(3) If $\lim\limits_{x\to x_0}g(x)\neq 0$, $\lim\limits_{x\to x_0}\dfrac{f(x)}{g(x)}$ is existent and

$$\lim_{x\to x_0}\frac{f(x)}{g(x)}=\frac{\lim\limits_{x\to x_0}f(x)}{\lim\limits_{x\to x_0}g(x)}.$$

Proof Let $\lim\limits_{x\to x_0}=A$ and $\lim\limits_{x\to x_0}g(x)=B$, and let ε be an arbitrarily selected positive number. Then there exist δ_1, $\delta_2>0$ such that

$$|f(x)-A|<\varepsilon,\ \text{whenever}\ 0<|x-x_0|<\delta_1;$$
$$|f(x)-B|<\varepsilon,\ \text{whenever}\ 0<|x-x_0|<\delta_2.$$

(1) Take $\delta=\min\{\delta_1,\delta_2\}$. Then, if $0<|x-x_0|<\delta$,

$$|(f(x)+g(x))-(A+B)|\leqslant|f(x)-A|+|g(x)-B|<2\varepsilon.$$

(2) Since $\lim\limits_{x\to x_0}g(x)=B$, for $\varepsilon_0=1$, there exists $\delta_3>0$ such that

$$|f(x)-B|<\varepsilon_0=1\ \text{whenever}\ 0<|x-x_0|<\delta_3,$$

and so $|g(x)|\leqslant|g(x)-B|+|B|<1+|B|$. Now take $\delta=\min\{\delta_1,\delta_2,\delta_3\}$. Then, if $0<|x-x_0|<\delta$,

$$\begin{aligned}|f(x)g(x)-AB|&\leqslant|f(x)-A||g(x)|+|A||g(x)-B|\\&<\varepsilon|g(x)|+|A|\varepsilon<\varepsilon(1+|B|+|A|).\end{aligned}$$

(3) Let $\lim\limits_{x\to x_0}g(x)=B\neq 0$. First we show

$$\lim_{x\to x_0}\frac{1}{g(x)}=\frac{1}{B}.$$

Noting that there exists $\delta_3 > 0$ such that

$$|g(x) - B| < \frac{|B|}{2} \text{ whenever } 0 < |x - x_0| < \delta_3,$$

we have

$$|B| = |B - g(x) + g(x)| \leqslant |B - g(x)| + |g(x)| < \frac{|B|}{2} + |g(x)|,$$

which implies that

$$\frac{1}{|g(x)|} < \frac{2}{|B|},$$

and so

$$\frac{1}{|Bg(x)|} < \frac{1}{|B|} \cdot \frac{2}{|B|} = \frac{2}{B^2}.$$

Take $\delta = \min\{\delta_2, \delta_3\}$. Then

$$\left|\frac{1}{g(x)} - \frac{1}{B}\right| = \frac{|B - g(x)|}{|Bg(x)|} < \varepsilon\left(\frac{2}{B^2}\right) \text{whenever } 0 < |x - x_0| < \delta,$$

which implies that

$$\lim_{x \to x_0} \frac{1}{g(x)} = \frac{1}{B} = \frac{1}{\lim\limits_{x \to x_0} g(x)}.$$

Thus by (2),

$$\lim_{x \to x_0} \frac{f(x)}{g(x)} = \lim_{x \to x_0}\left(f(x) \cdot \frac{1}{g(x)}\right) = \lim_{x \to x_0} f(x) \cdot \lim_{x \to x_0} \frac{1}{g(x)} = \frac{\lim\limits_{x \to x_0} f(x)}{\lim\limits_{x \to x_0} g(x)}. \quad \square$$

Remark The following formulas may be deduced directly by the preceding theorem:

(1) Let c_1 and c_2 be constants. Then

$$\lim_{x \to x_0}(c_1 f(x) + c_2 g(x)) = c_1 \lim_{x \to x_0} f(x) + c_2 \lim_{x \to x_0} g(x),$$

in particular,

$$\lim_{x \to x_0}(f(x) - g(x)) = \lim_{x \to x_0} f(x) - \lim_{x \to x_0} g(x);$$

(2)

$$\lim_{x \to x_0}(f(x))^n = (\lim_{x \to x_0} f(x))^n,$$ where n is a positive integer;

(3) Let $p(x)$ be a polynomial function, and x_0 any number. Then

$$\lim_{x \to x_0} p(x) = p(x_0);$$

(4) Let $r(x) = p(x)/q(x)$ be a rational function [i. e. $p(x)$, $q(x)$ are polynomials], and x_0 a number with $q(x_0) \neq 0$. Then

$$\lim_{x \to x_0} r(x) = r(x_0).$$

These formulas can be used effectively to determine new limits from those already established, and the basic $\varepsilon - \delta$ method is reserved for the special situations that require a return to fundamentals. For most purposes, then, the $\varepsilon - \delta$ approach is relegated to a distinctly secondary position, although as the foundation it supports everything else.

Examples (1) Prove $\lim_{x \to 1} e^{1/(x-1)}$ does not exist.

Proof $$\lim_{x \to 1^-} e^{1/(x-1)} = \lim_{x \to 1^-} e^{1/(x-1)} = e^{-\infty} = 0,$$

however,

$$\lim_{x \to 1^+} e^{1/(x-1)} = \lim_{x \to 1^+} e^{1/(x-1)} = e^{+\infty} = +\infty. \square$$

(2) Find

① $\lim_{x \to -\infty} x(\sqrt{x^2+100}+x)$; ② $\lim_{x \to +\infty} \dfrac{a^x + a^{-x}}{a^x - a^{-x}}$ $(a > 0, a \neq 1)$.

Solution ① $\lim\limits_{x\to-\infty} x(\sqrt{x^2+100}+x)=\lim\limits_{x\to-\infty}\dfrac{100x}{\sqrt{x^2+100}-x}$

$$=\lim_{x\to-\infty}\frac{100x}{|x|\sqrt{1+\frac{100}{x^2}}-x}$$

$$=\lim_{x\to-\infty}\frac{100x}{(-x)\sqrt{1+\frac{100}{x^2}}-x}$$

$$=\lim_{x\to-\infty}\frac{100}{-\sqrt{1+\frac{100}{x^2}}-1}$$

$$=\frac{100}{-2}=-50.$$

② If $a>1$, then $\lim\limits_{x\to+\infty} a^{-x}=0$, and so

$$\lim_{x\to+\infty}\frac{a^x+a^{-x}}{a^x-a^{-x}}=\lim_{x\to+\infty}\frac{1+(a^{-x})^2}{1-(a^{-x})^2}=1;$$

if $0<a<1$, then $\lim\limits_{x\to+\infty} a^x=0$, and so

$$\lim_{x\to+\infty}\frac{a^x+a^{-x}}{a^x-a^{-x}}=\lim_{x\to+\infty}\frac{(a^x)^2+1}{(a^x)^2-1}=-1.$$

(3) Find

$$\lim_{x\to0}\left[\frac{2+e^{\frac{1}{x}}}{1+e^{\frac{2}{x}}}+\frac{x}{|x|}\right].$$

Solution Noting that $\lim\limits_{x\to0^-} e^{\frac{1}{x}}=0$, $\lim\limits_{x\to0^-} e^{\frac{2}{x}}=0$ and $\lim\limits_{x\to0^-}\dfrac{x}{|x|}=-1$,

$$\lim_{x\to0^-}\left[\frac{2+e^{\frac{1}{x}}}{1+e^{\frac{2}{x}}}+\frac{x}{|x|}\right]=2-1=1;$$

while $\lim\limits_{x\to0^+} e^{\frac{1}{x}}=+\infty$, $\lim\limits_{x\to0^+} e^{\frac{2}{x}}=+\infty$ and $\lim\limits_{x\to0^+}\dfrac{x}{|x|}=1$,

$$\lim_{x\to 0^+}\left[\frac{2+e^{\frac{1}{x}}}{1+e^{\frac{2}{x}}}+\frac{x}{|x|}\right]=\lim_{x\to 0^+}\left[\frac{(2/e^{\frac{1}{x}})+1}{(1/e^{\frac{1}{x}})+e^{\frac{1}{x}}}+\frac{x}{|x|}\right]=0+1=1.$$

So

$$\lim_{x\to 0}\left[\frac{2+e^{\frac{1}{x}}}{1+e^{\frac{2}{x}}}+\frac{x}{|x|}\right]=1.$$

(4) Determine the constants λ and μ such that

$$\lim_{x\to\infty}(\sqrt[3]{1-x^3}-\lambda x-\mu)=0.$$

Solution Note

$$\lim_{x\to\infty} x\left(\sqrt[3]{\frac{1}{x^3}-1}-\lambda-\frac{\mu}{x}\right)=\lim_{x\to\infty}(\sqrt[3]{1-x^3}-\lambda x-\mu)=0.$$

$$\text{So } \lim_{x\to\infty}\left(\sqrt[3]{\frac{1}{x^3}-1}-\lambda-\frac{\mu}{x}\right)=0,$$

which implies $-1-\lambda=0$, i. e. $\lambda=-1$. Then

$$\mu=\lim_{x\to\infty}(\sqrt[3]{1-x^3}+x)=\lim_{x\to\infty}\frac{(1-x^3)+x^3}{(\sqrt[3]{1-x^3})^2-x\sqrt[3]{1-x^3}+x^2}$$

$$=\lim_{x\to\infty}\left[\frac{1}{\left(\sqrt[3]{\frac{1}{x^3}-1}\right)^2-\sqrt[3]{\frac{1}{x^3}-1}+1}\cdot\frac{1}{x^2}\right]=\frac{1}{3}\times 0=0.$$

The following theorem establishes a connection between limits of sequences and functions.

Theorem 3. 2. 7 (Heine theorem) Suppose the function f is defined on a free-center neighborhood of point x_0, $U^o(x_0)$. Then

$$\lim_{x\to x_0} f(x)=A\Leftrightarrow \lim_{n\to+\infty} f(x_n)=A$$

for any sequence $\{x_n\}\subset U^o(x_0)$ with $\lim\limits_{x\to+\infty} x_n=x_0\,(x_n\neq x_0)$.

Proof Necessity. Suppose $\lim\limits_{x\to x_0} f(x)=A$. Then for any $\varepsilon>0$, there exists $\delta > 0$ such that $|f(x)-A|<\varepsilon$ if $0<|x-x_0|<\delta$. Since $\lim\limits_{x\to+\infty} x_n = x_0 (x_n \neq x_0)$, for this δ, there exists a positive integer N such that $0<|x_n-x_0|<\delta$ whenever $n>N$, and so $|f(x_n)-A|<\varepsilon$, which implies $\lim\limits_{n\to+\infty} f(x_n) = A$.

Suffciency. Assume $\lim\limits_{x\to x_0} f(x)\neq A$. Then there exists $\varepsilon_0>0$ such that for any $\delta>0$, there is a x with $0<|x-x_0|<\delta$ and $|f(x)-A|\geqslant \varepsilon_0$. In particular, for $\delta_1 = 1$, there is x_1 with $0<|x_1-x_0|<\delta$ and $|f(x_1)-A|\geqslant\varepsilon_0$; for $\delta_2=1/2$, there is x_2 with $0<|x_2-x_0|<\delta$ and $|f(x_2)-A|\geqslant\varepsilon_0$;... ; for $\delta_n=1/n$, there is x_n with $0<|x_n-x_0|<\delta$ and $|f(x_n)-A|\geqslant\varepsilon_0$;.... Thus, we have a sequence $\{x_n\}$ with $0<|x_n-x_0|<1/n$ but $|f(x_n)-A|\geqslant\varepsilon_0$, which implies $\lim\limits_{n\to\infty} x_n=x_0 (x_n\neq x_0)$ but $\lim\limits_{n\to+\infty} f(x_n)\neq A$.

Remark By Heine theorem, if there is a sequence $\{x_n\}$ with $\lim\limits_{n\to+\infty} x_n = x_0 (x_n\neq x_0)$ but $\lim\limits_{n\to+\infty} f(x_n)$ is non-existent, then $\lim\limits_{x\to x_0} f(x)$ is nonexistent.

Example Let $f(x)=\sin(1/x)$. Prove that $\lim\limits_{x\to 0} f(x)$ does not exist.

Proof Let

$$x_n = \frac{1}{n\pi+(\pi/2)},$$

then $\lim\limits_{n\to+\infty} x_n = 0$. However, since

$$\sin\frac{1}{x_n} = \sin\frac{1}{1/(n\pi+(\pi/2))} = \sin\left(n\pi+\frac{\pi}{2}\right)= (-1)^n,$$

$\lim\limits_{n\to+\infty} \sin(1/x_n)$ does not exist. So, by Heine theorem, $\lim\limits_{x\to 0} f(x)$ does not exist. □

If a function $f(x)$ converges to a limit a as $x\to x_0$, the function values must become closer and closer to a as $x\to x_0$ and hence close to each other. This property is stated more formally in the next theorem, just as the similar result for the limit of a sequence.

Theorem 3.2.8 (Cauchy criterion for convergence) The sufficient and necessary condition for the existence of $\lim\limits_{x\to x_0} f(x)$ is:

for any $\varepsilon > 0$, there exists $\delta > 0$ such that $|f(x_1) - f(x_2)| < \varepsilon$ whenever $0 < |x_1 - x_0| < \delta$ and $0 < |x_2 - x_0| < \delta$.

Proof Necessity. Let $\lim\limits_{x\to x_0} f(x) = a$. Then, for any $\varepsilon > 0$, there exists $\delta > 0$ such that $|f(x) - a| < \varepsilon/2$ whenever $0 < |x - x_0| < \delta$. So, for any x_1 and x_2 with $0 < |x_1 - x_0| < \delta$ and $0 < |x_2 - x_0| < \delta$,

$$|f(x_1) - f(x_2)| \leqslant |f(x_1) - a| + |f(x_2) - a| < \frac{\varepsilon}{2} + \frac{\varepsilon}{2} = \varepsilon.$$

Sufficiency. Assume that for any $\varepsilon > 0$, there exists $\delta > 0$ such that $|f(x_1) - f(x_2)| < \varepsilon$ whenever $0 < |x_1 - x_0| < \delta$ and $0 < |x_2 - x_0| < \delta$.

Let $\{x_n\}$ $(x_n \neq x_0)$ be a sequence with $\lim\limits_{n\to+\infty} x_n = x_0$ (clearly such sequence is existent). Then, for the given $\delta > 0$, there is a positive integer N such that for any n, $m > N$, $0 < |x_n - x_0| < \delta$ and $0 < |x_m - x_0| < \delta$, and so $|f(x_n) - f(x_m)| < \varepsilon$. Then by Cauchy criterion for convergence of sequence, $\{f(x_n)\}$ is convergent.

Let $\lim\limits_{n\to+\infty} f(x_n) = a$. It remains to show $\lim\limits_{x\to x_0} f(x) = a$. Since $\lim\limits_{n\to\infty} x_n = x_0$ $(x_n \neq x_0)$ and $\lim\limits_{n\to\infty} f(x_n) = a$, there exists $x_{n_0} \in \{x_n\}$ such that $0 < |x_{n_0} - x_0| < \delta$ and $|f(x_{n_0}) - a| < \varepsilon$. Then for any x with $0 < |x - x_0| < \delta$,

$$|f(x) - a| \leqslant |f(x) - f(x_{n_0})| + |f(x_{n_0}) - a| < \varepsilon + \varepsilon = 2\varepsilon,$$

which implies $\lim\limits_{x\to x_0} f(x) = a$. □

Remark The sufficient and necessary condition for the existence of one-sided limits $\lim\limits_{x\to x_0^+} f(x)$ and $\lim\limits_{x\to x_0^-} f(x)$ can be similarly described.

§ 3.3 Two important limits

We now turn to two limits, the importance of which will become clear in later chapters.

Proposition The first important limit:

$$\lim_{x \to 0} \frac{\sin x}{x} = 1$$

Proof First we prove

$$\lim_{x \to 0^+} \frac{\sin x}{x} = 1.$$

Without loss of generality, we may suppose $0 < x < \pi/2$ as shown in a unit circle (cf. Figure 3 - 2). Denote by DD', $\overset{\frown}{DD'}$ and BB' the lengths of the line segment DD', the arc $\overset{\frown}{DD'}$ and the line segment BB' respectively.

Then, by the knowledge of plane geometry we have $DD' \leqslant DD' \leqslant BB'$. Since $\sin x = (DD'/2)$, $x = (\overset{\frown}{DD'}/2)$ and $\tan x = (BB'/2)$, we see that $\sin x \leqslant x \leqslant \tan x$ and so

$$1 \geqslant \left(\frac{\sin x}{x}\right) \geqslant \cos x.$$

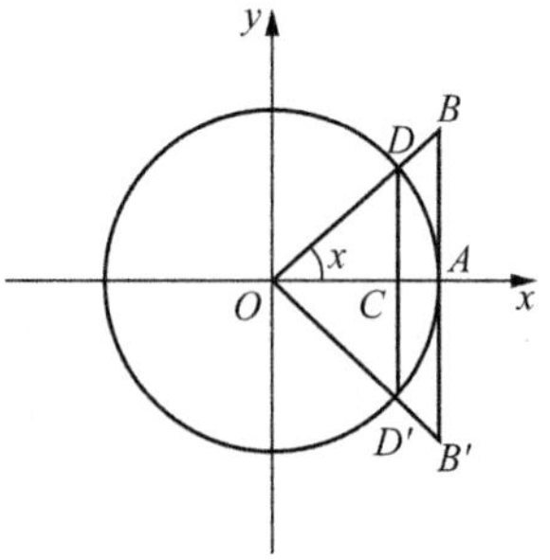

Figure 3 - 2

Because $\lim\limits_{x \to 0^+} \cos x = 1$, by squeezing law

$$\lim_{x \to 0^+} \frac{\sin x}{x} = 1.$$

It can be similarly shown that

$$\lim_{x \to 0^-} \frac{\sin x}{x} = 1. \square$$

Example Find the following limits:

① $\lim\limits_{x \to \pi} \dfrac{\sin x}{\pi - x}$; ② $\lim\limits_{x \to 0} \dfrac{1 + \cos x}{x^2}$; ③ $\lim\limits_{x \to 1} \dfrac{\sin^2 \dfrac{x-1}{2}}{x^2 - 2x + 1}$

Solution ① $\lim\limits_{x \to \pi} \dfrac{\sin x}{\pi - x} = \lim\limits_{x \to \pi} \dfrac{\sin(\pi - x)}{\pi - x} \xlongequal{(t = \pi - x)} \lim\limits_{t \to 0} \dfrac{\sin t}{t} =$ 1.

$$② \lim_{x \to 0} \frac{1 - \cos x}{x^2} = \lim_{x \to 0} \frac{2\sin^2 \dfrac{x}{2}}{x^2} = \lim_{x \to 0} \frac{1}{2} \left(\frac{\sin \dfrac{x}{2}}{\dfrac{x}{2}} \right)^2 = \frac{1}{2} \times 1^2 = \frac{1}{2}.$$

$$③ \lim_{x \to 1} \frac{\sin^2 \dfrac{x-1}{2}}{x^2 - 2x + 1} = \lim_{x \to 1} \frac{\sin^2 \dfrac{x-1}{2}}{4\left(\dfrac{x-1}{2}\right)^2} = \frac{1}{4} \lim_{x \to 1} \left(\frac{\sin \dfrac{x-1}{2}}{\dfrac{x-1}{2}} \right)^2 = \frac{1}{4}.$$

Proposition The second important limit:

$$\lim_{x \to \infty} \left(1 + \frac{1}{x}\right)^x = \mathrm{e}.$$

Proof First we prove

$$\lim_{x \to +\infty} \left(1 + \frac{1}{x}\right)^x = \mathrm{e}.$$

Let $x > 0$ and let $[x] = n$. Then $n \leqslant x < n + 1$ and $1 + \dfrac{1}{n+1} < 1 + \dfrac{1}{x} < 1 + \dfrac{1}{n}$. Thus

$$\left(1 + \frac{1}{n+1}\right)^n < \left(1 + \frac{1}{x}\right)^x < \left(1 + \frac{1}{n}\right)^{n+1}.$$

Note (cf. §2.3)

$$\lim_{x \to +\infty} \left(1 + \frac{1}{n+1}\right)^n = \lim_{x \to +\infty} \left(1 + \frac{1}{n}\right)^{n+1} = \mathrm{e}$$

and $n \to +\infty$ if and only if $x \to +\infty$. Hence, by squeezing law

$$\lim_{x\to+\infty}\left(1+\frac{1}{x}\right)^x = \mathrm{e}.$$

In a similar manner, we can prove that

$$\lim_{x\to-\infty}\left(1+\frac{1}{x}\right)^x = \mathrm{e}. \ \square$$

Corollary $\lim_{x\to 0}(1+x)^{\frac{1}{x}} = \mathrm{e}$.

Proof Take a transformation $x = 1/y$ and notice that $x \to 0$ if and only if $y \to \infty$. Then the limit follows from the second important limit. $\square$

Example Find the following limits:

① $\lim_{x\to\infty}\left(\frac{x-1}{x+1}\right)^x$; ② $\lim_{x\to 0}\left(\frac{1+\tan x}{1+\sin x}\right)^{\frac{-3}{2\sin x}}$; ③ $\lim_{x\to 0}\frac{\ln(1+x)}{x}$;

④ $\lim_{x\to 0}\frac{a^x-1}{x}$ $(a>0)$; ⑤ $\lim_{x\to 0}\left(\frac{a^x+b^x}{2}\right)^{\frac{1}{x}}$ $(a, b>0)$.

Solution ① $$\lim_{x\to\infty}\left(\frac{x-1}{x+1}\right)^x = \lim_{x\to\infty}\frac{\left(1-\frac{1}{x}\right)^x}{\left(1+\frac{1}{x}\right)^x}$$

$$= \frac{\lim_{x\to\infty}\left[\left(1+\frac{1}{-x}\right)^{-x}\right]^{-1}}{\lim_{x\to\infty}\left(1+\frac{1}{x}\right)^x} = \frac{\mathrm{e}^{-1}}{\mathrm{e}}$$

$$= \mathrm{e}^{-2};$$

or $$\lim_{x\to\infty}\left(\frac{x-1}{x+1}\right)^x = \lim_{x\to\infty}\left[\frac{(x+1)-2}{x+1}\right]^x$$

$$= \lim_{x\to\infty}\left[\left(1+\frac{-2}{x+1}\right)^{\frac{x+1}{-2}}\right]^{\frac{-2}{x+1}x} = \mathrm{e}^{-2}.$$

② $$\lim_{x\to 0}\left(\frac{1+\tan x}{1+\sin x}\right)^{\frac{-3}{2\sin x}} = \frac{\lim_{x\to 0}\left[(1+\tan x)^{\frac{1}{\tan x}}\right]^{\tan x\frac{-3}{2\sin x}}}{\lim_{x\to 0}\left[(1+\sin x)^{\frac{1}{\sin x}}\right]^{\sin x\frac{-3}{2\sin x}}} = \frac{\mathrm{e}^{-\frac{3}{2}}}{\mathrm{e}^{-\frac{3}{2}}} = 1.$$

③ $$\lim_{x\to 0}\frac{\ln(1+x)}{x} = \lim_{x\to 0}\ln[(1+x)^{1/x}]$$

$$= \ln[\lim_{x\to 0}(1+x)^{1/x}] = \ln e = 1.$$

④ If $a = 1$, $\lim_{x\to 0}\frac{a^x - 1}{x} = \lim_{x\to 0}\frac{0}{x} = 0 = \ln a$.

Now Assume $a \neq 1$. Let $u = a^x - 1$. Then $x \ln a = \ln(1+u)$ and so $x = \ln(1+u)/\ln a$. Note $x \to 0$ if and only if $u \to 0$. Thus

$$\lim_{x\to 0}\frac{a^x - 1}{x} = \lim_{u\to 0}\frac{u}{\frac{\ln(1+u)}{\ln a}} = \ln a \lim_{u\to 0}\frac{u}{\ln(1+u)}$$

$$= \ln a \cdot 1 = \ln a.$$

Thus, in any case $\lim_{x\to 0}\frac{a^x - 1}{x} = \ln a \quad (a > 0)$.

⑤ $$\lim_{x\to 0}\left(\frac{a^x + b^x}{2}\right)^{\frac{1}{x}} = \lim_{x\to 0}\left[\left(1+\frac{a^x + b^x - 2}{2}\right)^{\frac{2}{a^x+b^x-2}}\right]^{\frac{a^x+b^x-2}{2}\cdot\frac{1}{x}}$$

$$= e^{\lim_{x\to 0}\left(\frac{a^x-1}{2x}+\frac{b^x-1}{2x}\right)} = e^{\frac{1}{2}\ln a + \frac{1}{2}\ln b} = \sqrt{ab}.$$

§ 3.4 Infinitesimal and infinity

Functions or sequences which possess 0 as a limit may be treated as "ideal" numbers which are smaller in absolute value than any ordinary positive number and possessed of all the usual properties of arithmetic. Hence they are of particular significance in the development of calculus, and we will discuss them in this section.

§ 3.4.1 Infinitesimal

Definition If a function $f(x)$ or a sequence $\{x_n\}$ has 0 as a limit, then f or $\{x_n\}$ is called an *infinitesimal*.

Some simple facts, which can be proved just by definition, are listed below as a proposition.

Proposition (1) Suppose f is an infinitesimal. Then f is bounded.

(2) Suppose f_1, f_2, …, f_n are all infinitesimals. Then $f_1 + f_2 + \dots + f_n$ and $f_1 f_2 \dots f_n$ are both infinitesimals.

(3) Suppose f is an infinitesimal and $g(x)$ is bounded. Then fg is an infinitesimal.

(4) The following two statements are equivalent:

ⓘ Number A is the limit of $f(x)$ as x approaches x_0;

ⓘⓘ $f(x) - A$ is an infinitesimal as x approaches x_0.

Definition Suppose $f(x)$ and $g(x)$ are both infinitesimals.

(1) if

$$\lim_{x \to x_0} \frac{f(x)}{g(x)} = 0,$$

then $f(x)$ is called a *higher order infinitesimal* over $g(x)$, or equivalently, $g(x)$ is called a *lower order infinitesimal* of $f(x)$, and it is denoted by

$$f(x) = o(g(x)) \quad (x \to x_0).$$

(2) if there exist a, $b > 0$ such that $a \leqslant | f(x)/g(x) | \leqslant b$ on some free center neighborhood $U^o(x_0)$, then $f(x)$ and $g(x)$ are said to be a pair of *same order infinitesimals* (or one is a *same order infinitesimal* of the other). In particular, if

$$\lim_{x \to x_0} \frac{f(x)}{g(x)} = c \neq 0,$$

then $f(x)$ and $g(x)$ are a pair of same order infinitesimals.

If two infinitesimals $f(x)$ and $g(x)$ satisfy $| f(x)/g(x) | \leqslant b$ ($b > 0$), then they are denoted by

$$f(x) = O(g(x)).$$

By this notation, if $f(x)$ and $g(x)$ are a pair of same order infinitesimals, then $f(x) = O(g(x))$; if $f(x) = o(g(x))(x \to x_0)$, we also have $f(x) = O(g(x))(x \to x_0)$. Furthermore, we may use $f(x) = O(g(x))$ to indicate $f(x)/g(x)$ is bounded. In particular $f(x) = O(1)$ indicates that $f(x)$ is bounded while $f(x) = o(1)$ indicates that $f(x)$ is an infinitesimal.

(3) if $\lim\limits_{x\to x_0} \dfrac{f(x)}{g(x)} = 1$,

then $f(x)$ and $g(x)$ are called a pair of *equivalent infinitesimals* (or one is an *equivalent infinitesimal* of the other), and denoted by

$$f(x) \sim g(x)(x \to x_0).$$

In particular,

$$\text{if } \lim_{x\to x_0} \frac{f(x)}{g(x)} = a \neq 0, \text{ then } f(x) \sim ag(x)(x \to x_0).$$

(4) $\qquad$ if $f(x) = O(g^k(x)) \quad (k > 0)$,

then $f(x)$ is called a *k-order infinitesimal* of $g(x)$. In particular,

if $\dfrac{f(x)}{g^k(x)} \to a \neq 0$, then $f(x)$ is a k-order infinitesimal of $g(x)$.

If $f(x) = O(x^k)(k > 0)(x \to 0)$ then $f(x)$ is called a *k-order infinitesimal*.

In above definitions, $x \to x_0$ can also be replaced by $x \to x_0^{(\pm)}$ or $x \to (\pm)\infty$.

Examples (1) Since

$$\lim_{x\to 0} \frac{1-\cos x}{x} = 0; \qquad \lim_{x\to 0} \frac{1-\cos x}{x^2} = \frac{1}{2}; \qquad \lim_{x\to 0} \frac{1-\cos x}{\frac{1}{2}x^2} = 1,$$

so as $x \to 0$,

$$1-\cos x = o(x); \qquad 1-\cos x = O(x^2); \qquad 1-\cos x \sim \frac{1}{2}x^2.$$

(2) As $x \to 0$, compare the infinitesimal $f(x) = \sqrt{x} - 3x^3 + x^5$ with x.

Solution Since $f(x) = x^{1/2}(1 - 3x^{5/2} + x^{9/2})$, it is easy to see that

$$\lim_{x \to 0} \frac{f(x)}{x^{1/2}} = 1.$$

So, $f(x) \sim x^{1/2}$ as $x \to 0$, i. e. $f(x)$ is a $(1/2)$- order infinitesimal.

Similarly, we have

$2x^{1/3} + x^{3/10} + x^2 = O(x^{3/10})(x \to 0)$;

$x^{1/2} + 5x^{1/4} + x = O(x^{1/4})(x \to 0)$;

or more generally, let a_i, $b_i > 0$ $(i = 1, 2, \dots, n)$, then

$$b_1 x^{a_1} + b_2 x^{a_2} + \dots + b_n x^{a_n} = O(x^{\min\{a_1, a_2 \dots, a_n\}})(x \to 0).$$

(3) Suppose $x \to 0$.

① compare the infinitesimal $f(x) = \sqrt{1+x} - \sqrt{1-x}$ with x;

② compare the infinitesimal $f(x) = \tan x - \sin x$ with x.

Solution ① Note $f(x) = 2x/(\sqrt{1+x} + \sqrt{1-x})$. So, $\lim\limits_{x \to 0} f(x)/x = 1$, i. e. $f(x) \sim x$.

② Note $f(x) = \tan x(1 - \cos x)$. Since $\tan x \sim x$ and $1 - \cos x \sim (1/2)x^2$, $f(x) \sim (1/2)x^3$.

(4) As $x \to \infty$, compare the infinitesimal $f(x) = (x+1)/(x_2+1)$ with $1/x$.

Solution Since

$$\lim_{x \to \infty} \frac{f(x)}{1/x} = \lim_{x \to \infty} \frac{x+1}{x+1/x} = 1,$$
$$f(x) \sim 1/x(x \to \infty).$$

Remark (1) Usually, $x(x \to 0)$ or $1/x(x \to \infty)$ is called a *basic infinitesimal*.

(2) Some typical equivalent infinitesimals as $x \to 0$ are:

ⓘ $x \sim \sin x \sim \tan x \sim \arcsin x \sim \arctan x \sim \ln(1 + x) \sim$

$(e^x - 1)$;

ⅱ $[(1+x)^a - 1] \sim ax$ $(a \neq 0)$;

ⅲ $(1-\cos x) \sim \frac{1}{2}x^2$;

ⅳ $(a^x - 1) \sim x \ln a (a > 0)$;

ⅴ $(\sqrt[n]{1+x} - 1) \sim \frac{x}{n}$.

(3) It is easy to check that if $f(x) \sim g(x)$, then $f(x) - g(x) = o(f(x))$ and $f(x) - g(x) = o(g(x))$.

(4) In above definitions regarding comparison of infinitesimals, $x \to x_0$ can also be replaced by $x \to x_0^{(\pm)}$ or $x \to (\pm)\infty$.

Theorem 3.4.1 Suppose $f(x) \sim g(x) (x \to x_0)$.

(1) If $\lim\limits_{x \to x_0} f(x)h(x) = A$, $\lim\limits_{x \to x_0} g(x)h(x) = A$;

(2) If $\lim\limits_{x \to x_0} \frac{f(x)}{h(x)} = A$, $\lim\limits_{x \to x_0} \frac{g(x)}{h(x)} = A$.

Proof (1) $\lim\limits_{x \to x_0} g(x)h(x) = \lim\limits_{x \to x_0} \left(\frac{g(x)}{f(x)} f(x)h(x)\right) = \lim\limits_{x \to x_0} \frac{g(x)}{f(x)} \cdot \lim\limits_{x \to x_0} f(x)h(x) = 1 \cdot A = A$;

(2) $\lim\limits_{x \to x_0} \frac{g(x)}{h(x)} = \lim\limits_{x \to x_0} \left(\frac{g(x)}{f(x)} \frac{f(x)}{h(x)}\right) = \lim\limits_{x \to x_0} \frac{g(x)}{f(x)} \cdot \lim\limits_{x \to x_0} \frac{f(x)}{h(x)} = 1 \cdot A = A$. □

In the light of this theorem, the processing of finding a limit can be facilitated by substitution of equivalent infinitesimals in a product or a quotient.

Examples (1) $\lim\limits_{x \to 0} \frac{x^4 + x^3}{\left(\sin \frac{1}{2}x\right)^3} = \lim\limits_{x \to 0} \frac{x^4 + x^3}{\left(\frac{1}{2}x\right)^3} = 8$,

since $\sin \frac{1}{2}x \sim \frac{1}{2}x$.

(2) Find

$$\lim_{x\to+\infty} \ln(1+2^x)\ln\left(1+\frac{3}{x}\right).$$

Solution Note $\ln[1+(3/x)] \sim 3/x$ and $\ln(1+2^{-x}) \sim 2^{-x}$ as $x\to+\infty$. Then $\lim\limits_{x\to+\infty} \ln(1+2^x)\ln\left(1+\frac{3}{x}\right)$

$$= \lim_{x\to+\infty}\left[\ln(1+2^x)\cdot\frac{3}{x}\right] = \lim_{x\to+\infty}\frac{3\ln\left[2^x\left(1+\frac{1}{2^x}\right)\right]}{x}$$

$$= \lim_{x\to+\infty}\frac{3[\ln 2^x + \ln(1+2^{-x})]}{x}$$

$$= \lim_{x\to+\infty} 3\left[\ln 2 + \frac{\ln(1+2^{-x})}{x}\right]$$

$$= 3\ln 2 + 3\lim_{x\to+\infty}\frac{2^{-x}}{x} = 3\ln 2.$$

(3) $\lim\limits_{x\to 0}\dfrac{\tan x - \sin x}{x^3} = \lim\limits_{x\to 0}\dfrac{\sin x(1-\cos x)}{x^3\cos x} = \lim\limits_{x\to 0}\dfrac{x\cdot\frac{1}{2}x^2}{x^3\cos x} = \dfrac{1}{2}.$

(4) Find

$$\lim_{x\to 0}\frac{e^{\alpha x} - e^{\beta x}}{\sin\alpha x - \sin\beta x} \quad (\alpha\neq\beta,\ \alpha\neq 0,\ \beta\neq 0).$$

Solution $\lim\limits_{x\to 0}\dfrac{e^{\alpha x} - e^{\beta x}}{\sin\alpha x - \sin\beta x} = \lim\limits_{x\to 0}\dfrac{(e^{\alpha x}-1)-(e^{\beta x}-1)}{2\cos\left(\frac{\alpha+\beta}{2}x\right)\sin\left(\frac{\alpha-\beta}{2}x\right)}$

$$= \lim_{x\to 0}\frac{(e^{\alpha x}-1)-(e^{\beta x}-1)}{2\sin\left(\frac{\alpha-\beta}{2}x\right)}$$

$$= \lim_{x\to 0}\frac{(e^{\alpha x}-1)-(e^{\beta x}-1)}{2\cdot\frac{\alpha-\beta}{2}x}$$

$$= \lim_{x\to 0}\frac{e^{\alpha x}-1}{(\alpha-\beta)x} - \lim_{x\to 0}\frac{e^{\beta x}-1}{(\alpha-\beta)x}$$

$$= \lim_{x\to 0}\frac{\alpha x}{(\alpha-\beta)x} - \lim_{x\to 0}\frac{\beta x}{(\alpha-\beta)x}$$

$$= \frac{\alpha}{\alpha - \beta} - \frac{\beta}{\alpha - \beta} = 1.$$

(5) Find

$$\lim_{x \to 0} \left(\frac{1}{\sin^2 x} - \frac{1}{x^2} \right).$$

Solution $\lim_{x \to 0} \left(\frac{1}{\sin^2 x} - \frac{1}{x^2} \right) = \lim_{x \to 0} \frac{x^2 - \sin^2 x}{\sin^2 x \cdot x^2} = \lim_{x \to 0} \frac{x^2 - \sin^2 x}{x^2 x^2}$

$$= \lim_{x \to 0} \left(\frac{x - \sin x}{x^3} \cdot \frac{x + \sin x}{x} \right)$$

$$= \lim_{x \to 0} \left(\frac{x - \sin x}{x^3} \cdot \lim_{x \to 0} \frac{x + \sin x}{x} \right)$$

$$= 2 \lim_{x \to 0} \frac{x - \sin x}{x^3}.$$

Let

$$A = \lim_{x \to 0} \frac{x - \sin x}{x^3},$$

then

$$A = \lim_{x \to 0} \frac{x\left(1 - \cos \frac{x}{2}\right) - 2\sin \frac{x}{2} \cos \frac{x}{2} + x\cos \frac{x}{2}}{x^3}$$

$$= \lim_{x \to 0} \left[\frac{1 - \cos \frac{x}{2}}{x^2} + \frac{2\cos \frac{x}{2}\left(\frac{x}{2} - \sin \frac{x}{2}\right)}{x^3} \right]$$

$$= \lim_{x \to 0} \frac{2\sin^2 \frac{x}{4}}{x^2} + 2 \lim_{x \to 0} \frac{\frac{x}{2} - \sin \frac{x}{2}}{8\left(\frac{x}{2}\right)^3}$$

$$\overset{(t = x/2)}{=\!=\!=\!=} \frac{1}{8} + \frac{1}{4} \lim_{t \to 0} \frac{t - \sin t}{t^3} = \frac{1}{8} + \frac{1}{4} A,$$

which implies $A = 1/6$ and so

$$\lim_{x \to 0} \left(\frac{1}{\sin^2 x} - \frac{1}{x^2} \right) = \frac{1}{3}.$$

Note However, in the case which is not a product or a quotient, the substitution of equivalent infinitesimals will be invalid usually.

Example In spite of $1/x \sim 1/(x+1)(x \to \infty)$,

$$\lim_{x\to\infty}\frac{\dfrac{1}{x}-\dfrac{1}{x+1}}{\dfrac{1}{x^2}}=1,$$

while

$$\lim_{x\to\infty}\frac{\dfrac{1}{x}-\dfrac{1}{x}}{\dfrac{1}{x^2}}=0.$$

§ 3.4.2 Infinity

Definition (1) Suppose a function $f(x)$ is defined on a free center neighborhood $U^o(x_0)$. If for any number $M>0$, there is $\delta>0$ such that $f(x)>M$ (resp. $f(x)<-M$, $|f(x)|>M$) for any $x\in U^o(x_0,\delta)$, then f is said to have the *improper limit* $+\infty$ (resp. $-\infty$, ∞) as x approaches x_0, which is denoted by

$\lim\limits_{x\to x_0} f(x)=+\infty$ (resp. $\lim\limits_{x\to x_0} f(x)=-\infty$, $\lim\limits_{x\to x_0} f(x)=\infty$)

Similarly, we may define one-sided limits $\lim\limits_{x\to x_0^+} f(x)=(\pm)\infty$ and $\lim\limits_{x\to x_0^-} f(x)=(\pm)\infty$.

(2) Suppose a function $f(x)$ is defined on (a,∞). If for any number $M>0$, there is $N>0$ such that $f(x)>M$ (resp. $f(x)<-M$, $|f(x)|>M$) for any $x>N$, then f is said to have the *improper limit* $+\infty$ (resp. $-\infty$, ∞) as x approaches $+\infty$, which is denoted by

$\lim\limits_{x\to+\infty} f(x)=+\infty$(resp. $\lim\limits_{x\to+\infty} f(x)=-\infty$, $\lim\limits_{x\to+\infty} f(x)=\infty$)

Similarly, we may define $\lim\limits_{x\to-\infty} f(x)=(\pm)\infty$ and $\lim\limits_{x\to\infty} f(x)=$

$(\pm)\infty$.

(3) A function whose limit is $(\pm)\infty$ is called an *infinity*.

(4) Suppose $f(x)$ and $g(x)$ are both infinities as $x \to s$ where s may be $x_0^{(\pm)}$ or $(\pm)\infty$. If there is a real number $A \neq 0$ such that

$$\lim_{x \to s} \frac{f(x)}{g(x)} = A,$$

then f and g are called a pair of *same-order-infinities* as $x \to s$.

The relationship between infinitesimal and infinity is stated as follows, the proof of which is just by their definitions and so omitted.

Theorem 3.4.2 (1) If $f(x)$ is an infinitesimal as $x \to x_0$ and $f(x) \neq 0$ on some $U^o(x_0)$, then $1/f(x)$ is an infinity as $x \to x_0$;

(2) If $f(x)$ is an infinity as $x \to x_0$, then $1/f(x)$ is an infinitesimal as $x \to x_0$.

By the above theorem, the study of infinity can be converted to the study of infinitesimal.

§ 3.5 Concept of continuity

In ordinary language, to say that a certain process is "unbroken" is to say that it goes on without interruption and without abrupt change. We perceive the path of a moving object as an unbroken curve, without gaps, breaks, or holes. This is an intuitive understanding of what is meant by a continuous curve, and a verbal description would probably use such words as "connected," "without breaks", etc. The purpose of this section is to clarify the intuitive notion of continuity. We will present the definitions of continuity, one-sided continuity and discontinuity in a precise mathematical formulation.

§ 3.5.1 Continuity of a function at a point

A precise analytical definition of a continuous function will be given now in terms of limits. The motivation for such a mathematical statement is drawn from nature, where very slight changes usually produce correspondingly small effects. By analogy, a continuous function $f(x)$ must be defined as one that experiences a small variation for a slight change in the independent variable x. The continuity of $f(x)$ at $x = x_0$ therefore depends on its local behavior in some small interval about this point. Specifically, if x is any point near x_0 with $|x - x_0|$ being very small, then the corresponding values of the function, $f(x_0)$ and $f(x)$, should be almost equal. In other words, continuity requires the difference $|f(x) - f(x_0)|$ to be small when $|x - x_0|$ is also small, which in effect results in the following definition:

Definition A function $f(x)$ is *continuous* at x_0 if

$$\lim_{x \to x_0} f(x) = f(x_0).$$

In this case, x_0 is called a *continuity* or a *continuous point* of $f(x)$.

If $f(x)$ is not *continuous* at x_0, we also say that $f(x)$ is *discontinuous* at x_0, and x_0 is called a *discontinuity* or a *discontinuous point* of $f(x)$.

Remark By the definition, a function $f(x)$ is continuous at a point x_0 if and only if all the following three conditions are satisfied:

(1) $f(x_0)$ exists, i. e. the function $f(x)$ is defined at x_0;

(2) The limit $\lim\limits_{x \to x_0} f(x)$ exists;

(3) The above two are equal to each other.

Examples (1) The function

$$f(x) = \begin{cases} x \sin \dfrac{1}{x} & (\text{if } x \neq 0), \\ 0 & (\text{if } x = 0) \end{cases}$$

is continuous at $x = 0$, since

$$\lim_{x \to 0} x \sin \frac{1}{x} = 0 = f(0).$$

(2) The function

$$f(x) = \frac{1}{x}(x \neq 0)$$

is continuous at any point x_0 except 0, since for any $x_0 \neq 0$ we have

$$\lim_{x \to x_0} f(x) = \lim_{x \to x_0} \frac{1}{x} = \frac{1}{x_0} = f(x_0).$$

But for $x = 0$, the function has no definition, and so of course $f(x)$ is discontinuous at 0.

The essential feature of $f(x)$ being continuous at x_0 is that the assigned value of $f(x_0)$ must be consistent with the value obtained from the limit, and this will assure a smooth variation. The equivalent statement of continuity or discontinuity of $f(x)$ at a point x_0 in the $\varepsilon - \delta$ terminology is as follows:

$f(x)$ is continuous at x_0 :$\Leftrightarrow$for any $\varepsilon > 0$ there is $\delta > 0$ such that $|f(x) - f(x_0)| < \varepsilon$ whenever $|x - x_0| < \delta$.

$f(x)$ is discontinuous at $x_0 \Leftrightarrow$there is an $\varepsilon > 0$ such that for any $\delta > 0$

$|f(x) - f(x_0)| \geqslant \varepsilon$ for some x with $|x - x_0| < \delta$.

The general procedure to prove that a function is continuous at a point is the same as that used to show the existence of a limit. Start with $|f(x) - f(x_0)|$ and manipulate this quantity to determine the magnitude of δ which guarantees that the functional difference is less than the preassigned, arbitrarily small ε. Most of the examples of limits in this section also demonstrate that the functions involved are continuous.

Examples (1) Prove that $f(x) = x^4 - 5x^3 + 6$ is continuous at $x = 3$ by definition.

Analysis: Note $f(3) = -48$ and $|f(x) - f(3)| = |(x^4 - 5x^3 + 6) + 48| = |(x - 3)(x^3 - 2x^2 - 6x - 18)|$. Assuming we take

$\delta_1 = 1$, then if $|x-3| < \delta_1$, $2 < x < 4$ and furthermore

$|x^3 - 2x^2 - 6x - 18| \leqslant |x^3| + 2|x^2| + 6|x| + 18 < 4^3 + 2 \times 4^2 + 6 \times 4 + 18 = 138.$

Thus $|f(x) - f(3)| = |(x-3)||(x^3 - 2x^2 - 6x - 18)| \leqslant 138|x-3| < \varepsilon$, which implies $|x-3| < \varepsilon/138$.

Proof For any $\varepsilon > 0$, take $\delta = \min\{1, \varepsilon/138\}$, then

$$|f(x) - f(3)| < \varepsilon \text{ whenever } |x-3| < \delta.$$

So, $f(x)$ is continuous at $x = 3$. □

(2) Show that $H(x) = \begin{cases} 1 & (x \geqslant 0), \\ 0 & (x < 0) \end{cases}$ is discontinuous at $x = 0$ but that $xH(x)$ is continuous at $x = 0$.

Proof The piecewise function does not have a unique limit at $x = 0$, and this proves the first part of the statement. However,

$$\lim_{x \to 0} xH(x) = 0$$

and zero is the value of $xH(x)$ no matter what finite value is assigned arbitrarily to $H(0)$. Since the limit agrees with the value at $x = 0$, the product function $xH(x)$ is continuous. □

In order to introduce another form of the definition of continuity, we denote $\Delta x = x - x_0$, called an *increment* of the independent variable x at x_0, and $\Delta y = f(x) - f(x_0)$, called an *increment* of the function $y = f(x)$ at x_0. Then the definition of continuity can be presented in another form as:

Definition A function $y = f(x)$ is *continuous* at x_0 if

$$\lim_{\Delta x \to 0} \Delta y = 0.$$

Example Prove that $y = \sin x$ and $y = \cos x$ are both continuous at any $x \in (-\infty, +\infty)$.

Proof We only prove $\sin x$ is continuous at any point x, and for $\cos x$ it can be similarly proved. Let $x \in (-\infty, +\infty)$. Notice

$$\begin{aligned}|\Delta y| &= |\sin(x+\Delta x)-\sin x| \\ &= 2\left|\sin\frac{\Delta x}{2}\cos\left(x+\frac{\Delta x}{2}\right)\right| \leqslant 2\left|\sin\frac{\Delta x}{2}\right| < |\Delta x|.\end{aligned}$$

So, for any $\varepsilon > 0$, take $\delta = \varepsilon$. Then

$$|\Delta y| < \varepsilon \text{ whenever } |\Delta x| < \delta,$$

i. e. $\lim\limits_{\Delta x \to 0} \Delta y = 0$. □

Remark The following facts can be easily checked:

(1) $f(x)$ is continuous at $x_0 \Rightarrow |f(x)|$ or $f^2(x)$ is continuous at x_0; but usually the converse statement is not true.

(2) Let $f^{-1}(x)$ be an inverse function of $f(x)$. Then $f(x)$ is continuous at $x_0 \Leftrightarrow f^{-1}(x)$ is continuous at $f(x_0)$.

Just like one-sided limit, we can also define one-sided continuity of a function.

Definition (1) Suppose $f(x)$ is defined on a right-hand neighborhood of a point x_0. If $\lim\limits_{x \to x_0^+} f(x) = f(x_0)$, then f is said to be *continuous at x_0 from the right*.

(2) Suppose $f(x)$ is defined on a left-hand neighborhood of a point x_0. If $\lim\limits_{x \to x_0^-} f(x) = f(x_0)$, then f is said to be *continuous at x_0 from the left*.

The continuity of f at x_0 from the right or from the left is also called *one-sided continuity* of f at x_0. By the above definition we have immediately the following

Theorem 3.5.1 A function $f(x)$ is continuous at x_0 if and only if $f(x)$ is continuous at x_0 from both the right and the left.

Examples (1) Discuss the continuity of the function $f(x) = |x|$ at $x = 0$.

Solution Since

$$\lim_{x \to 0^+} f(x) = \lim_{x \to 0^+} x = 0,$$

$$\lim_{x \to 0^-} f(x) = \lim_{x \to 0^-} (-x) = 0,$$

$f(x)$ is continuous at $x = 0$ from both the right and the left, and so the function is continuous at $x = 0$.

(2) Let

$$f(x) = \begin{cases} \frac{1}{\pi} \arctan \frac{1}{x} + \frac{a + b\mathrm{e}^{1/x}}{1 + \mathrm{e}^{1/x}} & (\text{if } x \neq 0), \\ 1 & (\text{if } x = 0). \end{cases}$$

Determine the constants a and b such that $f(x)$ is continuous at $x = 0$.

Solution $\lim_{x \to 0^-} f(x) = \frac{1}{\pi}\left(-\frac{\pi}{2}\right) + \frac{a+0}{1+0} = 1,$

$$\lim_{x \to 0^+} f(x) = \frac{1}{\pi}\left(\frac{\pi}{2}\right) + \frac{0+b}{0+1} = 1.$$

Then we have $a = 3/2$ and $b = 1/2$.

§ 3. 5. 2 Discontinuities and their classification

The term discontinuity is used in two ways. The first refers to a point at which the function is not defined. In its second usage, the term is applied to a point where the function is defined but not continuous. More specifically, we have the following

Remark A function $f(x)$ is discontinuous at a point x_0, if and only if one of the following conditions is satisfied:

(1) $f(x)$ is not defined at x_0;

(2) $f(x)$ is defined at x_0 but $\lim_{x \to x_0} f(x)$ does not exist;

(3) $f(x)$ is defined at x_0 and $\lim_{x \to x_0} f(x)$ exists but $\lim_{x \to x_0} f(x) \neq f(x_0)$.

Discontinuities can be further classified as two classes according to their characteristics

Definition Let x_0 be a discontinuity of a function $f(x)$.

(1) If $\lim\limits_{x \to x_0} f(x_0)$ exists, the point x is called a *removable discontinuity* of $f(x)$.

That is, a discontinuity x is removable if and only if $\lim\limits_{x \to x_0} f(x_0)$ exists but $f(x_0)$ is not defined or $\lim\limits_{x \to x_0} f(x_0)$ exists but $\lim\limits_{x \to x_0} f(x_0) \neq f(x_0)$. If x_0 is a removable discontinuity of $f(x)$, then $f(x)$ can be modified as being continuous at x_0 if only the value of f at x_0 is redefined as the limit of f at x_0, i. e. define $f(x_0) = \lim\limits_{x \to x_0} f(x)$.

(2) If both $\lim\limits_{x \to x_0^+} f(x)$ and $\lim\limits_{x \to x_0^-} f(x)$ exist, but $\lim\limits_{x \to x_0^+} f(x) \neq \lim\limits_{x \to x_0^-} f(x)$, then x_0 is called a *jump discontinuity* of f. In this case, the number $\left| \lim\limits_{x \to x_0^+} f(x) - \lim\limits_{x \to x_0^-} f(x) \right|$ is called the *jump* of f at x_0.

(3) The removable discontinuity or the jump discontinuity is called a *discontinuity of the first class*. Clearly, a discontinuity x_0 is of the first class if and only if both of the right-hand and left-hand limits are existent.

(4) A continuity x_0 of f which does not belong to the first class is called a *discontinuity of the second class*.

Thus, a discontinuity x_0 of f is of the second class if and only if at least one of the right-hand and left-hand limits of f at x_0 is non-existent.

Examples (1) Let

$$f(x) = \begin{cases} \dfrac{\sin x}{x} & (\text{if } x \neq 0), \\ 0 & (\text{if } x = 0). \end{cases}$$

Then $f(x)$ has a removable discontinuity $x = 0$, since

$$\lim_{x \to 0} f(x) = 1 \neq f(0).$$

(2) Let

$$f(x)=\begin{cases}\arctan\dfrac{1}{x} & (\text{if } x\neq 0),\\ 0 & (\text{if } x=0).\end{cases}$$

Then, since

$$\lim_{x\to 0^+}f(x)=\frac{\pi}{2} \text{ and } \lim_{x\to 0^-}f(x)=-\frac{\pi}{2},$$

$x=0$ is a jump discontinuity of $f(x)$ with a jump π.

(3) The function $f(x)=1/x^2$ has a discontinuity $x=0$ which is of the second class (no matter whether or not $f(0)$ is defined), since

$$\lim_{x\to 0}f(x)=+\infty.$$

Such a discontinuity is also called an *infinite discontinuity*.

(4) Find the discontinuities of the following functions and determine their classes:

① $f(x)=\dfrac{\cos\frac{\pi}{2}x}{x^2(x-1)}$;

② $f(x)=\dfrac{1}{1-e^{x/(1-x)}}$;

③ $f(x)=\dfrac{x}{\tan x}$;

④ $f(x)=x[x]$.

Solution ① Discontinuities are $x=0$ and $x=1$. Note $\lim\limits_{x\to 0}\dfrac{\cos\frac{\pi}{2}x}{x^2(x-1)}=-\infty$. So, $x=0$ is a discontinuitiy of the second class, or exactly an infinite discontinuity. Note

$$\lim_{x\to 1}f(x)=\lim_{x\to 1}\frac{\sin\left[\frac{\pi}{2}(1-x)\right]}{-x^2(1-x)}=-\lim_{x\to 1}\frac{\frac{\pi}{2}(1-x)}{(1-x)}=-\frac{\pi}{2}.$$

So, $x = 1$ is a discontinuity of the first class, or exactly a removable discontinuity.

② Clearly, discontinuities are $x = 0$ and $x = 1$. Note $\lim_{x \to 0}(1 - e^{x/(1-x)}) = 0$ and then $\lim_{x \to 0} f(x) = \infty$. So, $x = 0$ is a discontinuity of the second class, or exactly an infinite discontinuity. Note $\lim_{x \to 1^+}(1 - e^{x/(1-x)}) = 1$ and then $\lim_{x \to 1^+} f(x) = 1$; $\lim_{x \to 1^-}(1 - e^{x/(1-x)}) = -\infty$ and then $\lim_{x \to 1^-} f(x) = 0$. So, $x = 1$ is a discontinuity of the first class, or exactly a jump discontinuity with a jump 1.

③ Note that the discontinuities of the function $\tan x$ are $n\pi + (\pi/2)$ and the roots of $\tan x$ are $n\pi (n = 0, \pm 1, \pm 2, \ldots)$. Since

$$\lim_{x \to 0} \frac{x}{\tan x} = 1,$$

$$\lim_{x \to n\pi + \frac{\pi}{2}} \frac{x}{\tan x} = 0 (n = 0, \pm 1, \pm 2, \ldots) \text{ and}$$

$$\lim_{x \to n\pi} \frac{x}{\tan x} = \infty (n = \pm 1, \pm 2, \ldots),$$

$x = 0$, $x = n\pi + \frac{\pi}{2} (n = 0, \pm 1, \pm 2, \ldots)$ are removable discontinuities, and $x = n\pi (n = \pm 1, \pm 2, \ldots)$ are infinite discontinuities.

④ Notice that the discontinuities of $[x]$ are $x = 0, \pm 1, \pm 2, \cdots$. Since

$$\lim_{x \to 0} f(x) = \lim_{x \to 0} x[x] = 0 = f(0),$$

and for $n = \pm 1, \pm 2, \ldots$, $\lim_{x \to n^-} x[x] = n(n-1)$, $\lim_{x \to n^+} x[x] = n^2$,

the discontinuities of $f(x)$ are $x = \pm 1, \pm 2, \ldots$, they are all jump discontinuities, and therefore of the first class.

§ 3.5.3 Continuous functions on an interval

Definition Let $f(x)$ be defined on an interval $I(=(a, b)$, $(a, b]$, $[a, b)$, $[a, b]$ or $(-\infty, +\infty))$. If f is continuous at any point $x \in I$, then f is said to be *continuous on* I.

At the endpoint of the interval $I = [a, b]$ we understand continuity to mean continuity from one side, i. e. the continuity of f at a is understood as continuity from the right; the continuity of f at b is understood as continuity from the left.

Thus, f is continuous on $[a, b]$ if and only if f is continuous on (a, b) and, in addition, $\lim_{x \to a^+} f(x) = f(a)$ and $\lim_{x \to b^-} f(x) = f(b)$.

Examples (1) As we have proved the functions $\sin x$ and $\cos x$ are both continuous on $(-\infty, +\infty)$.

(2) The function $f(x) = x^n (n \in \mathbb{N})$ is continuous on $(-\infty, +\infty)$.

Proof Let $x_0 \in (-\infty, +\infty)$. We show that $f(x)$ is continuous at x_0. Note $|x^n - x_0^n| \leqslant |x - x_0| (|x|^{n-1} + |x_0| |x|^{n-2} + \cdots + |x_0|^{n-1})$. Suppose $x_0 \neq 0$. If $|x - x_0| < |x_0|$, $|x| < 2|x_0|$ and so

$$|x^n - x_0^n| \leqslant |x - x_0| |x_0|^{n-1} (2^{n-1} + 2^{n-2} + \ldots + 2)$$

$$\leqslant 2^n |x_0|^{n-1} |x - x_0|.$$

So, for any $\varepsilon > 0$, take $\delta = \min\{|x_0|, \varepsilon/(2^n |x_0|^{n-1})\}$, then

$$|f(x) - f(x_0)| < \varepsilon \text{ whenever } |x - x_0| < \delta,$$

which implies $f(x)$ is continuous at x_0. Suppose $x_0 = 0$. Since $|x^n - x_0^n| = |x - x_0|^n$, we may take $\delta = \varepsilon^{1/n}$, and clearly the result still holds.

In fact, generally the power function $f(x) = x^a (a \in \mathbb{R})$ is continuous on $(0, +\infty)$.

Definition Let $f(x)$ be defined on interval $[a, b]$. If f has only a finite number of discontinuities of the first class, then f is called a *piecewise continuous function* on $[a, b]$.

For example, for any $L > 0$, the functions $f(x) = [x]$ and $g(x) = x - [x]$ are piecewise continuous functions on $[-L, L]$.

Continuous functions will be the main objects we discuss later, and we will see that primary functions are continuous on their domain intervals. However, there are some functions that are discontinuous everywhere on their domain intervals, for which Dirichlet function is one of the examples.

The following example shows that the continuity of $f(x)$ at a point x_0 does not necessarily implies the continuity of $f(x)$ at a point x which is sufficiently close to x_0, which, though, seems to be counterintuitive. Or briefly,

$f(x)$ is continuous at a point $x_0 \not\Rightarrow$ there exists a neighborhood of x_0, $U(x_0)$, such that $f(x)$ is continuous on $U(x_0)$.

Example Let

$$f(x) = \begin{cases} x^2 & (\text{if } x \text{ is rational}), \\ 0 & (\text{if } x \text{ is irrational}). \end{cases}$$

Prove ① $f(x)$ is continuous at $x = 0$; ② for any $a > 0$, there is $x \in (-a, a)$ such that $f(x)$ is discontinuous at x.

Proof ① For any $\varepsilon > 0$, Take $\delta = \sqrt{\varepsilon}$. Then if $|x - 0| = |x| < \delta$,

$$|f(x) - f(0)| = |f(x)| \leqslant |x^2| < \varepsilon,$$

which implies $f(x)$ is continuous at $x = 0$;

② Now let $x_0 \in (-a, a)\setminus\{0\}$ be a rational number. We show that $f(x)$ is discontinuous at x_0. Let $\varepsilon = x_0^2/2 > 0$. For any $\delta > 0$, take an irrational number $x \in (x_0 - \delta, x_0 + \delta)$. Then we see that though $|x - x_0| < \delta$, $|f(x) - f(x_0)| = |0 - x_0^2| = |x_0^2| > \varepsilon (= x_0^2/2)$. So, $f(x)$ is discontinuous at x_0. □

Example Riemann function (cf. Chapter 1)

$$f(x) = \begin{cases} \dfrac{1}{q} & \left(\text{if } x = \dfrac{p}{q} \in [0, 1] \text{ where } p, q \in \mathbb{N} \text{ and } p/q \text{ is irreducible fraction}\right), \\ 0 & (\text{if } x \in [0, 1] \text{ is irrational, or } x = 0, 1) \end{cases}$$

is continuous at any irrational point in $(0, 1)$, and discontinuous at at any rational point in $(0, 1)$.

Proof Let $\xi \in (0, 1)$ be an irrational. Then $|f(x) - f(\xi)| = f(x)$. We only need to show: for any $\varepsilon > 0$, there is $\delta > 0$ such that $f(x) < \varepsilon$ whenever $|x - \xi| < \delta$. If x is an irrational, $f(x) = 0 < \varepsilon$; if x is a rational $\dfrac{p}{q}$, in order to have $f(x) = \dfrac{1}{q} < \varepsilon$, merely take $n > \dfrac{1}{\varepsilon}$. There are only a finite number of irreducible fractions in $[0, 1]$, of which the denominator are less than or equal to n. Thus, we may select one such fraction, say x', which is closest to ξ, and take $\delta = |\xi - x'|$. Hence, $\dfrac{1}{q} < \dfrac{1}{n} < \varepsilon$ if $\left|\dfrac{p}{q} - \xi\right| < \delta$. This proves that $f(x)$ is continuous at any irrational point in $(0, 1)$. □

Now suppose ξ is a rational. Then, $f(\xi) = f\left(\dfrac{p}{q}\right) = \dfrac{1}{q}$. By the density of the irrationals, there are irrationals in any neighborhood of ξ. Clearly, at each of those irrationals the values of $f(x)$ is 0, while the difference between 0 and $\dfrac{1}{q}$ can not be arbitrarily small. So, $f(x)$ is discontinuous at at any rational point in $(0, 1)$. □

§ 3.6 Properties of continuous functions

In this section, we will discuss the properties of continuous functions from two points of view, namely, that of the local behavior and the global behavior.

§ 3.6.1 Local properties of continuous functions

Theorem 3.6.1 (Local boundedness) If $f(x)$ is continuous at point x_0, then there is a neighborhood of x_0, $U(x_0)$, such that $f(x)$ is bounded on $U(x_0)$.

Theorem 3.6.2 (Locally sign-preserving) If $f(x)$ is continuous at point x_0 and $f(x_0) \neq 0$, then there is a neighborhood of x_0, $U(x_0)$, such that $f(x)$ has a same sign as $f(x_0)$ on $U(x_0)$ and $|f(x)| \geqslant r > 0$ for some $r > 0$ at any $x \in U(x_0)$.

These two theorems can be proved in a similar manner as in the proof of the corresponding theorems for limits.

Corollary Suppose $f(x)$ and $g(x)$ are both continuous at x_0. (1) if $f(x_0) > g(x_0)$, then there is a neighborhood of x_0, $U(x_0)$, such that $f(x) > g(x)$ for any $x \in U(x_0)$; (2) if $f(x_0) < g(x_0)$, then there is a neighborhood $U(x_0)$ such that $f(x) < g(x)$ for any $x \in U(x_0)$.

Proof Let $F(x) = f(x) - g(x)$. Then using the locally sign-preserving theorem, we can derive the result immediately. □

Theorem 3.6.3 (1) Suppose $f(x)$ and $g(x)$ are both defined on an interval I and both continuous at $x_0 \in I$. Then $f(x) \pm g(x)$, $f(x) \cdot g(x)$ and $f(x)/g(x)(g(x_0) \neq 0)$ are all continuous at x_0.

(2) The function $\sqrt[n]{x}$ is continuous on $[0, +\infty)$ if n is a positive even number, and continuous on $(-\infty, +\infty)$ if n is a positive odd number.

This theorem follows from the theorem of operations for function limits.

Corollary Let $P_n(x) = a_0x^n + a_1x^{n-1} + \ldots + a_n$ be a polynomial function and let $R(x) = P_n(x)/Q_m(x)$ be a rational function. Then $P_n(x)$ and $R(x)$ are continuous at any point in their domains.

Proof Clearly $y = c$ (constant function) and $y = x$ are continuous at any point. Then we obtain the result by using Theorem 3.6.3(1) repeatedly. □

Theorem 3.6.4 Suppose $f(x)$ is continuous at x_0 and $g(u)$ is continuous at $u_0 = f(x_0)$. Then the composite function $g(f(x))$ is continuous at x_0, i.e.

$$\lim_{x\to x_0} g(f(x)) = g(\lim_{x\to x_0} f(x)) = g(f(\lim_{x\to x_0} x)) = g(f(x_0)).$$

Proof By the condition, for any $\varepsilon > 0$, there is $\delta_1 > 0$ such that

$$|g(u) - g(u_0)| < \varepsilon \text{ whenever } |u - u_0| < \delta_1.$$

Also for the above $\delta_1 > 0$, there is $\delta > 0$ such that

$$|u - u_0| = |f(x) - f(x_0)| < \delta_1 \text{ whenever } |x - x_0| < \delta.$$

Hence, for any $\varepsilon > 0$, there is $\delta > 0$ such that

$$|g(f(x)) - g(f(x_0))| < \varepsilon \text{ whenever } |x - x_0| < \delta. \ \square$$

Remark (1) It can be similarly proved that the limit expression in Theorem 3.6.4 remains valid for $x \to x_0^{\pm}$ and $x \to (\pm)\infty$.

(2) The above theorem can be expressed verbally as *a composite function of two continuous functions is still a continuous function.*

(3) Note that $\sin x$ and $\cos x$ are both continuous functions, and

$$\tan x = \frac{\sin x}{\cos x}; \quad \cot x = \frac{\cos x}{\sin x}; \quad \sec x = \frac{1}{\cos x}; \quad \csc x = \frac{1}{\sin x}.$$

Then by the above theorem, these trigonometric functions are all continuous on their domains.

(4) If $\lim_{x \to x_0} f(x) = a \neq f(x_0)$ (i. e. $x = x_0$ is a removable discontinuity of f), and $g(u)$ is continuous at $u = a$, then the preceding theorem remains valid in the following sense:

$$\lim_{x \to x_0} g(f(x)) = g(\lim_{x \to x_0} f(x)) = g(a).$$

Examples (1) Find $\lim_{x \to 1} \cos(1 - x^3)$.

Solution Note $\cos(1 - x^3)$ can be regarded as a composite function of $y = \cos u$ and $u = 1 - x^3$. Since the function $1 - x^3$ is continuous at $x = 1$ with function value 0, and the function $\cos u$ is also continuous at $u = 0$, then

$$\lim_{x \to 1} \cos(1 - x^3) = \cos(\lim_{x \to 1}(1 - x^3)) = \cos 0 = 1.$$

(2) Find

$$\lim_{x \to 0} \sqrt{2 - \frac{\sin x}{x}} \text{ and } \lim_{x \to \infty} \sqrt{2 - \frac{\sin x}{x}}.$$

Solution Since $\lim_{x \to 0}(\sin x / x) = 1$, $\lim_{x \to \infty}(\sin x / x) = 0$ and $\sqrt{2 - u}$ is continuous at $u = 0$, 1,

$$\lim_{x \to 0} \sqrt{2 - \frac{\sin x}{x}} = \sqrt{2 - \lim_{x \to 0} \frac{\sin x}{x}} = \sqrt{2 - 1} = 1,$$

$$\lim_{x \to \infty} \sqrt{2 - \frac{\sin x}{x}} = \sqrt{2 - \lim_{x \to \infty} \frac{\sin x}{x}} = \sqrt{2 - 0} = \sqrt{2}.$$

§ 3.6.2 Properties of a continuous function on a closed interval

So far our results have followed because f is continuous at a particular point. In fact we get the best results from assuming rather more, i. e. f is continuous on a closed interval. The results in this subsection are precisely why we are interested in discussing continuity in the first place. Although some of the results are intuitive, they only follow from the continuity property, and indeed we present counterexamples whenever that fails.

Definition Let $f(x)$ be a function with domain D. If there is $x_0 \in D$ such that $f(x_0) \geqslant f(x)$ $(f(x_0) \leqslant f(x))$ for any $x \in D$, then $f(x_0)$ is called the *absolute maximum* (*absolute minimum*) of $f(x)$ on D. The absolute maximum or the absolute minimum of $f(x)$ on D is also called the *absolute extremum* of $f(x)$ on D. The absolute maximum and the absolute minimum of $f(x)$ on D are respectively denoted as $\max\limits_{x \in D} f(x)$ and $\min\limits_{x \in D} f(x)$.

Theorem 3.6.5 (Boundedness theorem) Suppose a function $f(x)$ is continuous on a closed interval $[a, b]$. Then $f(x)$ is bounded on $[a, b]$.

Proof We prove this theorem by contradiction. Suppose that $f(x)$ is unbounded on $[a, b]$. Then, for any $n \in \mathbb{N}$, there is $x_n \in [a, b]$ such that $f(x_n) > n(n = 1, 2, \dots)$. Thus, we have a sequence $\{x_n\} \subseteq [a, b]$, and so it has a convergent subsequence $\{x_{i_n}\}$ $(a \leqslant x_{i_n} \leqslant b)$ by Weierstrass theorem. Let $\lim\limits_{n \to +\infty} x_{i_n} = x_0$. By Heine theorem $\lim\limits_{n \to +\infty} f(x_{i_n}) = \lim\limits_{x \to x_0} f(x)$. Furthermore, since $x_0 \in [a, b]$, $f(x)$ is continuous at x_0, i. e. $\lim\limits_{x \to x_0} f(x) = f(x_0)$. Hence, $\lim\limits_{n \to +\infty} f(x_{i_n}) = f(x_0)$. However, since $f(x_{i_n}) > i_n$, $\lim\limits_{n \to +\infty} f(x_{i_n}) = +\infty$, which yields a contradiction. □

The following theorem gives a sufficiency for the existence of an absolute extremum value of a continuous function on a closed interval.

Theorem 3.6.6 (Absolute extremum value theorem) Suppose a function $f(x)$ is continuous on a closed interval $[a, b]$. Then $f(x)$ has both absolute maximum value and absolute minimum value on

$[a, b]$, i. e. there exists $x_0 \in [a, b]$ such that $f(x) \leqslant f(x_0)$ ($f(x) \geqslant f(x_0)$) for any $x \in [a, b]$.

Proof By Theorem 3. 6. 5, the number set $A = \{f(x) \mid x \in [a, b]\}$ is bounded, and so A has a supremum, denoted by $M = \sup A$. We assert that there exists $x_0 \in [a, b]$ such that $f(x_0) = M$. Because otherwise, let $g(x) = \frac{1}{M - f(x)}$. Then $g(x)$ is well-defined on $[a, b]$, and also $g(x)$ is continuous on $[a, b]$ since $f(x)$ is continuous on $[a, b]$. Thus, by Theorem 3. 6. 5, $g(x)$ is bounded on $[a, b]$. However, on the other hand, by the definition of supremum we see that for any $n \in \mathbb{N}$, there is $x_n \in [a, b]$ such that $0 < M - f(x_n) < \frac{1}{n}$, i. e. $\frac{1}{M - f(x_n)} > n (n = 1, 2, \dots)$, which contradicts the boundedness of $g(x)$ on $[a, b]$. Therefore, $f(x)$ has the absolute maximum value on $[a, b]$. Similarly, we can show that $f(x)$ has the absolute minimum value on $[a, b]$. □

Theorem 3. 6. 7 (Existence of the root) Suppose a function $f(x)$ is continuous on a closed interval $[a, b]$ with $f(a)f(b) < 0$. Then there exists $x_0 \in (a, b)$ such that $f(x_0) = 0$, i. e. there is a root of the equation $f(x) = 0$ in (a, b).

Proof Divide $[a, b]$ into two subintervals $[a, x_1]$ and $[x_1, b]$ where $x_1 = \frac{a+b}{2}$, the middle point of a and b. If $f(x_1) = 0$, x_1 is what we seek; if $f(x_1) \neq 0$, then either $f(x_1)f(a) < 0$ or $f(x_1)f(b) < 0$. If $f(x_1)f(a) < 0$, denote $[a_1, b_1] = [a, x_1]$; otherwise, denote $[a_1, b_1] = [x_1, b]$. Thus, $f(a_1)f(b_1) < 0$.

The same approach works for $[a_1, b_1]$: divide $[a_1, b_1]$ into two subintervals $[a_1, x_2]$ and $[x_2, b_2]$ where $x_2 = \frac{a_1 + b_1}{2}$. If $f(x_2) = 0$, x_2 is what we require; if $f(x_2) \neq 0$, then either $f(x_2)f(a_1) < 0$ or $f(x_2)f(b_1) < 0$. For the former case, denote $[a_2, b_2] = [a_1, x_2]$;

otherwise, denote $[a_2, b_2] = [x_2, b_1]$. Thus, $f(a_2)f(b_2) < 0$.

As we repeat the same operation in this way, we have two possible cases:

(1) For some middle point x_i, $f(x_i) = 0$. Then the result holds true;

(2) For any middle point x_i, $f(x_i) \neq 0$. In this case, we obtain a series of closed intervals:

$[a, b] \supseteq [a_1, b_2] \supseteq [a_3, b_3] \supseteq \dots [a_n, b_n] \supseteq \dots$ with $|b_n - a_n| = \frac{b-a}{2^n} \to 0$ as $n \to +\infty$. By the nested interval theorem, there is uniquely a point x_0 belonging to each of these intervals.

We now assert that $f(x_0) = 0$. Because, if $f(x_0) \neq 0$, noting that $a_n \to x_0$ and $b_n \to x_0$ as $n \to +\infty$, and also $f(x)$ is continuous at x_0, $f(a_n)$ and $f(b_n)$ have a same sign as $f(x_0)$ when n is large enough by the locally sign-preserving theorem, which is in contradiction with the fact $f(a_n)f(b_n) < 0$ for any n. □

Examples (1) Prove the equation $x = a\sin x + b$ $(a, b > 0)$ has a root in $(0, a+b]$.

Proof Let $f(x) = x - a\sin x - b$. Clearly $f(x)$ is continuous on $[0, a+b]$ with $f(0) = -b < 0$ and $f(a+b) = a(1 - \sin(a+b)) \geqslant 0$. If $f(a+b) = 0$, then $a+b$ is a root required; if $f(a+b) \neq 0$, then $f(a+b) > 0$ and so $f(0)f(a+b) < 0$. Thus by Theorem 3. 6. 7, the equation has a root in $(0, a+b]$. □

(2) Let $f(x) = a_0x^n + a_1x^{n-1} + \dots + a_n (a_0 \neq 0)$. Prove that $f(x)$ has a real root if n is an odd number.

Proof As $x \to \infty$, $a_0 + a_1x^{-1} + \dots + a_nx^{-n} \to a_0$. Thus, for $\varepsilon = \frac{|a_0|}{2} > 0$, there is $M > 0$ such that $|a_0 + a_1x^{-1} + \dots + a_nx^{-n} - a_0| < \frac{|a_0|}{2}$ whenever $|x| > M$, i. e. $a_0 - \frac{|a_0|}{2} < a_0 + a_1x^{-1} + \dots + a_nx^{-n} <$

$a_0 + \frac{|a_0|}{2}$. So, if $|x| > M > 0$, $a_0 + a_1 x^{-1} + \ldots + a_n x^{-n}$ has a same sign with a_0. Hence, $f(M+1) \cdot f(-M-1) = -(M+1)^{2n}[a_0 + a_1(M+1)^{-1} + \cdots + a_n(M+1)^{-n}][a_0 + a_1(-M-1)^{-1} + \cdots + a_n(-M-1)^{-n}] < 0$. Then, by Theorem 3. 6. 7 the result holds true. □

Theorem 3. 6. 8 (The intermediate value theorem) Suppose a function $f(x)$ is continuous on a closed interval $[a, b]$ with $f(a) \neq f(b)$. Then for any number ξ between $f(a)$ and $f(b)$, there exists $x_0 \in (a, b)$ such that $f(x_0) = \xi$.

Proof Without loss of generality, suppose $f(a) > f(b)$ and then $f(a) > \xi > f(b)$. Let $g(x) = f(x) - \xi$. Then clearly $g(x)$ is continuous on $[a, b]$ with $g(a) > 0$ and $g(b) < 0$. Thus by Theorem 3. 6. 7 there is $x_0 \in (a, b)$ such that $g(x_0) = 0$, which implies $f(x_0) = \xi$. □

Example A Tibetan monk leaves the monastery at 7:00 a. m. and takes his usual path to the top of the mountain, arriving at 7:00 p. m. The following morning, he starts at 7:00 a. m. at the top and takes the same path back, arriving at the monastery at 7:00 p. m. Show that there is a point on the path that the monk will cross at exactly the same time of day on both days.

Proof Let l be the length of the path. Suppose $s_1 = s_1(t)$ $(t: 0 \to 12)$ is the distance function of the monk walking along the path from the monastery to the top of the mountain, and $s_2 = s_2(t)$ $(t: 0 \to 12)$ is the distance function from the top of the mountain back to the monastery. Then clearly $s_1(0) = s_2(0) = 0$ and $s_1(12) = s_2(12) = l$. Let $f(t) = s_1(t) + s_2(t)$ $(t: 0 \to 12)$. Thus $f(0) = 0$ and $f(12) = 2l$. So by the intermediate value theorem, for $l \in (0, 2l)$, there is a point $t_0 \in (0, 12)$ such that $f(t_0) = l$, i. e. $s_1(t_0) + s_2(t_0) = l$. Let P be the point on the path with a distance $s_1(t_0)$ from the monastery (also a distance $s_2(t_0)$ from the top). Hence, the monk will cross the point

P at exactly the same time of day on both days. □

The next corollary is a direct follower from Theorem 3. 6. 8.

Corollary Suppose a non-constant function $f(x)$ is continuous on an interval I. Then the range of f, $f(I)$, is also an interval; moreover, if $I = [a, b]$, then

$$f([a, b]) = [\alpha, \beta],$$

where $\alpha = \min_{x \in [a, b]} f(x)$ and $\beta = \max_{x \in [a, b]} f(x)$.

§ 3.6.3 Continuity of inverse functions

Theorem 3. 6. 9 Suppose $y = f(x)$ is continuous and strictly increasing (strictly decreasing) on $[a, b]$, then its inverse function $f^{-1}(x)$ is continuous on the corresponding domain $[f(a), f(b)]$ ($[f(b), f(a)]$).

Proof We only prove the case that $y = f(x)$ is strictly increasing on $[a, b]$ (The proof is similar for the case that $y = f(x)$ is strictly decreasing on $[a, b]$). It is easy to see that the range of f is $[f(a), f(b)]$, and so the domain of f^{-1} is $[f(a), f(b)]$.

Let $y_0 \in (f(a), f(b))$ and $x_0 = f^{-1}(y_0)$. Then $x_0 \in (a, b)$. For any $\varepsilon > 0$, we may take $x_1 \in (a, x_0)$, and $x_2 \in (x_0, b)$ such that $\max\{x_2 - x_0, x_0 - x_1\} < \varepsilon$. Let $y_1 = f(x_1)$ and $y_2 = f(x_2)$. Since f is strictly increasing, $y_1 < y_0 < y_2$ and $x = f^{-1}(y) \in (x_1, x_2)$ whenever $y \in (y_1, y_2)$, i. e. the distance between $f^{-1}(y)$ and $f^{-1}(y_0)$ is less than ε whenever $y \in (y_1, y_2)$, which implies $x = f^{-1}(y)$ is continuous at y_0.

Using the definition of right-hand and left-hand continuity, it can be similarly proved that $x = f^{-1}(y)$ is right-hand continuous at $f(a)$ and left-hand continuous at $f(b)$. □

Examples (1) Since $y = \sin x$ is continuous and strictly monotone on the interval $[-\pi/2, \pi/2]$, its inverse function $y =$

arcsin x is continuous on the interval $[-1, 1]$. Similarly, we can see that other anti-trigonometric functions are also continuous on their domains, e. g., $y = \arccos x$ is continuous on $[-1, 1]$, $y = \operatorname{arccot} x$ is continuous on $(-\infty, +\infty)$.

(2) Since $y = x^2$ is continuous and strictly monotone on $[0, +\infty)$, its inverse function $y = \sqrt{x}$ is continuous on $[0, +\infty)$.

(3) Since $y = x^n$ (n is any positive integer) is continuous and strictly monotone on $[0, +\infty)$, its inverse function $y = \sqrt[n]{x}$ is continuous on $[0, +\infty)$. Furthermore, since $y = x^{-1/n}$ is a composite function of $y = \sqrt[n]{u}$ and $u = 1/x$, by the continuity of the composite function $y = x^{-1/n}$ is continuous on $(0, +\infty)$, i. e. the function $y = x^{1/p}$ (p is a non-zero integer) is continuous on its domain interval.

(4) The function $y = x^\alpha$ is continuous on its domain interval, where α is any rational number.

Proof Let $\alpha = q/p (p \neq 0)$, where p and q are integers. Since $y = u^{1/p}$ and $u = x^q$ are both continuous on their domain intervals, then the composite function

$$y = (x^q)^{1/p} = x^{\frac{q}{p}} = x^\alpha$$

is a continuous function on its domain interval.

§ 3.7 Continuity of primary functions

In this section we will see that all primary functions are continuous on their domain intervals.

Theorem 3.7.1 The exponential function $f(x) = a^x (a > 0, a \neq 1)$ is continuous on $(-\infty, +\infty)$.

Proof We only need to prove that for any $x_0 \in (-\infty, +\infty)$, $\lim\limits_{x \to x_0} a^x = a^{x_0}$. First suppose $a > 1$. Since $\lim\limits_{n \to +\infty} a^{1/n} = 1$, for any $\varepsilon > 0$

there is $N > 0$ such that $0 < a^{1/N} - 1 < \varepsilon$. Since a^x is strictly increasing, $1 < a^x < a^{1/N}$ if $0 < x < 1/N$, and so $0 < a^x - 1 < \varepsilon$ if $0 < x < 1/N$, which implies $\lim\limits_{x \to 0^+} a^x = 1$. By substitution $x = -y$, we can see that

$$\lim_{x \to 0^-} a^x = \lim_{y \to 0^+} a^{-y} = \frac{1}{\lim\limits_{y \to 0^+} a^y} = 1.$$

This means if $x_0 = 0$, then $\lim\limits_{x \to x_0} a^x = a^{x_0}$.

Now let $x_0 \neq 0$. Note

$$\begin{aligned} \lim_{x \to x_0} (a^x - a^{x_0}) &= \lim_{x \to x_0} [a^{x_0}(a^{x - x_0} - 1)] \\ &= a^{x_0} \cdot \lim_{(x - x_0) \to 0} (a^{x - x_0} - 1) = a^{x_0} \cdot 0 = 0, \end{aligned}$$

i. e. $\lim\limits_{x \to x_0} a^x = a^{x_0}$.

Now suppose $a < 1$. Let $a = 1/b$, and then $b > 1$. Thus

$$\lim_{x \to x_0} a^x = \lim_{x \to x_0} \left(\frac{1}{b}\right)^x = \frac{1}{\lim\limits_{x \to x_0} b^x} = \frac{1}{b^{x_0}} = a^{x_0}.$$

If $a = 1$, trivially $\lim\limits_{x \to x_0} a^x = a^{x_0}$. □

Corollary The logarithmic function is continuous on the domain $(0, +\infty)$.

Proof Since the exponential function $f(x) = a^x (a > 0, a \neq 1)$ is strictly monotone and continuous on $(-\infty, +\infty)$, its inverse function, the logarithmic function is also continuous on the domain $(0, +\infty)$. □

Corollary The power function $f(x) = x^a (a \in \mathbb{R})$ is continuous on $(0, +\infty)$.

Proof Since the power function $y = x^a$ with $a \in \mathbb{R}$ is a composite function of $y = \mathrm{e}^u$ and $u = a \ln x$, which are both continuous, then x^a is also continuous on its domain $(0, +\infty)$. □

Theorem 3. 7. 2 All fundamental primary functions are continuous on their domain intervals.

Proof Note that we have proved the constant function, the trigonometric function, the anti-trigonometric function, the exponential function, the logarithmic function and the power function are all continuous on their domain intervals. The theorem follows immediately. □

More generally, we can derive the following:

Theorem 3. 7. 3 All primary functions are continuous on their domain intervals.

Proof Noticing that the fundamental operations and the composition keep the continuity of functions, we obtain the consequence directly in the light of the definition of the primary function and the preceding theorem. □

Using the continuity of primary functions, we can find limits of functions in a convenient way.

Examples (1) Find the following limits

$$① \lim_{x\to 3}\frac{e^x \sin x+5}{x^2+\ln x};\ ② \lim_{x\to 0}\frac{\ln(1+x)}{x}.$$

Solution ① Since the function is primary and $x=3$ is in its domain, so the limit is exactly the function value at $x=3$, i. e.

$$\lim_{x\to 3}\frac{e^x \sin x+5}{x^2+\ln x}=\frac{e^3 \sin 3+5}{3^2+\ln 3}.$$

$$② \lim_{x\to 0}\frac{\ln(1+x)}{x}=\lim_{x\to 0}\ln(1+x)^{1/x}$$
$$=\ln\left[\lim_{x\to 0}(1+x)^{1/x}\right]=\ln e=1.$$

(2) Let $f(x)$ be a polynomial with degree 3 such that

$$\lim_{x\to 2a}\frac{f(x)}{x-2a}=\lim_{x\to 4a}\frac{f(x)}{x-4a}=1(a\neq 0).\ \text{Find}\ \lim_{x\to 3a}\frac{f(x)}{x-3a}.$$

Solution Note that

$$\lim_{x\to 2a}\frac{f(x)}{x-2a}=1 \text{ and } \lim_{x\to 2a}(x-2a)=0.$$

Then $\lim_{x\to 2a}f(x)=0$, which implies $f(2a)=0$ since $f(x)$ is continuous. Similarly, we have $f(4a)=0$. Considering $f(x)$ is a polynomial with degree 3 and $2a\neq 4a$, we may set $f(x)=b(x-2a)(x-4a)(x-c)$ where b and c are to be determined. By the given limits, we see that $\lim_{x\to 2a}[b(x-4a)(x-c)]=b(-2a)(2a-c)=1$ and $\lim_{x\to 4a}[b(x-2a)(x-c)]=b(2a)(4a-c)=1$. Thus, $b=1/(2a^2)$ and $c=3a$, and

$$f(x)=\frac{1}{2a^2}(x-2a)(x-4a)(x-3a).$$

Therefore,

$$\lim_{x\to 3a}\frac{f(x)}{x-3a}=\frac{1}{2a^2}(3a-2a)(3a-4a)=-\frac{1}{2}.$$

Remark Let $f(u)$ be a primary function and let $u(x)$ be any function. By Theorem 3. 7. 3 and the remark in Section 3. 6. 1, if $\lim_{x\to x_0}u(x)$ exists and belongs to the domain interval of $f(x)$, then

$$\lim_{x\to x_0}f(u(x))=f(\lim_{x\to x_0}u(x)).$$

§ 3. 8 Uniform continuity

We now make a closer study of the concept of continuity of a function, particularly with respect to the way the behavior of a function near one point may differ from that near another. In fact, in the notion of the continuity of a function $f(x)$ at a point x_0, we notice that the δ we selected depended not only upon ε but also upon

x_0. Suppose $f(x)$ is continuous on some interval I such that we can select the δ depending only upon ε. This special phenomenon of being able to pick δ independent of x_0 is one to which a special name will be given. As we shall see, the contrast here is between properties that are local and properties that are global.

Definition We say a function f is *uniformly continuous* on an interval I, if corresponding to each $\varepsilon > 0$, a number $\delta > 0$ can be found such that $|f(x_1) - f(x_2)| < \varepsilon$ whenever x_1 and x_2 are in I and $|x_1 - x_2| < \delta$.

Remark Let $f(x)$ be a function defined on an interval I. Then $f(x)$ is not uniformly continuous on I: $\Leftrightarrow$there is some $\varepsilon > 0$ such that for any $\delta > 0$, there exist points x_1 and x_2 in I with $|x_1 - x_2| < \delta$, however $|f(x_1) - f(x_2)| \geqslant \varepsilon$.

To emphasize the difference between continuity and uniform continuity on an interval, we consider the following examples.

Examples (1) Prove: $f(x) = \sin x$ is uniformly continuous on $(-\infty, +\infty)$.

Proof For any $\varepsilon > 0$, take $\delta = \varepsilon$, then for any $x_1, x_2 \in (-\infty, +\infty)$ with $|x_1 - x_2| < \delta$, we have

$$|\sin x_1 - \sin x_2| = 2\left|\sin\frac{x_1 - x_2}{2}\cos\frac{x_1 + x_2}{2}\right|$$
$$\leqslant 2\left|\sin\frac{x_1 - x_2}{2}\right| \leqslant |x_1 - x_2| < \delta = \varepsilon.$$

So, $f(x) = \sin x$ is uniformly continuous on $(-\infty, +\infty)$. $\square$

(2) Let $f(x) = 1/x$ for $x > 0$ and take $I = (0, 1]$. (Clearly, $f(x)$ is continuous on I.) Prove $f(x)$ is not uniformly continuous on I.

Proof Let $\varepsilon = 1$. For any $\delta > 0$, take $x_1 = \min\{\delta, 1/2\}$ and $x_2 = x_1/2$. Then $x_1, x_2 \in I$ with $|x_1 - x_2| = |x_1 - x_1/2| = x_1/2 \leqslant \delta/2 < \delta$. However,

$$|f(x_1)-f(x_2)| = \left|\frac{1}{x_1}-\frac{1}{x_2}\right| = \left|\frac{1}{x_1}-\frac{2}{x_1}\right| = \frac{1}{x_1} \geqslant 1 = \varepsilon.$$

Hence, by the remark, $f(x)$ is not uniformly continuous on $(0, 1]$. □

(3) Let $f(x) = x^2$. ① Prove $f(x)$ is uniformly continuous on the interval $I = (0, 1]$. ② Prove $f(x)$ is not uniformly continuous on the interval $I = [1, +\infty)$.

① **Proof** Observe that for any $x, x_0 \in I$,

$$|f(x)-f(x_0)| = |x^2 - x_0^2| = |(x-x_0)(x+x_0)| < 2|x-x_0|.$$

If $|x-x_0| < \delta$, then $|f(x)-f(x_0)| < 2\delta$. Hence, if ε is given, we need only take $\delta = \varepsilon/2$ to guarantee that $|f(x)-f(x_0)| < \varepsilon$ for every pair x and x_0 with $|x-x_0| < \delta$. This shows that f is uniformly continuous on $I = (0, 1]$.

② For the proof of the second result, we give three different versions:

Proof 1 Take $\varepsilon = 1/2$. For any $\delta > 0$, there exist $x_1 = \sqrt{n}$ and $x_2 = \sqrt{n+1}$ where n is large enough such that

$$|x_1 - x_2| = \sqrt{n+1} - \sqrt{n} = \frac{1}{\sqrt{n+1}+\sqrt{n}} < \delta.$$

However, $|f(x_1)-f(x_2)| = |n-(n+1)| = 1 > 1/2 = \varepsilon$. Hence, $f(x)$ is not uniformly continuous on $[1, +\infty)$.

Proof 2 Take $\varepsilon = 1$. For any $\delta > 0$, there exist $x_1 = n$ and $x_2 = n + \frac{1}{n}$ where n is large enough such that $|x_1 - x_2| = \frac{1}{n} < \delta$. However,

$$|f(x_1)-f(x_2)| = \left|n^2 - \left(n+\frac{1}{n}\right)^2\right| = 2+\frac{1}{n^2} > 2 > 1 = \varepsilon.$$

So, $f(x)$ is not uniformly continuous on $[1, +\infty)$.

Proof 3 Take $\varepsilon = 1$. For any $\delta > 0$ (Without loss of generality, we may suppose $\delta < 1$), there exist $x_1 = \frac{1}{8}$ and $x_2 = \frac{1}{\delta} + \frac{\delta}{2}$. Thus, $|x_1 - x_2| = \frac{\delta}{2} < \delta$. However,

$$|f(x_1) - f(x_2)| = \left|\left(\frac{1}{\delta}\right)^2 - \left(\frac{1}{\delta} + \frac{\delta}{2}\right)^2\right| = 1 + \frac{\delta^2}{4} > 1 = \varepsilon,$$

which implies that $f(x)$ is not uniformly continuous on $[1, +\infty)$. □

(4) Show that $f(x) = ax + b$ ($a \neq 0$) is uniformly continuous on $(-\infty, \infty)$.

Proof For any $\varepsilon > 0$, since $|f(x_1) - f(x_2)| = |a||x_1 - x_2|$, we may take $\delta = \varepsilon / |a|$. Then

$$|f(x_1) - f(x_2)| < \varepsilon,$$

whenever $x_1, x_2 \in (-\infty, \infty)$ with $|x_1 - x_2| < \delta$. □

(5) Show that $f(x) = \sin \frac{1}{x}$ is not uniformly continuous on $(0, 1)$.

Proof Take $\varepsilon = 1$. Then for any $\delta > 0$, there exist

$$x_1 = \frac{1}{\left(\left[\frac{1}{\delta}\right] + 1\right)\pi} \in (0, 1), \quad x_2 = \frac{1}{\left(\left[\frac{1}{\delta}\right] + 1\right)\pi + \frac{\pi}{2}} \in (0, 1),$$

such that

$$\begin{aligned}
|x_1 - x_2| &= \left|\frac{1}{\left(\left[\frac{1}{\delta}\right] + 1\right)\pi} - \frac{1}{\left(\left[\frac{1}{\delta}\right] + 1\right)\pi + \frac{\pi}{2}}\right| \\
&= \frac{\pi/2}{\left\{\left(\left[\frac{1}{\delta}\right] + 1\right)\pi\right\}\left\{\left(\left[\frac{1}{\delta}\right] + 1\right)\pi + \frac{\pi}{2}\right\}} \\
&< \frac{1}{2\left(\left[\frac{1}{\delta}\right] + 1\right)} < \frac{\delta}{2} < \delta,
\end{aligned}$$

however,

$$\left|\sin\frac{1}{x_1}-\sin\frac{1}{x_2}\right|=\left|\sin\left(\left[\frac{1}{\delta}\right]+1\right)\pi-\sin\left\{\left(\left[\frac{1}{\delta}\right]+1\right)\pi+\frac{\pi}{2}\right\}\right|$$
$$=\left|\sin\left\{\left(\left[\frac{1}{\delta}\right]+1\right)\pi+\frac{\pi}{2}\right\}\right|=1.$$

Hence, $f(x)=\sin\frac{1}{x}$ is not uniformly continuous on $(0, 1)$. □

From the above examples, we observe that the uniform continuity is a global property of a function on some interval. Clearly, if a function is uniformly continuous on an interval I, then it is continuous everywhere on I, but conversely it is usually not true. However, if a function is continuous on a closed interval I, then it is also uniformly continuous on I. We now present this result, which will be proved in two versions, as follows:

Theorem 3. 8. 1 (Cantor theorem) If a function $f(x)$ is continuous on $[a, b]$, then $f(x)$ is uniformly continuous on $[a, b]$.

Proof 1 We prove this theorem by contradiction. Assume that $f(x)$ is not uniformly continuous on $[a, b]$, i. e. there is some $\varepsilon_0>0$ such that for any $\delta>0$, there exist $x', x''\in[a, b]$, though $|x'-x''|<\delta$, $|f(x')-f(x'')|\geqslant\varepsilon_0$. We will show that this assumption is in contradiction with the fact that $f(x)$ is continuous on $[a, b]$.

Divide $[a, b]$ into two subintervals $[a, c]$ and $[c, b]$ where $c=\frac{a+b}{2}$. Then at least one of the subintervals possesses the following property:

For this $\varepsilon_0>0$ and for any $\delta>0$, there exist x', x'' in this subinterval such that $|x'-x''|<\delta$, but $|f(x')-f(x'')|\geqslant\varepsilon_0$.

Because otherwise, suppose that both of these two subintervals do not possess this property. Then, for this $\varepsilon_0>0$, there are $\delta_1>0$

and $\delta_2 > 0$ respectively such that for any x'_1, $x''_1 \in [a, c]$ and any x'_2, $x''_2 \in [c, b]$, $|f(x'_1) - f(x''_1)| < \varepsilon_0$ and $|f(x'_2) - f(x''_2)| < \varepsilon_0$ if only $|x'_1 - x''_1| < \delta_1$ and $|x'_2 - x''_2| < \delta_2$. Since $f(x)$ is continuous at c, for this $\varepsilon_0 > 0$, there is $\delta_3 > 0$ such that $|f(x) - f(c)| < \frac{\varepsilon_0}{2}$ if only $|x - c| < \delta_3$. Now, we take $\delta = \min\{\delta_1, \delta_2, \delta_3\}$. Then, for any x', $x'' \in [a, b]$, no matter whether these two points are both in $[a, c]$, or both in $[c, b]$, or one is in $[a, c]$ and the other is in $[c, b]$, $|f(x') - f(x'')| < \varepsilon_0$ if only $|x' - x''| < \delta$. However, this is a contradiction to our initial assumption. So, at least one of the subintervals possesses the above property.

We select a subinterval which possesses the property and denote it by $[a_1, b_1]$ (if both subintervals possess the property, select either of them). Then, $[a_1, b_1] \subseteq [a, b]$ with $b_1 - a_1 = \dfrac{b-a}{2}$.

Repeat this process with $[a_1, b_1]$ to generate a subinterval $[a_2, b_2] \subseteq [a_1, b_1]$ such that $[a_2, b_2]$ possesses the property and $b_2 - a_2 = \dfrac{b-a}{2^2}$. In this manner, we may continue the process and obtain a nested sequence of closed intervals $\{[a_n, b_n]\}$ such that $[a_{n+1}, b_{n+1}] \subseteq [a_n, b_n] (n = 1, 2, \cdots)$, $b_n - a_n = \dfrac{b-a}{2^n}$, and each $[a_n, b_n]$ possesses the above property. Thus, by the nested interval theorem, there is uniquely a point $\xi \in [a_n, b_n] \subseteq [a, b]$.

Since $f(x)$ is continuous at $\xi \in [a, b]$, for the above $\varepsilon_0 > 0$, there exists $\delta_0 > 0$ such that $|f(x) - f(\xi)| < \frac{\varepsilon_0}{2}$ whenever $x \in U(\xi, \delta_0)$. Noticing that $[a_n, b_n] \subseteq U(\xi, \delta_0)$ if n is large enough, we see that for any x', $x'' \in [a_n, b_n]$, since $|x' - \xi| < \delta_0$ and $|x'' - \xi| < \delta_0$, $|f(x') - f(\xi)| < \frac{\varepsilon_0}{2}$ and $|f(x'') - f(\xi)| < \frac{\varepsilon_0}{2}$. Thus,

$$|f(x')-f(x'')|\leqslant|f(x')-f(\xi)|+|f(x'')-f(\xi)|<\frac{\varepsilon_0}{2}+\frac{\varepsilon_0}{2}=\varepsilon_0.$$

However, this contradicts to the above property each subinterval $[a_n, b_n]$ possesses: for this $\varepsilon_0>0$ and for any $\delta>0$, there exist x', x'' in this subinterval such that $|x'-x''|<\delta$, but $|f(x')-f(x'')|\geqslant\varepsilon_0$. Therefore, it is impossible that a function $f(x)$ is continuous on $[a, b]$ but $f(x)$ is not uniformly continuous on $[a, b]$. □

Proof 2 We now use Heine-Borel finite open covering theorem to prove this theorem. Extend $f(x)$ to $g(x)$ as follows:

$$g(x)=\begin{cases} f(a) & (x<a), \\ f(x) & (a\leqslant x\leqslant b), \\ f(b) & (b<x). \end{cases}$$

Clearly, $g(x)$ is continuous on $(-\infty, +\infty)$. Then, for any $\varepsilon>0$ and any $x_0\in[a, b]$, there exists $\delta_{(x_0,\varepsilon)}>0$ such that $|g(x)-g(x_0)|<\frac{\varepsilon}{2}$ whenever $|x-x_0|<\delta_{(x_0,\varepsilon)}$.

Let $U_{x_0}=\{x\mid |x-x_0|<\frac{1}{2}\delta_{(x_0,\varepsilon)}\}$. Then, $[a, b]\subseteq\bigcup\limits_{x_0\in[a, b]}U_{x_0}$. By Heine-Borel finite open covering theorem, there are a finite number of open intervals: U_{x_1}, U_{x_2}, $\cdots$, U_{x_n} such that

$$[a, b]\subseteq\bigcup_{j=1}^{n}U_{x_j}.$$

Take $\delta=\frac{1}{4}\min\{\delta_{(x_1,\varepsilon)}, \delta_{(x_2,\varepsilon)}, \cdots, \delta_{(x_n,\varepsilon)}\}$. Then, for any x', $x''\in[a, b]$ with $|x'-x''|<\delta$, there is x_j such that $x'\in U_{x_j}$, and

$$|x''-x_j|\leqslant|x''-x'|+|x'-x_j|\leqslant\frac{\delta_{(x_j,\varepsilon)}}{4}+\frac{1}{2}\delta_{(x_j,\varepsilon)}<\delta_{(x_j,\varepsilon)}.$$

So, $|g(x')-g(x'')|\leqslant|g(x')-g(x_j)|+|g(x_j)-g(x'')|<\frac{\varepsilon}{2}+$

$\frac{\varepsilon}{2}=\varepsilon$. Considering that $f(x)=g(x)$ if $x\in[a, b]$, we complete the proof. □

Theorem 3. 8. 2 A function $f(x)$ is uniformly continuous on (a, b) if and only if $f(x)$ is continuous on (a, b) and the limits

$$\lim_{x\to a^+} f(x) \text{ and } \lim_{x\to b^-} f(x).$$

are both existent.

Proof Necessity. Clearly, we merely need to show the existence of the limits

$$\lim_{x\to a^+} f(x) \text{ and } \lim_{x\to b^-} f(x).$$

For any $\varepsilon>0$, since $f(x)$ is uniformly continuous on (a, b), there exists a positive number $\delta<b-a$ such that for any $x_1, x_2\in(a, b)$ with $|x_1-x_2|<\delta$, we have $|f(x_1)-f(x_2)|<\varepsilon$. Therefore, for any x_1, x_2 with $0<x_1-a<\delta$ and $0<x_2-a<\delta$, we have $x_1, x_2\in(a, b)$ with $|x_1-x_2|<\delta$, and so it holds that

$$|f(x_1)-f(x_2)|<\varepsilon.$$

Thus, by Cauchy criterion for convergence (Theorem 3. 2. 8), $\lim_{x\to a^+} f(x)$ exists. Similarly, we may show $\lim_{x\to b^-} f(x)$ exists.

Sufficiency. Construct an auxiliary function:

$$F(x)=\begin{cases}\lim_{x\to a^+} f(x) & (x=a),\\ f(x) & (a<x<b),\\ \lim_{x\to b^-} f(x) & (x=b).\end{cases}$$

Then, $F(x)$ is continuous on $[a, b]$. By Theorem 3. 8. 1, $F(x)$ is uniformly continuous on $[a, b]$, and so clearly $F(x)$ is uniformly continuous on (a, b), i. e. $f(x)$ is uniformly continuous on (a, b).

Remark By an analogous argument, we may derive the following statements:

(1) A function $f(x)$ is uniformly continuous on $[a, b)$ if and only if $f(x)$ is continuous on $[a, b)$ and the limit $\lim\limits_{x \to b^-} f(x)$ exists;

(2) A function $f(x)$ is uniformly continuous on $(a, b]$ if and only if $f(x)$ is continuous on $(a, b]$ and the limit $\lim\limits_{x \to a^+} f(x)$ exists.

Example Since $\ln x$ is continuous on $(0, \mathrm{e}]$, but $\lim\limits_{x \to 0^+} = -\infty$, $\ln x$ is not uniformly continuous on $(0, \mathrm{e}]$.

For the uniform continuity of a function defined on an infinite interval, we have the following sufficient condition.

Theorem 3.8.3 If a function $f(x)$ is continuous on $(-\infty, +\infty)$ and the limits

$$\lim_{x \to +\infty} f(x) \text{ and } \lim_{x \to -\infty} f(x)$$

are both existent, then $f(x)$ is uniformly continuous on $(-\infty, +\infty)$.

Proof Since the limits

$$\lim_{x \to +\infty} f(x) \text{ and } \lim_{x \to -\infty} f(x)$$

are both existent, for any $\varepsilon > 0$, there exists $l > 0$ such that

$$| f(x_1) - f(x_2) | < \varepsilon$$

whenever $x_1 < -l$ and $x_2 < -l$, and

$$| f(x_3) - f(x_4) | < \varepsilon$$

whenever $x_3 > l$ and $x_4 > l$. Evidently, $f(x)$ is continuous at $x = -l$ and $x = l$, and also $f(x)$ is continuous on $[-l, l]$. So, by theorem 8.3.1, it is uniformly continuous on $[-l, l]$. Thus, there exists $\delta > 0$ such that

$$| f(x) - f(-l) | < \frac{\varepsilon}{2},$$

whenever $| x - (-l) | < \delta$, and

$$| f(x) - f(l) | < \frac{\varepsilon}{2},$$

whenever $| x - l | < \delta$; for any x_5, $x_6 \in [-l, l]$ with $| x_5 - x_6 | < \delta$,

$$| f(x_5) - f(x_6) | < \varepsilon.$$

Now, let x', $x'' \in (-\infty, +\infty)$ with $| x' - x'' | < \delta$. There are five cases to be considered:

(1) x', $x'' \in (-\infty, -l)$. Then, $| f(x') - f(x'') | < \varepsilon$ holds.

(2) One of x', x'' belongs to $(-l-\delta, -l)$, the other belongs to $[-l, -l+\delta)$. Then, $| x' + l | < \delta$ and $| x'' + l | < \delta$, and so $| f(x') - f(-l) | < \frac{\varepsilon}{2}$ and $| f(x'') - f(-l) | < \frac{\varepsilon}{2}$, which implies $| f(x') - f(x'') | < \varepsilon$.

(3) x', $x'' \in [-l, l]$. Then, $| f(x') - f(x'') | < \varepsilon$ holds.

(4) One of x', x'' belongs to $(l-\delta, l]$, the other belongs to $(l, l+\delta)$. Then, $| x' - l | < \delta$ and $| x'' - l | < \delta$, and so $| f(x') - f(l) | < \frac{\varepsilon}{2}$ and $| f(x'') - f(l) | < \frac{\varepsilon}{2}$, which implies $| f(x') - f(x'') | < \varepsilon$.

(5) x', $x'' \in (l, +\infty)$. Then, $| f(x') - f(x'') | < \varepsilon$ holds.

In summary, for any $\varepsilon > 0$, there exists $\delta > 0$ such that $| f(x') - f(x'') | < \varepsilon$ whenever x', $x'' \in (-\infty, +\infty)$ with $| x' - x'' | < \delta$. Hence, $f(x)$ is uniformly continuous on $(-\infty, +\infty)$. □

Remark Analogously, we may prove the following statements:

(1) If a function $f(x)$ is continuous on $(a, +\infty)$ and the limits $\lim_{x \to a^+} f(x)$ and $\lim_{x \to +\infty} f(x)$ are both existent, then $f(x)$ is uniformly continuous on $(a, +\infty)$.

(2) If a function $f(x)$ is continuous on $[a, +\infty)$ and the limit

$\lim\limits_{x \to +\infty} f(x)$ is existent, then $f(x)$ is uniformly continuous on $[a, +\infty)$.

(3) If a function $f(x)$ is continuous on $(-\infty, b)$ and the limits $\lim\limits_{x \to -\infty} f(x)$ and $\lim\limits_{x \to b^-} f(x)$ are both existent, then $f(x)$ is uniformly continuous on $(-\infty, b)$.

(4) If a function $f(x)$ is continuous on $(-\infty, b]$ and the limit $\lim\limits_{x \to -\infty} f(x)$ is existent, then $f(x)$ is uniformly continuous on $(-\infty, b]$.

Example Since $\frac{\sin x}{x}$ is continuous on $(0, +\infty)$ and $\lim\limits_{x \to 0^+} \frac{\sin x}{x} = 1$ and $\lim\limits_{x \to +\infty} \frac{\sin x}{x} = 0$, $\frac{\sin x}{x}$ is uniformly continuous on $(0, +\infty)$.

Note that the above mentioned conditions are just sufficient, but not necessary. For example, $\sin x$ is uniformly continuous on $(-\infty, +\infty)$, but $\lim\limits_{x \to -\infty} \sin x$ and $\lim\limits_{x \to +\infty} \sin x$ do not exist.

For an arbitrary interval, we have the following sufficiency.

Theorem 3. 8. 4 Suppose I is an interval and there exists a number $L > 0$ such that

$$| f(x_1) - f(x_2) | \leqslant L | g(x_1) - g(x_2) |$$

for any $x_1, x_2 \in I$. If the function $g(x)$ is uniformly continuous on I, then the function $f(x)$ is also uniformly continuous on I.

Proof By the condition, for any $\varepsilon > 0$, there is $\delta > 0$ such that

$$| g(x_1) - g(x_2) | < \frac{\varepsilon}{L}$$

whenever $x_1, x_2 \in I$ with $| x_1 - x_2 | < \delta$. Thus, for any $x_1, x_2 \in I$ with $| x_1 - x_2 | < \delta$, we have

$$| f(x_1) - f(x_2) | \leqslant L | g(x_1) - g(x_2) | < L \frac{\varepsilon}{L} = \varepsilon,$$

from which the result follows. □

Definition A function $f(x)$ is said to satisfy *Lipschitz condition* on an interval I, if there is a number $L > 0$ such that for any x_1, $x_2 \in I$,

$$|f(x_1) - f(x_2)| \leqslant L |x_1 - x_2|.$$

Corollary If function $f(x)$ satisfies Lipschitz condition on an interval I, $f(x)$ is uniformly continuous on I.

Proof Only notice that the function $g(x) = x$ is uniformly continuous on I. Then the result follows directly from Theorem 3. 8. 4. □

For a function defined on a finite non-closed interval, we have the necessity for the uniform continuity as follows:

Theorem 3. 8. 5 If function $f(x)$ is uniformly continuous on a finite non-closed interval I, then $f(x)$ is continuous and bounded on I.

Proof We merely need to show the boundedness.

If $I = (a, b)$, by Theorem 3. 8. 2 the limits

$$\lim_{x \to a^+} f(x) \text{ and } \lim_{x \to b^-} f(x)$$

exist. Let

$$F(x) = \begin{cases} \lim\limits_{x \to a^+} f(x) & (x = a), \\ f(x) & (a < x < b), \\ \lim\limits_{x \to b^-} f(x) & (x = b). \end{cases}$$

Then, $F(x)$ is continuous on $[a, b]$, and so bounded on $[a, b]$. Thus, $F(x)$ is bounded on (a, b), i. e. $f(x)$ is bounded on I.

If $I = (a, b]$ or $I = [a, b)$, we may similarly show $f(x)$ is bounded on I. □

Example Since $\dfrac{1}{x}$ is unbounded on $(0, 1)$, $\dfrac{1}{x}$ is not uniformly continuous on (0, 1).

Note (1) The condition in Theorem 3. 8. 5 is not sufficient. For example, $\sin\dfrac{1}{x}$ is continuous and bounded on (0, 1), but we have proved that it is not uniformly continuous on (0, 1).

(2) Theorem 3. 8. 5 will not hold true if the interval is infinite. For example, $f(x)=x$ is uniformly continuous on $(-\infty, +\infty)$, but it is unbounded on $(-\infty, +\infty)$.

Exercises

1. Prove by definition the following function limits:

(1) $\lim\limits_{x\to\infty}\dfrac{1}{x}=0$;　　(2) $\lim\limits_{x\to 0}e^{\frac{-1}{x^2}}=0$;

(3) $\lim\limits_{x\to 1}(3x-1)=2$;　　(4) $\lim\limits_{x\to 5}\dfrac{x^2-6x+3}{x-5}=4$;

(5) $\lim\limits_{x\to 0}a^x=1$;　　(6) $\lim\limits_{x\to a}\cos x=\cos a$;

(7) $\lim\limits_{x\to x_0}\sin x=\sin x_0$;　　(8) $\lim\limits_{x\to\infty}\dfrac{x^2-5}{x^2-1}=1$;

(9) $\lim\limits_{x\to 2}\sqrt{x^2+5}=3$.

2. Identify the infinitesimals and the infinities of the following:

(1) $f(x)=\dfrac{x-1}{x^2-4}$, as $x\to -2$;

(2) $f(x)=\ln x$, as $x\to 0^+$;

(3) $f(x)=e^{\frac{1}{x}}$, as $x\to 0^-$;

(4) $f(x)=\dfrac{\pi}{2}-\arctan x$, as $x\to +\infty$.

3. Prove by definition the following limits:

(1) $\lim\limits_{x\to 0}\dfrac{1+x}{x}=\infty$;　　(2) $\lim\limits_{x\to +\infty}a^x=+\infty\quad(a>1)$.

4. Prove the following statements:

(1) $x\cos x$ is unbounded but not an infinity as $x\to\infty$;

(2) $\frac{1}{x}\sin\frac{1}{x}$ is unbounded but not an infinity as $x \to 0$.

5. Find the following limits:

(1) $\lim\limits_{x\to 0}\frac{x^2+2x+1}{x^3+2x}$; (2) $\lim\limits_{x\to\frac{\pi}{2}}\frac{\sin x+1}{\cos x}$; (3) $\lim\limits_{x\to e}\frac{x-1}{\ln x-1}$;

(4) $\lim\limits_{x\to-\infty}\frac{\arctan x}{e^x}$; (5) $\lim\limits_{x\to 2}\frac{x^3-8}{x-2}$; (6) $\lim\limits_{x\to 1}\frac{2x^2-x-1}{x^2-1}$;

(7) $\lim\limits_{x\to 3}\frac{x^2-5x+6}{x^2+2x-15}$; (8) $\lim\limits_{x\to\frac{\pi}{2}}\frac{\sin 2x}{\cos x}$; (9) $\lim\limits_{x\to\pi}\frac{\tan x}{\sin 2x}$;

(10) $\lim\limits_{x\to\frac{\pi}{4}}\frac{\sin x-\cos x}{\cos 2x}$; (11) $\lim\limits_{h\to 0}\frac{(x+h)^3-x^3}{h}$;

(12) $\lim\limits_{h\to 0}\left(\frac{1}{x+h}-\frac{1}{x}\right)\frac{1}{h}$; (13) $\lim\limits_{t\to 1}\left(\frac{1}{1-t}-\frac{1}{1-t^2}\right)$;

(14) $\lim\limits_{x\to 1}\frac{x^2-1}{\sqrt{3-x}-\sqrt{1+x}}$; (15) $\lim\limits_{x\to 1}\frac{\sqrt[3]{x}-1}{\sqrt{x}-1}$;

(16) $\lim\limits_{x\to a}\frac{\sqrt[m]{x}-\sqrt[m]{a}}{x-a}$ $(a>0)$; (17) $\lim\limits_{x\to 4}\frac{\sqrt{2x+1}-3}{\sqrt{x}-2}$;

(18) $\lim\limits_{x\to 1}\frac{x^m-1}{x^n-1}$ (m, n are positive integers);

(19) $\lim\limits_{x\to\infty}\frac{x+\sin x}{x-\sin x}$; (20) $\lim\limits_{x\to\infty}\frac{x-\arctan x}{x+\arctan x}$;

(21) $\lim\limits_{x\to\infty}\left(\frac{x^3}{2x^2-1}-\frac{x^2}{2x+1}\right)$; (22) $\lim\limits_{x\to 1}\frac{x+x^2+\cdots+x^n-n}{x-1}$;

(23) $\lim\limits_{x\to\infty}(\sqrt{x^2+1}-\sqrt{x^2-1})$;

(24) $\lim\limits_{x\to+\infty}(\sqrt{x^2+x}-\sqrt{x^2-x})$; (25) $\lim\limits_{x\to+\infty}\frac{\sqrt{x^2+1}}{x+1}$;

(26) $\lim\limits_{x\to+\infty}\frac{\sqrt{x^2-3}}{\sqrt[3]{8x^3+1}}$; (27) $\lim\limits_{x\to-1}\left(\frac{1}{x+1}-\frac{3}{x^3+1}\right)$;

(28) $\lim\limits_{x\to+\infty}(\sqrt{x^2+x}-\sqrt{x^2+1})$.

6. Find the following limits:

(1) $\lim\limits_{x\to 0}\dfrac{\tan kx}{x}\quad (k\neq 0)$; (2) $\lim\limits_{x\to 0}\dfrac{\sin\alpha x}{\sin\beta x}\quad (\beta\neq 0)$;

(3) $\lim\limits_{x\to 0}\dfrac{\tan 2x}{\sin 5x}$; (4) $\lim\limits_{x\to \pi}\dfrac{\sin x}{\pi - x}$;

(5) $\lim\limits_{x\to \pi}\dfrac{\sin 3x}{\tan 5x}$; (6) $\lim\limits_{n\to +\infty} 2^n \sin\dfrac{\alpha}{2^n}\quad (\alpha\neq 0)$;

(7) $\lim\limits_{x\to 0} x\cot 5x$; (8) $\lim\limits_{x\to 0}\dfrac{\sqrt{1-\cos x}}{|x|}$;

(9) $\lim\limits_{x\to 0}\dfrac{\cos mx - \cos nx}{x^2}$; (10) $\lim\limits_{x\to 0}\dfrac{1-\cos 2x}{x\sin 2x}$;

(11) $\lim\limits_{x\to 0}\dfrac{\sin 2x\tan x}{x^2}$; (12) $\lim\limits_{x\to\infty}\left(1+\dfrac{1}{x}\right)^{kx}$;

(13) $\lim\limits_{t\to +\infty}\left(1-\dfrac{1}{t}\right)^{t}$; (14) $\lim\limits_{x\to\infty}\left(\dfrac{x}{1+x}\right)^{x}$;

(15) $\lim\limits_{x\to 0}(1-2x)^{\frac{1}{x}}$; (16) $\lim\limits_{x\to 0}(1+\sin x)^{\cot x}$;

(17) $\lim\limits_{x\to 0}\dfrac{a^x-1}{x}$; (18) $\lim\limits_{x\to 0}\dfrac{a^x-a^{-x}}{x}$;

(19) $\lim\limits_{x\to\infty}\dfrac{\ln(x+1)-\ln x}{x}$; (20) $\lim\limits_{x\to\infty}\left(\dfrac{3-2x}{2-2x}\right)^{x}$;

(21) $\lim\limits_{x\to 0}\dfrac{\sin x^3}{(\sin x)^2}$; (22) $\lim\limits_{x\to\infty}\dfrac{(3x+6)^{70}(8x-5)^{20}}{(5x-1)^{90}}$;

(23) $\lim\limits_{x\to 0}\left[\dfrac{a_1^x+a_2^x+\cdots+a_n^x}{n}\right]^{1/x}\quad (a_i>0)$;

(24) $\lim\limits_{x\to\frac{\pi}{2}}\dfrac{\cos x}{x-\dfrac{\pi}{2}}$; (25) $\lim\limits_{x\to 0}\dfrac{\tan x - \sin x}{x^3}$;

(26) $\lim\limits_{x\to\infty} x\sin\dfrac{1}{x}$; (27) $\lim\limits_{x\to 0}\dfrac{\sin 4x}{\sqrt{x+1}-1}$;

(28) $\lim\limits_{x\to 0}(1+\tan x)^{\cot x}$; (29) $\lim\limits_{x\to\infty}\left(\dfrac{3x+2}{3x-1}\right)^{2x-1}$.

7. (1) As $n\to +\infty$, $u_n = 1/n$ and $v_n = 1/n!$ are both infinitesimals. Which is the higher-order infinitesimal?

(2) As $x \to 1$, the functions $\frac{1-x}{1+x}$ and $1-\sqrt{x}$ are both infinitesimals. Are they equivalent infinitesimals?

8. Using equivalent infinitesimal to find the following limits:

(1) $\lim\limits_{x \to 0} \frac{\tan \alpha x}{\sin \beta x} \quad (\beta \neq 0)$; (2) $\lim\limits_{x \to 0} \frac{1-\cos mx}{x^2}$;

(3) $\lim\limits_{x \to 0} \frac{\ln(1+x)}{\sqrt{1+x}-1}$; (4) $\lim\limits_{x \to 0} \frac{\sqrt{2}-\sqrt{1+\cos x}}{\sqrt{1+x^2}-1}$.

9. As $x \to 0$, determine the order of the following infinitesimals compared with x:

(1) x^3+100x^2; (2) $x^{2/3}-x^{1/2}$; (3) $e^{\sqrt{x}}-1$; (4) $e^x-\cos x$; (5) $\cos x-\cos 2x$.

10. Let

$$f(x)=\begin{cases} 2(1-\cos x) & (x<0), \\ x^2+x^3 & (x \geqslant 0). \end{cases} \quad \text{Find } \lim_{x \to 0} \frac{f(x)}{x^2}.$$

11. Judge whether the following statements are correct. Give the proof if it is correct, and set a counter example otherwise.

(1) If $\lim\limits_{x \to x_0} f(x)=A$, then there exists a number $\delta>0$ such that either $f(x)>A$ or $f(x)<A$ if $0<|x-x_0|<\delta$;

(2) If $\lim\limits_{x \to \infty} f(x) \cdot g(x)=0$ and $g(x)$ is bounded as $x \to \infty$, then $\lim\limits_{x \to \infty} f(x)=0$.

(3) If $f(x) \cdot g(x) \to 0$ and $g(x) \to 0$, then $f(x) \to \infty$.

12. Prove by definition that the following functions are continuous on their domains.

(1) $y=\sqrt{x}$; (2) $y=\begin{cases} x^2 \sin \frac{1}{x} & (x \neq 0), \\ 0 & (x=0). \end{cases}$

13. Let

$$f(x)=\begin{cases}\dfrac{x^2-4}{x-2} & (x\neq 2),\\ a & (x=2).\end{cases}$$

Evaluate a such that $f(x)$ is continuous at 2.

14. Locate and classify the discontinuities of the following functions:

(1) $f(x)=\dfrac{\sin x}{x}$;　(2) $f(x)=e^{1/x}$;

(3) $f(x)=\sin\dfrac{1}{x}+e^{1/x}$;　(4) $f(x)=\dfrac{1}{1-e^{1/x}}$;

(5) $f(x)=\dfrac{1}{x-2}$;　(6) $f(x)=(1+x)^{1/x}$;

(7) $f(x)=\dfrac{x^2-1}{x^3-1}$;　(8) $f(x)=x\cos^2\dfrac{1}{x}$;

(9) $f(x)=\dfrac{1}{\sin\pi x}$;　(10) $f(x)=\arctan\dfrac{1}{1+x}$;

(11) $f(x)=\dfrac{1-\cos x}{x^2}$.

15. Evaluate the limits of the following functions:

(1) $\lim\limits_{x\to 7}\dfrac{2-\sqrt{x-3}}{x^2-49}$;　(2) $\lim\limits_{x\to\infty}2x(\sqrt{x^2+1}-x)$;

(3) $\lim\limits_{x\to+\infty}x^{1/3}\cdot[(x+1)^{2/3}-(x-1)^{2/3}]$;　(4) $\lim\limits_{x\to\infty}x(e^{\frac{1}{x}}-1)$;

(5) $\lim\limits_{x\to+\infty}\arcsin(\sqrt{x^2+x}-x)$;　(6) $\lim\limits_{x\to+\infty}\dfrac{x^a-x^{-a}}{x^a+x^{-a}}$;

(7) $\lim\limits_{x\to 0^+}\dfrac{2^{1/x}-1}{2^{1/x}+1}$;　(8) $\lim\limits_{x\to 0^-}\dfrac{2^{1/x}-1}{2^{1/x}+1}$;

(9) $\lim\limits_{x\to+\infty}\dfrac{x\sin x}{x^2-4}$;

(10) $\lim\limits_{x\to 0^+}\sqrt{\dfrac{1}{x}+\sqrt{\dfrac{1}{x}+\sqrt{\dfrac{1}{x}}}}-\sqrt{\dfrac{1}{x}-\sqrt{\dfrac{1}{x}+\sqrt{\dfrac{1}{x}}}}$;

(11) $\lim\limits_{x\to\infty}(\sqrt{x+\sqrt{x+\sqrt{x}}}-\sqrt{x})$.

16. Prove the following statements:

(1) the equation $x2^x = 1$ has at least one root $r \in (0, 1)$;

(2) the equation $x^5 - 3x = 1$ has at least one root $r \in [1, 2]$.

17. Give functions in the following cases respectively:

(1) $f(x)$ is defined on $[0, 2]$ such that it has no limit at any point in $[0, 2]$ except $x=1$;

(2) $g(x)$ satisfies the following conditions: (i) $\lim\limits_{x \to 0} |g(x)| = 1$; (ii) $\lim\limits_{x \to 0^-} g(x) = -1$; (iii) $\lim\limits_{x \to 0^+} g(x)$ is non-existent.

18. Prove the following limits are non-existent using Heine theorem and Cauchy criterion respectively:

(1) $\lim\limits_{x \to 0} \cos \dfrac{1}{x}$;

(2) $\lim\limits_{x \to \infty} \sin x$.

19. Let $f(x) > 0$. If $\lim\limits_{x \to a} f(x) = A$, show that $\lim\limits_{x \to a} \sqrt{f(x)} = \sqrt{A}$.

20. Suppose $f(x)$ is a periodical function and $\lim\limits_{x \to \infty} f(x) = 0$. Prove $f(x) \equiv 0$.

21. Prove:

$$\lim_{n \to +\infty} \left(\cos x \cos \frac{x}{2} \cos \frac{x}{2^2} \cdots \cos \frac{x}{2^n} \right) = \frac{\sin 2x}{2x}.$$

22. Let $x \to 0$. Prove:

(1) $2x - x^2 = O(x)$;

(2) $x \sin \sqrt{x} = O(x^{\frac{3}{2}})$;

(3) $\sqrt{x + \sqrt{x + \sqrt{x}}} \sim \sqrt[8]{x}$;

(4) $(1+x)^n = 1 + nx + o(x)$.

23. Let $x \to \infty$. Prove:

(1) $2x^3 + 3x^2 = O(x^3)$; (2) $\sqrt{x + \sqrt{x + \sqrt{x}}} \sim \sqrt{x}$.

24. Prove that the sum of an infinity and a bounded magnitude is still an infinity.

25. Answer and explain:

(1) Must an infinity be bounded?

(2) Must a bounded magnitude be an infinity?

26. Determine the value of α such that the following functions are equivalent infinitesimals to x^{α} as $x \to 0$:

(1) $\sin 2x - 2\sin x$; (2) $\dfrac{1}{1+x} - (1-x)$;

(3) $\sqrt{1+\tan x} - \sqrt{1-\sin x}$; (4) $\sqrt[5]{3x^2 - 4x^3}$.

27. Determine the value of α such that the following functions are equivalent infinities to x^{α} as $x \to \infty$:

(1) $\sqrt{x^2 + x^5}$; (2) $x + x^2(2 + \sin x)$;

(3) $(1+x)(1+x^2)\cdots(1+x^n)$.

28. Determine the constants a and b such that

(1) $\lim\limits_{x \to \infty}\left(\dfrac{x^2+1}{x+1} - ax - b\right) = 0$;

(2) $\lim\limits_{x \to -\infty}(\sqrt{x^2 - x + 1} - ax - b) = 0$;

(3) $\lim\limits_{x \to \infty}(\sqrt{x^2 - x + 1} - ax - b) = 0$.

29. Give a function $f(x)$ which satisfies the following conditions respectively:

(1) $\lim\limits_{x \to -2} f(x) = 2$;

(2) $\lim\limits_{x \to -2}$ does not exist;

(3) $\lim\limits_{x \to -2} \neq f(-2)$.

30. Give a function such that $f(x) > 0$ for any x, but $\lim\limits_{x \to x_0} f(x) = 0$ for some x_0. Is this a contradiction to the local sign-preserving property of the limit?

31. Suppose $\lim\limits_{x \to a} f(x) = A$ and $\lim\limits_{x \to A} g(x) = B$. Can we infer that $\lim\limits_{x \to a} g(f(x)) = B$? Why?

32. Assume $f(x)$ is defined on $(a, +\infty)$, and for any $b > a$, $f(x)$ *is bounded on* (a, b). If $\lim_{x\to+\infty}(f(x+1)-f(x))=A$, prove that $\lim_{x\to+\infty}\frac{f(x)}{x}=A$.

33. Let $f(x) = x\cos x$. Construct a sequence

(1) $\{x_n\}$ such that $f(x_n)\to 0$ as $x_n \to +\infty$;

(2) $\{y_n\}$ such that $f(y_n) \to +\infty$ as $y_n \to +\infty$;

(3) $\{z_n\}$ such that $f(z_n) \to -\infty$ as $z_n \to +\infty$.

34. Determine the discontinuities and their classes of the following functions:

(1) $f(x) = [|\cos x|]$;

(2) $f(x) = \operatorname{sgn}|x|$;

(3) $f(x) = \operatorname{sgn}(\cos x)$;

(4) $f(x) = \begin{cases} x+\dfrac{1}{x} & (x \neq 0), \\ 0 & (x = 0); \end{cases}$

(5) $f(x) = \begin{cases} \dfrac{\sin x}{|x|} & (x \neq 0), \\ 1 & (x = 0); \end{cases}$

(6) $f(x) = \begin{cases} \dfrac{1}{x+7} & (-\infty < x < -7), \\ x & (-7 \leqslant x \leqslant 1), \\ (x-1)\sin\dfrac{1}{x-1} & (1 < x < +\infty). \end{cases}$

35. Suppose $f(x)$ is continuous. Prove that for any $c > 0$, the following function is also continuous:

$$F(x) = \begin{cases} -c & (\text{if } f(x) < -c), \\ f(x) & (\text{if } |f(x)| \leqslant c), \\ c & (\text{if } f(x) > c). \end{cases}$$

36. If $f(x)$ is continuous at x_0, are $|f(x)|$ and $f^2(x)$ also continuous at x_0? If $|f(x)|$ and $f^2(x)$ are both continuous at x_0, is $f(x)$ *also continuous at* x_0?

37. Assume $f(0) \neq g(0)$ and $f(x) = g(x)$ if $x \neq 0$. Prove that at most one of the functions $f(x)$ and $g(x)$ is continuous at $x = 0$.

38. Suppose $f(x)$ is monotone on an interval I. If $x_0 \in I$ is a discontinuity of $f(x)$, prove that it must be a discontinuity of the first class.

39. Let $f(x)$ and $g(x)$ are both continuous at x_0. Prove:

(1) If $f(x_0) > g(x_0)$, then there exists a δ-neighborhood of x_0, $U(x_0, \delta)$, such that $f(x) > g(x)$ for any $x \in U(x_0, \delta)$;

(2) If $x \neq x_0$ and $f(x) > g(x)$, then $f(x_0) \geqslant g(x_0)$.

40. Discuss the continuity of composite functions $f(g(x))$ and $g(f(x))$ where:

(1) $f(x) = \mathrm{sgn}(x)$, $g(x) = 1 + x^2$;

(2) $f(x) = \mathrm{sgn}(x)$, $g(x) = (1 - x^2)x$.

41. Assume $f(x)$ is continuous on $[a, +\infty)$ and $\lim\limits_{x \to +\infty} f(x)$ exists. Prove that $f(x)$ is bounded on $[a, +\infty)$. Does $f(x)$ have its maximum and minimum on $(a, +\infty)$?

42. If $f(x)$ is continuous on $[a+\varepsilon, b-\varepsilon]$ for any $\varepsilon > 0$, can we infer that $f(x)$ is continuous on (a, b)?

43. Prove: if $f(x)$ is continuous on $[a, b]$ and $f(x) \neq 0$ for any $x \in [a, b]$, then $f(x)$ is identically positive or identically negative on $[a, b]$.

44. (1) Prove: The function $f(x) = \sqrt{x}$ is uniformly continuous on $[0, +\infty)$;

(2) Judge whether the functions $\ln x$ and $\sin \dfrac{1}{x}$ are uniformly continuous on $(0, 1)$.

45. Suppose a function $f(x)$ is continuous on $[a, +\infty)$ and

$\lim\limits_{x\to+\infty} f(x)$ exists. Prove that $f(x)$ must be uniformly continuous on $[a, +\infty)$.

46. Suppose a function $f(x)$ is continuous on $[0, 2a]$ and $f(0) = f(2a)$. Prove that there exists $x \in [0, a]$ such that $f(x) = f(x+a)$.

47. Let $f(x)$ be an increasing function on $[a, b]$ with range $[f(a), f(b)]$. Prove that $f(x)$ is continuous on $[a, b]$.

48. Suppose $f(x)$ is continuous on $[a, b]$ and $a < x_1 < x_2 < \cdots < x_n < b$. Prove that there exists $\xi \in [x_1, x_n]$ such that

$$f(\xi) = \frac{f(x_1) + f(x_2) + \cdots + f(x_n)}{n}.$$

49. Let $f(x) = \sin x$ and

$$g(x) = \begin{cases} x - \pi & (x \leqslant 0), \\ x + \pi & (x > 0). \end{cases}$$

Prove that $f(g(x))$ is continuous at $x = 0$, but $g(x)$ is discontinuous at $x = 0$.

50. Suppose $\lim\limits_{n\to+\infty} x_n = x > 0$ and $\lim\limits_{n\to+\infty} y_n = y$. Prove that $\lim\limits_{n\to+\infty} x_n^{y_n} = x^y$.

51. Let $f(x)$ be a continuous function on $[a, b]$. Prove that the functions

$$m(x) = \min_{a\leqslant\xi\leqslant x}\{f(\xi)\} \text{ and } M(x) = \max_{a\leqslant\xi\leqslant x}\{f(\xi)\}$$

are both continuous on $[a, b]$.

52. Let $f(x)$ and $g(x)$ be continuous functions on $[a, b]$. Prove that the functions

$$m(x) = \min\{f(x), g(x)\} \text{ and } M(x) = \max\{f(x), g(x)\}$$

are both continuous on $[a, b]$.

53. Prove: The equation

$$\frac{a_1}{x_1-\lambda_1}+\frac{a_2}{x_2-\lambda_2}+\frac{a_3}{x_3-\lambda_3}=0$$

(where $a_1, a_2, a_3>0$ and $\lambda_1<\lambda_2<\lambda_3$) has one root on (λ_1, λ_2) and (λ_2, λ_3) respectively.

54. Let $f(x)$ be a continuous function on $[a, b]$. Suppose $a<c<d<b$ and $k=f(c)+f(d)$. Prove:

(1) There exists $\xi \in (a, b)$ such that $k=2f(\xi)$;

(2) There exists $\xi \in (a, b)$ such that $mf(c)+nf(d)=(m+n)f(\xi)$, where $m, n>0$.

55. Let $f(x)$ be a continuous function on $[a, b]$. Prove the following statements:

(1) If $f(r)=0$ for any rational number $r \in [a, b]$, then $f(x) \equiv 0$ on $[a, b]$;

(2) If $f(r_1)<f(r_2)$ for any two rational numbers $r_1, r_2 \in [a, b]$ with $r_1<r_2$, then $f(x)$ is an increasing function on $[a, b]$.

56. Prove: If $f(x)$ and $g(x)$ are both uniformly continuous on interval I, then $f(x) \pm g(x)$ is also uniformly continuous on I.

57. Prove: If $f(x)$ and $g(x)$ are both uniformly continuous and bounded on interval I, then $f(x)g(x)$ is also uniformly continuous and bounded on I.

58. Prove: If $f(x)$ and $g(x)$ are both uniformly continuous on a finite interval I, then $f(x)g(x)$ is also uniformly continuous on I.

59. Suppose $f(x)$ is uniformly continuous on an interval I, and there is a number $\alpha>0$ such that $|f(x)| \geqslant \alpha$ for any $x \in I$. Then $\frac{1}{f(x)}$ is uniformly continuous and bounded on I.

60. Suppose $f(x)$ is uniformly continuous on an interval I, and c is a constant. Then $cf(x)$ is also uniformly continuous on I.

61. Suppose $f(x)$ and $g(x)$ are both uniformly continuous on an interval I, $f(x)$ is bounded on I, and there is a number $\alpha > 0$ such that $|g(x)| \geqslant \alpha$ for any $x \in I$. Then $\dfrac{f(x)}{g(x)}$ is uniformly continuous and bounded on I. (Hint: using Problems 56 and 58)

62. Suppose $f(x)$ and $g(x)$ are both uniformly continuous on a finite interval I, and there is a number $\alpha > 0$ such that $|g(x)| \geqslant \alpha$ for any $x \in I$. Then $\dfrac{f(x)}{g(x)}$ is uniformly continuous on I.

Chapter 4 Derivatives and differentials

Many real-world phenomena involve changing quantities - the speed of a rocket, the inflation of currency, the shock intensity of an earthquake and so forth. This chapter treats the derivative, the central concept of differential calculus, which is a mathematical tool that is used to study rates at which quantities change.

§ 4.1 Concept of derivatives

§ 4.1.1 The origin of the concept of the derivative

Two important different types of problems - the geometrical problem of finding the tangent line to a curve at a given point, and the physical problem of finding the instantaneous velocity of a moving particle-both lead quite naturally to the notion of derivative.

(1) The problem of the tangent line of a curve

From plane geometry, we know that a tangent to a circle is a line that passes through one and only one point on the circle, but how do we define and find a tangent line to a graph of a function at a point? The concept of the slope of a straight line will play a central role in the process. If we pass a straight line through two points $P_1 = (x_1, y_1)$ and $P_2 = (x_2, y_2)$ on the graph of $y = f(x)$, as in Figure 4 - 1, we obtain a secant line. Given the coordinates of the two points, we can find the slope of the secant line using the point-slope formula:

$$\text{Secant line slope} = \frac{y_2 - y_1}{x_2 - x_1} = \frac{f(x_2) - f(x_1)}{x_2 - x_1} = \frac{f(x_1 + \Delta x) - f(x_1)}{\Delta x}.$$

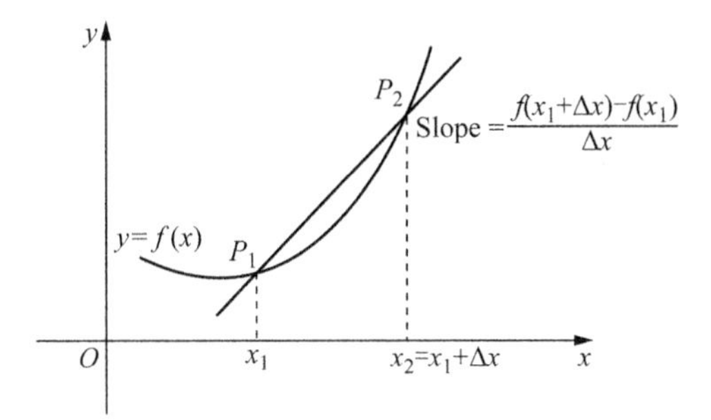

Figure 4 - 1

As we let Δx tend to 0, P_2 will approach P_1, and this appears that the secant lines will approach a limiting position and the secant slopes will approach a limiting value (cf. Figure 4 - 2).

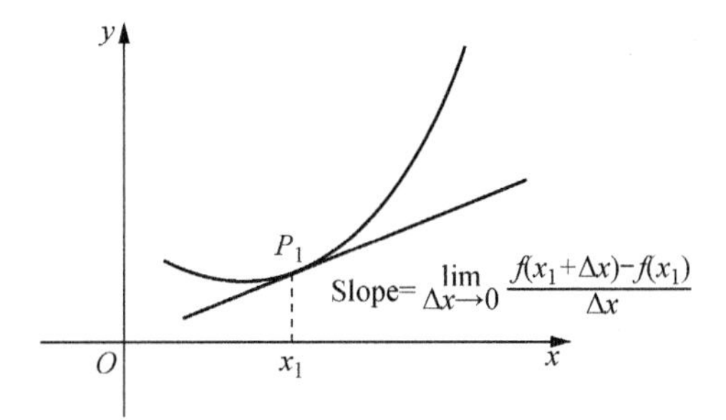

Figure 4 - 2

If they do, then we will call the line that the secant lines approach the *tangent line* to the graph at $(x_1, f(x_1))$, and the limiting slope will be the slope of the tangent line. This leads to the following definition of a tangent line:

Definition Given the graph of $y = f(x)$, then the *tangent line* at $(x_1, f(x_1))$ is the line that passes through this point with slope

$$k = \lim_{\Delta x \to 0} \frac{f(x_1 + \Delta x) - f(x_1)}{\Delta x}$$

if the limit exists.

(2) The problem of the velocity

How can we measure the velocity of a moving object at a given instant in time? Or, more fundamentally, what do we mean by the term *velocity*?

Suppose $s(t)$ is the position of an object at time t, then the *average velocity* of the object over the interval $a \leqslant t \leqslant b$ is

$$\text{Average velocity} = \frac{\text{Change in position}}{\text{Change in time}} = \frac{s(b)-s(a)}{b-a}.$$

in words, the *average velocity* of the object over an interval is the net change in position during the interval divided by the change in time. The average velocity is a useful concept since it gives a rough idea of the behavior of the object. But the average velocity over an interval doesn't solve the problem of measuring the velocity of the object at exactly $t = t_0 \in (a, b)$. To get closer to an answer to that problem, we have to look at what happen near $t = t_0$ in more detail, i. e. the average velocity over small interval on either side of $t = t_0$. We are convinced the smaller and smaller intervals will provide average velocities that come arbitrarily close to the velocity at $t = t_0$. This process is just *taking the limit*. This is written more compactly using limit notation as follows:

Definition Let $s(t)$ give the position of an object at time t. Then the *instantaneous velocity* at $t = t_0$, denoted by $v(t_0)$, is

$$v(t_0) = \lim_{h \to 0} \frac{s(t_0 + h) - s(t_0)}{h}.$$

In words, the Instantaneous velocity of an object at time $t = t_0$ is given by the limit of the average velocity over an interval, as the interval shrinks around t_0.

§ 4.1.2 The definition of the derivative

We can also define the *instantaneous rate of change* of a function at a point in the same way as a special limit that we define the slope of a tangent line and the instantaneous velocity: we look at the average rate of change over smaller and smaller intervals. This instantaneous rate of change is so important that it is given its own name, the derivative of the function at a point.

Suppose f is defined on an open interval (a, b), then for two distinct points x and x_0 in (a, b) we can form the difference quotient

$$\frac{f(x) - f(x_0)}{x - x_0}.$$

We keep x_0 fixed and study the behavior of this quotient as $x \to x_0$.

Definition Let $y = f(x)$ be defined on an open interval (a, b), and assume that $x_0 \in (a, b)$. Then f is said to be *derivable* at x_0 whenever the limit

$$\lim_{x \to x_0} \frac{f(x) - f(x_0)}{x - x_0}$$

exists. If f is not derivable at x_0, then f is said to be *underivable* at x_0. The limit, denoted by $f'(x_0)$, is called the *derivative* of f at x_0. The calculation of a derivative is called *differentiation* or *derivation.*

Let $\Delta x = x - x_0$ (the increment of the independent variable at x_0) and $\Delta y = f(x) - f(x_0)$ (the increment of the function at x_0). The definition of the derivative can also be expressed as

$$f'(x_0) = \left.\frac{\mathrm{d}y}{\mathrm{d}x}\right|_{x=x_0} = \lim_{\Delta x \to 0} \frac{f(x_0 + \Delta x) - f(x_0)}{\Delta x} = \lim_{\Delta x \to 0} \frac{\Delta y}{\Delta x}.$$

Examples (1) Find the derivative of $y = 1/x$ at $x = x_0 (\neq 0)$ by definition:

$$\text{As } \Delta y = \frac{1}{x_0 + \Delta x} - \frac{1}{x_0} = \frac{-\Delta x}{x_0(x_0 + \Delta x)},$$

$$\frac{\Delta y}{\Delta x} = -\frac{1}{x_0(x_0 + \Delta x)}.$$

So,

$$f'(x_0) = \lim_{\Delta x \to 0} \frac{\Delta y}{\Delta x} = -\frac{1}{x_0^2}.$$

(2) Find the derivative of $y = x^3$ at $x = 1$ by definition:

$$f'(1) = \lim_{\Delta x \to 0} \frac{(1 + \Delta x)^3 - 1^3}{\Delta x} = \lim_{\Delta x \to 0} \frac{1 + 3\Delta x + 3\Delta x^2 + \Delta x^3 - 1}{\Delta x}$$
$$= \lim_{\Delta x \to 0}(3 + 3\Delta x + \Delta x^2) = 3.$$

(3) Show that the derivative of the constant function $y = c$ at any point x_0 is 0:

$$f'(0) = \lim_{x \to x_0} \frac{f(x) - f(x_0)}{x - x_0} = \lim_{x \to x_0} \frac{c - c}{x - x_0} = \lim_{x \to x_0} 0 = 0.$$

(4) Suppose $\psi(x) = f(\varphi(x))$ where $\varphi(x) = \begin{cases} x^2 \sin x & (x \neq 0), \\ 0 & (x = 0), \end{cases}$

and $f'(0)$ exists. Find $\psi'(0)$:

$$\psi'(0) = \lim_{x \to 0} \frac{\psi(x) - \psi(0)}{x} = \lim_{x \to 0} \frac{f(\varphi(x)) - f(\varphi(0))}{x}$$
$$= \lim_{x \to 0} \frac{f(x^2 \sin x) - f(0)}{x}$$
$$= \lim_{x \to 0}\left(\frac{f(x^2 \sin x) - f(0)}{x^2 \sin x} \cdot x \sin x\right)$$
$$= f'(0) \cdot 0 = 0.$$

(5) Show that the function $f(x) = |x|$ is underivable at $x = 0$:

Since

$$\frac{\Delta y}{\Delta x}=\frac{f(0+\Delta x)-f(0)}{\Delta x}=\frac{|\Delta x|}{\Delta x}=\begin{cases}1 & (\Delta x>0),\\ -1 & (\Delta x<0),\end{cases}$$

$$\lim_{\Delta x\to 0^+}\frac{\Delta y}{\Delta x}\neq\lim_{\Delta x\to 0^-}\frac{\Delta y}{\Delta x},$$

and so $\lim\limits_{\Delta x\to 0}(\Delta y/\Delta x)$ does not exist.

(6) Suppose $f(x)$ is derivable at a. Find

$$I=\lim_{h\to 0}\frac{f(a+h^4)-f(a)}{1-\cos(h^2)};$$

$$\begin{aligned}I&=\lim_{h\to 0}\left[\frac{f(a+h^4)-f(a)}{h^4}\cdot\frac{h^4}{2\sin^2\left(\frac{h^2}{2}\right)}\right]\\&=\lim_{h\to 0}\frac{f(a+h^4)-f(a)}{h^4}\cdot\frac{h^4}{2\left(\frac{h^2}{2}\right)^2}\\&=f'(a)\cdot 2=2f'(a).\end{aligned}$$

Remark By the definition of the derivative, if we let

$$\frac{\Delta y}{\Delta x}-f'(x_0)=\varepsilon,$$

then ε is an infinitesimal as $\Delta x\to 0$ and so $\varepsilon\Delta x=o(\Delta x)$ as $\Delta x\to 0$. Hence, we have

$$\Delta y=f'(x_0)\Delta x+o(\Delta x),$$

which is called *the finite increment formula*.

Example Suppose $f(x)$ is a function, $f(0)=0$ and $f'(0)$ exists. Let

$$x_n=f\left(\frac{1}{n^2}\right)+f\left(\frac{2}{n^2}\right)+\cdots+f\left(\frac{n}{n^2}\right),$$

where n is a positive integer. Find $\lim\limits_{n\to+\infty}x_n$.

Solution By the finite increment formula,

$$\frac{f(x_0+\Delta x)-f(x_0)}{\Delta x}=f'(x_0)+\frac{o(\Delta x)}{\Delta x}\quad (\Delta x\to 0).$$

Then,

$$\frac{f\left(\frac{k}{n^2}\right)-f(0)}{\frac{k}{n^2}}=f'(0)+\alpha_k\quad (k=1,2,\cdots,n;\ \alpha_k\to 0 \text{ as } n\to+\infty).$$

Hence,

$$\begin{aligned}
x_n&=\frac{\left(f\left(\frac{1}{n^2}\right)-f(0)\right)+\left(f\left(\frac{2}{n^2}\right)-f(0)\right)+\cdots+\left(f\left(\frac{n}{n^2}\right)-f(0)\right)}{\frac{1}{n^2}}\cdot\frac{1}{n^2}\\
&=\frac{(f'(0)+\alpha_1)+2(f'(0)+\alpha_2)+\cdots+n(f'(0)+\alpha_n)}{n^2}\\
&=\frac{(1+2+\cdots+n)f'(0)}{n^2}+\frac{\alpha_1+2\alpha_2+\cdots+n\alpha_n}{n^2}\\
&=\frac{\frac{1}{2}n(1+n)}{n^2}f'(0)+\frac{\alpha_1+2\alpha_2+\cdots+n\alpha_n}{n^2}.
\end{aligned}$$

By Stolz theorem,

$$\begin{aligned}
\lim_{n\to+\infty}\frac{\alpha_1+2\alpha_2+\cdots+n\alpha_n}{n^2}&=\lim_{n\to+\infty}\frac{n\alpha_n}{n^2-(n-1)^2}\\
&=\lim_{n\to+\infty}\frac{n}{2n-1}\alpha_n=0,
\end{aligned}$$

and so $\lim\limits_{n\to+\infty}x_n=\frac{1}{2}f'(0)$.

By the finite increment formula, we can reveal the relationship between the derivability and continuity of a function as follows:

Theorem 4. 1. 1 If $f(x)$ is derivable at x_0, then $f(x)$ is continuous at x_0.

Proof By the finite increment formula,

$$\lim_{\Delta x \to 0} \Delta y = \lim_{\Delta x \to 0} (f'(x_0)\Delta x + o(\Delta x)) = 0,$$

which implies $f(x)$ is continuous at x_0. □

Note The converse statement is not true, i. e. if $f(x)$ is continuous at x_0, then it is not necessarily true that $f(x)$ is derivable at x_0. For example, the function $y = |x|$ is underivable at $x = 0$. However, it is continuous at $x = 0$.

A more general result concerning the relationship between derivability and continuity is given as follows:

Theorem 4. 1. 2 If f is defined on (a, b) and derivable at $x_0 \in (a, b)$, then there is a function $g(x)$ (depending on f and x_0) which is continuous at x_0 and which satisfies the equation

$$f(x) - f(x_0) = (x - x_0)g(x) \qquad (*)$$

for any $x \in (a, b)$, and $g(x_0) = f'(x_0)$.

Conversely, if there is a function g, continuous at x_0, which satisfies the above equation $(*)$, then f is derivable at x_0 and $f'(x_0) = g(x_0)$.

Proof If $f'(x_0)$ exists, let g be defined on (a, b) as follows:

$$g(x) = \begin{cases} \dfrac{f(x) - f(x_0)}{x - x_0} & (\text{if } x \neq x_0), \\ f'(x) & (\text{if } x = x_0), \end{cases}$$

then g is continuous at x_0 and equation $(*)$ holds for any $x \in (a, b)$

Conversely, dividing both sides of the equality $(*)$ by $x - x_0$ and letting $x \to x_0$, we see that $f'(x_0)$ exists and equals $g(x_0)$. □

Note. Theorem 4. 1. 2 makes it possible to reduce Theorem 4. 1. 1 as a special case just by letting $x \to x_0$ in the equality $(*)$.

§ 4. 1. 3 One-sided derivatives and improper derivatives

Up to this point, the statement that f has a derivative at a point

x_0 has meant that f was defined on some neighborhood of x_0 and that the limit defining $f'(x_0)$ was finite. It is convenient to extend the scope of our ideas somewhat in order to discuss derivatives at endpoints of intervals on which $f(x)$ is defined. It is also desirable to introduce infinite derivatives, so that the usual geometric interpretation of a derivative as the slope of a tangent line will still be valid in case the tangent line happens to be vertical. In such a case we cannot prove that f is continuous at x_0. Therefore, we explicitly require it to be so.

Definition Let f be defined on a closed interval S and assume that f continuous at the point x_0 in S. Then f is said to have a *right-hand derivative* at x_0 if the right-hand limit

$$\lim_{x \to x_0^+} \frac{f(x) - f(x_0)}{x - x_0}$$

exists as a finite value. This limit is denoted by $f'_+(x_0)$; *Left-hand derivatives*, denoted by $f'_-(x_0)$, are similarly defined, i. e.

$$f'_-(x_0) = \lim_{x \to x_0^-} \frac{f(x) - f(x_0)}{x - x_0}.$$

The right-hand derivative and the left-hand derivative are also called *one-sided derivative*.

Definition Let $y = f(x)$ be a function. If

$$\lim_{x \to x_0} \frac{f(x) - f(x_0)}{x - x_0} = (\pm)\infty,$$

then we say the function is *improperly derivable* at $x = x_0$, and its *improper derivative* at $x = x_0$ is $(\pm)\infty$, denoted by $f'(x_0) = (\pm)\infty$. One-sided improper derivative is similarly defined.

For example, let $y = \sqrt[3]{x}$. Since

$$\lim_{x\to 0}\frac{f(x)-f(0)}{x-0}=\lim_{x\to 0}\frac{\sqrt[3]{x}}{x}=\lim_{x\to 0}\frac{1}{(\sqrt[3]{x})^2}=+\infty,$$

the function has an improper derivative at $x=0$.

Analogous to the case of the limit and the one-sided limit, the relationship between the derivative and the one-sided derivative is as follows:

Theorem 4.1.3 Suppose $f(x)$ is defined on a neighborhood $U(x_0)$. Then f has a derivative at x_0 if and only if $f'_+(x_0)$ and $f'_-(x_0)$ exist, and

$$f'_+(x_0)=f'_-(x_0).$$

In this case, $f'_+(x_0)=f'_-(x_0)=f'(x_0)$.

Examples (1) Discuss the derivative of $f(x)=x^2\operatorname{sgn} x$ at $x=0$: Notice that

$$f(x)=x^2\operatorname{sgn} x=\begin{cases} x^2 & (x\geqslant 0),\\ -x^2 & (x<0),\end{cases}$$

and $f'_+(0)=\lim\limits_{x\to 0^+}\dfrac{x^2-0}{x-0}=0$, $f'_-(0)=\lim\limits_{x\to 0^-}\dfrac{-x^2-0}{x-0}=0$.

By the above theorem, $f'(0)=0$.

(2) Let

$$f(x)=\begin{cases} 1+\sin 2x & (x\leqslant 0),\\ a+be^x & (x>0).\end{cases}$$

Determine the numbers a and b such that $f(x)$ is derivable at $x=0$, and find $f'(0)$ in this case.

Solution Notice that

$$\lim_{x\to 0^-}f(x)=\lim_{x\to 0^-}(1+\sin 2x)=1(=f(0)),$$

$$\lim_{x\to 0^+}f(x)=\lim_{x\to 0^+}(a+be^x)=a+b.$$

If $f(x)$ is derivable at $x=0$, $f(x)$ is continuous at $x=0$, and so $a+b=1$ must be satisfied. Furthermore,

$$f'_{-}(0)=\lim_{\Delta x\to 0^{-}}\frac{f(\Delta x)-f(0)}{\Delta x}=\lim_{\Delta x\to 0^{-}}\frac{\sin(2\Delta x)}{\Delta x}=2;$$

$$f'_{+}(0)=\lim_{\Delta x\to 0^{+}}\frac{f(\Delta x)-f(0)}{\Delta x}=\lim_{\Delta x\to 0^{+}}\frac{(a+be^{\Delta x})-1}{\Delta x}$$

$$=\lim_{\Delta x\to 0^{+}}\frac{b(e^{\Delta x}-1)}{\Delta x}=b.$$

Hence, only if $b=2$, $f'_{-}(0)=f'_{+}(0)$. In summary, as $a=-1$ and $b=2$, $f(x)$ is derivable at $x=0$ and in this case, $f'(0)=2$.

(3) Suppose $f(x)$ is continuous on $[a, b]$, $f(a)=f(b)=0$ and $f'_{+}(a)f'_{-}(b)>0$. Prove that there exists $x_0\in(a, b)$ such that $f(x_0)=0$.

Proof The condition $f'_{+}(a)f'_{-}(b)>0$ implies that $f'_{+}(a)$ and $f'_{-}(b)$ have a same sign, i. e. $f'_{+}(a)>0$ and $f'_{-}(b)>0$ or $f'_{+}(a)<0$ and $f'_{-}(b)<0$. We only assume the first case; the statement can be similarly proved for the second case.

Suppose $f(x)\neq 0$ for any $x\in(a, b)$. So by the continuity of $f(x)$, $f(x)>0$ for any $x\in(a, b)$ or $f(x)<0$ for any $x\in(a, b)$. If $f(x)<0$, $\dfrac{f(x)-f(a)}{x-a}=\dfrac{f(x)}{x-a}<0$ $(x\in(a, b))$, and so

$f'_{+}(a)=\lim\limits_{x\to a^{+}}\dfrac{f(x)-f(a)}{x-a}\leqslant 0$, which contradicts $f'_{+}(a)>0$; If

$f(x)>0$, $\dfrac{f(x)-f(b)}{x-b}=\dfrac{f(x)}{x-b}<0$ $(x\in(a, b))$, and so $f'_{-}(b)=$

$\lim\limits_{x\to b^{-}}\dfrac{f(x)-f(b)}{x-b}\leqslant 0$, which contradicts $f'_{-}(b)>0$. □

It is still possible to reduce the continuity from the existence of one-sided derivatives.

Theorem 4.1.4 If both of the right-hand and left-hand derivatives of $f(x)$ at x_0 are existent, then $f(x)$ is continuous at x_0. In particular, if $f(x)$ is derivable at x_0, then $f(x)$ is continuous at x_0.

Proof Note that

$$\begin{aligned}\lim_{x \to x_0^+}(f(x) - f(x_0)) &= \lim_{x \to x_0^+} \frac{f(x) - f(x_0)}{x - x_0}(x - x_0) \\ &= \lim_{x \to x_0^+} \frac{f(x) - f(x_0)}{x - x_0} \lim_{x \to x_0^+}(x - x_0) \\ &= f'_+(x_0) \cdot 0 = 0.\end{aligned}$$

Similarly, we have $\lim_{x \to x_0^+}(f(x) - f(x_0)) = 0$. Thus $f(x)$ is continuous at x_0 both from the right and from the left. Thus $f(x)$ is continuous at x_0. In particular, if $f(x)$ is derivable at x_0, then both of the right-hand and left-hand derivatives of $f(x)$ at x_0 must be existent, and so $f(x)$ is continuous at x_0. □

§ 4. 1. 4 Derivative functions

If a function $y = f(x)$ is derivable at any $x \in (a, b)$, then we say f is *derivable on* (a, b) or f is a *derivable function on* (a, b). In fact the limit process in the definition of derivative defines a new function f', whose domain consists of those points in (a, b) at which f is derivable. The function f' is also called the *derivative function* (or just the *derivative*) of f, i. e.

$$f'(x) = \lim_{\Delta x \to 0} \frac{f(x + \Delta x) - f(x)}{\Delta x}, \quad x \in (a, b).$$

Note that in this limit process, Δx is a variable while x is regarded as a constant.

Other common notations of the derivative are

$$y', \frac{\mathrm{d}y(x)}{\mathrm{d}x}, \frac{\mathrm{d}f(x)}{\mathrm{d}x}, \frac{\mathrm{d}y}{\mathrm{d}x}, \frac{\mathrm{d}f}{\mathrm{d}x}, \text{ etc.}$$

Remark If a function f is derivable at $x = x_0$, there does not necessarily exist a neighborhood of x_0, $U(x_0)$, such that f is derivable on $U(x_0)$. For example, let

$$f(x) = \begin{cases} 0 & (\text{if } x \text{ is a rational number}), \\ x^2 & (\text{if } x \text{ is an irrational number}). \end{cases}$$

Then f is derivable at $x = 0$. In fact,

$$0 \leqslant \left| \frac{f(\Delta x) - f(0)}{\Delta x} \right| \leqslant \left| \frac{(\Delta x)^2}{\Delta x} \right| = |\Delta x| \to 0 \text{ as } \Delta x \to 0,$$

and so by squeezing law,

$$f'(0) = \lim_{\Delta x \to 0} \frac{f(\Delta x) - f(0)}{\Delta x} = 0.$$

However, at any $x \neq 0$, $f(x)$ is discontinuous (cf. Example in Section 3. 5. 3), and of course it is underivable.

Examples (1) Find the derivative of $y = x^{1/2}$ by definition:

$$\frac{dy}{dx} = \lim_{\Delta x \to 0} \frac{f(x + \Delta x) - f(x)}{\Delta x} = \lim_{\Delta x \to 0} \frac{(x + \Delta x)^{1/2} - x^{1/2}}{\Delta x}$$
$$= \lim_{\Delta x \to 0} \frac{1}{(x + \Delta x)^{1/2} + x^{1/2}} = \frac{1}{2\sqrt{x}}.$$

(2) Find the derivative of $y=1/x$ by definition:

$$\frac{dy}{dx} = \lim_{\Delta x \to 0} \frac{1}{\Delta x}\left(\frac{1}{x + \Delta x} - \frac{1}{x}\right) = \lim_{\Delta x \to 0} \frac{-1}{(x + \Delta x)x} = -\frac{1}{x^2}.$$

(3) Find the derivative of $y = x^n$ (n is positive integer) by definition:

$$\frac{dy}{dx} = \lim_{\Delta x \to 0} \frac{(x + \Delta x)^n - x^n}{\Delta x}$$
$$= \frac{(x^n + n\Delta x \cdot x^{n-1} + \cdots + (\Delta x)^n) - x^n}{\Delta x} = nx^{n-1}.$$

(4) Find the derivatives of $y=\cos kx$ and $y=\sin kx$ by definition:

$$\frac{\mathrm{d}y}{\mathrm{d}x}=\lim_{\Delta x\to 0}\frac{\cos k(x+\Delta x)-\cos kx}{\Delta x}$$

$$=\lim_{\Delta x\to 0}\frac{1}{\Delta x}[(\cos kx\ \cos k\cdot\Delta x-\sin kx\ \sin k\cdot\Delta x)-\cos kx]$$

$$=\cos kx\cdot\lim_{\Delta x\to 0}\frac{\cos k\Delta x-1}{\Delta x}-\sin kx\cdot\lim_{\Delta x\to 0}\frac{\sin k\Delta x}{\Delta x}$$

$$=\cos kx\cdot 0-\sin kx\cdot k=-k\sin kx.$$

Similarly, we can derive $\frac{\mathrm{d}}{\mathrm{d}x}(\sin kx)=k\cos kx$. In particular,

$$(\sin x)'=\cos x \text{ and } (\cos x)'=-\sin x.$$

§ 4.1.5 Geometric interpretation of derivatives

As we see in the beginning of this chapter, the derivative of a function f at $x=x_0$ represents the slope of the tangent line of the curve $y=f(x)$ at the point $(x_0, f(x_0))$. For clarity, we restate it in the version of Δx and Δy as follows:

Let $y=f(x)$ be a function whose graph is a smooth curve like that shown in Figure 4-3. If P denotes point (x_0, y_0) and Q is $(x_0+\Delta x, y_0+\Delta y)$, where

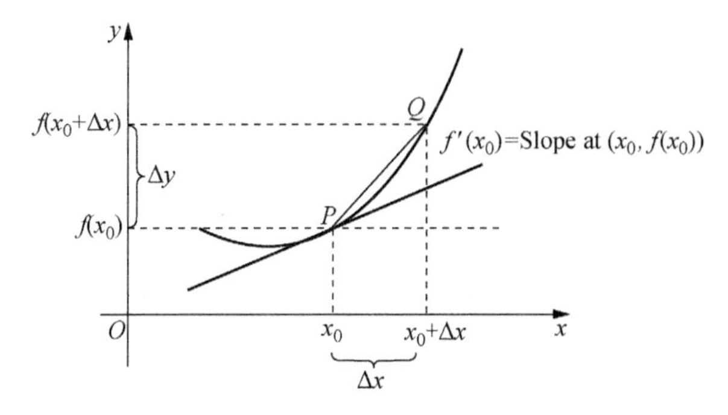

Figure 4-3

$$\Delta y = f(x_0 + \Delta x) - f(x_0),$$

then the slope of the tangent line of $f(x)$ at P is determined from the limit, as $\Delta x \to 0$, of the ratio

$$\frac{\Delta y}{\Delta x} = \frac{f(x_0 + \Delta x) - f(x_0)}{\Delta x}.$$

The slope of the tangent line and the derivative of the function are one and the same at the point in question.

Let the angle between this tangent line and the positive direction of x-axis be α. Then $f'(x_0) = \tan \alpha$. Thus, if $f'(x_0) > 0$, α is an acute angle; if $f'(x_0) < 0$, α is an obtuse angle; if $f'(x_0) = 0$, the tangent line is parallel to the x-axis (cf. Figure 4 - 4).

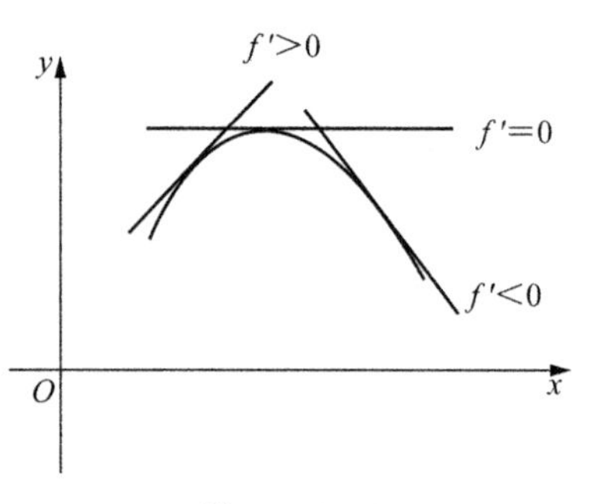

Figure 4 - 4

The geometric interpretation of derivative helps us gain insight into its meaning, by which we may derive immediately the equations of the tangent line and the normal line of a curve at some point. In fact, the equations of the tangent line and the normal line of a curve $y = f(x)$ at the point $(x_0, f(x_0))$ are respectively:

$$y - f(x_0) = f'(x_0)(x - x_0);$$

$$y - f(x_0) = -\frac{1}{f'(x_0)}(x - x_0)(f'(x_0) \neq 0).$$

If $f'(x_0) = 0$, then the equations of the tangent line and the normal line are $y = f(x_0)$ and $x = x_0$ respectively.

Example Let C be a curve with equation $y = \ln x$. Find the equations of the tangent line and the normal line of C at the point

(x_0, y_0).

Solution Since $(\ln x)'|_{x=x_0}=1/x_0$, the equations of the tangent line and the normal line at (x_0, y_0) are respectively

$$y-y_0=\frac{1}{x_0}(x-x_0);\ y-y_0=-x_0(x-x_0).$$

According to the derivative derived, to construct the tangent line of the curve C at the point $P(x_0, y_0)$, we only need to fix the point $Q(x_0, 1)$ on the straight line connecting P and the point $(x_0, 0)$, then draw a line through P parallel to the line segment OQ. This is exactly the tangent line required (cf. Figure 4 - 5).

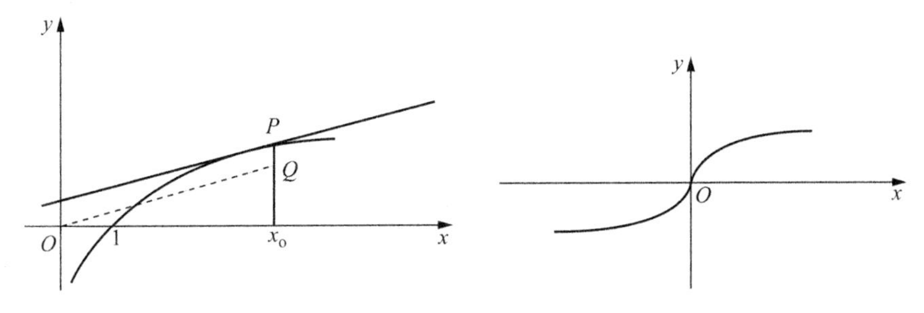

Figure 4 - 5　　Figure 4 - 6

If $f(x)$ is improperly derivable at $x=x_0$, i. e. $f'(x_0)=(\pm)\infty$, geometrically the curve $y=f(x)$ has a tangent line at the point $(x_0, f(x_0))$ which is perpendicular to the x-axis. Recall the example $y=\sqrt[3]{x}$ in the section 4. 1. 3. Since $y'(0)=+\infty$, the curve $y=\sqrt[3]{x}$ has a tangent line at the point $(0, 0)$, which is perpendicular to the x-axis (cf. Figure 4 - 6).

§ 4. 2 Computation of derivatives

This section is concerned almost exclusively with the development of technique and analytical skill. The differentiation of complicated functions is greatly simplified by using general formulas.

§ 4. 2. 1 Fundamental operations for derivation

We now develop some important theorems that will enable us to calculate derivatives more efficiently.

Theorem 4. 2. 1 If $u(x)$ and $v(x)$ are derivable functions and c is any constant, then $u \pm v$, uv, cu and $\frac{u}{v}$ $(v(x) \neq 0)$ are also derivable and

(1) (addition rule) $\frac{\mathrm{d}}{\mathrm{d}x}(u \pm v) = \frac{\mathrm{d}u}{\mathrm{d}x} \pm \frac{\mathrm{d}v}{\mathrm{d}x}$, i. e. $(u \pm v)' = u' \pm v'$

(2) (constant rule) $\frac{\mathrm{d}}{\mathrm{d}x}(cu) = c\frac{\mathrm{d}u}{\mathrm{d}x}$, i. e. $(cu)' = cu'$

(3) (product rule) $\frac{\mathrm{d}}{\mathrm{d}x}(uv) = u\frac{\mathrm{d}v}{\mathrm{d}x} + v\frac{\mathrm{d}u}{\mathrm{d}x}$, i. e. $(uv)' = uv' + u'v$

(4) (quotient rule) $\frac{\mathrm{d}}{\mathrm{d}x}\left(\frac{u}{v}\right) = \frac{1}{v^2}\left(v\frac{\mathrm{d}u}{\mathrm{d}x} - u\frac{\mathrm{d}v}{\mathrm{d}x}\right)$ $(v(x) \neq 0)$,

i. e. $\left(\frac{u}{v}\right)' = \frac{vu' - uv'}{v^2}$

(5) (reciprocal rule) $\frac{\mathrm{d}}{\mathrm{d}x}\left(\frac{1}{v}\right) = -\frac{1}{v^2}\frac{\mathrm{d}v}{\mathrm{d}x}$, i. e. $\left(\frac{1}{v}\right)' = \frac{-v'}{v^2}$ $(v(x) \neq 0)$

Proof (1) and (2). The first two results are simple operations in manipulating limits.

(3) To prove the product rule requires some algebraic rearrangement:

$$\frac{\mathrm{d}}{\mathrm{d}x}(u(x)v(x)) = \lim_{h \to 0}\frac{u(x+h)v(x+h) - u(x)v(x)}{h}$$

$$= \lim_{h \to 0}\left[u(x+h)\left(\frac{v(x+h) - v(x)}{h}\right) + v(x)\left(\frac{u(x+h) - u(x)}{h}\right)\right]$$

$$= u(x)\frac{\mathrm{d}v}{\mathrm{d}x} + v(x)\frac{\mathrm{d}u}{\mathrm{d}x}.$$

(4) The quotient rule can be proved by evaluating a limit or, more simply, by a clever application of the product rule. Since the

function u is the product of the functions v and u/v, we have

$$\frac{\mathrm{d}u}{\mathrm{d}x}=\frac{\mathrm{d}}{\mathrm{d}x}\left(v\frac{u}{v}\right)=v\frac{\mathrm{d}}{\mathrm{d}x}\left(\frac{u}{v}\right)+\frac{u}{v}\frac{\mathrm{d}v}{\mathrm{d}x}.$$

Solving for the derivative of the quotient,

$$\frac{\mathrm{d}}{\mathrm{d}x}\left(\frac{u}{v}\right)=\frac{1}{v^2}\left(v\frac{\mathrm{d}u}{\mathrm{d}x}-u\frac{\mathrm{d}v}{\mathrm{d}x}\right).$$

This formula is valid if $v(x)\neq 0$.

(5) follows directly from (4) as setting $u=1$. □

The product rule can be easily generalized to the product of a finite number of functions, e. g.

$$(uvw)'=u'vw+uv'w+uvw'.$$

These rules allow the derivatives of many functions to be calculated with very little efforts.

Examples (1) Differentiate $y=x^4\cos x+5x^3\sin x$:

$$\begin{aligned}\frac{\mathrm{d}y}{\mathrm{d}x}&=\frac{\mathrm{d}}{\mathrm{d}x}(x^4\cos x)+\frac{\mathrm{d}}{\mathrm{d}x}(5x^3\sin x)\\&=\frac{\mathrm{d}}{\mathrm{d}x}(x^4)\cos x+x^4\frac{\mathrm{d}}{\mathrm{d}x}(\cos x)+\frac{\mathrm{d}}{\mathrm{d}x}(5x^3)\sin x+5x^3\frac{\mathrm{d}}{\mathrm{d}x}(\sin x)\\&=4x^3\cos x-x^4\sin x+15x^2\sin x+5x^3\cos x.\end{aligned}$$

(2) Show that $(\tan x)'=\sec^2 x$:

$$\begin{aligned}\frac{\mathrm{d}}{\mathrm{d}x}\tan x&=\frac{\mathrm{d}}{\mathrm{d}x}\left(\frac{\sin x}{\cos x}\right)=\frac{\cos x\frac{\mathrm{d}}{\mathrm{d}x}(\sin x)-\sin x\frac{\mathrm{d}}{\mathrm{d}x}(\cos x)}{\cos^2 x}\\&=\frac{\cos x\cos x-\sin x(-\sin x)}{\cos^2 x}=\frac{1}{\cos^2 x}=\sec^2 x.\end{aligned}$$

§ 4. 2. 2 The chain rule

We now derive a formula, called *chain rule*, that expresses the derivative of a composition $u \circ v$ in terms of the derivatives of functions u and v. This formula will enable us to differentiate complicated functions using known derivatives of simpler functions.

Theorem 4. 2. 2 (chain rule) Suppose the composite function $u(v(x))$ is composed of functions $u(v)$ and $v(x)$. If $v(x)$ is derivable at x and $u(v)$ is derivable at $v=v(x)$, then $u(v(x))$ is derivable at x with

$$\frac{\mathrm{d}}{\mathrm{d}x}(u(v(x))) = \frac{\mathrm{d}u(v)}{\mathrm{d}v}\frac{\mathrm{d}v(x)}{\mathrm{d}x}.$$

Proof In the function $u(v(x))$, u is a function of v while v is a function of x. We know that

$$\frac{\mathrm{d}u}{\mathrm{d}v} = \lim_{\Delta v \to 0}\frac{u(v+\Delta v)-u(v)}{\Delta v}$$

and

$$\frac{\mathrm{d}v}{\mathrm{d}x} = \lim_{h \to 0}\frac{v(x+h)-v(x)}{h}.$$

From the definition, the derivative of $u(v(x))$ is

$$\begin{aligned}
\frac{\mathrm{d}}{\mathrm{d}x}u(v(x)) &= \lim_{h\to 0}\frac{u(v(x+h))-u(v(x))}{h} \\
&= \lim_{h\to 0}\left(\frac{u(v(x+h))-u(v(x))}{v(x+h)-v(x)}\cdot\frac{v(x+h)-v(x)}{h}\right) \\
&= \lim_{\Delta v\to 0}\frac{u(v+\Delta v)-u(v)}{\Delta v}\cdot\lim_{h\to 0}\frac{v(x+h)-v(x)}{h} \\
&= \frac{\mathrm{d}u}{\mathrm{d}v}\frac{\mathrm{d}v}{\mathrm{d}x},
\end{aligned}$$

where we have used the identity $\Delta v = v(x+h)-v(x)$ and the fact that, as h approaches 0, Δv also approaches 0. □

Remark (1) Let $u=u(v)$ and $v=v(x)$. Then the chain rule can be simplified as

$$\frac{\mathrm{d}u}{\mathrm{d}x} = \frac{\mathrm{d}u}{\mathrm{d}v}\frac{\mathrm{d}v}{\mathrm{d}x}.$$

(2) If a function is a composition of more than two functions, the chain rule can be similarly presented, e. g. let $u=u(v)$, $v=v(w)$ and $w=w(t)$, then

$$\frac{\mathrm{d}u}{\mathrm{d}t} = \frac{\mathrm{d}u}{\mathrm{d}v}\frac{\mathrm{d}v}{\mathrm{d}w}\frac{\mathrm{d}w}{\mathrm{d}t}.$$

Examples (1) Differentiate $u(x)=(2+x^5)^8$:
Let $v(x)=2+x^5$ and $u(v)=v^8$. Then

$$\begin{aligned}\frac{\mathrm{d}u}{\mathrm{d}x} &= \frac{\mathrm{d}u}{\mathrm{d}v}\frac{\mathrm{d}v}{\mathrm{d}x} = 8v^7 \cdot (0+5x^4)\\ &= 8(2+x^5)^7 \cdot 5x^4 = 40x^4(2+x^5)^7.\end{aligned}$$

(2) Differentiate $y(x)=(\sin x+\cos^2 x)^3$:

$$\begin{aligned}\frac{\mathrm{d}y}{\mathrm{d}x} &= \frac{\mathrm{d}}{\mathrm{d}x}(\sin x+\cos^2 x)^3\\ &= 3(\sin x+\cos^2 x)^2\frac{\mathrm{d}}{\mathrm{d}x}(\sin x+\cos^2 x)\\ &= 3(\sin x+\cos^2 x)^2\left(\frac{\mathrm{d}}{\mathrm{d}x}\sin x+\frac{\mathrm{d}}{\mathrm{d}x}\cos^2 x\right)\\ &= 3(\sin x+\cos^2 x)^2(\cos x+2\cos x\frac{\mathrm{d}}{\mathrm{d}x}\cos x)\\ &= 3(\sin x+\cos^2 x)^2(\cos x+2\cos x(-\sin x))\\ &= 3\cos x(\sin x+\cos^2 x)^2(1-2\sin x).\end{aligned}$$

(3) Let $y=|\sin x|$. Find y'.

Solution Note $(|x|)'=\operatorname{sgn} x=|x|/x(x\neq 0)$. Then,

$$\begin{aligned}y' &= \frac{|\sin x|}{\sin x}\cos x\\ &= |\sin x|\cot x\ (x\neq 0, \pm n\pi, \text{ where } n=1, 2, \cdots).\end{aligned}$$

In finding the derivative of a mixed expression of products, quotients and powers, it is usually advisable to take logarithm first, which is called *method of logarithmic derivation*.

Examples (1) Find the derivative of

$$y=\frac{(x+5)^2(x-4)^{\frac{1}{3}}}{(x+2)^5(x+4)^{\frac{1}{2}}}.$$

Solution Take logarithm on both sides first:

$$\ln y=2\ln(x+5)+\frac{1}{3}\ln(x-4)-5\ln(x+2)-\frac{1}{2}\ln(x+4),$$

then by differentiating both sides with respect to x, we have

$$\frac{1}{y}y'=\frac{2}{x+5}+\frac{1}{3}\frac{1}{x-4}-\frac{5}{x+2}-\frac{1}{2}\frac{1}{x+4}.$$

Therefore,

$$f'=\frac{(x+5)^2(x-4)^{\frac{1}{3}}}{(x+2)^5(x+4)^{\frac{1}{2}}}\left(\frac{2}{x+5}+\frac{1}{3}\frac{1}{x-4}-\frac{5}{x+2}-\frac{1}{2}\frac{1}{x+4}\right).$$

(2) Find the derivative of

$$y=(\ln x)^{\cos x}\quad(x>1).$$

Solution Note

$$\ln y=\cos x\ln(\ln x).$$

Then

$$\frac{1}{y}y'=(\cos x)'\ln(\ln x)+\cos x(\ln(\ln x))'$$

$$=-\sin x\ln(\ln x)+\cos x\frac{1}{\ln x}\frac{1}{x}.$$

Thus

$$y'=(\ln x)^{\cos x}\left(-\sin x\ln(\ln x)+\cos x\frac{1}{x\ln x}\right).$$

(3) Find the derivative of

$$y = x^{x^x}.$$

Solution $\ln y = x^x \ln x$, and so $\frac{1}{y}y' = (x^x)'\ln x + x^x(\ln x)' =$ $(x^x)'\ln x + \frac{1}{x}x^x$.

To find $(x^x)'$, let $z=x^x$. Then

$$\ln z = x \ln x, \text{ and so } \frac{1}{z}z' = \ln x + x\frac{1}{x} = \ln x + 1.$$

Then $(x^x)' = z' = z(\ln x + 1) = x^x(\ln x + 1)$, and finally we have

$$y' = y\left[(x^x)'\ln x + \frac{1}{x}x^x\right] = x^{x^x}[x^x(\ln x + 1)\ln x + x^{x-1}].$$

(4) Let

$$y = \sqrt[\varphi(x)]{\psi(x)} \text{ where } \varphi(x) \neq 0, \psi(x) > 0. \text{ Find } \frac{\mathrm{d}y}{\mathrm{d}x}.$$

Solution Since

$$\ln y = \frac{1}{\varphi(x)} \ln \psi(x).$$

Differentiating both sides of above equality, we have

$$\frac{1}{y}y' = \frac{\varphi(x)\frac{1}{\psi(x)}\psi'(x) - \ln \psi(x) \cdot \varphi'(x)}{\varphi^2(x)}.$$

Thus,

$$\frac{\mathrm{d}y}{\mathrm{d}x} = \sqrt[\varphi(x)]{\psi(x)}\left(\frac{\varphi(x)\frac{1}{\psi(x)}\psi'(x) - \ln \psi(x) \cdot \varphi'(x)}{\varphi^2(x)}\right).$$

§4.2.3 Derivative of implicit function

In earlier sections we were concerned with differentiating

functions that were given by the explicit form $y = f(x)$. However, functional dependence can also be described without necessarily solving for one variable in terms of the other by writing

$$F(x, y) = 0.$$

In this case, the function $y = f(x)$ defined by the equation $F(x, y) = 0$ is called an *implicit function*, while the former given by the explicit form $y = f(x)$ is called an *explicit function*.

A typical example of such a relationship is

$$x^2 + y^2 - 1 = 0.$$

This defines y implicitly as a function of x (or vice versa) but not uniquely. There are two continuous functions, $y = \sqrt{1 - x^2}$ and $y = -\sqrt{1 - x^2}$, defined by this implicit equation when $-1 < x < 1$.

The derivative of an implicit function is calculated in a completely straightforward fashion using the general rules for differentiation, especially the chain rule. To find $y'(x)$ in $x^2 + y^2 - 1 = 0$, for example, each side of the equation is differentiated term by term with respect to x. The differentiation of the right-hand side obviously gives zero, and since

$$\frac{d}{dx}(x^2 + y^2 - 1) = \frac{d}{dx}x^2 + \frac{d}{dx}y^2 - \frac{d}{dx}1 = 2x + 2y\frac{dy}{dx},$$

it follows that

$$x + y\frac{dy}{dx} = 0,$$

and so the derivative of y with respective to x is

$$\frac{dy}{dx} = -\frac{x}{y}.$$

This technique for finding the derivative of an implicit function is called *implicit differentiation*. Other examples are as follows:

Examples (1) Find the slope of the curve $x^3 - 3xy^2 + y^3 = 1$ at the point $(2, -1)$:

Differentiation gives

$$3x^2 - 3\left[x\left(2y\frac{dy}{dx}\right) + y^2\right] + 3y^2\frac{dy}{dx} = 0.$$

At $x = 2$ and $y = -1$, the equation becomes

$$12 + 12\frac{dy}{dx} - 3 + 3\frac{dy}{dx} = 0.$$

Thus

$$\frac{dy}{dx} = -\frac{3}{5},$$

i. e. the slope is $-\frac{3}{5}$.

(2) Let C be a curve with equation $x^2 + 2xy + y^2 - 4x - 5y + 3 = 0$ and let l be a straight line with equation $2x + 3y = 0$. Find the equation of a tangent line T of C which is parallel to l.

Solution Suppose T is the tangent line of C at the point (x_0, y_0). Differentiation on both sides of the equation of C gives

$$2x + 2y + 2xy' + 2yy' - 4 - 5y' = 0, \text{ and so } y' = \frac{4 - 2(x + y)}{2(x + y) - 5}.$$

Since T is required to be parallel to l, $y' = -2/3$, which implies

$$\frac{4 - 2(x_0 + y_0)}{2(x_0 + y_0) - 5} = -\frac{2}{3}, \text{ i. e. } x_0 + y_0 - 1 = 0.$$

Noticing (x_0, y_0) is on C, $x_0^2 + 2x_0y_0 + y_0^2 - 4x_0 - 5y_0 + 3 = 0$. Then we have $x_0 = 1$ and $y_0 = 0$. Hence, the equation of T is

$$y - y_0 = -\frac{2}{3}(x - x_0), \text{ i. e. } 2x + 3y - 2 = 0.$$

(3) Let C be a parabola with the equation $x^{1/2}+y^{1/2}=a^{1/2}$ where a is a positive constant. Suppose T is a tangent line of C at any point on C. Prove that the sum of the intersection of T with x-axis and that of T with y-axis is always a constant.

Proof Assume (x_0, y_0) is any point on C and T is a tangent line of C at (x_0, y_0). Differentiating both sides of the equation of C, we have

$$\frac{1}{2}x^{-\frac{1}{2}}+\frac{1}{2}y^{-\frac{1}{2}}y'=0, \text{ and so } y'=-\sqrt{\frac{y}{x}}.$$

Thus, the tangent line of C through (x_0, y_0) is

$$y-y_0=-\sqrt{\frac{y_0}{x_0}}(x-x_0),$$

$$\text{i. e. } \frac{x}{\sqrt{x_0}(\sqrt{x_0}+\sqrt{y_0})}+\frac{y}{\sqrt{y_0}(\sqrt{x_0}+\sqrt{y_0})}=1.$$

So, the sum of the intersection of T with x-axis and that of T with y-axis is

$$\begin{aligned}\sqrt{x_0}(\sqrt{x_0}+\sqrt{y_0})+\sqrt{y_0}(\sqrt{x_0}+\sqrt{y_0}) &= (\sqrt{x_0}+\sqrt{y_0})^2 \\ &= (a^{1/2})^2=a. \square\end{aligned}$$

§ 4. 2. 4 Derivative of inverse function

Theorem 4. 2. 3 Suppose $y=f(x)$ is the inverse function of $x=g(y)$. If $g(y)$ is continuous and strictly increasing on a neighborhood of y_0, $U(y_0)$, and $g(y)$ is derivable at y_0 with $g'(y_0)\neq 0$, then $f(x)$ is derivable at $x_0(=g(y_0))$, and

$$f'(x_0)=\frac{1}{g'(y_0)},$$

or simply

$$\frac{\mathrm{d}y}{\mathrm{d}x}=\frac{1}{\mathrm{d}x/\mathrm{d}y}.$$

Proof Let $\Delta x = x - x_0\,(\neq 0)$. Since f is strictly increasing, $\Delta y = f(x_0+\Delta x)-f(x_0)\neq 0$, and so

$$\frac{\Delta y}{\Delta x}=\frac{1}{\Delta x/\Delta y}.$$

Noticing $g(y)$ is continuous at y_0, by the continuity of the inverse function, $f(x)$ is continuous at x_0. So, $\Delta y\to 0$ as $\Delta x\to 0$. Then as $g'(y_0)\neq 0$,

$$f'(x_0)=\lim_{\Delta x\to 0}\frac{\Delta y}{\Delta x}=\frac{1}{\lim\limits_{\Delta y\to 0}(\Delta x/\Delta y)}=\frac{1}{g'(y_0)}.$$

We note the identity

$$\frac{\mathrm{d}y}{\mathrm{d}x}=\frac{1}{\mathrm{d}x/\mathrm{d}y}$$

indicates a relationship that further enhances the interpretation of $\mathrm{d}x$ and $\mathrm{d}y$ as real infinitesimal elements which are manipulated like ordinary numbers. From this viewpoint, the above identity represents a generally valid algebraic manipulation in which numerator and denominator are divided by $\mathrm{d}y$:

$$\frac{\mathrm{d}y}{\mathrm{d}x}=\frac{\mathrm{d}y/\mathrm{d}y}{\mathrm{d}x/\mathrm{d}y}=\frac{1}{\mathrm{d}x/\mathrm{d}y},$$

in exact analogy with the fraction $\dfrac{p}{q}$ written as $\dfrac{1}{q/p}$.

Examples (1) Find derivative of $y=\arcsin x$:

Noticing the inverse function of $y=\arcsin x\,(-1<x<1)$ is $x=\sin y\ (-\pi/2<y<\pi/2)$ and $\cos y>0$ as $-\pi/2<y<\pi/2$,

$$(\arcsin x)'=\frac{1}{(\sin y)'}=\frac{1}{\cos y}=\frac{1}{\sqrt{1-\sin^2 y}}=\frac{1}{\sqrt{1-x^2}}.$$

Similarly, we can derive

$$(\arccos x)' = -\frac{1}{\sqrt{1-x^2}};$$

$$(\arctan x)' = \frac{1}{1+x^2};$$

$$(\operatorname{arccot} x)' = -\frac{1}{1+x^2}.$$

(2) Find derivative of $y=a^x (a>0, a\neq 1)$:

Since $y = a^x$ $(-\infty < x < \infty)$ is the inverse function of the function $x = \log_a y$ $(0 < y < +\infty)$,

$$(a^x)' = \frac{1}{(\log_a y)'} = \frac{1}{\frac{1}{y}\log_a \mathrm{e}} = \frac{y}{\log_a \mathrm{e}} = a^x \ln a.$$

In particular, $(\mathrm{e}^x)' = \mathrm{e}^x$.

§ 4.2.5 Derivative of parametric function

The functional relationship between two variables x and y is often conveniently described in terms of a parameter or an auxiliary variable, as for example in the *parametric equations*

$$\begin{cases} x = x(t), \\ y = y(t), \end{cases} \tag{1}$$

where $\alpha \leqslant t \leqslant \beta$.

Here the corresponding values of x and y are calculated directly from t, a procedure that may be much simpler than having an explicit statement $y = f(x)$. This explicit form can be obtained by solving the first equation in (1) for t as a function of x, $t = t(x)$, and then substituting this in the second:

$$y = y(t(x)) = f(x).$$

The derivative $\mathrm{d}y/\mathrm{d}x$ is determined by implicit differentiation and use of the chain rule, with x assigned the role of the independent variable and y and t the dependent variables. If both expressions in (1)are differentiated with respect to x, the results are

$$\begin{cases} 1 = \dfrac{\mathrm{d}x}{\mathrm{d}t}\dfrac{\mathrm{d}t}{\mathrm{d}x}, \\ \dfrac{\mathrm{d}y}{\mathrm{d}x} = \dfrac{\mathrm{d}y}{\mathrm{d}t}\dfrac{\mathrm{d}t}{\mathrm{d}x}, \end{cases}$$

and by eliminating $\mathrm{d}t/\mathrm{d}x$ between these we obtain

$$\frac{\mathrm{d}y}{\mathrm{d}x} = \frac{\mathrm{d}y}{\mathrm{d}t}\bigg/\frac{\mathrm{d}x}{\mathrm{d}t}. \tag{2}$$

The derivative $y'(x)$ is given as the ratio of derivatives with respect to the parametric variable t.

Another way of obtaining (2) is to divide the formulas

$$\mathrm{d}x = x'(t)\mathrm{d}t,\ \mathrm{d}y = y'(t)\mathrm{d}t,$$

thereby eliminating $\mathrm{d}t$. Once again we see that it is perfectly valid to handle $\mathrm{d}t$ like real numbers.

Example Suppose a curve is given by

$$l: \begin{cases} x = r(\theta - \sin\theta), \\ y = r(1 - \cos\theta). \end{cases}$$

① find $y'(x)$;

② find the equation of the tangent line of l at $\theta = \pi/3$;

③ determine the points at which l has horizontal tangent line or vertical tangent line respectively.

Solution ① $\dfrac{\mathrm{d}y}{\mathrm{d}x} = \dfrac{\mathrm{d}y}{\mathrm{d}t}\bigg/\dfrac{\mathrm{d}x}{\mathrm{d}t} = \dfrac{r\sin\theta}{r(1-\cos\theta)}$

$$= \frac{2\sin\dfrac{\theta}{2}\cos\dfrac{\theta}{2}}{2\sin^2\dfrac{\theta}{2}} = \cot\frac{\theta}{2}.$$

② as $\theta=\pi/3$, $x=r\left(\frac{\pi}{3}-\frac{\sqrt{3}}{2}\right)$, $y=r/2$ and so $y'(x)=\cot(\pi/6)=\sqrt{3}$. Thus the equation of the tangent line is

$$y-\frac{r}{2}=\sqrt{3}\left[x-r\left(\frac{\pi}{3}-\frac{\sqrt{3}}{2}\right)\right],\ \text{i. e. } \sqrt{3}-y=r\left(\frac{\pi}{\sqrt{3}}-2\right).$$

③ Let $dy/dx=\cot(\theta/2)=0$. Then $\theta=(2n+1)\pi$, where n is any integer, i. e. the curve l has horizontal tangent lines at points $((2n+1)\pi,\ 2r)$.

As $x=\varphi(\theta)$ is continuous, $\theta=\varphi^{-1}(x)$ is continuous. Also, $y=\psi(\theta)$ is continuous. Then $y=\psi(\varphi^{-1}(x))$ is continuous. Moreover,

$$\lim_{x\to 2n\pi r}\left|\frac{dy}{dx}\right|=\lim_{\theta\to 2n\pi}\left|\frac{dy}{dx}\right|=\lim_{\theta\to 2n\pi}\left|\cot\frac{\theta}{2}\right|=+\infty.$$

Therefore, the curve l has vertical tangent lines at points $(2n\pi r,\ 0)$ $(n=0,\ \pm 1,\ \pm 2,\cdots)$.

§ 4. 2. 6 Basic laws and formulas of derivation

For convenience of use we list the basic laws and formulas of derivation as follows.

Ⅰ. Basic laws of derivation

(1) $(u\pm v)'=u'\pm v'$;

(2) $(uv)'=u'v+uv'$, $(cu)'=cu'$;

(3) $\left(\frac{u}{v}\right)'=\frac{u'v-uv'}{v^2}$, $\left(\frac{1}{v}\right)'=-\frac{v'}{v^2}$;

(4) (derivative of inverse function) $\frac{dy}{dx}=-\frac{1}{\frac{dx}{dy}}$;

(5) (derivative of composite function) $\frac{dy}{dx}=\frac{dy}{du}\frac{du}{dx}$;

(6) (derivative of parameter function) $\frac{dy}{dx}=\frac{dy}{dt}\Big/\frac{dx}{dt}$.

Ⅱ. Formulas for derivatives of fundamental primary functions

(1) $(c)'=0$ (c is a constant);

(2) $(x^r)' = rx^{r-1}$ (r is a real number);

(3) $(\sin x)'=\cos x$; $(\cos x)'=-\sin x$;

$(\tan x)'=\sec^2 x$; $(\cot x)'=-\csc^2 x$;

$(\sec x)'=\sec x \tan x$; $(\csc x)'=-\csc x \cot x$;

(4) $(\arcsin x)'=-(\arccos x)'=\dfrac{1}{\sqrt{1-x^2}}$ $(|x|<1)$;

$(\arctan x)'=-(\text{arccot}\, x)'=\dfrac{1}{1+x^2}$;

(5) $(a^x)' = a^x \ln a$; $(e^x)' = e^x$;

(6) $(\log_a x)' = \dfrac{1}{x \ln a}$; $(\ln x)' = \dfrac{1}{x}$.

§4.2.7 Related rates

We now study related rates problems. In such problems one tries to find the rate at which some quantity is changing by relating the quantity to other quantities whose rates of change are known.

In many practical problems in which two variables are related by an equation, both of the variables change with time. So they can be regarded as functions of a third variable, time, denoted t. In this case the derivatives of the variables with respect to t measure the rate of change of the variables with respect to t. Since the variables are related by an equation, implicit differentiation can be used to obtain a relationship between their derivatives. If the value of one of these derivatives is known at a particular moment, it may be possible to use this relationship to find the value of the other. Such derivatives are called related rates.

Example Assume that oil spilled from a ruptured tanker spreads in a circular pattern whose radius increases at a constant rate of

2 ft/sec. How fast is the area of the spill increasing when the radius of the spill is 60 ft?

Solution Let

t=number of seconds elapsed from the time of the spill

r=radius of the spill in feet after t seconds

A=area of the spill in square feet after t seconds

We know the rate at which the radius is increasing, and we want to find the rate at which the area is increasing at the instant when $r=60$; that is, we want to find

$$\left.\frac{\mathrm{d}A}{\mathrm{d}t}\right|_{r=60}$$

given that $\mathrm{d}r/\mathrm{d}t=2\text{ft/sec}$. From the formula for the area of a circle we obtain $A=\pi r^2$. Because A and r are functions of t, we can differentiate both sides of $A=\pi r^2$ with respect to t to obtain

$$\frac{\mathrm{d}A}{\mathrm{d}t}=2\pi r\frac{\mathrm{d}r}{\mathrm{d}t}.$$

Thus, when $r=60$ the area of the spill is increasing at the rate of

$$\left.\frac{\mathrm{d}A}{\mathrm{d}t}\right|_{r=60}=2\pi\times 60\times 2=240\pi(\text{ft})^2/\text{sec}.$$

With only minor variations, the methods used in above example can be applied to a variety of related rates problems. The method consists of five steps:

Strategy for Solving Related Rates Problems

Step 1 Identify the rates of change that are known and the rate of change that is to be found. Interpret each rate as a derivative of a variable with respect to time, and provide a description of each variable involved.

Step 2 Find an equation relating those quantities whose rates are identified in Step 1. In a geometric problem, this is aided by drawing an appropriately labelled figure that illustrates a relationship involving these quantities.

Step 3 Obtain an equation involving the rates in Step 1 by differentiating both sides of the equation in Step 2 with respect to the time variable.

Step 4 Evaluate the equation found in Step 3 using the known values for the quantities and their rates of change at the moment in question.

Step 5 Solve for the value of the remaining rate of change at this moment.

Note Do not substitute prematurely; that is, always perform the differentiation in Step 3 before performing the substitution in Step 4.

We now apply this strategy in the following examples:

Examples (1) An airplane at an altitude of 3 000 m, flying horizontally at 300 km/hr, passes directly over an observer. Evaluate the rate at which it is approaching the observer when it is 5 000 m away from the observer.

Solution Refer to Figure 4 - 7. The airplane is at point A flying towards point B. The distance OB is labelled as 3 000 (Step 1). Note that this distance does not change in the course of the problem. Let x be the distance between A and B where B is directly over the observer O, and s be the distance from the airplane to the observer. We want to find $\mathrm{d}s/\mathrm{d}t$ at $s = 5\,000$. We are given $\mathrm{d}x/\mathrm{d}t = -300$ km/hr. The negative sign is used because x is decreasing. A change of units gives

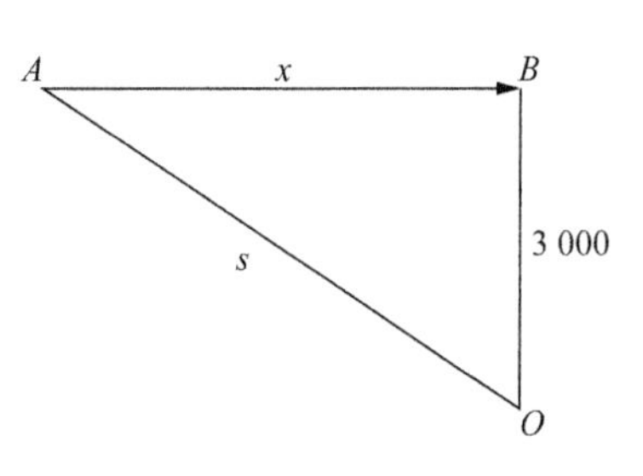

Figure 4 - 7

us $dx/dt=-83\frac{1}{3}$ m/sec. Thus we have (Step 2)

$$s^2 = x^2 + (3\,000)^2,$$

and differentiating both sides with respect to t (Step 3), we get

$$s\frac{ds}{dt} = x\frac{dx}{dt}.$$

Noting that $x=4\,000$ when $s=5\,000$ (Step 4), we have

$$5\,000\frac{ds}{dt} = 4\,000\left(-83\frac{1}{3}\right),$$

i. e. (Step 5) $\frac{ds}{dt}=-66\frac{2}{3}$ m/sec $=-240$ km/hr.

(2) One airplane flew over an airport at the rate of 300 km/hr. Ten minutes later another airplane flew over the airport at 240 km/hr. If the first airplane was flying west and the second flying south (both at the same altitude), determine the rate at which they were separating 20 minutes after the second plane flew over the airport. (We assume the airplanes are travelling at constant speed.)

Solution Let s be the distance between the airplanes at time t, let x and y respectively be the distances the westbound airplane and the southbound airplane have travelled at time t. We know that $dx/dt=5$ km/min and $dy/dt=4$ km/min. We want to find ds/dt when $t=20$ min, measured from the time the second airplane passes over the airport. We have the relation

$$s^2 = x^2 + y^2.$$

By Taking derivatives on both sides with respect to t, we obtain:

$$s\frac{ds}{dt} = x\frac{dx}{dt} + y\frac{dy}{dt}.$$

When $t=20$, $y=80$, $x=150$ and $s=170$. Substituting these values, we get

$$170 \cdot \frac{\mathrm{d}s}{\mathrm{d}t} = 150 \times 5 + 80 \times 4, \text{ i. e. } \frac{\mathrm{d}s}{\mathrm{d}t} = \frac{107}{17} \text{ km/min.}$$

§ 4. 3 Differentials

Up to now we have not defined the symbols $\mathrm{d}y$ and $\mathrm{d}x$ as separate entities, and we have treated $\mathrm{d}y/\mathrm{d}x$ as a single entity but not a ratio. In this section we will define the quantities $\mathrm{d}y$ and $\mathrm{d}x$, called differentials, independently and use them to interpret the derivative $\mathrm{d}y/\mathrm{d}x$ as a ratio of differentials. Differentials will play an important role in finding approximate values of functions.

§ 4. 3. 1 The concept of differential

First we investigate a concrete problem. Let S be a square with edge length x. Then its area $A = x^2$, a function of x. Suppose the original edge length is x_0 and it is added by Δx, the corresponding area will be added by

$$\Delta A = (x_0 + \Delta x)^2 - x_0^2 = 2x_0 \Delta x + \Delta x^2,$$

which is composed of two parts: the first part $2x_0 \Delta x$ is a linear function of Δx (Figure 4 - 8), and the second part Δx^2 is a higher order infinitesimal over Δx, i. e. $\Delta x^2 = o(\Delta x)$.

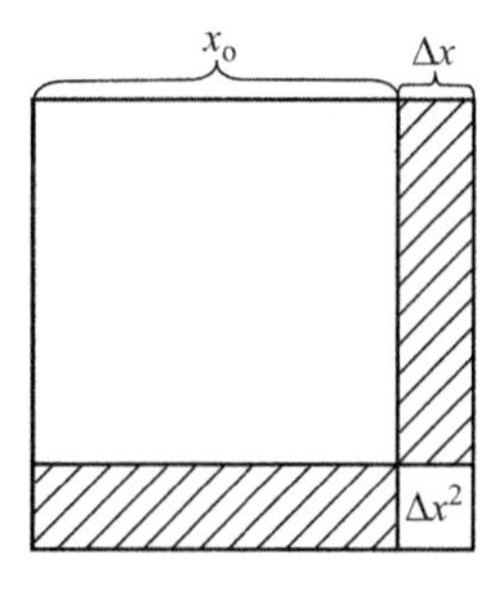

Figure 4 - 8

From this we see that as the edge length x_0 is given a tiny increment Δx, the corresponding increment of the area of the square, ΔA, can be approximated by the first part $2x_0 \Delta x$, the linear function of Δx. The error induced in this way is Δx^2, a higher order infinitesimal over Δx. Such a linear function of Δx will be defined specifically as follows.

Definition Let $y = f(x)$ and let $\Delta y = f(x_0 + \Delta x) - f(x_0)$ be the increment of y at x_0. If Δy can be expressed as a sum of $A\Delta x$ where A is a constant (i. e. a linear function of Δx) and a higher-order infinitesimal over Δx:

$$\Delta y = A\Delta x + o(\Delta x),$$

then $f(x)$ is said to be *differentiable* at x_0. The $A\Delta x$ is called the *differential* of $f(x)$ at x_0, which is denoted by

$$\mathrm{d}y \,|_{x=x_0} = \mathrm{d}f(x) \,|_{x=x_0} = A\Delta x.$$

Theorem 4. 3. 1 The function $f(x)$ is differentiable at x_0 if and only if $f(x)$ is derivable at x_0, and in this case, $f'(x_0) = A$.

Proof Necessity. If f is differentiable at x_0,

$$\Delta y = A\Delta x + o(\Delta x).$$

Hence

$$f'(x_0) = \lim_{\Delta x \to 0} \frac{\Delta y}{\Delta x} = \lim_{\Delta x \to 0} \left(A + \frac{o(\Delta x)}{\Delta x} \right) = A.$$

Sufficiency. If f is derivable at x_0, then by the finite increment formula:

$$\Delta y = f'(x_0)\Delta x + o(\Delta x),$$

which implies that $f(x)$ is differentiable at x_0 and

$$\mathrm{d}y \,|_{x=x_0} = f'(x_0)\Delta x.$$

Note This theorem shows for single variable function, differentiability is equivalent to derivability, and the differential of $f(x)$ at x_0 can be evaluated by

$$dy\,|_{x=x_0} = f'(x_0)\Delta x.$$

We will see later that differentiability is not equivalent to derivability for multivariable functions.

The geometric interpretation of differential

The geometric interpretation of differential is referred to Figure 4 - 9. Let $y = f(x)$. As the independent variable changes from x_0 to $x_0 + \Delta x$, the corresponding increment of the function f is

$$\Delta y = f(x_0 + \Delta x) - f(x_0) = AB,$$

while the increment of the y-coordinate of the tangent line of f at point $P(x_0, y_0)$ is exactly

$$dy\,|_{x=x_0} = f'(x_0)\Delta x = AC.$$

Then the difference between Δy and dy, CB, tends to 0 as Δx tends to 0, and moreover, CB is a higher order infinitesimal over Δx as $\Delta x \to 0$. Therefore, in a sufficiently small neighborhood of x_0, the curve segment at $x = x_0$ can be approximated by a corresponding tangent segment at $x = x_0$.

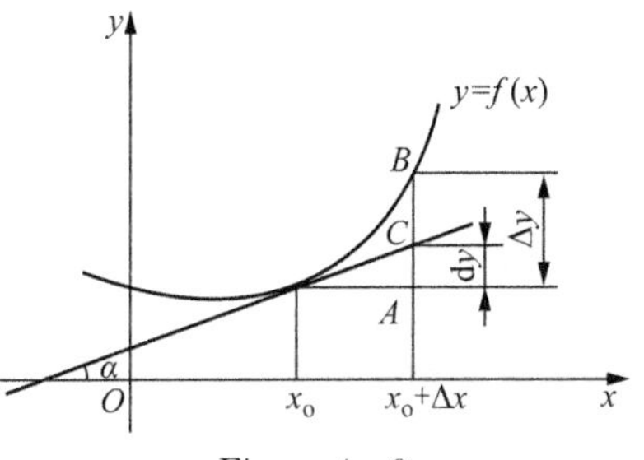

Figure 4 - 9

Definition If for any point x of an interval I, $y = f(x)$ is differentiable at x, then $f(x)$ is said to be *differentiable on* I, and the *differential* of $f(x)$ on I is denoted by

$$dy = f'(x)\Delta x,$$

which depends on not only Δx but also x.

Note (1) Let $y = f(x) = x$. Then $dy = f'(x)\Delta x = 1 \cdot \Delta x = \Delta x$. So, for independent variable x, the increment of x is exactly the differential of the function $y = x$, which is, for convenience, also called the differential of the independent variable x, denoted as $\Delta x = \mathrm{d}x$.

(2) Let $y = f(x)$. From (1) it follows straight that

$$\mathrm{d}y = f'(x)\mathrm{d}x,$$

i. e. the differential of a function is exactly the product of the derivative of the function and the differential of the independent variable.

(3) Another expression in (2) is

$$f'(x) = \frac{\mathrm{d}y}{\mathrm{d}x},$$

i. e. the derivative of a function is exactly the quotient of the differential of the function and the differential of the independent variable. So, the derivative is also called the *differential quotient*, and the denotation $\frac{\mathrm{d}y}{\mathrm{d}x}$ can be regarded as fraction instead of an inseparable symbol.

§4.3.2 Laws of differentiation

The basic laws for derivation presented in Section 4.2.6 can be interpreted in terms of differentials which relate the various infinitesimal increments of change. In each case, the differential $\mathrm{d}x$ appearing in these rules is simply cancelled throughout like a common algebraic factor. Suppose $u(x)$ and $v(x)$ are differentiable functions. Then

$$\mathrm{d}(u \pm v) = \mathrm{d}u \pm \mathrm{d}v$$

$$\mathrm{d}(cu) = c\,\mathrm{d}u$$

$$\mathrm{d}(uv) = v\,\mathrm{d}u + u\,\mathrm{d}v$$

$$d\left(\frac{u}{v}\right)=\frac{v\,du-u\,dv}{v^2}$$

$$du(v)=\left(\frac{du}{dv}\right)dv.$$

Note Suppose $y=f(x)$ and $x=\varphi(t)$. Then $y=f(\varphi(t))$ is a composite function of t, and so

$$\frac{dy}{dt}=\frac{dy}{dx}\frac{dx}{dt}=f'(x)\varphi'(t),$$

i. e.

$$dy=f'(x)\varphi'(t)dt=f'(x)dx,$$

which implies the expression

$$dy=f'(x)dx$$

is always true whether x is an independent variable or a function of another variable. This property is called the *form-invariance of first differentials*. However, we will see that there is no such property for higher differentials later.

Examples Find the differentials of the following functions:

$$① \ y=x^3\ln(x^2)+\sin x; \quad ② \ y=e^{\cos(ax+b)}.$$

Solution ① $dy=d(x^3\ln(x^2))+d(\sin x)$

$$\begin{aligned}&=d(x^3)\ln(x^2)+x^3d(\ln(x^2))+d(\sin x)\\&=3x^2\ln(x^2)dx+2x^2dx+\cos x\,dx\\&=(3x^2\ln(x^2)+2x^2+\cos x)dx;\end{aligned}$$

② $dy=e^{\cos(ax+b)}d(\cos(ax+b))$

$$\begin{aligned}&=e^{\cos(ax+b)}(-\sin(ax+b))d(ax+b)\\&=(-a)e^{\cos(ax+b)}\sin(ax+b)dx.\end{aligned}$$

§ 4. 3. 3 Approximations using differentials

In light of the relationship of the increment and the differential:

$$\Delta y = f'(x_0)\Delta x + o(\Delta x) = \mathrm{d}y + o(\Delta x),$$

if Δx is very small, the term $o(\Delta x)$ can be neglected, and so $\Delta y \approx \mathrm{d}y$, or

$$f(x_0 + \Delta x) \approx f(x_0) + f'(x_0)\Delta x.$$

For the practical purpose of approximating a value $f(x)$, three factors should be determined beforehand: (1) the function $f(x)$; (2) the point x_0 which is "near" x and it is "easy" to calculate $f(x_0)$ and $f'(x_0)$; (3) $\Delta x = x - x_0$. The remaining work is just to apply the approximation expression directly.

Examples (1) Use differential to approximate $\sqrt[3]{27.54}$.

Solution Form the function $y = f(x) = \sqrt[3]{x} = x^{\frac{1}{3}}$; Take $x_0 = 27$ and so $\Delta x = x - x_0 = 27.54 - 27 = 0.54$. Thus,

$$\sqrt[3]{27.54} = f(x + \Delta x) \approx f(x_0) + f'(x_0)\Delta x = f(27) + f'(27)(0.54)$$
$$= 27^{\frac{1}{3}} + \frac{1}{3 \times (27)^{2/3}} \times 0.54 = 3.02.$$

(2) Find the approximate value of $\sin 29^\circ$.

Solution Let $y = f(x) = \sin x$, $x_0 = 30^\circ = \pi/6$ and $x = x_0 + \Delta x = 29^\circ$ with $\Delta x = -1^\circ \approx -0.0175$ (radian). Then

$$\sin 29^\circ = f(x + \Delta x) \approx f(x_0) + f'(x_0)\Delta x$$
$$= \sin\frac{\pi}{6} + \cos\frac{\pi}{6} \cdot (-0.0175)$$
$$= \frac{1}{2} - \frac{\sqrt{3}}{2} \times 0.0175$$
$$\approx \frac{1}{2} - \frac{1}{2} \times 1.732 \times 0.0175 \approx 0.485.$$

Remark Some common-used approximations: Let $|x| \ll 1$. Then

$\sin x \approx x$, $\tan x \approx x$, $\arcsin x \approx x$,

$\arctan x \approx x$, $\sinh x \approx x$, $\tanh x \approx x$,

$\operatorname{arcsinh} x \approx x$, $\operatorname{arctanh} x \approx x$, $\ln(1+x) \approx x$,

$e^x \approx 1+x$, $(1+x)^a \approx 1+ax$,

$\ln \dfrac{1+x}{1-x} \approx 2x$, $\sqrt{1+x} \approx 1+\dfrac{1}{2}x$.

§ 4. 4 Derivatives and differentials of higher order

Since the derivative is itself a function, we can also consider its derivative, which will be called the second derivative. In this way we can introduce the concept the derivative of higher order (or higher derivative) as well as the concept the differential of higher order (or higher differential).

§ 4. 4. 1 Higher derivatives

Definition Suppose $y = f(x)$ with $f'(x)$ being derivable at x_0. Then the derivative of $f'(x)$ at x_0 is called the *second derivative* of f at x_0, denoted by $f''(x_0)$, i. e.

$$f''(x_0) = \lim_{x \to x_0} \frac{f'(x) - f'(x_0)}{x - x_0}.$$

In this case, $y=f(x)$ is also said to be *secondly derivable* at x_0.

Moreover, if $y=f(x)$ is derivable at any point $x \in I$ where I is an interval, then f is said to be *secondly derivable* on I, and in this case $f''(x)$ $(x \in I)$ is called the *second derivative* of f on I, also denoted as

$$\frac{d}{dx}\left(\frac{d}{dx}f(x)\right) = \frac{d^2}{dx^2}f(x).$$

Obviously, the definition can be generalized to even higher orders, e. g. the third derivative $f'''(x)$, the fourth derivative

$f^{(4)}(x)$, …, and the n-th derivative $f^{(n)}(x)$, which can also be denoted as

$$\frac{d^n}{dx^n}f(x) \text{ or } \frac{d^n f}{dx^n} \text{ or } \frac{d^n y}{dx^n}$$

which is read as "the n-th derivative of function $y=f(x)$ with respect to x".

The rules established in the last section apply at each stage of higher-order differentiation, and it is rather easy to prove, for example, that the n-th derivative of a sum functions is just the sum of the n-th derivatives:

$$\frac{d^n}{dx^n}(u_1(x)+\cdots+u_m(x))=\frac{d^n}{dx^n}u_1(x)+\cdots+\frac{d^n}{dx^n}u_m(x),$$

or, written more succinctly using the summation notation,

$$\frac{d^n}{dx^n}\sum_{i=1}^{m}u_i(x)=\sum_{i=1}^{m}\frac{d^n}{dx^n}u_i(x).$$

Formulas for product differentiation are also simply derived but cumbersome as a typical calculation illustrates:

$$\begin{aligned}\frac{d^2}{dx^2}(u(x)v(x)) &= \frac{d}{dx}\left(\frac{d}{dx}(uv)\right)=\frac{d}{dx}\left(u\frac{dv}{dx}+v\frac{du}{dx}\right)\\ &= \frac{d}{dx}\left(u\frac{dv}{dx}\right)+\frac{d}{dx}\left(v\frac{du}{dx}\right)\\ &= u\frac{d^2v}{dx^2}+2\frac{du}{dx}\frac{dv}{dx}+v\frac{d^2u}{dx^2}.\end{aligned}$$

The third derivative of a product has a similar formula:

$$\frac{d^3}{dx^3}(uv)=u\frac{d^3v}{dx^3}+3\frac{du}{dx}\frac{d^2v}{dx^2}+3\frac{d^2u}{dx^2}\frac{dv}{dx}+\frac{d^3u}{dx^3}v.$$

The general result, known as *Leibniz formula*, is presented in the following.

Theorem 4.4.1 $\frac{d^n}{dx^n}(uv) = \sum_{k=0}^{n} C_n^k \frac{d^k u}{dx^k}\frac{d^{n-k}v}{dx^{n-k}}$,

i.e. $(uv)^{(n)} = u^{(n)}v^{(0)} + C_n^1 u^{(n-1)}v' + C_n^2 u^{(n-2)}v'' + \cdots + C_n^k u^{(n-k)}v^{(k)} + \cdots + u^{(0)}v^{(n)}$, where $u^{(0)} = u$, $v^{(0)} = v$ and $C_n^k = \frac{n!}{k!(n-k)!}$.

Noting $C_n^k + C_n^{k-1} = C_{n+1}^k$, this theorem can be proved just by mathematical induction. It is not necessary to commit such formulas to memory; it is important instead to understand the basic procedure well enough so that the result can be derived whenever needed.

Examples (1) Let $y = x^2 e^{2x}$. Find $f^{(6)}$.

Solution Note $(x^2)^{(n)} = 0$ for any $n \geqslant 3$ and $(e^{2x})^{(n)} = 2^n e^{2x}$. Thus

$$\begin{aligned} y^{(6)} &= (e^{2x})^{(6)}x^2 + C_6^1(e^{2x})^{(5)} \cdot 2x + C_6^2(e^{2x})^{(4)} \cdot 2 \\ &= 2^6 e^{2x}x^2 + 6 \times 2^6 e^{2x}x + \left(6 \times \frac{5}{2}\right)2^5 e^{2x} \\ &= 32e^{2x}(2x^2 + 12x + 15). \end{aligned}$$

(2) Let $y = 1/(1-x^2)$. Find $f^{(n)}$.

Solution Since $y = (1/2)[(1+x)^{-1} + (1-x)^{-1}]$, and $[(1+x)^{-1}]^{(n)} = (-1)^n n! \cdot (1+x)^{-n-1}$

$$[(1-x)^{-1}]^{(n)} = n! \cdot (1-x)^{-n-1},$$

so,

$$y^{(n)} = \frac{n!}{2}\left[\frac{(-1)^n}{(1+x)^{n+1}} + \frac{1}{(1-x)^{n+1}}\right].$$

(3) Let $y = \cos x$, $y = \cos kx$, $y = \sin x$, $y = \sin kx$ (k is a constant). Find $y^{(n)}$ respectively.

Solution Note

$$y' = -\sin x = \cos\left(x + \frac{\pi}{2}\right);\ y'' = -\cos x = \cos\left(x + 2\cdot\frac{\pi}{2}\right),$$

$$y''' = \sin x = \cos\left(x + 3\cdot\frac{\pi}{2}\right);\ y^{(4)} = \cos x = \cos\left(x + 4\cdot\frac{\pi}{2}\right).$$

so, we can see that

$$(\cos x)^{(n)} = \cos\left(x + \frac{n\pi}{2}\right);\ \text{and } (\cos kx)^{(n)} = k^n\cos\left(kx + \frac{n\pi}{2}\right).$$

In a analogous manner, we may derive that

$$(\sin x)^{(n)} = \sin\left(x + \frac{n\pi}{2}\right);\ \text{and } (\sin kx)^{(n)} = k^n\sin\left(kx + \frac{n\pi}{2}\right).$$

(4) Let

$$y = \frac{ax+b}{cx+d}(a,\ b,\ c,\ d \text{ are costants, } c \neq 0).\ \text{Find } \frac{\mathrm{d}^n y}{\mathrm{d}x^n}.$$

Solution Note

$$y = \frac{acx+bc}{c(cx+d)} = \frac{acx+ad-ad+bc}{c(cx+d)}$$
$$= \frac{a(cx+d)}{c(cx+d)} + \frac{bc-ad}{c(cx+d)} = \frac{a}{c} + \frac{bc-ad}{c}(cx+d)^{-1}.$$

Let $a/c = \alpha$ and let $(bc-ad)/c = \beta$. Then $y = \alpha + \beta(cx+d)^{-1}$. Thus

$$y' = (-1)\beta c(cx+d)^{-2},\ y'' = (-1)^2\beta\cdot 2!\cdot c^2(cx+d)^{-3},$$
$$y''' = (-1)^3\beta\cdot 3!\cdot c^3(cx+d)^{-4},\cdots,$$

from which it follows that

$$\frac{\mathrm{d}^n y}{\mathrm{d}x^n} = (-1)^n\beta\cdot n!\cdot c^n(cx+d)^{-(n+1)}$$
$$= (-1)^n\frac{bc-ad}{c}\cdot n!\cdot c^n(cx+d)^{-(n+1)}.$$

We can also find higher derivatives of implicit functions and parameter functions in a similar manner as in finding the first derivative of those functions:

(5) Find

$$\frac{d^2y}{dx^2} \text{ given that } y^3 - x^2 = 4:$$

Differentiation with respect to x gives

$$3y^2\frac{dy}{dx} - 2x = 0. \tag{1}$$

Differentiating again we have

$$3y^2\left(\frac{d^2y}{dx^2}\right) + 6y\left(\frac{dy}{dx}\right)^2 - 2 = 0. \tag{2}$$

From (1) we know that

$$\frac{dy}{dx} = \frac{2x}{3y^2}. \tag{3}$$

Substituting (3) in (2) we have

$$3y^2\left(\frac{d^2y}{dx^2}\right) + 6y\left(\frac{2x}{3y^2}\right)^2 - 2 = 0.$$

which, as you can check, gives

$$\frac{d^2y}{dx^2} = \frac{6y^3 - 8x^2}{9y^5}.$$

(6) Find

$$\frac{d^2y}{dx^2} \text{ given that } e^{x+y} = xy.$$

By taking logarithm we see $x + y = \ln x + \ln y$. Differentiating both sides with respect to x gives

$$1 + y' = \frac{1}{x} + \frac{y'}{y} \text{ and } y'' = -\frac{1}{x^2} + \frac{y''}{y} - \frac{(y')^2}{y^2}.$$

Thus

$$y' = \frac{1-x}{x} \cdot \frac{y}{y-1},$$

and so

$$y'' \cdot \left(1 - \frac{1}{y}\right) = -\frac{1}{x^2} - \frac{1}{y^2}\left[\frac{1-x}{x} \cdot \frac{y}{y-1}\right]^2 = -\frac{(y-1)^2 + (1-x)^2}{x^2(1-y)^2}.$$

Consequently, we have

$$\frac{d^2 y}{dx^2} = \frac{y[(1-y)^2 + (1-x)^2]}{x^2(1-y)^3}.$$

(7) Find the highest order of the derivatives of the following function at $x = 0$:

$$f(x) = \begin{cases} x^5 \sin \frac{1}{x} & (x \neq 0), \\ 0 & (x = 0). \end{cases}$$

Solution Let $x \neq 0$. Then,

$$f'(x) = 5x^4 \sin \frac{1}{x} + x^5 \cos \frac{1}{x}\left(-\frac{1}{x^2}\right) = 5x^4 \sin \frac{1}{x} - x^3 \cos \frac{1}{x}.$$

Let $x = 0$. Then

$$f'(0) = \lim_{x \to 0} \frac{f(x) - f(0)}{x} = \lim_{x \to 0} \frac{x^5 \sin \frac{1}{x}}{x} = 0.$$

Thus

$$f''(0) = \lim_{x \to 0} \frac{f'(x) - f'(0)}{x} = \lim_{x \to 0}\left(5x^3 \sin \frac{1}{x} - x^2 \cos \frac{1}{x}\right) = 0.$$

If $x \neq 0$,

$$f''(x) = 20x^3 \sin \frac{1}{x} + 5x^4 \cos \frac{1}{x}\left(-\frac{1}{x^2}\right) - 3x^2 \cos \frac{1}{x},$$

$$-x^3\left(-\sin \frac{1}{x}\right)\left(-\frac{1}{x^2}\right) = 20x^3 \sin \frac{1}{x} - 8x^2 \cos \frac{1}{x} - x \sin \frac{1}{x}.$$

Furthermore,

$$\frac{f''(x)-f''(0)}{x} = 20x^2 \sin\frac{1}{x} - 8x\cos\frac{1}{x} - \sin\frac{1}{x}. \tag{4}$$

As $x \to 0$, the limits of the preceding two terms on the right side of Expression (4) are both 0 while the limit of the third term does not exist. So, by the definition of the derivative, the highest order of the derivatives of $f(x)$ at $x = 0$ is 2.

§4.4.2 Higher differentials

Higher differential of $y = f(x)$ can be introduced in an analogous fashion as for derivative. Let $\mathrm{d}x(=\Delta x)$ be the increment of independent variable x. If $\mathrm{d}x$ is fixed, then the (first) differential $\mathrm{d}y = f'(x)\mathrm{d}x$ can be regarded as a function of x. So, by differentiating $\mathrm{d}y$, we have

$$\mathrm{d}(\mathrm{d}y) = \mathrm{d}(f'(x)\mathrm{d}x) = f''(x)\mathrm{d}x^2,$$

where $\mathrm{d}x^2 = (\mathrm{d}x)^2$. We call it *second differential* of $y = f(x)$ and denote it as

$$\mathrm{d}^2 y = f''(x)\mathrm{d}x^2.$$

Note that $\mathrm{d}x^2$ refers to the square of $\mathrm{d}x$, while $\mathrm{d}^2 x$ is the second differential of x and $\mathrm{d}(x^2)$ is the differential of the function $y = x^2$.

In general, the differential of the $(n-1)$-th differential of $y = f(x)$ is defined as the *n-th differential* of $f(x)$, denoted by

$$\mathrm{d}^n y = \mathrm{d}(\mathrm{d}^{(n-1)} y) = f^{(n)}(x)\mathrm{d}x^n.$$

Notice that the denotation is conformable to the n-th derivative

$$\frac{\mathrm{d}^n y}{\mathrm{d}x^n} = f^{(n)}(x).$$

Note Form-invariance property (cf. Section 4.3.2) is not valid any longer for higher differentials. For example, let $y = f(x)$, if x is an independent variable, then $d^2y = f''(x)dx^2$; however, if $y = f(x)$ and $x = \varphi(t)$, since $dy = f'(x)dx$ and $dx = \varphi'(t)dt$ is a function of t,

$$\begin{aligned} d^2y &= d(dy) = d(f'(x)dx) = d(f'(x)) \cdot dx + f'(x) \cdot d(dx) \\ &= f''(x)dx \cdot dx + f'(x)d^2x = f''(x)dx^2 + f'(x)d^2x, \end{aligned}$$

with an additional term $f'(x)d^2x$ compared with $d^2y = f''(x)dx^2$.

Examples (1) $y = \sqrt{1+x^2}$, find d^2y.

Solution Since

$$y' = \frac{x}{\sqrt{1+x^2}} \text{ and } y'' = \frac{1}{\sqrt{(1+x^2)^3}},$$

and so

$$d^2y = \frac{dx^2}{\sqrt{(1+x^2)^3}}.$$

(2) $y = x^x$, find d^2y.

Solution Since

$$\begin{aligned} y' &= x^x(\ln x + 1) \text{ and } y'' = x^x\left[(\ln x + 1)^2 + \frac{1}{x}\right] \\ &= x^{x-1}[x(\ln x + 1)^2 + 1], \end{aligned}$$

and so

$$d^2y = x^{x-1}[x(\ln x + 1)^2 + 1]dx^2.$$

(3) $y = x^n e^x$, find $d^n y$.

Solution Let $u = x^n$ and $v = e^x$. Then

$$u^{(k)} = n(n-1)\cdots(n-k+1)x^{n-k} = \frac{n!}{(n-k)!}x^{n-k},\ v^{(n-k)} = e^x.$$

So,

$$\begin{aligned} d^n y &= \left(\sum_{k=0}^{n} C_n^k u^{(k)} v^{(n-k)}\right)dx^n \\ &= \left(\sum_{k=0}^{n} \frac{n!}{k!(n-k)!} \cdot \frac{n!}{(n-k)!} \cdot x^{n-k}e^x\right)dx^n \end{aligned}$$

$$= \left[\sum_{k=0}^{n} \frac{(n!)^2}{k![(n-k)!]^2} \cdot x^{n-k} e^x \right] dx^n.$$

(4) Let u and v be secondly differentiable functions of x, find $d^2 y$ of the following functions:

① $y = uv$;

② $y = u/v$.

Solution ① $d^2 y = u d^2 v + 2 du dv + v d^2 u$.

② $d^2 y = d\left(d\left(\frac{u}{v}\right)\right) = d\left(\frac{v\,du - u\,dv}{v^2}\right)$

$$= \frac{v^2 d(v\,du - u\,dv) - (v\,du - u\,dv) d(v^2)}{v^4}$$

$$= \frac{v^2 (dv du + v\,d^2 u - du dv - u\,d^2 v) - 2v dv (v\,du - u\,dv)}{v^4}$$

$$= \frac{v(v\,d^2 u - u\,d^2 v) - 2dv(v\,du - u\,dv)}{v^3}.$$

Exercises

1. Find the derivatives of the following functions at the assigned points by definition of the derivative.

(1) $f(x) = \sqrt{x}$, $x_0 = 1$;　　(2) $f(x) = \frac{1}{x}$, $x_0 = -1$;

(3) $f(x) = \tan x$, $x_0 = \frac{\pi}{4}$;　　(4) $f(x) = \ln x$, $x_0 = e$.

2. Discuss the existence of the derivatives of following functions at $x_0 = 0$:

(1) $f(x) = \sqrt[3]{x}$;　　(2) $f(x) = |\sin x|$;

(3) $f(x) = \begin{cases} \ln(1+x) & (x \geqslant 0), \\ x & (x < 0); \end{cases}$

(4) $f(x) = \begin{cases} x^2 \sin \frac{1}{x} & (x \neq 0), \\ 0 & (x = 0). \end{cases}$

3. Let

$$f(x)=\begin{cases} x^2 & (x\leqslant 1), \\ ax+b & (x>1). \end{cases}$$

Determine a and b such that $f(x)$ is derivable at $x=1$.

4. Suppose a curve is expressed by

$$f(x)=\begin{cases} \ln x & (x\geqslant 1), \\ x-1 & (x<1). \end{cases}$$

(1) Sketch the graph of the curve and its tangent line at $x=1$;

(2) Present the equation of the tangent line.

5. Find the derivatives of the following functions:

(1) $\dfrac{mx^2+nx+p}{a+b}$;

(2) $2\sqrt[3]{x^2}-\dfrac{1}{x^2}+1$;

(3) $(\sqrt{x}+1)\left(\dfrac{1}{\sqrt{x}-1}\right)$;

(4) $\dfrac{1-x}{1+x}$;

(5) $\dfrac{2+3x+4x^2}{x^2}$;

(6) $x\sin x$;

(7) $2\tan x+\sec x-1$;

(8) $\dfrac{2}{\tan x}+\dfrac{\cot x}{3}$;

(9) $\dfrac{\tan x}{x}+\tan\dfrac{\pi}{3}$;

(10) $\dfrac{t}{1-\cos t}$;

(11) $(2+\sec t)\sin t$;

(12) a^x+e^x;

(13) $e^x\ln x$;

(14) $e^x\sin x$;

(15) $\dfrac{10^x-1}{10^x+1}$;

(16) $e^x(\sin x+\cos x)+e^{-x}$;

(17) $\dfrac{\ln x}{x^n}$;

(18) $\dfrac{1-\ln t}{1+\ln t}$;

(19) x^{-3x};

(20) $\ln(1-2x)$;

(21) $\dfrac{1}{(1-x)^2}$;

(22) $\sqrt{3-2x}$;

(23) $\cos(4-3x)$;

(24) $\tan(1-x)+\sec(1+x)$;

(25) $\tan\frac{x}{2}-\cot\frac{x}{2}$；

(26) $(10-x)^{10}$；

(27) $\arcsin(1-2x)$；

(28) $\arctan(1+x^2)$；

(29) $\sin(1-x^2)$；

(30) e^{1-2x^2}；

(31) $\sin\ln(1-x)$；

(32) $\frac{x}{\sqrt{1-x^2}}$；

(33) $e^{-\sin x}+\sin^2 x$；

(34) $\ln(x+\sqrt{a^2+x^2})$；

(35) $x\arcsin\sqrt{x}$；

(36) $x^2\sin\frac{1}{x}$；

(37) $\sqrt{\tan\frac{1}{x}}$；

(38) $\ln[\ln(\ln x)]$；

(39) $\tan(e^{-2x}+1)$；

(40) $10^{-\sin^3 2x}$；

(41) $\sqrt{x+\sqrt{x+\sqrt{x}}}$；

(42) $\arcsin\sqrt{1-4x}$；

(43) $\arctan\sqrt{6x-1}$；

(44) $\arccos\frac{1}{x}$；

(45) $\arcsin\frac{x+1}{x}$；

(46) $\cosh 2x$；

(47) $\sinh(1-x)+\cosh(1+x)$；

(48) $\sinh^2 x-\cosh^2 x$；

(49) $\sinh^2 x+\cosh^2 x$；

(50) $\tanh(1-x^2)$；

(51) $\sqrt{1+\sinh^2 x}$；

(52) $\frac{\cosh x}{1+\cosh^2 x}$；

(53) $\arctan\sqrt{\frac{1-x}{1+x}}$；

(54) $\tan x+\frac{1}{3}\tan^3 x+\frac{1}{5}\tan^5 x$；

(55) $\frac{x}{2}\sqrt{a^2-x^2}+\frac{a^2}{2}\arcsin\frac{x}{a}$；

(56) $e^{\tan^3\frac{1}{x}}$；

(57) $x^2\sqrt{1+\sqrt{x}}$；

(58) $x\arcsin\ln x$；

(59) $\sin[\sin(\sin x)]$；

(60) $\ln\frac{u+\sqrt{1+u^2}}{u}$.

6. Using logarithmic differentiation to find the derivatives of the following functions:

(1) x^{2x}；

(2) x^{e^x}；

(3) $(2x)^{\sqrt{x}}$；

(4) $(\sin x)^{\ln x}$; (5) $\sqrt{\dfrac{x(x-1)}{x^2+1}}$; (6) $\ln\dfrac{\sqrt{1-x}}{\sqrt{1+x}}$;

(7) $\sqrt{(x\sin x)\sqrt{1-e^x}}$; (8) $e^{\sqrt{\frac{1-x}{1+x}}}$;

(9) $\ln\left(x\sqrt{x+\sqrt{x}}\right)$; (10) $(x\sin x)^x$;

(11) $(\sin x)^{\cos x}$; (12) x^{x^a};

(13) a^{x^x}; (14) $\dfrac{x^2}{1-x}\sqrt{\dfrac{x+1}{1+x+x^2}}$.

7. Suppose the function $y=y(x)$ is given as follows. Find $\dfrac{dy}{dx}$.

(1) $\begin{cases} x=\dfrac{t-1}{t+1}, \\ y=\dfrac{t^2}{t+1}; \end{cases}$ (2) $\begin{cases} x=2e^t, \\ y=e^{-t}; \end{cases}$ (3) $\begin{cases} x=\sin t, \\ y=\cos 2t; \end{cases}$

(4) $\begin{cases} x=a(t-\sin t), \\ y=a(1-\cos t); \end{cases}$ (5) $\begin{cases} x=\cos^4 t, \\ y=\sin^4 t; \end{cases}$ (6) $\begin{cases} x=at\cos t, \\ y=at\sin t. \end{cases}$

8. Find $\dfrac{d^2y}{dx^2}$ of the following functions.

(1) $\begin{cases} x=2t-t^2, \\ y=3t-t^3; \end{cases}$ (2) $\begin{cases} x=e^t\cos t, \\ y=e^t\sin t; \end{cases}$ (3) $\begin{cases} x=f'(t), \\ y=tf'(t)-f(t). \end{cases}$

9. Suppose a curve C is given by

$$\begin{cases} x=a(\cos t+t\sin t), \\ y=a(\sin t-t\cos t). \end{cases}$$

Prove that the distance of the normal line at any point on C to the origin $(0,0)$ is identical.

10. Find the derivatives y'_x of the following implicit functions:

(1) $xy=1$; (2) $y=\cos(x+y)$;

(3) $x^{2/3}+y^{2/3}=a^{2/3}$; (4) $x=y+\arctan y$;

(5) $x^3+y^3-3a^2xy=0$; (6) $e^y-e^{-x}+xy=0$;

(7) $\arctan\dfrac{y}{x}=\ln\sqrt{x^2+y^2}$; (8) $x^y=y^x$.

11. Prove:

(1) $(\sinh x)' = \cosh x$;　　(2) $(\cosh x)' = \sinh x$;

(3) $(\tanh x)' = \dfrac{1}{\cosh^2 x}$;　　(4) $(\coth x)' = -\dfrac{1}{\sinh^2 x}$.

12. Find the derivatives of the following functions:

(1) $y = \sinh^3 x$;　　(2) $y = \cosh(\sinh x)$;

(3) $y = \ln(\cosh x)$;　　(4) $y = \arctan(\operatorname{arctanh} x)$;

(5) $y = \operatorname{arcsinh} x$;　　(6) $y = \operatorname{arccosh} x$;

(7) $y = \operatorname{arccoth} x$;　　(8) $y = \operatorname{arcsinh}(\tan x)$.

13. Suppose $f(x)$ is derivable at x_0. Prove:

$$\lim_{h \to 0} \frac{f(x_0 + \alpha h) - f(x_0 - \beta h)}{h} = (\alpha + \beta) f'(x_0).$$

14. Prove:

(1) The derivative function of a derivable even function is an odd function;

(2) The derivative function of a derivable odd function is an even function;

(3) The derivative function of a derivable periodical function is still a periodical function.

15. (1) Find the equations of the tangent line and the normal line of the curve l: $y = x(\ln x - 1)$.

(2) On the curve l: $y = x^2$, at which point the tangent line of l is parallel to the line $y = 4x - 5$? at which point the normal line of l is parallel to the line $2x - 6y = 5$? at which point the angle between the tangent line of l and the line $3x - y + 1 = 0$ is $\pi/4$?

(3) Determine the value of the parameter a such that the curves $y = ax^2$ and $y = \ln x$ are tac to each other. Locate the tac-point and give the equation of the tangent line.

(4) Suppose the $y = 2x$ is the tangent line of the curve $y = x^2 +$

$ax+b$ at the point (2, 4). Find a and b.

16. Find the differentials dy of the following functions:

(1) $y=2\sqrt{x}$; (2) $y=\sqrt{1+x^2}$; (3) $y=\cos ax$;

(4) $y=e^{-ax^2}$; (5) $y=\sinh 2x$; (6) $y=x\ln x$;

(7) $y=5^{\ln(\tan x)}$; (8) $y=\dfrac{\sin x}{2\cos^2 x}+\dfrac{1}{2}\ln\left|\tan\left(\dfrac{x}{2}+\dfrac{\pi}{4}\right)\right|$.

17. Find the differentials dy of the following implicit functions:

(1) $x^2+xy+y^2=3$; (2) $e^x\sin y-e^{-y}\cos x=0$.

18. Find the first and second order derivatives of the following functions:

(1) $y=e^{-x^2}$; (2) $y=\arcsin x$; (3) $y=\cos^2 x$;

(4) $y=x^3\ln x$; (5) $y=\dfrac{x}{\sqrt{1-x^2}}$; (6) $y=x\ln x$;

(7) $y=f(\ln x)$.

19. Find the n-th order derivatives of the following functions and determine $y^{(n)}(0)$:

(1) $y=\ln(1-x)$; (2) $y=\ln\left(\dfrac{1+x}{1-x}\right)$;

(3) $y=a^x\ (a>0)$; (4) $y=\sin^2 x$;

(5) $y=x^a\ (a\in\mathbb{R})$; (6) $y=e^{ax}\sin bx\ (a, b\in\mathbb{R})$.

20. Using Leibniz formula to find the higher order derivatives of the following functions:

(1) $y=x^2\sin x$, find $y^{(10)}$; (2) $y=x^3e^x$, find $y^{(50)}$;

(3) $y=\dfrac{\ln x}{x}$, find $y^{(n)}$.

21. Let $y=\arctan x$. Prove that it satisfies the equation $(1+x^2)y''+2xy'=0$, and find $y^{(n)}(0)$.

22. Let $y=\arcsin x$. Prove that it satisfies the equality $(1-x^2)y^{(n+2)}-(2n+1)xy^{(n+1)}-n^2y^{(n)}=0\ (n\geqslant 0)$, and find $y^{(n)}(0)$.

23. Let the function

$$f(x)=\begin{cases} e^{-\frac{1}{x^2}} & (x\neq 0), \\ 0 & (x=0). \end{cases}$$

Prove that any order derivative of $f(x)$ is existent at $x=0$ and it is identically equal to 0.

24. Find the higher order derivatives of the following implicit functions:

(1) $x^2+y^2=a^2$, find y'''; (2) $y=1+xe^y$, find y'';

(3) $x^2-xy+y^2=1$, find y''; (4) $\cos(x+y)=y$, find y''.

25. Find the higher order differentials of the following functions:

(1) $f(x)=\ln x$, $g(x)=e^x$, find $d^3(f(x)g(x))$, $d^3\left(\frac{f(x)}{g(x)}\right)$;

(2) $f(x)=e^{\frac{x}{2}}$, $g(x)=\cos 2x$, find $d^3(f(x)g(x))$, $d^3\left(\frac{f(x)}{g(x)}\right)$.

26. Find differentials of the following functions:

(1) $y=x+2x^2-\frac{1}{3}x^3+x^4$; (2) $y=x\ln x-x$;

(3) $y=x^2\sin x$; (4) $y=\frac{x}{1-x^2}$;

(5) $y=e^a x\sin bx$; (6) $y=\arcsin\sqrt{1-x^2}$.

27. Using differential to find approximate values:

(1) $\sqrt[3]{1.03}$; (2) $\lg 11$; (3) $\tan 45°10'$; (4) $\sqrt{26}$.

28. Find the higher order derivatives of the following functions given by parameter equations:

(1) $\begin{cases} x=at+b, \\ y=\frac{1}{2}at^2+bt, \end{cases}$ find y'_x, y''_x; (2) $\begin{cases} x=\frac{1}{1+t}, \\ y=\frac{t}{1+t}, \end{cases}$ find y''_x;

(3) $\begin{cases} x=a\cos t, \\ y=a\sin t, \end{cases}$ find y'''_x; (4) $\begin{cases} x=a\cos^3 t, \\ y=a\sin^3 t, \end{cases}$ find y''_x;

(5) $\begin{cases} x = a(t - \sin t), \\ y = a(1 - \cos t), \end{cases}$ find y''_x;

(6) $\begin{cases} x = at\ \cos t, \\ y = at\ \sin t, \end{cases}$ find y'_x, y''_x at $t = \frac{\pi}{2}$.

29. Let $x = \mathrm{e}^t$.

Express $x^2 \frac{\mathrm{d}^2 y}{\mathrm{d}x^2} + x \frac{\mathrm{d}y}{\mathrm{d}x} + y$ in terms of $\frac{\mathrm{d}^2 y}{\mathrm{d}t^2}$, $\frac{\mathrm{d}y}{\mathrm{d}t}$ and y.

30. Prove the following statements:

(1) Suppose $f(x)$ and $g(x)$ are both derivable at $x = 0$, and $g'(0) \neq 0$, $g(x) \neq 0 (x \neq 0)$, $f(0) = g(0) = 0$. Then

$$\lim_{x \to 0} \frac{f(x)}{g(x)} = \frac{f'(0)}{g'(0)}.$$

(2) Suppose $f(x)$ is defined on $(-\infty, +\infty)$ such that $f(x+y) = f(x) \cdot f(y)$ for any x, $y \in (-\infty, +\infty)$, $f(x) = 1 + xg(x)$ and $\lim_{x \to 0} g(x) = 1$. Then $f(x)$ is derivable on $(-\infty, +\infty)$.

(3) Suppose $y = f(x)$ and $y = \varphi(x)$ are both derivable at x_0. Then the curves $y = f(x)$ and $y = \varphi(x)$ are tac to each other at x_0 if and only if $f(x) - \varphi(x)$ is a higher order infinitesimal over $x - x_0$ as $x \to x_0$.

31. Suppose $g(x)$ is derivable at $x = 0$, and

$$f(x) = \begin{cases} g(x) \sin \frac{1}{x} & (x \neq 0), \\ 0 & (x = 0). \end{cases}$$

Determine the values $g(0)$ and $g'(0)$ such that $f(x)$ is derivable at $x = 0$.

32. A manufacturer's production output P for a certain product is given by $P = 64x^{0.75} y^{0.25}$, where x and y denote the number of units of labor and capital, respectively, required as production inputs.

Suppose that the number of units of labor and capital required as production inputs are increasing by 1/4 and 1/3 units per year, respectively. How fast will the production output be changing when 81 units of labor and 16 units of capital are required as inputs?

33. At noon a ship sailed east from a Caribbean port at 32 knots per hour. One hour later another ship left this port sailing north at 30 knots per hour. How fast will the distance between the ships be increasing at 5 p. m?

34. A metal cube expands uniformly when heated. If the length of a side is increasing at the rate of 0. 2 inches per hour (1 inch = 2. 54 cm), how fast is the volume of the cube changing when the length of a side is 10 inches?

35. A man 6 feet tall is walking away from a lamppost at a constant speed of 5 feet per second (1 foot = 30. 48 cm). If the light at the top of the lamppost is 18 feet above the sidewalk, how fast is the length of the man's shadow increasing?

36. A metal disk with a 9-inch radius expands when heated. If its radius increases at the rate of 0. 01 inch per second, how fast is the area of its top face increasing when the radius is 9. 1 inches?

37. Two spotlights, each 60 feet high, are located on poles 100 feet apart. One of the lights is working, and the other is not. A workman repairing the second light lowers his toolkit from the top of the second pole at the rate of 5 feet per second. How fast is the shadow of the toolkit moving when it is 30 feet above the ground?

38. A certain piece of computer software is expected to sell approximately 30, 000 - $150p$ copies at a price of p dollars. If the price of the software is increased by 1 dollar per month, how will the revenue change when the price of the software reaches 80 dollars per copy?

39. A child's kite was flying 90 feet above the ground when it got caught in a wind gust that blew the kite horizontally at a rate of 15 feet per second. How fast will the child be letting out kite-string when there are 150 feet of string already out? (Assume that the kite-string is straight.)

40. An American coffee distributor can sell approximately 10,000 −1250p pounds of coffee each week at a price of p dollars per pound. Because of a severe frost in South America, the cost of imported coffee beans is increasing at the rate of $0.10 per pound each week. If the distributor passes the entire cost increase along to his customers, at what rate will the distributor's revenue be changing when the price of coffee reaches $5.00 per pound?

41. A farmer wants to raise a bale of hay into the loft of her barn. To do so, she has run a 59-foot rope through a pulley attached to the side of the barn at a height of 25 feet above the ground. One end of the rope is attached to the hay and the other end to the hitch of a tractor 1 foot above the ground. If the farmer drives the tractor away from the barn at a speed of 10 feet per second, how fast will the hay be rising when the tractor is 32 feet from the barn?

42. Prove the formula:

$$\left(\frac{ax+b}{cx+d}\right)' = \frac{\begin{vmatrix} a & b \\ c & d \end{vmatrix}}{(cx+d)^2}.$$

43. Let

$$f(x) = \begin{cases} x^n \sin\dfrac{1}{x} & (\text{if } x \neq 0), \\ 0 & (\text{if } x = 0). \end{cases}$$

How large should the positive integer n be such that $f''(0)$ exists? In this case find the value of $f''(0)$.

44. Suppose $\varphi(x)$ is continuous at point a and $f(x) =$

$|x-a|\varphi(x)$. Find $f'_+(a)$ and $f'_-(a)$. Under which condition is $f'(a)$ existent?

45. Assume $f(x)$ is derivable. Find y' of the following functions:

(1) $y=f(x^2)$; (2) $y=f(e^x)e^{f(x)}$; (3) $y=f(f(f(x)))$.

46. Assume $\varphi(x)$ and $\psi(x)$ are both derivable. Find y' of the following functions:

(1) $y=\sqrt{\varphi^2(x)+\psi^2(x)}$; (2) $y=\arctan\dfrac{\varphi(x)}{\psi(x)}(\psi(x)\neq 0)$;

(3) $y=\log_{\varphi(x)}\psi(x)$ $(\varphi, \psi>0)$.

47. Suppose $x=g(y)$ is the inverse function of $y=f(x)$. Express $g'''(y)$ in terms of $f(x)$, $f'(x)$ and $f'''(x)$.

48. Suppose $f_{ij}(x)(i, j=1, 2, \cdots, n)$ are derivable functions. Prove that

$$\frac{d}{dx}\begin{vmatrix} f_{11}(x) & f_{12}(x) & \cdots & f_{1n}(x) \\ f_{21}(x) & f_{22}(x) & \cdots & f_{2n}(x) \\ \vdots & \vdots & & \vdots \\ f_{n1}(x) & f_{n2}(x) & \cdots & f_{nn}(x) \end{vmatrix} = \sum_{k=1}^{n}\begin{vmatrix} f_{11}(x) & f_{12}(x) & \cdots & f_{1n}(x) \\ f_{21}(x) & f_{22}(x) & \cdots & f_{2n}(x) \\ \vdots & \vdots & & \vdots \\ f'_{k1}(x) & f'_{k2}(x) & \cdots & f'_{kn}(x) \\ \vdots & \vdots & & \vdots \\ f_{n1}(x) & f_{n2}(x) & \cdots & f_{nn}(x) \end{vmatrix},$$

and using this result to find $F'(x)$ of the following functions $F(x)$:

(1) $F(x)=\begin{vmatrix} x-1 & 1 & 2 \\ -3 & x & 3 \\ -2 & -3 & x+1 \end{vmatrix}$;

(2) $F(x)=\begin{vmatrix} x & x^2 & x^3 \\ 1 & 2x & 3x^2 \\ 0 & 2 & 6x \end{vmatrix}$.

Chapter 5 Mean value theorems and applications of derivative

In this chapter, we will concentrate on results known as the mean value theorems. First stated by the French mathematician Joseph Louis Lagrange (1736 - 1813), these theorems have come to permeate the theoretical structure of the theory of calculus.

§ 5. 1 Mean value theorems

§ 5. 1. 1 Fermat theorem

Definition Suppose $f(x)$ is defined on $U(x_0)$, a neighborhood of the point x_0. If

$$f(x_0) \geqslant f(x)(f(x_0) \leqslant f(x))$$

for any $x \in U(x_0)$, then we say $f(x)$ obtains its *local maximum* (*local minimum*) $f(x_0)$ at x_0, and the point x_0 is called a *local maximal point* (*local minimal point*); the term *local extremum* may be used to refer to either of local maximum and local minimum, and the term *local extremal point* may be used to refer to either of local maximal point and local minimal point. Obviously, the nature of a local extremum is dictated by the local behavior of the function.

Theorem 5. 1. 1 (Fermat theorem) Suppose $f(x)$ is defined on some neighborhood of a point x_0. If $f(x)$ obtains its local extremum at x_0 and $f(x)$ is differentiable at x_0, then $f'(x_0) = 0$.

Proof Without loss of generality, assume $f(x_0)$ is a local maximum. Then there is $U(x_0)$, a neighborhood of x_0, such that $f(x) \leqslant f(x_0)$ for any $x \in U(x_0)$. Thus

$$\frac{f(x)-f(x_0)}{x-x_0} \geqslant 0 \text{ if } x < x_0;$$

and

$$\frac{f(x)-f(x_0)}{x-x_0} \leqslant 0 \text{ if } x > x_0.$$

Then

$$f'(x_0) = f'_-(x_0) = \lim_{x \to x_0^-} \frac{f(x)-f(x_0)}{x-x_0} \geqslant 0;$$

$$f'(x_0) = f'_+(x_0) = \lim_{x \to x_0^+} \frac{f(x)-f(x_0)}{x-x_0} \leqslant 0.$$

Therefore, $f'(x_0) = 0$. □

Note Geometric interpretation of Fermat theorem:

From a geometrical viewpoint, the Fermat theorem can be interpreted as: If a curve $y = f(x)$ obtains its local extremum at x_0, and the curve has a tangent line at x_0, then this tangent line must be horizontal (cf. Figure 5 - 1).

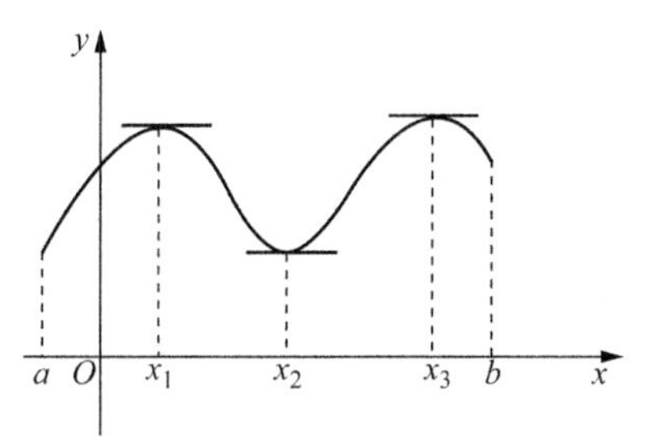

Figure 5 - 1

Note that $f'(x)$ represents the rate of the change of $f(x)$ at x. Hence, $f'(x_0) = 0$ explains that the rate of the change of $f(x)$ at x_0 is zero. Such a point x_0 is called a *stationary point* of $f(x)$. A point x_0 at which $f'(x_0)$ does not exist is called a *singular point* of $f(x)$. The Fermat theorem can be further interpreted as: The necessary condition for a differentiable function $f(x)$ to obtain its local extremum at x_0 is that x_0 is a stationary point of $f(x)$.

Example Suppose $f(x)$ is continuous on $[a, b]$ and differentiable on (a, b). Prove: If $f'_+(a)f'_-(b) < 0$, then there exists $x_0 \in (a, b)$ such that $f'(x_0) = 0$.

Proof Without loss of generality, we may assume $f'_+(a) > 0$ and $f'_-(b) < 0$, i. e.

$$\lim_{x \to a^+} \frac{f(x) - f(a)}{x - a} > 0 \text{ and } \lim_{x \to b^-} \frac{f(x) - f(b)}{x - b} < 0.$$

By the locally sign-preserving property of the limit, there exist $x_1, x_2 \in (a, b)$ such that

$$\frac{f(x_1) - f(a)}{x_1 - a} > 0, \frac{f(x_2) - f(b)}{x_2 - b} < 0,$$

which implies $f(x_1) > f(a)$ and $f(x_2) > f(b)$, and so $f(a) \neq \max_{[a, b]} f(x)$ and $f(b) \neq \max_{[a, b]} f(x)$. Hence, there is $x_0 \in (a, b)$ with $f(x_0) = \max_{[a, b]} f(x)$ as f is continuous on $[a, b]$. Notice $f(x_0)$ must be a local maximum and f is differentiable at $x = x_0$, by Fermat theorem $f'(x_0) = 0$. □

Suppose $f(x)$ is differentiable on an interval (a, b). If the one-sided derivatives $f'_+(a)$ and $f'_-(b)$ are both existent, we also say *$f(x)$ is differentiable on $[a, b]$*. As an application of Fermat theorem, we show the following

Theorem 5. 1. 2 (Darboux theorem) Suppose $f(x)$ is differentiable on $[a, b]$ with $f'_+(a) \neq f'_-(b)$. Then for any number p which is strictly between $f'_+(a)$ and $f'_-(b)$, there exists $c \in (a, b)$ such that $f'(c) = p$.

Proof First we assume $f'_+(a)f'_-(b) < 0$. By above example, there is a point $c \in (a, b)$ such that $f'(c) = 0$.

Now we let $f'_+(a) \neq f'_-(b)$, say, $f'_+(a) > f'_-(b)$ and $f'_+(a) > p > f'_-(b)$. Setting $\varphi(x) = f(x) - px$, we have $\varphi'(a) = f'_+(a) - p >$

0 and $\varphi'(b) = f'_{-}(b) - p < 0$. So, there exists $c \in (a, b)$ with $\varphi'(c) = 0$, i.e. $f'(c) = p$. □

The next corollary follows directly from Darboux theorem.

Corollary Let I be an open interval and suppose $f(x)$ is differentiable on I. Then the image of $f'(x)$ is still an interval; that is, if a, $b \in I$ and p is strictly between $f'(a)$ and $f'(b)$, then there exists $c \in I$ such that $f'(c) = p$.

Example Let

$$f(x) = \begin{cases} 1 & (\text{if } x \in [0, +\infty)), \\ 0 & (\text{if } x \in (-\infty, 0)). \end{cases}$$

Then the image of $f(x)$ consists of two points, so the image is certainly not an interval. It follows from the above corollary that there does not exist a differentiable function $F(x)$ such that $F'(x) = f(x)$ for all $x \in (-\infty, +\infty)$.

§5.1.2 Rolle theorem, Lagrange theorem and Cauchy theorem

One immediate consequence of Fermat theorem has acquired a special name as follows.

Theorem 5.1.3 (Rolle theorem) Let $f(x)$ be continuous on the interval $[a, b]$ and differentiable on (a, b). If $f(a) = f(b)$, there is a point $x_0 \in (a, b)$ with $f'(x_0) = 0$.

Proof If f is a constant function, any choice of the point x_0 will do. If f is not constant, then it must have either a local minimum or a local maximum at some point $x_0 \in (a, b)$; and since f is differentiable there, $f'(x_0) = 0$ in the light of Fermat theorem. □

Note Geometric interpretation of Rolle theorem:

Rolle theorem points out that a suffciently smooth curve defined on $[a, b]$ which takes on same values at endpoints must have a horizontal tangent line at a point somewhere between a and b. (cf.

Figure 5-2)

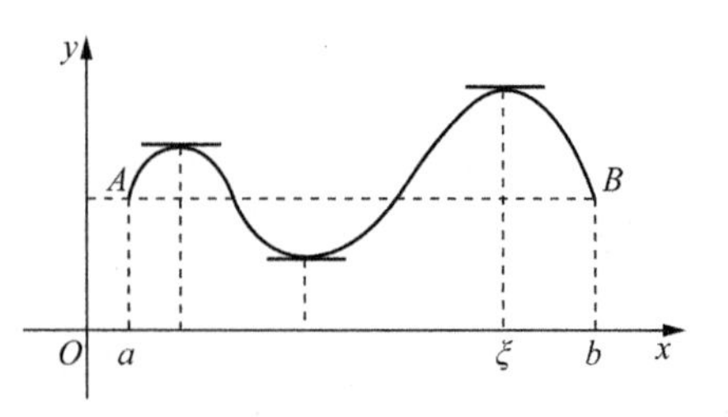

Figure 5-2

Corollary 1 If $f(x)$ is differentiable on the interval (a, b), then the zeros of f on (a, b) are separated by zeros of f'.

Corollary 2 Let f and g be continuous on $[a, b]$ and differentiable on (a, b). Suppose that $f(a)=g(a)$ and $f(b)=g(b)$. Then, there is at least one point $x_0 \in (a, b)$ such that $f'(x_0) = g'(x_0)$.

Corollary 1 follows directly from Rolle theorem and Corollary 2 follows from Corollary 1 if we consider the function $f-g$.

Examples (1) Suppose $f(x)$ is continuous on $[0, 1]$ and differentiable on $(0, 1)$. If $0<f(x)<1$ on $[0, 1]$ and $f'(x)\neq 1$ on $(0, 1)$, show that there exists a unique point $x_0 \in (0, 1)$ such that $f(x_0)=x_0$.

Proof Let $g(x)=f(x)-x$. Then $g(x)$ is continuous on $[0, 1]$ and differentiable on $(0, 1)$ with $g(0)=f(0)>0$ and $g(1)=f(1)-1<0$. By the intermediate value theorem there exists a point $x \in (0, 1)$ such that $g(x_0)=0$, i. e. $f(x_0)=x_0$. Assume there is another $x_1 \in (0, 1)$ such that $f(x_1)=x_1$. Since $g(x_0)=g(x_1)(=0)$, by Rolle theorem there exists a point c between x_0 and x_1 with $g'(c)=0$, i. e. $f'(c)=1$, which yields a contradiction. □

(2) Suppose $f(x)$ is continuous on $[0, 1]$, differentiable on $(0, 1)$, and $f(1)=0$. Prove:

① for any positive integer n there is $\xi \in (0, 1)$ such that $nf(\xi)+\xi f'(\xi)=0$;

② there is $\xi \in (0, 1)$ such that $f'(\xi)=-\dfrac{f(\xi)}{\xi}$.

Proof ① Let $h(x) = x^n f(x)$. Then $h(0) = h(1)$. So by Rolle theorem there is $\xi \in (0, 1)$ with $h'(\xi) = 0$, i. e. $n\xi^{n-1} f(\xi) + \xi^n f'(\xi) = 0$ and so $nf(\xi) + \xi f'(\xi) = 0$ as $\xi \neq 0$.

② Let $n = 1$, and then the second result follows. □

(3) Suppose $f(x)$ is continuous on $[0, 1]$, $f''(x)$ exists on $(0, 1)$ and $f(0) = f(1) = f'(1) = 1$. Prove that there exists $x_0 \in (0, 1)$ such that $f''(x_0) = 2$.

Proof Let $F(x) = f(x) - x(x-1)$. Then $F(0) = F(1)$ and $F'(x) = f'(x) - 2x + 1$. So by Rolle theorem there exists $c \in (0, 1)$ such that $F'(c) = 0$. Note $F'(1) = 0$. There is $x_0 \in (c, 1)$ such that $F''(x_0) = 0$, i. e. $f''(x_0) = 2$. □

(4) Let f and g be functions such that $f''(x)$ and $g''(x)$ both exist on $[a, b]$ where $g''(x) \neq 0$ on $[a, b]$. Suppose $f(a) = f(b) = g(a) = g(b) = 0$. Prove that

① for any $x \in (a, b)$ $g(x) \neq 0$;

② there exists $x_0 \in (a, b)$ such that $f(x_0)/g(x_0) = f''(x_0)/g''(x_0)$.

Proof ① Assume there is a point $c \in (a, b)$ with $g(c) = 0$. Since $g(a) = g(b) = 0$, by Rolle theorem there exist $x_1 \in (a, c)$ with $g'(x_1) = 0$ and $x_2 \in (c, b)$ with $g'(x_2) = 0$. Using Rolle theorem again we see that there exist $x_3 \in (x_1, x_2)$ with $g''(x_3) = 0$, which yields a contradiction.

② Let $h(x) = f(x)g'(x) - f'(x)g(x)$. Then $h(a) = h(b) = 0$, and so there exists $x_0 \in (a, b)$ such that $h'(x_0) = 0$, i. e. $f(x_0)g''(x_0) - f''(x_0)g(x_0) = 0$, which directly implies the result as $g(x_0) \neq 0$ and $g''(x_0) \neq 0$. □

(5) Suppose $f(x)$ is continuous on $[a, b]$ and differentiable on (a, b). If $f(x) \neq 0$ on (a, b) and $f(a) = f(b) = 0$, show that there exists $x_0 \in (a, b)$ such that $f'(x_0) = kf(x_0)$, where k is a real number.

Proof Let $g(x) = e^{-kx}f(x)$. Then $g(a) = g(b) = 0$, and so there is $x_0 \in (a, b)$ with $g'(x_0) = 0$, i. e. $e^{-kx_0}(f'(x_0) - kf(x_0)) = 0$, from which the result clearly follows. □

(6) Suppose $p(x)$ and $q(x)$ are both continuous on $[a, b]$ and differentiable on (a, b). If there are two different points $x_1, x_2 \in (a, b)$ with $p(x_1) = p(x_2) = 0$, show that there exists a point x_0 between x_1 and x_2 such that $p'(x_0) + p(x_0)q'(x_0) = 0$.

Proof Let $x_1 < x_2$ and let $g(x) = p(x)e^{q(x)}$. Then $g(x)$ is continuous on $[x_1, x_2]$ and differentiable on (x_1, x_2). Since $p(x_1) = p(x_2) = 0$, $g(x_1) = g(x_2) = 0$, and so there is $x_0 \in (x_1, x_2)$ such that $g'(x_0) = 0$, i. e. $e^{q(x_0)}(p'(x_0) + p(x_0)q'(x_0)) = 0$, from which the result follows. □

(7) Suppose $f(x)$ and $g(x)$ are both continuous on $[a, b]$ and differentiable on (a, b). If $g'(x) \neq 0$ for any $x \in (a, b)$, show that there exists $x_0 \in (a, b)$ such that

$$\frac{f(a) - f(x_0)}{g(x_0) - g(b)} = \frac{f'(x_0)}{g'(x_0)}.$$

Proof Applying Rolle theorem to function $F(x) = (f(a) - f(x))g(x) + f(x)g(b)$ on $[a, b]$, we see that there exists $x_0 \in (a, b)$ with $F'(x_0) = (f(a) - f(x_0))g'(x_0) - (g(x_0) - g(b))f'(x_0) = 0$. Noticing $g'(x) \neq 0$ implies $g(x_0) - g(b) \neq 0$, we have the result immediately. □

(8) Suppose $f(x)$ and $g(x)$ are differentiable on $(-\infty, +\infty)$ such that $f(x)g'(x) - f'(x)g(x)$ has no root on $(-\infty, +\infty)$. Prove that between any two neighboring roots of $f(x)$, there exists a root of $g(x)$.

Proof Let x_1 and x_2 $(x_1 < x_2)$ be two neighboring roots of $f(x)$, i. e. $f(x_1) = f(x_2) = 0$. If $g(x)$ has no root in (x_1, x_2), i. e. $g(x) \neq 0$ for any $x \in (x_1, x_2)$, noting also $g(x_1) \neq 0$ and $g(x_2) \neq 0$

(because otherwise, x_1 or x_2 will be a root of $f(x)g'(x) - f'(x)g(x)$), we may define $h(x) = \dfrac{f(x)}{g(x)}$ on $[x_1, x_2]$. Since $h(x_1) = h(x_2) (= 0)$, by Rolle theorem, there is $c \in (x_1, x_2)$ with $h'(c) = 0$, which implies c is a root of $f(x)g'(x) - f'(x)g(x)$, contradicting the condition. □

(9) Suppose $f(x)$ is continuous on $[a, b]$ and $f''(x)$ is existent on (a, b). If $f'_+(a) < 0$, $f'_-(b) < 0$ and $f(a) = f(b) = 0$, prove that there exists $c \in (a, b)$ such that $f''(c) = 0$.

Proof Since

$$\lim_{x \to a^+} \frac{f(x)}{x-a} = \lim_{x \to a^+} \frac{f(x) - f(a)}{x-a} = f'_+(a) < 0,$$

by the Sign-preserving theorem there exists $c_1 \in (a, b)$ with

$$\frac{f(c_1)}{c_1 - a} < 0, \text{ and so } f(c_1) < 0;$$

similarly, since

$$\lim_{x \to b^-} \frac{f(x)}{x-b} = \lim_{x \to b^-} \frac{f(x) - f(b)}{x-a} = f'_-(b) < 0,$$

there exists $c_2 \in (c_1, b)$ with

$$\frac{f(c_2)}{c_2 - b} < 0, \text{ and so } f(c_2) > 0.$$

Thus, by the intermediate value theorem there is $c_3 \in (c_1, c_2)$ such that $f(c_3) = 0$.

Then by Rolle theorem, there exists $c_4 \in (a, c_3)$ with $f'(c_4) = 0$, and there exists $c_5 \in (c_3, b)$ with $f'(c_5) = 0$. Now, using Rolle theorem on f', we see there exists $c \in (c_4, c_5)$ such that $f''(c) = 0$. □

The next theorem can be regarded as a generalization of Rolle theorem.

Theorem 5.1.4 Let a, $b \in \mathbb{R}$.

(1) If $f(x)$ is differentiable on (a, b) and there is $r \in \mathbb{R}$ such that $\lim_{x \to a^+} f(x) = \lim_{x \to b^-} f(x) = r$, then there exists $x_0 \in (a, b)$ with $f'(x_0) = 0$;

(2) If $f(x)$ is differentiable on $(a, +\infty)$ and there is $r \in \mathbb{R}$ such that $\lim_{x \to a^+} f(x) = \lim_{x \to +\infty} f(x) = r$, then there exists $x_0 \in (a, +\infty)$ with $f'(x_0) = 0$;

(3) If $f(x)$ is differentiable on $(-\infty, b)$ and there is $r \in \mathbb{R}$ such that $\lim_{x \to -\infty} f(x) = \lim_{x \to b^-} f(x) = r$, then there exists $x_0 \in (-\infty, b)$ with $f'(x_0) = 0$;

(4) If $f(x)$ is differentiable on $(-\infty, +\infty)$ and there is $r \in \mathbb{R}$ such that $\lim_{x \to -\infty} f(x) = \lim_{x \to +\infty} f(x) = r$, then there exists $x_0 \in (-\infty, +\infty)$ with $f'(x_0) = 0$.

Proof We merely prove the statements (1) and (4). The statements (2) and (3) can be similarly proved.

(1) Define $f(a) = f(b) = r$. Then f is continuous on $[a, b]$, and so from Rolle theorem the result follows directly.

(4) If for any $x \in \mathbb{R}$, $f(x) = r$, the result holds clearly. Now let $b \in \mathbb{R}$ with $f(b) \neq r$, say, $f(b) > r$. Thus, for $\varepsilon_0 = f(b) - r$, there is $M_1 > 0$ such that $|f(x) - r| < \varepsilon_0$ whenever $x < -M_1$ and there is $M_2 > 0$ such that $|f(x) - r| < \varepsilon_0$ whenever $x > M_2$, which implies that $f(x) < f(b)$ on $(-\infty, -M_1) \cup (M_2, +\infty)$. Clearly, $f(x)$ is continuous on $[M_1, M_2]$, and so there exists $x_0 \in [M_1, M_2]$ with $f(x_0) = \max_{x \in [M_1, M_2]} f(x)$. Notice $b \in [M_1, M_2]$ and $f(b) \leqslant f(x_0)$. Then $f(x) \leqslant f(x_0)$ for any $x \in \mathbb{R}$, and then by Fermat theorem, we have $f'(x_0) = 0$. □

Example Suppose $f(x)$ is differentiable on $(0, +\infty)$ and $0 \leqslant f(x) \leqslant x/(1 + x^2)$. Show that there is a point $x_0 > 0$ such that

$$f'(x_0) = \frac{1-x_0^2}{(1+x_0^2)^2}.$$

Proof Construct a new function

$$g(x) = \frac{x}{1+x^2} - f(x).$$

Then $g(x)$ is differentiable on $(0, +\infty)$. Since $0 \leqslant f(x) \leqslant x/(1+x^2)$,

$$0 \leqslant \lim_{x\to 0^+} f(x) \leqslant \lim_{x\to 0^+} \frac{x}{1+x^2} = 0, \text{ and so } \lim_{x\to 0^+} f(x) = 0,$$

which implies $\lim_{x\to 0^+} g(x) = 0$. Similarly, as $0 \leqslant f(x) \leqslant x/(1+x^2)$ and $\lim_{x\to+\infty} \frac{x}{1+x^2} = 0$, $\lim_{x\to+\infty} f(x) = 0$, which implies $\lim_{x\to+\infty} g(x) = 0$. Hence, by Theorem 5.1.4 there exists $x_0 \in (0, +\infty)$ such that

$$g'(x_0) = 0, \text{ i.e. } f'(x_0) = \frac{1-x_0^2}{(1+x_0^2)^2}. \square$$

Theorem 5.1.5 (Lagrange theorem) Let $f(x)$ be continuous on the interval $[a, b]$, and let $f(x)$ be differentiable on (a, b). Then there is $x_0 \in (a, b)$ such that

$$f'(x_0) = \frac{f(b)-f(a)}{b-a}.$$

Proof Form a new function

$$F(x) = f(x) - \frac{f(b)-f(a)}{b-a}x.$$

It is routine to check that $F(x)$ satisfies all conditions of Rolle theorem, and so there is a point $x_0 \in (a, b)$ such that

$$F'(x_0) = f'(x_0) - \frac{f(b)-f(a)}{b-a} = 0. \square$$

Lagrange theorem provides us with a way to research the global property of a function from its local property.

Remark (1) Geometric interpretation of Lagrange theorem:

Lagrange theorem states that a suffciently smooth curve joining two points A and B has somewhere on the curve a tangent line with the same slope as the chord AB. (cf. Figure 5 - 3).

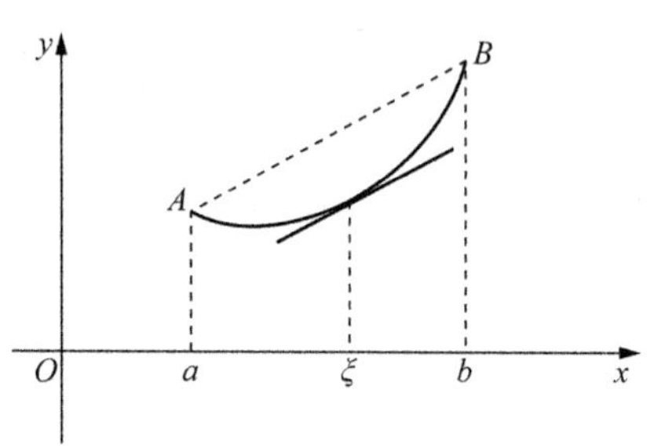

Figure 5 - 3

(2) The expression

$$f'(x_0)=\frac{f(b)-f(a)}{b-a}$$

is also called Lagrange formula, which can be presented as

$$f(b)-f(a)=f'(x_0)(b-a),\ a<x_0<b,$$

or

$$f(b)-f(a)=f'(a+\theta(b-a))(b-a)\quad (0<\theta<1).$$

(3) In Lagrange theorem, if we suppose $f(a)=f(b)$, then there exists $x_0\in(a,b)$ with

$$f'(x_0)=\frac{f(b)-f(a)}{b-a}=0.$$

This is just Rolle theorem. Thus, Lagrange theorem is a generalization of Rolle theorem.

Corollary 1 Suppose $f(x)$ is differentiable on the interval I and $f'(x)\equiv 0$. Then $f(x)$ is a constant function on I.

Proof Take $x_0\in I$. Then for any $x\in I\backslash\{x_0\}$, by Lagrange theorem

$$f(x)-f(x_0)=f'(\xi)(x-x_0)$$

where ξ is some point between x and x_0. Thus $f(x)-f(x_0)=0$ since $f'(x)\equiv 0$, which implies $f(x)=f(x_0)$, a constant. □

Corollary 2 Let $y=f(x)$ with $f'(x)$ being bounded on an interval I, then $f(x)$ satisfies Lipschitz condition on I (cf. Section 3.8), i.e. there is $M>0$ such that for any $x_1, x_2\in I$,

$$|f(x_1)-f(x_2)|\leqslant M|x_1-x_2|.$$

Proof Without loss of generality, let $x_1<x_2$. Applying Lagrange theorem on $[x_1, x_2]$, we have

$$f(x_1)-f(x_2)=f'(\xi)(x_1-x_2),$$

where ξ is between x_1 and x_2. Since $f'(x)$ is bounded on I, there is $M>0$ such that for any $x\in I$, $|f'(x)|<M$, and so $|f'(\xi)|<M$. Then the result follows clearly. □

Examples (1) Suppose $f''(x)$ exists on $[0, a]$ with $|f''(x)|\leqslant M$ on $(0, a)$. If $f(x)$ attains its maximum on $(0, a)$, prove that $|f'(0)|+|f'(a)|\leqslant Ma$.

Proof Let $c\in(0, a)$ with $f(c)=\max\limits_{[0, a]}f(x)$. By Fermat theorem, $f'(c)=0$. Then by Lagrange theorem on $[0, c]$, $f'(c)-f'(0)=f''(\xi_1)(c-0)$ where $\xi_1\in(0, c)$, i.e. $f'(0)=-cf''(\xi_1)$, which implies $|f'(0)|\leqslant Mc$. Now using Lagrange theorem on $[c, a]$, we obtain $f'(a)-f'(c)=f''(\xi_2)(a-c)$ where $\xi_2\in(c, a)$, i.e. $f'(a)=f''(\xi_2)(a-c)$, which implies $|f'(a)|\leqslant M(a-c)$. Hence, $|f'(0)|+|f'(a)|\leqslant Mc+M(a-c)=Ma$.

(2) Let $f(x)$ be continuous for $x\in[a, b]$. Assume $f''(x)$ exists on (a, b) and there exists $c\in(a, b)$ with $\max\{f(a), f(b)\}<f(c)$. Verify that there exists $\xi\in(a, b)$ such that $f''(\xi)<0$.

Proof The use of Lagrange theorem and the inequality $\max\{f(a), f(b)\}<f(c)$ give that

$$f'(\xi_1) = \frac{f(c) - f(a)}{c - a} > 0 \text{ for some } \xi_1 \in (a, c), \text{ and}$$

$$f'(\xi_2) = \frac{f(b) - f(c)}{b - c} < 0 \text{ for some } \xi_2 \in (c, b).$$

Thus, there exists $\xi \in (\xi_1, \xi_2) \subseteq (a, b)$ such that

$$f''(\xi) = \frac{f'(\xi_2) - f'(\xi_1)}{\xi_2 - \xi_1} < 0. \ \square$$

(3) Suppose $f'(x)$ is existent and decreasing on $[0, c]$. If $f(0) = 0$, show that $f(a+b) \leqslant f(a) + f(b)$ for $0 \leqslant a \leqslant b \leqslant a + b \leqslant c$.

Proof By Lagrange theorem, $f(a) - f(0) = f'(\xi_1)(a - 0) = af'(\xi_1)$ for some $\xi_1 \in (0, a)$ and $f(a+b) - f(b) = f'(\xi_2)(a+b-b) = af'(\xi_2)$ for some $\xi_2 \in (b, a+b)$. Since $f'(x)$ is decreasing, $f'(\xi_1) \geqslant f'(\xi_2)$ and so

$$f(a) = f(a) - f(0) \geqslant f(a+b) - f(b). \ \square$$

(4) Suppose $f(x)$ is continuous on $[0, 1]$, differentiable on $(0, 1)$ and $f(0) = 0$, $f(1) = \frac{1}{2}$. Prove that there exist $\xi, \eta \in (0, 1)$ $(\xi \neq \eta)$ such that $f'(\xi) + f'(\eta) = \xi + \eta$.

Proof Let $F(x) = f(x) - \frac{1}{2}x^2$. By Lagrange theorem,

$$F(0) - F\left(\frac{1}{2}\right) = F'(\xi)\left(0 - \frac{1}{2}\right) = -\frac{1}{2}(f'(\xi) - \xi)$$

for some $\xi \in \left(0, \frac{1}{2}\right)$, and

$$F(1) - F\left(\frac{1}{2}\right) = F'(\eta)\left(1 - \frac{1}{2}\right) = \frac{1}{2}(f'(\eta) - \eta)$$

for some $\eta \in \left(\frac{1}{2}, 1\right)$. Noticing $F(0) = F(1)(= 0)$, we have

$$-\frac{1}{2}(f'(\xi)-\xi)=\frac{1}{2}(f'(\eta)-\eta),$$

from which the result follows.

Theorem 5.1.6 (Cauchy theorem) Let f and g be continuous on $[a, b]$ with $g(a)\neq g(b)$. If f and g are both differentiable on (a, b) with $(f'(x))^2+(g'(x))^2\neq 0$, then there is $x_0\in(a, b)$ such that

$$\frac{f'(x_0)}{g'(x_0)}=\frac{f(b)-f(a)}{g(b)-g(a)}.$$

Proof Noticing $g(a)\neq g(b)$, we may form a new function

$$F(x)=\frac{f(b)-f(a)}{g(b)-g(a)}g(x)-f(x).$$

It is routine to check that $F(x)$ satisfies all the conditions of Lagrange theorem, and so there is $x_0\in(a, b)$ such that

$$F'(x_0)=\frac{f(b)-f(a)}{g(b)-g(a)}g'(x_0)-f'(x_0)=0.$$

Since $(f'(x_0))^2+(g'(x_0))^2\neq 0$, $g'(x_0)\neq 0$, from which the result follows.

Examples (1) Suppose $f(x)$ is continuous on $[a, b]$ and differentiable on (a, b) $(b>a>0)$. Prove that there exists $\xi\in(a, b)$ such that $2\xi(f(b)-f(a))=(b^2-a^2)f'(\xi)$.

Proof Only need to take $g(x)=x^2$. Then the result follows directly from Cauchy theorem. □

(2) Suppose $f(x)$ is continuous on $[a, b]$, differentiable on (a, b) $(0<a<b)$. Prove that there exist ξ, $\eta\in(a, b)$ such that

$$f'(\xi)=(a^2+ab+b^2)\frac{f'(\eta)}{3\eta^2}.$$

Proof Let $g(x)=x^3$. By Cauchy theorem, there is $\eta\in(a, b)$ such that

$$\frac{f(b)-f(a)}{g(b)-g(a)}=\frac{f(b)-f(a)}{b^3-a^3}=\frac{f'(\eta)}{3\eta^2}, \text{ i. e.}$$

$$\frac{f(b)-f(a)}{b-a}=(a^2+ab+b^2)\frac{f'(\eta)}{3\eta^2}.$$

On the other hand, by Lagrange theorem, there is $\xi\in(a, b)$ such that

$$\frac{f(b)-f(a)}{b-a}=f'(\xi),$$

and hence we have the result. □

Remark

(1) Geometric interpretation of Cauchy theorem:

The reader should interpret Cauchy theorem geometrically by referring to the curve in the xy - plane described by the parametric equations $x=g(t)$, $y=f(t)$, $(a\leqslant x\leqslant b)$. In this case $\frac{f(b)-f(a)}{g(b)-g(a)}$ represents the slope of the chord AB (cf. Figure 5 - 4), while $\frac{f'(x_0)}{g'(x_0)}$ represents the slope of a tangent line at some point P (corresponding to the parameter x_0) on the curve. Cauchy theorem states that, provided the conditions satisfied, the curve described by parametric equations has at least one point at which the tangent line of the curve is parallel to the chord AB. So, geometric interpretation of Cauchy theorem is the same as that of Lagrange theorem, merely, Cauchy theorem is not only applicable to a curve with equation $y=f(x)$ (i. e. each x is just allowed to be corresponding to one value of y), but also applicable to a curve which can be described by

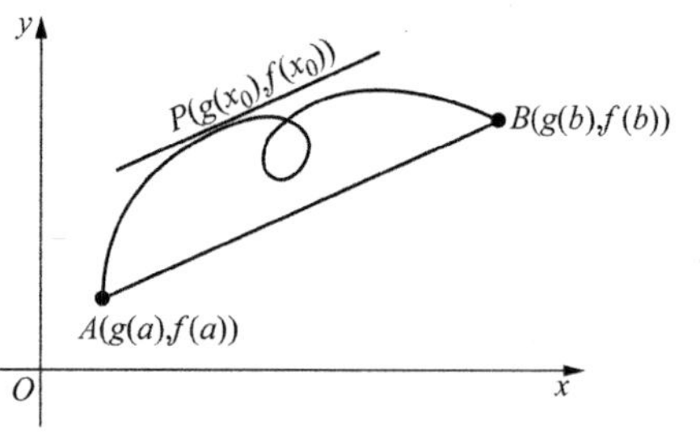

Figure 5 - 4

parametric equations.

(2) The relationship among Rolle theorem, Lagrange theorem and Cauchy theorem:

In Cauchy theorem , we suppose $g(x) = x$, then $g'(x) = 1$. Thus there is $x_0 \in (a, b)$ such that

$$f'(x_0) = \frac{f'(x_0)}{1} = \frac{f(b) - f(a)}{g(b) - g(a)} = \frac{f(b) - f(a)}{b - a}.$$

This is exactly Lagrange theorem. Hence, Cauchy theorem is a generalization of Lagrange theorem. Recalling that Lagrange theorem is a generalization of Rolle theorem, we see the following implications of these three theorems:

Cauchy theorem⇒Lagrange theorem⇒Rolle theorem.

§5.1.3 Taylor theorem

As noted earlier, if f is differentiable at a, then f is approximately a linear function near a. That is, the equation

$$f(x) = f(a) + f'(a)(x - a),$$

is approximately correct when $x - a$ is small. Taylor theorem will tell us that, more generally, f can be approximated by a polynomial of degree $n - 1$ if f has a derivative of order n. Moreover, Taylor theorem can yield a useful expression for the error made by this approximation.

Theorem 5.1.7 (Taylor theorem) Suppose $y = f(x)$ is a function such that

(1) $f(x)$, $f'(x)$, $f''(x)$, $\cdots$, $f^{(n)}(x)$ are continuous on $[a, b]$;

(2) $f^{(n+1)}(x)$ exists on (a, b).

Then for any $x \in (a, b)$, there is a point $\xi \in (a, x)$ such that

$$f(x)=f(a)+f'(a)(x-a)+\frac{f''(a)}{2!}(x-a)^2+\cdots$$
$$+\frac{f^{(n)}(a)}{n!}(x-a)^n+\frac{f^{(n+1)}(\xi)}{(n+1)!}(x-a)^{n+1}. \qquad (*)$$

Proof Let

$$p_n(x)=f(a)+f'(a)(x-a)+\frac{f''(a)}{2!}(x-a)^2+\cdots$$
$$+\frac{f^{(n)}(a)}{n!}(x-a)^n.$$

Then the expression (*) can be written as

$$f(x)=p_n(x)+\frac{f^{(n+1)}(\xi)}{(n+1)!}(x-a)^{n+1},$$

or

$$f(x)-p_n(x)=\frac{f^{(n+1)}(\xi)}{(n+1)!}(x-a)^{n+1}.$$

Denote $R_n(x)=f(x)-p_n(x)$ and $Q_n(x)=(x-a)^{n+1}$. Then $R_n(x)$ and $Q_n(x)$ both possess nth continuous derivative on $[a, b]$ and $(n+1)$th derivative on (a, b).

Since $R_n^{(i)}(a)=0$ and $Q_n^{(i)}(a)=0$ for any $i\in\{0, 1, \dots, n\}$, by using Cauchy theorem $n+1$ times on $[a, x]$ $(x\leqslant b)$, we have

$$\frac{R_n(x)}{Q_n(x)}=\frac{R_n(x)-Q_n(a)}{Q_n(x)-Q_n(a)}=\frac{R_n'(\xi_1)}{Q_n'(\xi_1)}=\frac{R_n'(\xi_1)-Q_n'(a)}{Q_n'(\xi_1)-Q_n'(a)}$$
$$=\frac{R_n''(\xi_2)}{Q_n''(\xi_2)}=\frac{R_n''(\xi_2)-Q_n''(a)}{Q_n''(\xi_2)-Q_n''(a)}=\frac{R_n'''(\xi_3)}{Q_n'''(\xi_3)}=\cdots$$
$$=\frac{R_n^{(n)}(\xi_n)}{Q_n^{(n)}(\xi_n)}=\frac{R_n^{(n)}(\xi_n)-R_n^{(n)}(a)}{Q_n^{(n)}(\xi_n)-Q_n^{(n)}(a)}=\frac{R_n^{(n+1)}(\xi)}{Q_n^{(n+1)}(\xi)},$$

where $a<\xi<\xi_n<\cdots<\xi_3<\xi_2<\xi_1<x\leqslant b$ and

$$R_n^{(n+1)}(\xi)=f^{(n+1)}(\xi);\ Q_n^{(n+1)}(\xi)=(n+1)!.$$

Thus

$$\frac{R_n(x)}{Q_n(x)} = \frac{f^{(n+1)}(\xi)}{(n+1)!} \qquad (a<\xi<x\leqslant b),$$

or

$$R_n(x) = \frac{1}{(n+1)!}f^{(n+1)}(\xi)(x-a)^{n+1} \quad (a<\xi<x\leqslant b). \ \square \tag{* *}$$

The expression (*) is called *(n-order) Taylor formula* of $f(x)$ (at point a), $p_n(x)$ is called *(n-order) Taylor polynomial* of $f(x)$, $R_n(x)$ is called the *remainder* of Taylor formula of $f(x)$, a remainder with the form (* *) is called *(n-order) remainder in Lagrange form*.

Examples (1) Find the 3-order Taylor formula of $f(x)=\sin x$ at $x=\pi/4$ with remainder in Lagrange form.

Solution Since

$$f\left(\frac{\pi}{4}\right)=\frac{\sqrt{2}}{2},\ f'\left(\frac{\pi}{4}\right)=\frac{\sqrt{2}}{2},\ f''\left(\frac{\pi}{4}\right)=-\frac{\sqrt{2}}{2},$$

$$f'''\left(\frac{\pi}{4}\right)=-\frac{\sqrt{2}}{2} \text{ and } f^{(4)}(\xi) = \sin\xi,$$

$$\sin x = \frac{\sqrt{2}}{2}+\frac{\sqrt{2}}{2}\left(x-\frac{\pi}{4}\right)-\frac{\sqrt{2}}{2}\frac{1}{2!}\left(x-\frac{\pi}{4}\right)^2 + \frac{\sqrt{2}}{2}\frac{1}{3!}\left(x-\frac{\pi}{4}\right)^3+\frac{1}{4!}(\sin\xi)\left(x-\frac{\pi}{4}\right)^4,$$

where ξ is between x and $\pi/4$, or

$$\sin x = \frac{\sqrt{2}}{2}\left[1+\left(x-\frac{\pi}{4}\right)-\frac{1}{2}\left(x-\frac{\pi}{4}\right)^2 + \frac{1}{6}\left(x-\frac{\pi}{4}\right)^3\right]+\frac{1}{24}\left(x-\frac{\pi}{4}\right)^4\sin\xi,$$

where ξ is between x and $\pi/4$.

(2) Suppose $f(x)$ is continuous on $[0, 1]$ with $f(0) = f(1) = 0$ and $f''(x)$ exists on $(0, 1)$. If $\max\limits_{[0,1]} f(x) = 2$, show that there exists $\xi \in (0, 1)$ such that $f''(\xi) \leqslant -16$.

Proof Let $x_0 \in [0, 1]$ with $f(x_0) = \max\limits_{[0,1]} f(x) = 2$. Since $f(0) = f(1) = 0$, $x_0 \in (0, 1)$ and $f'(x_0) = 0$ (Fermat theorem), so by Taylor theorem there is ξ between x and x_0 with

$$f(x) = f(x_0) + f'(x_0)(x - x_0) + \frac{f''(\xi)}{2!}(x - x_0)^2$$

$$= 2 + \frac{f''(\xi)}{2!}(x - x_0)^2,$$

from which, by setting $x = 1$ and $x = 0$ respectively, it follows

$$f(1) = 2 + \frac{f''(\xi_1)}{2!}(1 - x_0)^2 \quad (\xi_1 \in (x_0, 1)), \text{ and}$$

$$f(0) = 2 + \frac{f''(\xi_2)}{2!}(0 - x_0)^2 \quad (\xi_2 \in (0, x_0)).$$

Thus, we have

$$f''(\xi_1) = -\frac{4}{(1 - x_0)^2} \text{ and } f''(\xi_2) = -\frac{4}{x_0^2}.$$

If $x_0 \in (0, 1/2]$, $-4/x_0^2 \leqslant -4/(\frac{1}{2})^2 = -16$; if $x_0 \in (1/2, 1)$, $-4/(1 - x_0)^2 \leqslant -4/(1 - \frac{1}{2})^2 = -16$. Consequently, we see

$$f''(\xi_1) \leqslant -16 \text{ or } f''(\xi_2) \leqslant -16. \ \square$$

(3) Suppose $f(x)$ is continuous on $[a, b]$ with $f'(a) = f'(b) = 0$ and $f''(x)$ exists on (a, b). Prove that there exists $\xi \in (a, b)$ such that

$$| f''(\xi) | \geqslant \frac{4 | f(b) - f(a) |}{(b - a)^2}.$$

Proof Using Taylor theorem at $x=a$ and $x=b$ respectively, we have

$$f(x)=f(a)+f'(a)(x-a)+\frac{f''(\xi_1)}{2!}(x-a)^2$$
$$=f(a)+\frac{f''(\xi_1)}{2!}(x-a)^2 \quad (\xi_1\in(a,x)),$$

and

$$f(x)=f(b)+f'(b)(x-b)+\frac{f''(\xi_2)}{2!}(x-b)^2$$
$$=f(b)+\frac{f''(\xi_2)}{2!}(x-b)^2 \quad (\xi_2\in(x,b)).$$

Upon replacing x with $(a+b)/2$, these two equalities become

$$f\left(\frac{a+b}{2}\right)=f(a)+\frac{f''(\xi_1)}{2!}\left(\frac{a+b}{2}-a\right)^2 \quad \left(\xi_1\in\left(a,\frac{a+b}{2}\right)\right),$$

and

$$f\left(\frac{a+b}{2}\right)=f(b)+\frac{f''(\xi_2)}{2!}\left(\frac{a+b}{2}-b\right)^2 \quad \left(\xi_2\in\left(\frac{a+b}{2},b\right)\right),$$

which implies

$$f(a)+\frac{f''(\xi_1)}{2!}\left(\frac{a+b}{2}-a\right)^2=f(b)+\frac{f''(\xi_2)}{2!}\left(\frac{a+b}{2}-b\right)^2,$$

$$\text{i.e. } f(b)-f(a)=\frac{1}{2}\left(\frac{b-a}{2}\right)^2 | f''(\xi_1)-f''(\xi_2) |.$$

Let $| f''(\xi) |=\max\{| f''(\xi_1) |, | f''(\xi_2) |\}$. Then,

$$| f(b)-f(a) | \leqslant \frac{(b-a)^2}{8}(| f''(\xi_1) |+| f''(\xi_2) |)$$
$$\leqslant \frac{(b-a)^2}{4}. | f''(\xi) |. \square$$

Note (1) By remainder in Lagrange form, we may derive that $R_n(x)=o((x-a)^n)$, $(x\to a)$. So, if $f(x)$ is approximated by $p_n(x)$ for x near a, the error is a high-order infinitesimal over $(x-a)^n$ as

$x \to a$.

(2) When $n = 0$, Taylor theorem is exactly Lagrange theorem, and so Taylor theorem is a generalization of Lagrange theorem.

(3) The relationship of these theorems can be indicated as follows:

Cauchy theorem↘
Lagrange theorem⇒Rolle theorem
Taylor theorem↗

Theorem 5.1.8 (Peano theorem) Let $y = f(x)$ with $f^{(n)}$ being continuous at a. Then

$$f(x) = f(a) + f'(a)(x-a) + \frac{f''(a)}{2!}(x-a)^2 + \cdots \frac{f^{(n-1)}(a)}{(n-1)!}(x-a)^{n-1} + \frac{f^{(n)}(a)}{n!}(x-a)^n + o((x-a)^n).$$

Proof By Taylor theorem,

$$f(x) = f(a) + f'(a)(x-a) + \frac{f''(a)}{2!}(x-a)^2 + \cdots \frac{f^{(n-1)}(a)}{(n-1)!}(x-a)^{n-1} + \frac{f^{(n)}(\xi)}{n!}(x-a)^n,$$

where ξ is between x and a. So, we only need to show: as $x \to 0$

$$\frac{f^{(n)}(\xi)}{n!}(x-a)^n = \frac{f^{(n)}(a)}{n!}(x-a)^n + o((x-a)^n).$$

Because $f^{(n)}$ is continuous at a, and so we have

$$\lim_{x\to a}\left[\left(\frac{f^{(n)}(\xi)}{n!}(x-a)^n - \frac{f^{(n)}(a)}{n!}(x-a)^n\right)/(x-a)^n\right]$$
$$= \frac{1}{n!}\lim_{x\to a}(f^{(n)}(\xi) - f^{(n)}(a)) = \frac{1}{n!}\lim_{\xi\to a}(f^{(n)}(\xi) - f^{(n)}(a))$$
$$= \frac{1}{n!}(f^n(a) - f^n(a)) = 0.$$

i. e. as $x \to 0$

$$\frac{f^{(n)}(\xi)}{n!}(x-a)^n - \frac{f^{(n)}(a)}{n!}(x-a)^n = o((x-a)^n). \square$$

The expression in above theorem is called (n-*order*) *Peano formula*. The term $o((x-a)^n)$ in Peano formula is called (n-*order*) *remainder in Peano form*, also denoted by $R_n(x) = o((x-a)^n)$.

Definition The Taylor formula of $f(x)$ at point $a = 0$ is called *Maclaurin formula* of $f(x)$, i. e.

$$f(x) = f(0) + f'(0)x + \cdots + \frac{f^{(n)}(0)}{n!}x^n + R_n(x),$$

where the remainder $R_n(x)$ can be taken as that in Lagrange form, i. e.

$$R_n(x) = \frac{f^{(n+1)}(\xi)}{(n+1)!}x^{n+1} \quad (0 < \xi < x \leqslant b)$$

or as that in Peano form, i. e.

$$R_n(x) = o(x^n).$$

Remark Suppose $x \to 0$ and n, m are positive integers. Then

(1) $o(x^n) \cdot x^m = o(x^{n+m})$;

(2) If $m \geqslant n$, $o(x^n) + o(x^m) = o(x^n)$;

(3) $x^{n+1} + x^{n+2} + \cdots + x^{n+m} = o(x^n)$.

Proof (1) Only need to note

$$\lim_{x \to 0} \frac{o(x^n) \cdot x^m}{x^{n+m}} = \lim_{x \to 0} \frac{o(x^n)}{x^n} = 0.$$

(2) Let $m = n + n_1 (n_1 \geqslant 0)$. Then by (1),

$$\lim_{x \to 0} \frac{o(x^n) + o(x^{n+n_1})}{x^n} = \lim_{x \to 0} \frac{o(x^n) + o(x^n) \cdot x^{n_1}}{x^n} = 0.$$

(3) $\lim\limits_{x \to 0} \dfrac{x^{n+1} + x^{n+2} + \cdots + x^{n+m}}{x^n} = \lim\limits_{x \to 0}(x + x^2 + \cdots + x^m) = 0.$

Proposition Some basic Maclaurin formulas:

(1) $e^x = 1 + \frac{x}{1!} + \frac{x^2}{2!} + \cdots + \frac{x^n}{n!} + \frac{e^\xi}{(n+1)!}x^{n+1}$;

(2) $\sin x = x - \frac{x^3}{3!} + \frac{x^5}{5!} + \cdots$

$$+ \frac{x^n}{n!}\sin\left(\frac{n}{2}\pi\right) + \frac{x^{n+1}}{(n+1)!}\sin\left(\frac{n+1}{2}\pi + \xi\right);$$

(3) $\cos x = 1 - \frac{x^2}{2!} + \frac{x^4}{4!} + \cdots + \frac{x^n}{n!}\cos\left(\frac{n}{2}\pi\right)$

$$+ \frac{x^{n+1}}{(n+1)!}\cos\left(\frac{n+1}{2}\pi + \xi\right);$$

(4) $\ln(1+x) = x - \frac{x^2}{2} + \frac{x^3}{3} - \cdots + (-1)^{n-1}\frac{x^n}{n}$

$$+ (-1)^n \frac{x^{n+1}}{(n+1)} \cdot \frac{1}{(1+\xi)^{n+1}};$$

(5) Let α be any real number.

$$(1+x)^\alpha = 1 + \alpha x + \frac{\alpha(\alpha-1)}{2!}x^2 + \cdots + \frac{\alpha(\alpha-1)(\alpha-2)\cdots(\alpha-n+1)}{n!}x^n$$

$$+ \frac{\alpha(\alpha-1)(\alpha-2)\cdots(\alpha-n)}{(n+1)!}(1+\xi)^{\alpha-n-1}x^{n+1}.$$

In these Maclaurin formulas, the remainders are all in Lagrange form where ξ is between 0 and x. For the corresponding Maclaurin formulas with remainders in Peano form, we merely need to change the last term into $o(x^n)$.

Proof (1) As $f(0) = f'(0) = f''(0) = \cdots = f^{(n)}(0) = 1$ and $f^{(n+1)}(x) = e^x$, Maclaurin formula of $f(x) = e^x$ with remainder in Lagrange form is

$$e^x = 1 + \frac{x}{1!} + \frac{x^2}{2!} + \cdots + \frac{x^n}{n!} + \frac{e^\xi}{(n+1)!}x^{n+1},$$

where ξ is between 0 and x.

(2) In Section 4.4.1, we derived

$$(\sin x)^{(n)}=\sin\left(x+\frac{n\pi}{2}\right),$$

and so

$$f(0)=0,\ f'(0)=1,\ f''(0)=0,\ f'''(0)=-1,\ f^{(4)}=0,\ \cdots,$$

$$f^{(n)}(0)=\sin\left(\frac{n}{2}\pi\right),\ f^{(n+1)}(\xi)=\sin\left(\frac{n+1}{2}\pi+\xi\right).$$

Thus, Maclaurin formula of $f(x)=\sin x$ with remainder in Lagrange form is

$$\sin x=x-\frac{x^3}{3!}+\frac{x^5}{5!}+\cdots+\frac{x^n}{n!}\sin\left(\frac{n}{2}\pi\right)+\frac{x^{n+1}}{(n+1)!}\sin\left(\frac{n+1}{2}\pi+\xi\right),$$

where ξ is between 0 and x.

(3) Maclaurin formula of $f(x)=\cos x$ with remainder in Lagrange form can be similarly derived.

(4) It is easy to find

$$f^{(n)}(x)=(-1)^{n-1}\frac{(n-1)!}{(1+x)^n}.$$

Then, $f^{(n)}(0)=(-1)^{n-1}(n-1)!$ and $f^{(n+1)}(\xi)=(-1)^n\dfrac{n!}{(1+\xi)^{n+1}}$.

So, Maclaurin formula of $f(x)=\ln(1+x)$ with remainder in Lagrange form is

$$\ln(1+x)=x-\frac{x^2}{2}+\frac{x^3}{3}-\cdots+(-1)^{n-1}\frac{x^n}{n}$$
$$+(-1)^n\frac{x^{n+1}}{(n+1)}\cdot\frac{1}{(1+\xi)^{n+1}},$$

where ξ is between 0 and x.

(5) Since

$$f^{(n)}(x)=\alpha(\alpha-1)(\alpha-2)\cdots(\alpha-n+1)(1+x)^{\alpha-n},$$

$f^{(n)}(0) = \alpha(\alpha-1)(\alpha-2)\cdots(\alpha-n+1)$. Thus, Maclaurin formula of $f(x) = (1+x)^{\alpha}$ with remainder in Lagrange form is

$$(1+x)^{\alpha} = 1+\alpha x+\frac{\alpha(\alpha-1)}{2!}x^2+\cdots+\frac{\alpha(\alpha-1)(\alpha-2)\cdots(\alpha-n+1)}{n!}x^n$$
$$+\frac{\alpha(\alpha-1)(\alpha-2)\cdots(\alpha-n)}{(n+1)!}(1+\xi)^{\alpha-n-1}x^{n+1},$$

where ξ is between 0 and x. □

Note If $\alpha = m$ is a positive integer, from Maclaurin formula of $(1+x)^{\alpha}$ it follows immediately that

$$(1+x)^m = 1+mx+\frac{m(m-1)}{2!}x^2+\cdots+mx^{m-1}+x^m,$$

which is exactly the expansion of binomial. Hence, Maclaurin formula of $(1+x)^{\alpha}$ can be regarded as the generalization of the theorem of binomial expansion.

Remark Some other common-used Maclaurin formulas are listed as follows:

$$\arcsin x = x+\frac{x^3}{6}+o(x^3);$$

$$\arccos x = \frac{\pi}{2}-x-\frac{x^3}{6}+o(x^3);$$

$$\arctan x = x-\frac{x^3}{3}+o(x^3);$$

$$\tan x = x+\frac{x^3}{3}+o(x^3);$$

$$a^x = 1+x\ln a+o(x) \quad (a>0);$$

$$\sqrt[n]{1+x} = 1+\frac{x}{n}+o(x);$$

$$\ln(x+\sqrt{1+x^2}) = x-\frac{x^3}{6}+o(x^3).$$

Examples (1) Prove the number e is irrational.

Proof By setting $x = 1$ in Maclaurin formula of $f(x) = \mathrm{e}^x$, we see

$$\mathrm{e} = 1 + \frac{1}{1!} + \frac{1}{2!} + \cdots + \frac{1}{n!} + \frac{\mathrm{e}^{\xi}}{(n+1)!} \quad (\xi \in (0, 1)),$$

from which it follows that

$$n! \cdot \mathrm{e} = I + \frac{\mathrm{e}^{\xi}}{n+1} \quad (\xi \in (0, 1)),$$

where I is some integer. Assume e is a rational number, say, $\mathrm{e} = p/q$ for some integers p and q. By taking $n \geqslant q$, we see that $n! \cdot \mathrm{e}$ is an integer. However, $I + \dfrac{\mathrm{e}^{\xi}}{n+1}$ is a fraction if $n \geqslant 2$, which yields a contradiction. □

(2) Using Maclaurin formula to find limits of the following functions:

① $\lim\limits_{x \to 0} \dfrac{\sin x - x\cos x}{x - \sin x}$; ② $\lim\limits_{x \to 0} \dfrac{x\ln(1+x^2)}{\mathrm{e}^{x^2} - x - 1}$; ③ $\lim\limits_{x \to 0} \dfrac{\cos x - \mathrm{e}^{-x^2/2}}{x^4}$.

Solution

① Since

$$\sin x = x - \frac{1}{3!}x^3 + o(x^3) \text{ and } \cos x = 1 - \frac{1}{2!}x^2 + o(x^2),$$

Noting $o(x^2) \cdot x = o(x^3)$, we have

$$\lim_{x \to 0} \frac{\sin x - x\cos x}{x - \sin x} = \lim_{x \to 0} \frac{\left(x - \frac{1}{3!}x^3 + o(x^3)\right) - x\left(1 - \frac{1}{2!}x^2 + o(x^2)\right)}{x - \left(x - \frac{1}{3!}x^3 + o(x^3)\right)}$$

$$= \lim_{x \to 0} \frac{\left(x - \frac{1}{3!}x^3 + o(x^3)\right) - \left(x - \frac{1}{2!}x^3 + o(x^3)\right)}{x - \left(x - \frac{1}{3!}x^3 + o(x^3)\right)}$$

$$= \lim_{x \to 0} \frac{\frac{1}{3}x^3 + o(x^3)}{\frac{1}{6}x^3 - o(x^3)} = 2.$$

②

$$\lim_{x \to 0} \frac{x\ln(1+x^2)}{e^{x^2} - x - 1} = \lim_{x \to 0} \frac{x\left(x^2 - \frac{1}{2}x^4 + o(x^4)\right)}{(1 + x^2 + o(x^2)) - x - 1}$$

$$= \lim_{x \to 0} \frac{x^3 - \frac{1}{2}x^5 + o(x^5)}{x^2 - x + o(x^2)} = 0.$$

③ Note

$$\cos x = 1 - \frac{x^2}{2!} + \frac{x^4}{4!} + o(x^4),$$

$$e^{-x^2/2} = 1 + \left(-\frac{x^2}{2}\right) + \frac{1}{2!}\left(-\frac{x^2}{2}\right)^2 + o(x^4).$$

We have

$$\lim_{x \to 0} \frac{\cos x - e^{-x^2/2}}{x^4} = \lim_{x \to 0} \frac{-\frac{1}{12}x^4 + o(x^4)}{x^4} = -\frac{1}{12}.$$

(3) Using Maclaurin formula to find derivatives:

① $f(x) = \dfrac{x}{\sqrt{1+x^2}}$, find $f^{(5)}(0)$; ② $f(x) = x^2 \sin x$, find $f^{(99)}(0)$.

Solution ① Note that

$$\frac{1}{\sqrt{1+x}} = 1 - \frac{1}{2}x + \frac{\left(-\frac{1}{2}\right)\left(-\frac{1}{2} - 1\right)}{2!}x^2 + o(x^2).$$

Then

$$\frac{1}{\sqrt{1+x^2}}=1-\frac{1}{2}x^2+\frac{\left(-\frac{1}{2}\right)\left(-\frac{1}{2}-1\right)}{2!}x^4+o(x^4),$$

and so

$$\frac{x}{\sqrt{1+x^2}}=x-\frac{1}{2}x^3+\frac{\left(-\frac{1}{2}\right)\left(-\frac{1}{2}-1\right)}{2!}x^5+o(x^5).$$

Since Maclaurin formula is unique, by comparing coefficient of x^5 we have

$$\frac{f^{(5)}(0)}{5!}=\frac{\left(-\frac{1}{2}\right)\left(-\frac{1}{2}-1\right)}{2!},$$

from which it follows that $f^{(5)}(0)=45$.

② Since

$$\sin x=x-\frac{1}{3!}x^3+\cdots+\frac{\sin\left(\frac{n\pi}{2}\right)}{n!}x^n+o(x^n),$$

$$x^2\sin x=x^3-\frac{1}{3!}x^5+\cdots+\frac{\sin\left(\frac{n\pi}{2}\right)}{n!}x^{n+2}+o(x^{n+2}),$$

and so

$$\frac{f^{(99)}(0)}{99!}=\frac{1}{97!}\sin\left(\frac{97\pi}{2}\right).$$

i. e.

$$f^{(99)}(0)=99\times 98\sin\left(\frac{97\pi}{2}\right)=9\ 702.$$

(4) Determine the constants a and b such that

$$\lim_{x\to+\infty}(\sqrt{2x^2+4x-1}-ax-b)=0.$$

Solution Since

$$\sqrt{2x^2+4x-1}=\sqrt{2}x\sqrt{1+\left(\frac{2}{x}-\frac{1}{2x^2}\right)}=\sqrt{2}x+\sqrt{2}-\frac{\sqrt{2}}{4x}+\varepsilon,$$

where $\lim\limits_{x\to+\infty}\varepsilon=0$, we see that in order to make $\lim\limits_{x\to+\infty}(\sqrt{2x^2+4x-1}-ax-b)=\lim\limits_{x\to+\infty}\left[(\sqrt{2}-a)x+(\sqrt{2}-b)-\frac{\sqrt{2}}{4x}+\varepsilon\right]=0$, it must be $a=b=\sqrt{2}$.

(5) Determine the infinitesimal order of the following functions as compared with x $(x\to 0)$:

① $f(x)=\mathrm{e}^{-x}+\sin x-\cos x$;

② $f(x)=\cos^2 x\sin^2 x-x^2(1-x^2)^{\frac{4}{3}}$;

③ $\sqrt{b}\arctan\sqrt{\frac{x}{b}}-\sqrt{a}\arctan\sqrt{\frac{x}{a}}\quad(b>a>0)$.

Solution ① $f(x)=\left(1+(-x)+\frac{(-x)^2}{2!}+o(x^2)\right)$

$$+\left(x-\frac{x^3}{3!}+o(x^3)\right)+\left(1-\frac{x^2}{2!}+o(x^2)\right)$$

$$=x^2-\frac{x^3}{6}+o(x^2),$$

i. e. $f(x)=O(x^2)$, the order of $f(x)$ is 2 as compared with x.

② Since $\quad\cos^2 x\sin^2 x=\frac{1}{4}\sin^2 2x=\frac{1}{8}-\frac{1}{8}\cos 4x$

$$=\frac{1}{8}-\frac{1}{8}\left(1-\frac{1}{2!}(4x)^2+\frac{1}{4!}(4x)^4-\frac{1}{6!}(4x)^6+o(x^6)\right)$$

$$=x^2-\frac{4}{3}x^4+\frac{32}{45}x^6+o(x^6),$$

and

$$x^2(1-x^2)^{\frac{4}{3}}=x^2\left[1-\frac{4}{3}x^2+\frac{1}{2!}\frac{4}{3}\left(\frac{4}{3}-1\right)(-x^2)^2+o(x^4)\right]$$

$$= x^2 - \frac{4}{3}x^4 + \frac{2}{9}x^6 + o(x^6),$$

$$f(x) = \frac{22}{45}x^6 + o(x^6),$$

i. e. $f(x) = O(x^6)$, the order of $f(x)$ is 6 as compared with x.

③ Notice $\arctan x = x - \frac{x^3}{3} + o(x^3)$.

$$f(x) = \sqrt{b}\left(\sqrt{\frac{x}{b}} - \frac{1}{3}\left(\sqrt{\frac{x}{b}}\right)^3 + o(x^{3/2})\right)$$

$$-\sqrt{a}\left(\sqrt{\frac{x}{a}} - \frac{1}{3}\left(\sqrt{\frac{x}{a}}\right)^3 + o(x^{3/2})\right)$$

$$= \left(\sqrt{x} - \frac{1}{3b}(x^{3/2}) + o(x^{3/2})\right) - \left(\sqrt{x} - \frac{1}{3a}(x^{3/2}) + o(x^{3/2})\right)$$

$$= \frac{1}{3}\left(\frac{1}{a} - \frac{1}{b}\right)x^{3/2} + o(x^{3/2}),$$

i. e. $f(x) = O(x^{3/2})$, the order of $f(x)$ is $\frac{3}{2}$ as compared with x.

(6) Find the approximation of the following numbers:

① $\sqrt[3]{30}$; ② $\sqrt{e}$; ③ $\sin 18^\circ$; ④ $\ln 1.1$.

Solution ① $\sqrt[3]{30} = 3\left(1 + \frac{1}{9}\right)^{1/3}$

$$\approx 3\left[1 + \frac{1}{3}\,\frac{1}{9} + \frac{\frac{1}{3}\left(\frac{1}{3} - 1\right)}{2!}\left(\frac{1}{9}\right)^2\right]$$

$$\approx 3.107\,2;$$

② $\sqrt{e} = e^{1/2} \approx 1 + \frac{1}{2} + \frac{1}{2}\left(\frac{1}{2}\right)^2 + \frac{1}{6}\left(\frac{1}{2}\right)^3 \approx 1.645$;

③ $\sin 18^\circ = \sin\frac{\pi}{10} \approx \frac{\pi}{10} - \frac{1}{3!}\left(\frac{\pi}{10}\right)^3 + \frac{1}{5!}\left(\frac{\pi}{10}\right)^5 \approx 0.309\,017$;

④ $\ln 1.1 = \ln(1 + 0.1) \approx 0.1 - \frac{0.1^2}{2} + \cdots + \frac{0.1^5}{5} \approx 0.095\,3$.

§ 5.2 Monotony and extremum of functions

In many problems of economics, engineering and physics it is important to determine how large or how small a certain quantity can be. If a problem admits a mathematical formulation, it is often reducible to the problem of finding the maximum or minimum value of some function. In this section we generally consider maximum or minimum values for functions defined on an interval, and we begin with discussion of monotonicity of a function.

§ 5.2.1 Test of monotonicity of functions

Theorem 5.2.1 Let $f(x)$ be differentiable on (a, b). Then

(1) $f(x)$ is increasing on (a, b) if and only if $f'(x) \geqslant 0$, $x \in (a, b)$;

(2) $f(x)$ is decreasing on (a, b) if and only if $f'(x) \leqslant 0$, $x \in (a, b)$.

Proof We only prove (1) while (2) can be similarly proved.

Sufficiency. Let $x_1, x_2 \in (a, b)$ with $x_1 < x_2$. By Lagrange theorem and the condition,

$$f(x_2) - f(x_1) = f'(\xi)(x_2 - x_1) \geqslant 0 \quad (x_1 < \xi < x_2),$$

which means $f(x)$ is increasing on (a, b).

Necessity. Let $x \in (a, b)$ with $\Delta x > 0$ and $x + \Delta x \in (a, b)$. Then

$$\frac{f(x + \Delta x) - f(x)}{\Delta x} \geqslant 0.$$

Letting $\Delta x \to 0$, we have $f'(x) \geqslant 0$, $x \in (a, b)$. □

Theorem 5.2.2 Let $f(x)$ be a differentiable function on (a, b). Then $f(x)$ is strictly increasing (decreasing) on (a, b) if and only if the following conditions are both satisfied:

(1) for any $x \in (a, b)$, $f'(x) \geqslant 0$ ($f'(x) \leqslant 0$);

(2) $f'(x) \not\equiv 0$ on any subinterval $(a_1, b_1) \subseteq (a, b)$.

Proof We only prove the case of strictly increasing. It can be proved similarly for the case of strictly decreasing.

Sufficiency By Theorem 5.2.1, $f(x)$ is increasing on (a, b). Let $x_1, x_2 \in (a, b)$ with $x_1 < x_2$ and let $x \in (x_1, x_2)$. Then

$$f(x_1) \leqslant f(x) \leqslant f(x_2).$$

Suppose $f(x_1) = f(x_2)$. Then $f(x) = f(x_1) = f(x_2)$, i.e. $f(x)$ is a constant function on $[x_1, x_2]$, and so $f'(x) \equiv 0$ on (x_1, x_2), which yields a contradiction. Thus $f(x_1) < f(x_2)$.

Necessity. Suppose $f(x)$ is strictly increasing on (a, b). Then by Theorem 5.2.1, $f'(x) \geqslant 0$ for any $x \in (a, b)$, i.e. the condition (1) holds; Assume there is a subinterval $(a_1, b_1) \subseteq (a, b)$ such that $f'(x) \equiv 0$ on (a_1, b_2). Then by Corollary 1 of Lagrange theorem, $f(x)$ is constant on (a_1, b_1), which contradicts the property of strictly increasing of $f(x)$. □

Corollary Let $f(x)$ be a differentiable function on (a, b). If $f'(x) > 0$ ($f'(x) < 0$), then $f(x)$ is strictly increasing (decreasing) on (a, b).

This corollary is an immediate consequence of the precedent theorem. Note the corollary is merely a sufficient condition for f to be strictly increasing (decreasing). For example, $f(x) = x^3$ is strictly increasing on $(-\infty, +\infty)$, but $f'(0) = 0$.

Examples (1) Find the monotone intervals of the following functions:

① $y = \sqrt[3]{(2x-a)(a-x)^2}$ $(a > 0)$;

② $y = \dfrac{10}{4x^3 - 9x^2 + 6x}$.

Solution ① $y' = \dfrac{2(2a-3x)}{3\sqrt[3]{(2x-a)^2(a-x)}}$,

by which we see that $x=2a/3$ is a stationary point, and $x=a/2$, $x=a$ are two singular points. Note $y'>0$ on $(-\infty, a/2)$, $(a/2, 2a/3)$ and $(a, +\infty)$, $y'<0$ on $(2a/3, a)$. So, f is strictly increasing on $(-\infty, a/2)$, $(a/2, 2a/3)$ and $(a, +\infty)$, and f is strictly decreasing on $(2a/3, a)$;

② Note the domain of y is $(-\infty, 0)\cup(0, +\infty)$. Let

$$z=\frac{1}{y}=\frac{4x^3-9x^2+6x}{10}.$$

Then

$$z'(x)=\frac{3}{5}(2x^2-3x+1)=\frac{6}{5}\left(x-\frac{1}{2}\right)(x-1),$$

by which we see the stationary points of z are $x=1/2$ and $x=1$. Since $z'(x)>0$ on $(-\infty, 0)$, $(0, 1/2)$ and $(1, +\infty)$, and $z'(x)<0$ on $(1/2, 1)$, $z(x)$ is strictly increasing on $(-\infty, 0)$, $(0, 1/2)$ and $(1, +\infty)$, and strictly decreasing on $(1/2, 1)$, i. e. y is strictly decreasing on $(-\infty, 0)$, $(0, 1/2)$ and $(1, +\infty)$, and strictly increasing on $(1/2, 1)$.

(2) Show that $x>\ln(1+x)$ if $x>0$.

Proof Let $f(x)=x-\ln(1+x)$ $(x>0)$. Then

$$f'(x)=1-\frac{1}{1+x}=\frac{x}{1+x}>0,$$

which, since $f(x)$ is right-continuous at $x=0$, implies that $f(x)$ is strictly increasing on $[0, +\infty)$. So, if $x>0$, $f(x)=x-\ln(1+x)>f(0)=0$, i. e. $x>\ln(1+x)$. □

(3) Let $x, y>0$ and $\beta>\alpha>0$. Prove $(x^\alpha+y^\alpha)^{1/\alpha}>(x^\beta+y^\beta)^{1/\beta}$.

Proof Let $f(z)=(t^z+1)^{1/z}$ $(z>0)$. We only need to show $f(\alpha)>f(\beta)$.

Since $\ln f(z) = \frac{1}{z}\ln(t^z + 1)$,

$$\frac{1}{f(z)}f'(z) = \frac{1}{z} \cdot \frac{t^z \ln t}{t^z + 1} + \left(-\frac{1}{z^2}\right)\ln(t^z + 1), \text{ and so}$$

$$\begin{aligned} f'(z) &= f(z)\left[\frac{t^z \ln t}{z(t^z+1)} - \frac{\ln(t^z+1)}{z^2}\right] \\ &= f(z)\,\frac{zt^z \ln t - \ln(t^z+1)\cdot(t^z+1)}{z^2(t^z+1)} \\ &= \frac{f(z)}{z^2(t^z+1)}[\ln(t^z)^{t^z} - \ln(t^z+1)^{t^z+1}] < 0, \end{aligned}$$

which implies that $f(z)$ is strictly decreasing. Hence $f(\alpha) > f(\beta)$. □

(4) Show the function $f(x) = \left(1 + \frac{1}{x}\right)^x$ is strictly increasing on $(0, +\infty)$.

Proof We only need to show $f'(x) > 0$ on $(0, +\infty)$. Since $\ln f(x) = x \ln\left(1 + \frac{1}{x}\right)$, by differentiating both sides we have

$$\frac{f'(x)}{f(x)} = \ln\left(1 + \frac{1}{x}\right) - \frac{1}{1+x} = \ln(x+1) - \ln x - \frac{1}{x+1}.$$

By Lagrange theorem,

$$\ln(x+1) - \ln x = \frac{1}{\xi}[(x+1) - x] = \frac{1}{\xi} > \frac{1}{x+1},$$

where $x < \xi < x+1$. Thus $f'(x)/f(x) > 0$, and so $f'(x) > 0$ on $(0, +\infty)$. □

(5) Let

$$f(x) = 1 - x + \frac{x^2}{2} - \frac{x^3}{3} + \cdots + (-1)^n \frac{x^n}{n}.$$

Show that $f(x)$ has exactly one real root if n is odd, and that $f(x)$ has no real root if n is even.

Proof Note $f'(x) = -1 + x - x^2 + \cdots + (-1)^n x^{n-1}$. So, if

$x \neq -1$,

$$f'(x) = -\frac{1-(-x)^n}{1+x}.$$

Suppose n is odd. Then $f'(-1) = -n < 0$ and

$$f'(x) = -\frac{1+x^n}{1+x} < 0 \text{ if } x \neq -1,$$

which implies $f(x)$ is strictly decreasing on $(-\infty, +\infty)$. Considering $\lim_{x\to+\infty} f(x) = -\infty$ and $\lim_{x\to-\infty} f(x) = +\infty$, we see that $f(x)$ has exactly one real root.

Now suppose n is even. Then $f'(-1) = -n < 0$ and

$$f'(x) = -\frac{1-x^n}{1+x} \text{ if } x \neq -1,$$

which implies $f'(x) < 0$ on $(-\infty, 1)$, $f'(x) > 0$ on $(1, +\infty)$ and $f'(1)=0$. Thus $f(x)$ has its local minimum

$$\begin{aligned} f(1) &= 1 - 1 + \frac{1}{2} - \frac{1}{3} + \cdots + (-1)^n \frac{1}{n} \\ &= \left(\frac{1}{2} - \frac{1}{3}\right) + \left(\frac{1}{4} - \frac{1}{5}\right) + \cdots + \left(\frac{1}{n-2} - \frac{1}{n-1}\right) + \frac{1}{n} > 0. \end{aligned}$$

Furthermore, since $\lim_{x\to+\infty} f(x) = +\infty$ and $\lim_{x\to-\infty} f(x) = +\infty$, $f(x)$ has no real root. □

§5.2.2 Test of local extremum

Trivially, Fermat theorem implies the following:

Theorem 5.2.3 (Necessity of extremum) Suppose $f(x)$ is differentiable at x_0. If $f(x_0)$ is a local extremum, then $f'(x_0) = 0$ (i.e. x_0 is a stationary point of f).

Note that the above condition is usually not a sufficiency for $f(x_0)$ to be a local extremum. For example, as we already see, $x=0$

is a stationary point of $f(x)=x^3$, but $f(0)=0$ is not a local extremum of f.

Also note that the precondition of the theorem is that $f'(x_0)$ exists. However, a point, at which the derivative does not exist, is still possible to be a extremum point. For example, $f(x)=|x|$ is undifferentiable at $x=0$, but $f(0)=0$ is clearly a local minimum of $f(x)$.

As for how to ascertain the extremum points of a function, we have the following theorems:

Theorem 5.2.4 (Sufficiency I of extremum) Suppose $f(x)$ is continuous at x_0 and differentiable on $U^o(x_0, \delta)$, a free-center neighborhood of x_0.

(1) If $f'(x)\geqslant 0$ for $x\in(x_0-\delta, x_0)$ and $f'(x)\leqslant 0$ for $x\in(x_0, x_0+\delta)$, then $f(x_0)$ is a local maximum;

(2) If $f'(x)\leqslant 0$ for $x\in(x_0-\delta, x_0)$ and $f'(x)\geqslant 0$ for $x\in(x_0, x_0+\delta)$, then $f(x_0)$ is a local minimum.

Proof Also we only need to prove the first case. By the condition, $f(x)$ is increasing on $(x_0-\delta, x_0)$ and decreasing on $(x_0, x_0+\delta)$. Thus, considering the continuity of $f(x)$ at x_0, we see $f(x)\leqslant f(x_0)$ for $x\in U^o(x_0, \delta)$.

Remark The condition in Theorem 5.2.4 is not a necessary one, i.e.

$f(x_0)$ is a local maximum $\nRightarrow$ there exists $\delta>0$ such that $f'(x)\geqslant 0$ for $x\in(x_0-\delta, x_0)$ and $f'(x)\leqslant 0$ for $x\in(x_0, x_0+\delta)$;

$f(x_0)$ is a local minimum $\nRightarrow$ there exists $\delta>0$ such that $f'(x)\leqslant 0$ for $x\in(x_0-\delta, x_0)$ and $f'(x)\geqslant 0$ for $x\in(x_0, x_0+\delta)$.

For example, let

$$f(x)=\begin{cases} 2-x^2\left(2+\sin\dfrac{1}{x}\right) & (\text{if } x\neq 0), \\ 2 & (\text{if } x=0). \end{cases}$$

Clearly, for any $\delta > 0$, $f(0) = 2 \geqslant f(x)$ on $(-\delta, \delta)$, and so $f(0) = 2$ is a local maximum. However, $f'(x) = -2x\left(2 + \sin\frac{1}{x}\right) + \cos\frac{1}{x}$ $(x \neq 0)$. Then, for any $\delta > 0$, there exists $x_1 = -\frac{1}{2k\pi + \pi} \in (-\delta, 0)$ such that $f'(x_1) < 0$, and there exists $x_2 = \frac{1}{2k\pi} \in (0, \delta)$ such that $f'(x_2) > 0$ where k is a sufficiently large positive integer.

Theorem 5.2.5 (Sufficiency Ⅱ of extremum) Suppose $f(x)$ is differentiable on $U(x_0, \delta)$, a neighborhood of x_0, such that $f'(x_0) = 0$ and $f''(x_0)$ exists.

(1) If $f''(x_0) < 0$, then $f(x_0)$ is a local maximum;

(2) If $f''(x_0) > 0$, then $f(x_0)$ is a local minimum.

Proof We merely prove the first case. Since

$$\lim_{x \to x_0} \frac{f'(x)}{x - x_0} = \lim_{x \to x_0} \frac{f'(x) - f'(x_0)}{x - x_0} = f''(x_0) < 0,$$

by the property of sign-preserving of the limit, there is $\delta > 0$ such that if $0 < |x - x_0| < \delta$,

$$\frac{f'(x)}{x - x_0} < 0,$$

which implies that $f'(x) > 0$ for $x \in (x_0 - \delta, x_0)$ and $f'(x) < 0$ for $x \in (x_0, x_0 + \delta)$. Hence $f(x_0)$ is a local maximum by the foregoing theorem. □

Examples (1) Determine all the values of a such that the function $f(x) = x^3 + 3ax^2 - ax - 1$ possesses neither local maximum nor local minimum.

Solution Set $f'(x) = 3x^2 + 6ax - a = 0$, and we have $x = -a \pm (1/3)\sqrt{9a^2 + 3a}$. If $9a^2 + 3a < 0$, i.e. $a \in (-1/3, 0)$, $f(x)$ has no stationary points, and so $f(x)$ has no local extremum. (Note $f(x)$

clearly has no singular points.) If $9a^2 + 3a \geqslant 0$, then we have stationary points $x_0 = -a \pm (1/3)\sqrt{9a^2 + 3a}$. In this case, by the second sufficiency of extremum $f(x)$ has no local extremum only if $f''(x_0) = 0$. Since $f''(x) = 6x + 6a$, we may set $6[-a \pm (1/3)\sqrt{9a^2 + 3a}] + 6a = 0$, and so $a=0$ or $a=-1/3$. Hence, in summary $f(x)$ possesses neither local maximum nor local minimum if and only if $a \in [-1/3, 0]$.

(2) Find the local extremum of the following functions:

① $y = \cos x + (1/2)\cos 2x$; ② $f(x) = e^{-x}\cos x$.

Solutions

① Note $y' = -\sin x - \sin 2x = -\sin x(1+2\cos x)$. By letting $y' = 0$, we have

$$x = k\pi, \ 2k\pi + \frac{2\pi}{3}, \ 2k\pi + \frac{4\pi}{3},$$

where k is any integer. Since $y'' = -\cos x - 2\cos 2x$,

$$y''(k\pi) = -(-1)^k - 2 < 0,$$

$$y''\left(2k\pi + \frac{2\pi}{3}\right) = \frac{3}{2} > 0, \ y''\left(2k\pi + \frac{4\pi}{3}\right) = \frac{3}{2} > 0,$$

then we have:

$$\text{local maximum } y(k\pi) = (-1)^k + \frac{1}{2},$$

$$\text{local minimum } y\left(2k\pi + \frac{2\pi}{3}\right) = -\frac{3}{4},$$

$$\text{local minimum } y\left(2k\pi + \frac{4\pi}{3}\right) = -\frac{3}{4}.$$

② Note $f'(x) = -e^{-x}(\cos x + \sin x) = -\sqrt{2}e^{-x}\cos\left(x - \frac{\pi}{4}\right)$. By setting $f'(x) = 0$, we have the stationary points: $x_n = n\pi - \left(\frac{\pi}{4}\right)$

where n is any integer.

$$\text{As } n = 2k,\ f''(x_n) = 2e^{-x_n}\sin\left(2k\pi - \frac{\pi}{4}\right) < 0,$$

$$\text{As } n = 2k+1,\ f''(x_n) = 2e^{-x_n}\sin\left(2k\pi + \frac{3\pi}{4}\right) > 0.$$

Hence, as $n=2k$, f has its local maximum $f\left(2k\pi - \frac{\pi}{4}\right) = (1/\sqrt{2})e^{-2k\pi+(\pi/4)}$;

as $n=2k+1$, f has its local minimum $f\left(2k\pi + \frac{3\pi}{4}\right) = -(1/\sqrt{2})e^{-2k\pi-(3\pi/4)}$.

§5.2.3 Evaluation of absolute extremum

In Chapter 3, we concluded that the absolute extremum of a continuous function on a closed interval must be existent. Now we briefly describe the evaluation of the absolute extremum.

Procedure of finding the absolute extremum of f which is continuous on $[a, b]$:

(1) Find all of the stationary points of f on $[a, b]$: $a_1, a_2, \cdots, a_n$ (i. e. $f'(a_i) = 0$ ($i = 1, 2, \cdots, n$));

(2) Find all of the singular points of f on $[a, b]$: $b_1, b_2, \cdots, b_m$ (i. e. $f'(b_i)$ is non-existent ($i=1, 2, \cdots, m$));

(3) Compare all these values: $f(a_i)$ ($i=1, 2, \cdots, n$), $f(b_i)$ ($i=1, 2, \cdots, m$) and $f(a)$, $f(b)$;

(4) The greatest value is the absolute maximum, and the smallest value is the absolute minimum.

Examples (1) Let $f(x) = (x-1)^2(x-2)^3$. Find absolute extremum of f on $[0, 3]$.

Solution Noting $f'(x) = (x-1)(5x-7)(x-2)^2$. By letting $f'(x)=0$, we have stationary points: $a_1 = 1$, $a_2 = 7/5$, $a_3 = 2$. Clearly, f has no singular points. Then, by comparing $f(1) = 0$, $f(7/5) \approx -0.035$, $f(2) = 0$, $f(0) = -8$ and $f(3) = 4$, we have the absolute extremum of f:

$$\max_{[0,3]} f = f(3) = 4;\ \min_{[0,3]} f = f(0) = -8.$$

(2) Let $f(x) = |x^2 - 3x + 2|$. Find local extremum and absolute extremum of f on $[-10, 10]$.

Solution Since

$$f'(x) = \frac{x^2 - 3x + 2}{|x^2 - 3x + 2|}(x^2 - 3x + 2)' = \frac{x^2 - 3x + 2}{|x^2 - 3x + 2|}(2x - 3),$$

we have the stationary point $x = 3/2$ and singular points $x = 1$ and $x = 2$. By comparing $f(3/2) = 1/4$; $f(1) = f(2) = 0$, $f(-10) = 132$ and $f(10) = 72$, we see that the local maximum is $f(3/2) = 1/4$, the absolute maximum is $f(-10) = 132$ and the absolute minimum (also local minimum) is $f(1) = f(2) = 0$.

(3) Let $x \in [0, 1]$ and $p > 1$. Prove $\frac{1}{2^{p-1}} \leqslant x^p + (1-x)^p \leqslant 1$.

Proof Let $f(x) = x^p + (1-x)^p (x \in [0, 1])$. Then by setting $f'(x) = p[x^{p-1} - (1-x)^{p-1}] = 0$, we have $x = \frac{1}{2}$. Since

$$f''\left(\frac{1}{2}\right) = p(p-1)[x^{p-2} + (1-x)^{p-2}]\Big|_{x=1/2} > 0,$$

$f\left(\frac{1}{2}\right) = \frac{1}{2^{p-1}}$ is a local minimum. Noting $f(0) = f(1) = 1$, we see that $\frac{1}{2^{p-1}}$ is the absolute minimum and 1 is the absolute maximum of $f(x)$ on $[0, 1]$, i. e. $\frac{1}{2^{p-1}} \leqslant x^p + (1-x)^p \leqslant 1$ on $[0, 1]$. □

§ 5.3 Graph of a function

§ 5.3.1 Convexity of a curve

Definition Suppose $f(x)$ is differentiable on (a, b).

(1) If for any $x_1, x_2 \in (a, b)$,

$$f\left(\frac{x_1+x_2}{2}\right) \leqslant \frac{f(x_1)+f(x_2)}{2},$$

then $f(x)$ is said to be *convex downward* on (a, b);

(2) If for any $x_1, x_2 \in (a, b)$,

$$f\left(\frac{x_1+x_2}{2}\right) \geqslant \frac{f(x_1)+f(x_2)}{2},$$

then $f(x)$ is said to be *convex upward* on (a, b).

By the definition of convexity, it is easy to check the following:

Note Suppose $f(x)$ is differentiable on (a, b).

(1) $f(x)$ is convex downward on (a, b) if and only if the curve $y=f(x)$ lies above its tangent line at each point in (a, b);

(2) $f(x)$ is convex upward on (a, b) if and only if the curve $y=f(x)$ lies below its tangent line at each point in (a, b).

Theorem 5.3.1 Suppose $f(x)$ is differentiable on (a, b).

(1) $f(x)$ is convex downward on (a, b) if and only if $f'(x)$ is increasing on (a, b);

(2) $f(x)$ is convex upward on (a, b) if and only if $f'(x)$ is decreasing on (a, b).

Proof We only prove the first case, and the second case can be proved similarly. Sufficiency. Suppose $f'(x)$ is increasing on (a, b). Let $x, x_0 \in (a, b)$. By Lagrange theorem,

$$f(x)-f(x_0)=f'(\xi)(x-x_0),$$

where ξ is between x and x_0.

Since $f'(x)$ is increasing and ξ is between x and x_0, it is routine to check that $(f'(\xi)-f'(x_0))(x-x_0) \geqslant 0$, i.e.

$$f'(\xi)(x-x_0) \geqslant f'(x_0)(x-x_0).$$

Thus

$$f(x)-f(x_0)\geqslant f'(x_0)(x-x_0),$$

i. e.

$$f(x)\geqslant f'(x_0)(x-x_0)+f(x_0).$$

This means for any point $x_0\in(a, b)$, the curve $y=f(x)$ is above its tangent line $y-f(x_0)=f'(x_0)(x-x_0)$. So, by the note we see that $f(x)$ is convex downward on (a, b).

Necessity. Suppose $f(x)$ is convex downward on (a, b). Then the curve $y=f(x)$ is above its tangent line $y-f(x_0)=f'(x_0)(x-x_0)$, i. e.

$$f(x)\geqslant f'(x_0)(x-x_0)+f(x_0).$$

Notice that x and x_0 are arbitrary points in (a, b). Therefore, by a same reason we have

$$f(x_0)\geqslant f'(x)(x_0-x)+f(x).$$

Thus, by adding these two inequality we have

$$f'(x)(x_0-x)+f'(x_0)(x-x_0)=(f'(x)-f'(x_0))(x_0-x)\leqslant 0.$$

So, if $x_0<x$, then $f'(x_0)\leqslant f'(x)$, which implies f' is increasing on (a, b). □

Considering that the monotony of f' can be tested by f'', we derive immediately the following.

Theorem 5. 3. 2 Suppose $f(x)$ has second derivative $f''(x)$ on (a, b). Then

(1) $f(x)$ is convex downward on (a, b) if and only if $f''(x)\geqslant 0$ on (a, b);

(2) $f(x)$ is convex upward on (a, b) if and only if $f''(x)\leqslant 0$ on (a, b).

Example Discuss the convexity of the curve $y=\arctan x$.

Solution Note

$$y'' = -\frac{2x}{(1+x^2)^2}.$$

Thus, $y''>0$ if $x<0$, and $y''<0$ if $x>0$. Then $f(x)$ is convex downward on $(-\infty, 0)$, and $f(x)$ is convex upward on $(0, +\infty)$.

§5.3.2 Inflection points

Definition An *inflection point* of a curve $y=f(x)$ is a point on the curve at which the curve changes its convexity (from upward to downward or from downward to upward).

The following necessary condition for an inflection point is easy to prove.

Theorem 5.3.3 Suppose $f''(x)$ exists on (a, b). If $(x_0, f(x_0))$ is an inflection point of the curve $y=f(x)$, then $f''(x_0)=0$.

Note If a point $(x_0, f(x_0))$ is an inflection point of a curve $y=f(x)$, $f''(x_0)$ is not necessarily existent, i. e. a point $(x_0, f(x_0))$ on the curve $y=f(x)$ such that $f''(x_0)$ does not exist may also be an inflection point.

The procedure to determine inflection points and convexity of a curve $y=f(x)$:

(1) Find all x_i such that $f''(x_i)=0$ or $f''(x_i)$ does not exist.

(2) Test the sign of $f''(x)$ near each x_i. If $f''(x)$ changes its sign at x_i, then $(x_i, f(x_i))$ is an inflection point of the curve $y=f(x)$; otherwise, $(x_i, f(x_i))$ is not an inflection point of the curve $y=f(x)$.

(3) List all such x_i, e. g. $x_0<x_1<x_2<\cdots<x_n$, and then consider each interval (x_i, x_{i+1}). If $f''(x)\geqslant 0$ on (x_i, x_{i+1}), then the curve is convex downward on (x_i, x_{i+1}); if $f''(x)\leqslant 0$ on (x_i, x_{i+1}), then the curve is convex upward on (x_i, x_{i+1}).

Examples (1) Discuss the convexity and the inflection points of the curve $y=(x-1)\sqrt[3]{x^2}$.

Solution Note

$$y'(x)=\frac{5}{3}x^{\frac{2}{3}}-\frac{2}{3}x^{-\frac{1}{3}},\ y''(x)=\frac{10}{9}x^{-\frac{1}{3}}+\frac{2}{9}x^{-\frac{4}{3}}=\frac{2(5x+1)}{9x^{\frac{4}{3}}}.$$

So, if $x=-1/5$, $y''=0$; if $x=0$, y'' does not exist. Clearly, as $x<-1/5, y''<0$; as $x>-1/5$, $y''>0$. Thus, the point

$$\left(-\frac{1}{5},\ f\left(-\frac{1}{5}\right)\right)=\left(-\frac{1}{5},\ -\frac{6}{5}\sqrt[3]{\frac{1}{25}}\right)$$

is an inflection point, and the curve is convex upward on the left side of $x=-1/5$, and convex downward on the right side of $x=-1/5$. Since y'' does not change its sign at 0, the point $(0,\ f(0))=(0,\ 0)$ is not an inflection point.

(2) Discuss the convexity and the inflection points of the curve

$$y=x\sin(\ln x).$$

Solution Note

$$y'(x)=\sin(\ln x)+\cos(\ln x),\ y''(x)=\frac{1}{x}\cos(\ln x)-\frac{1}{x}\sin(\ln x).$$

By solving $y''(x)=0$, we have

$$x=\mathrm{e}^{k\pi+\frac{\pi}{4}}\quad(k=0,\ \pm 1,\ \pm 2,\ \cdots).$$

Then we see that on intervals $I_k=(\mathrm{e}^{2k\pi+\frac{\pi}{4}},\ \mathrm{e}^{2k\pi+\frac{5\pi}{4}})$, $y''(x)>0$, and so $f(x)$ is convex downward on I_k; on intervals $J_k=(\mathrm{e}^{2k\pi+\frac{5\pi}{4}},\ \mathrm{e}^{(2k+1)\pi+\frac{5\pi}{4}})$, $y''(x)<0$, and so $f(x)$ is convex upward on J_k, where $k=0,\ \pm 1,\ \pm 2,\ \cdots$.

The inflection points are

$$\begin{aligned}P_k&=(\mathrm{e}^{k\pi+\frac{\pi}{4}},\ f(\mathrm{e}^{k\pi+\frac{\pi}{4}}))\\&=\left(\mathrm{e}^{k\pi+\frac{\pi}{4}},\ (-1)^k\frac{\sqrt{2}}{2}\mathrm{e}^{k\pi+\frac{\pi}{4}}\right)\quad(k=0,\ \pm 1,\ \pm 2,\ \cdots).\end{aligned}$$

§5.3.3 Asymptote

Definition Suppose $C: y = f(x)$ is a curve and L is a line. If C approaches L as $x \to (\pm)\infty$ or $y \to (\pm)\infty$, then L is called a *asymptote* of C.

(1) The line $x=a$ is called a *vertical asymptote* of the curve $y=f(x)$ if at least one of the following statements is true:

$$\lim_{x \to a^-} f(x) = +\infty, \quad \lim_{x \to a^+} f(x) = +\infty,$$

$$\lim_{x \to a^-} f(x) = -\infty, \quad \lim_{x \to a^+} f(x) = -\infty.$$

(2) The line $y=b$ is called a *horizontal asymptote* of the curve $y=f(x)$ if either

$$\lim_{x \to -\infty} f(x) = b \text{ or } \lim_{x \to +\infty} f(x) = b.$$

(3) The line $y=kx+b$ $(k \neq 0)$ is called a *slant asymptote* (or *oblique asymptote*) of the curve $y=f(x)$ if

$$\lim_{x \to \infty} [f(x) - (kx + b)] = 0.$$

Remark Suppose the line $y = kx + b$ $(k \neq 0)$ is an slant asymptote of the curve $y=f(x)$. Then

$$\lim_{x \to \infty} \frac{f(x) - (kx + b)}{x} = 0.$$

Thus

$$k = \lim_{x \to \infty} \frac{f(x)}{x} \tag{1}$$

and

$$b = \lim_{x \to \infty} (f(x) - kx). \tag{2}$$

So, if a curve $y=f(x)$ has a slant asymptote $y=kx+b$, the

constant k and b can be determined by the above expression (1) and (2); conversely, it is easy to check that if we can find k and b by (1) and (2), the line $y=kx+b$ must be a slant asymptote of the curve $y=f(x)$.

Example Find asymptotes of the curve

$$y=\frac{x^3}{x^2+2x-3}.$$

Solution Since

$$y=f(x)=\frac{x^3}{x^2+2x-3}=\frac{x^3}{(x+3)(x-1)},$$

$\lim\limits_{x\to-3} f(x)=\lim\limits_{x\to 1} f(x)=\infty$. Therefore, the curve has the vertical asymptotes $x=-3$ and $x=1$.

Note that

$$k=\lim_{x\to\infty}\frac{f(x)}{x}=\lim_{x\to\infty}\frac{x^3}{x^3+2x^2-3x}=1.$$

Then

$$\begin{aligned} b &= \lim_{x\to\infty}(f(x)-kx) \\ &= \lim_{x\to\infty}\left(\frac{x^3}{x^2+2x-3}-x\right) \\ &= -2. \end{aligned}$$

So, the equation of the slant asymptote of the curve is $y=x-2$. (cf. Figure 5 - 5)

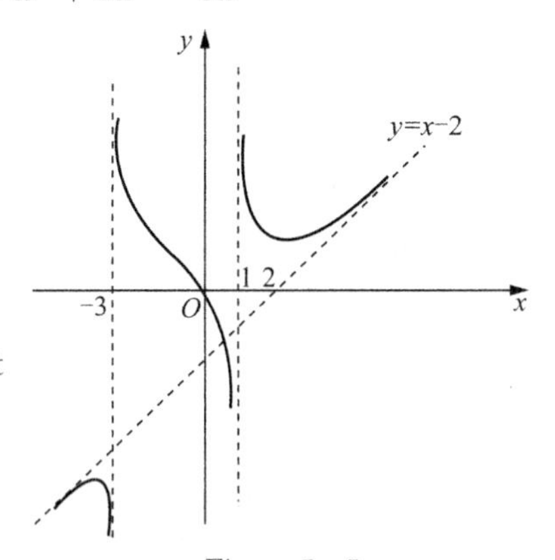

Figure 5 - 5

§5.3.4 Sketching graphs

We now present analytic strategy for graphing a function.

Strategy for Sketching the Graph of $y=f(x)$

(1) *Domain*: Determine the domain of $f(x)$.

(2) *Intercepts*: Find the x-intercept(s) and the y-intercept of $f(x)$.

(3) *Symmetry*: Determine whether the curve $y=f(x)$ is

symmetric about y-axis (i. e. f is an even function) or about the origin (i. e. f is an odd function).

(4) *Asymptotes*: Find the vertical, horizontal, and slant asymptotes of the curve $y=f(x)$.

(5) *Intervals of monotony*: Use the tests for monotone functions to determine the intervals on which $f(x)$ is increasing or decreasing.

(6) *Local extremum*: Use derivative tests to determine local maximum or local minimum.

(7) *Convexity and inflection points*: Find $f''(x)$ and use the derivative test to determine the convexity of $f(x)$. Inflection points occur where the convexity changes.

(8) *Sketch the graph*: Use the information from steps (1) through (7) to sketch the graph, plotting the characteristic points such as intercepts, maximum and minimum points and inflection points. Then make a curve smoothly pass through these points. If additional accuracy is required, you can plot more points arbitrarily on the curve, compute the values of the derivatives there. The tangent indicates the direction in which the curve proceeds.

Example Discuss the properties of the curve

$$y=\frac{(x-3)^2}{4(x-1)},$$

and sketch its graph.

Solution ① The domain of the function is $(-\infty, 1)\cup(1, +\infty)$.

② The x-intercept is $(3, 0)$ and the y-intercept is $(0, -9/4)$.

③ No Symmetry.

④ Since

$$\lim_{x\to 1}\left|\frac{(x-3)^2}{4(x-1)}\right|=+\infty,$$

the line $x=1$ is the vertical asymptote of the curve; as

$$k = \lim_{x \to \infty} \frac{f(x)}{x} = \frac{(x-3)^2}{4x(x-1)} = \frac{1}{4},$$

and

$$\lim_{x \to \infty}(f(x) - kx) = \lim_{x \to \infty}\left[\frac{(x-3)^2}{4(x-1)} - \frac{1}{4}x\right]$$

$$= \lim_{x \to \infty} \frac{-5x+9}{4(x-1)} = -\frac{5}{4},$$

the line

$$y = \frac{1}{4}x - \frac{5}{4}$$

is the slant asymptote of the curve.

⑤ Upon letting

$$f'(x) = \frac{(x-3)(x+1)}{4(x-1)^2} = 0,$$

the solutions are $x = -1$ and $x = 3$. $y' > 0$ if $x < -1$ or $x > 3$, and $y' < 0$ if $-1 < x < 1$ or $1 < x < 3$, i. e. f is strictly increasing on $(-\infty, -1)$ and $(3, +\infty)$, and strictly decreasing on $(-1, 1)$ and $(1, 3)$.

⑥ By (5) we see that the local maximum is $(-1, f(-1)) = (-1, -2)$ and the local minimum is $(3, f(3)) = (3, 0)$.

⑦ Note

$$y'' = \frac{2}{(x-1)^3}.$$

Then $y'' < 0$ if $x < 1$ and $y'' > 0$ if $x > 1$, and so the curve is convex downward on $(-\infty, 1)$ and convex upward on $(1, +\infty)$.

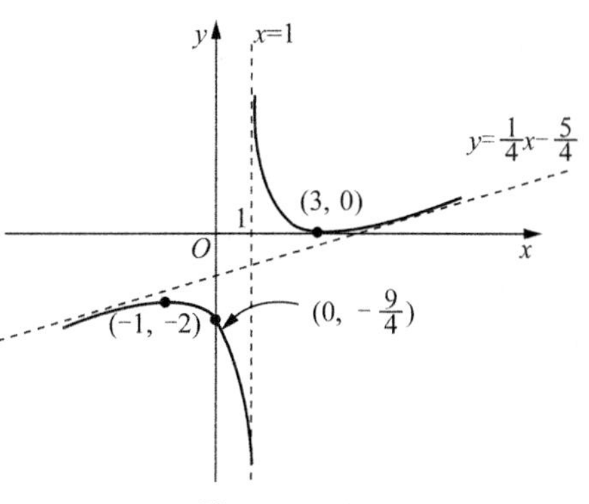

Figure 5 - 6

⑧ By the information

obtained from steps (1) through (7), the graph of the function can be drawn as in Figure 5-6:

§ 5.4 L'Hospital rules

The following types of limit operations are called *indeterminate types*:

$$\frac{0}{0}, \frac{+\infty}{+\infty}, \frac{+\infty}{-\infty}, \frac{-\infty}{+\infty}, \frac{-\infty}{-\infty}, \infty-\infty, 0\cdot\infty, 0^0, 1^\infty, \infty^0.$$

§ 5.4.1 The indeterminate type of $\frac{0}{0}$

Theorem 5.4.1 (L'Hospital rule Ⅰ) Suppose

(1) $\lim\limits_{x\to x_0} f(x) = \lim\limits_{x\to x_0} g(x) = 0$;

(2) $f(x)$ and $g(x)$ are differentiable on $U^o(x_0)$, a free-center neighborhood of x_0, and $g'(x)\neq 0$;

(3) $\lim\limits_{x\to x_0}(f'(x)/g'(x)) = A$, where A is a real number or $(\pm)\infty$. Then

$$\lim_{x\to x_0}\frac{f(x)}{g(x)} = \lim_{x\to x_0}\frac{f'(x)}{g'(x)} = A.$$

Proof Since the values of f and g at x_0 have no influence on the result, we may define $f(x_0) = g(x_0) = 0$, and so f and g are both continuous at x_0. Let $x\in U^o(x_0)$. Thus by Cauchy theorem,

$$\lim_{x\to x_0}\frac{f(x)}{g(x)} = \lim_{x\to x_0}\frac{f(x)-f(x_0)}{g(x)-g(x_0)} = \lim_{x\to x_0}\frac{f'(\xi)}{g'(\xi)} = \lim_{\xi\to x_0}\frac{f'(\xi)}{g'(\xi)} = A,$$

where ξ is between x and x_0. □

Note

(1) Theorem 5.4.1 remains valid if $x\to x_0^{(\pm)}$ or $x\to(\pm)\infty$, provided corresponding modifications are made in the second condition

of the theorem.

(2) If $f'(x)$, $g'(x)$ still satisfy the conditions of the theorem, L'Hospital rule Ⅰ can be used more than once, e. g.

$$\lim_{x \to x_0} \frac{f(x)}{g(x)} = \lim_{x \to x_0} \frac{f'(x)}{g'(x)} = \lim_{x \to x_0} \frac{f''(x)}{g''(x)}.$$

Examples (1) (cf. § 2. 7) Find

$$\lim_{x \to 0} \left(\frac{1}{\sin^2 x} - \frac{1}{x^2} \right).$$

Solution $\lim_{x \to 0} \left(\frac{1}{\sin^2 x} - \frac{1}{x^2} \right) = \lim_{x \to 0} \frac{x^2 - \sin^2 x}{\sin^2 x \cdot x^2}$

$$= \lim_{x \to 0} \frac{x^2 - \dfrac{1 - \cos 2x}{2}}{x^4}$$

$$= \lim_{x \to 0} \frac{2x^2 - 1 + \cos 2x}{2x^4}$$

$$= \lim_{x \to 0} \frac{4x - \sin 2x \cdot 2}{8x^3}$$

$$= \lim_{x \to 0} \frac{4 - 4\cos 2x}{24x^2} = \lim_{x \to 0} \frac{8\sin 2x}{48x}$$

$$= \lim_{x \to 0} \frac{8 \cdot 2x}{48x} = \frac{1}{3}.$$

(2) Suppose $f''(x)$ exists and

$g(x) = \dfrac{\sin 6x + xf(x)}{x^3}$. If $\lim_{x \to 0} g(x) = 0$, find $\lim_{x \to 0} \dfrac{6 + f(x)}{x^2}$.

Solution Since $\lim_{x \to 0} g(x) = 0$,

$$\lim_{x \to 0} (\sin 6x + xf(x)) = 0,$$

and so by L'Hospital rule Ⅰ,

$$\lim_{x \to 0} \frac{(\sin 6x + xf(x))'}{(x^3)'} = \lim_{x \to 0} \frac{6\cos 6x + xf'(x) + f(x)}{3x^2} = 0.$$

Since $\lim\limits_{x\to 0} 3x^2 = 0$,

$$\lim_{x\to 0}(6\cos 6x + xf'(x) + f(x)) = 0,$$

which implies $f(0) = -6$ and also

$$\lim_{x\to 0}\frac{(6\cos 6x + xf'(x) + f(x))'}{(3x^2)'}$$
$$= \lim_{x\to 0}\frac{-36\sin 6x + xf''(x) + 2f'(x)}{6x} = 0.$$

Also note $\lim\limits_{x\to 0} 6x = 0$. Then $\lim\limits_{x\to 0}(-36\sin 6x + xf''(x) + 2f'(x)) = 0$, which implies $f'(0) = 0$ and

$$\lim_{x\to 0}\frac{(-36\sin 6x + xf''(x) + 2f'(x))'}{(6x)'}$$
$$= \lim_{x\to 0}\frac{-6^3\cos 6x + xf'''(x) + 3f''(x)}{6} = 0.$$

Then $\lim\limits_{x\to 0} f''(x) = 72$. Thus we have

$$\lim_{x\to 0}\frac{6 + f(x)}{x^2} = \lim_{x\to 0}\frac{(6 + f(x))'}{(x^2)'} = \lim_{x\to 0}\frac{f'(x)}{2x} = \lim_{x\to 0}\frac{(f'(x))'}{(2x)'}$$
$$= \lim_{x\to 0}\frac{f''(x)}{2} = \frac{72}{2} = 36.$$

(3) Suppose $f''(x)$ exists on a neighborhood of 0, $U(0)$, and

$$\lim_{x\to 0}\left(1 + x + \frac{f(x)}{x}\right)^{1/x} = e^3.$$

Evaluate $f(0)$, $f'(0)$, $f''(0)$ and $\lim\limits_{x\to 0}\left(1 + \frac{f(x)}{x}\right)^{1/x}$.

Solution Since

$$\lim_{x\to 0}\left[\left(1 + x + \frac{f(x)}{x}\right)^{\frac{x}{x^2+f(x)}}\right]^{\frac{x^2+f(x)}{x}\cdot\frac{1}{x}} = e^3,$$

$$\lim_{x\to 0}\frac{x^2+f(x)}{x^2}=3.$$

Thus $\lim_{x\to 0}(x^2+f(x))=0$ and so $f(0)=\lim_{x\to 0}f(x)=0$. By L'Hospital rule Ⅰ,

$$\lim_{x\to 0}\frac{(x^2+f(x))'}{(x^2)'}=\lim_{x\to 0}\frac{2x+f'(x)}{2x}=3.$$

Then $\lim_{x\to 0}(2x+f'(x))=0$ and so $f'(0)=\lim_{x\to 0}f'(x)=0$. Furthermore,

$$\lim_{x\to 0}\frac{(2x+f'(x))'}{(2x)'}=\lim_{x\to 0}\frac{2+f''(x)}{2}=3,$$

which implies that $f''(0)=\lim_{x\to 0}f''(x)=4$. Note

$$\lim_{x\to 0}\frac{f(x)}{x}=\lim_{x\to 0}\frac{f'(x)}{1}=0.$$

We have

$$\lim_{x\to 0}\left(1+\frac{f(x)}{x}\right)^{1/x}=\lim_{x\to 0}\left[\left(1+\frac{f(x)}{x}\right)^{\frac{x}{f(x)}}\right]^{\frac{f(x)}{x}\cdot\frac{1}{x}}$$
$$=\mathrm{e}^{\lim_{x\to 0}\frac{f(x)}{x^2}}=\mathrm{e}^{\lim_{x\to 0}\frac{f'(x)}{2x}}=\mathrm{e}^{\lim_{x\to 0}\frac{f''(x)}{2}}=\mathrm{e}^2.$$

§5.4.2 The indeterminate type of $\frac{\infty}{\infty}$

Theorem 5.4.2 (L'Hospital rule Ⅱ) Suppose

(1) $\lim_{x\to x_0^+}f(x)=\infty$, $\lim_{x\to x_0^+}g(x)=\infty$;

(2) $f(x)$ and $g(x)$ are differentiable on $(x_0, x_0+\delta)$, a right neighborhood of x_0, and $g'(x)\neq 0$;

(3) $\lim_{x\to x_0^+}(f'(x)/g'(x))=A$. where A is a real number or $(\pm)\infty$. Then

$$\lim_{x\to x_0^+}\frac{f(x)}{g(x)}=\lim_{x\to x_0^+}\frac{f'(x)}{g'(x)}=A.$$

Proof We only consider the case in which A is a real number. For other cases it can be proved similarly. By condition (1), we may suppose $f(x)\neq 0$ and $g(x)\neq 0$ for any $x\in(x_0, x_0+\delta)$. Let $x_1\in(x_0, x_0+\delta)$ and let $x\in(x_0, x_1)$. Using Cauchy theorem on $[x, x_1]$, we have

$$\begin{aligned}\frac{f(x)}{g(x)}-A&=\frac{(f(x_1)-Ag(x_1))+(f(x)-Ag(x)-f(x_1)-Ag(x_1))}{g(x)}\\&=\frac{f(x_1)-Ag(x_1)}{g(x)}+\frac{g(x)-g(x_1)}{g(x)}\\&\quad\cdot\frac{(f(x)-f(x_1))-A(g(x)-g(x_1))}{g(x)-g(x_1)}\\&=\frac{f(x_1)-Ag(x_1)}{g(x)}+\left(1-\frac{g(x_1)}{g(x)}\right)\left(\frac{f(x)-f(x_1)}{g(x)-g(x_1)}-A\right)\\&=\frac{f(x_1)-Ag(x_1)}{g(x)}+\left(1-\frac{g(x_1)}{g(x)}\right)\left(\frac{f'(\xi)}{g'(\xi)}-A\right),\end{aligned}$$

where $x_0<x<\xi<x_1$.

For any $\varepsilon>0$, by Condition (3) we may take x_1 sufficiently close to x_0 such that $\left|\frac{f'(\xi)}{g'(\xi)}-A\right|<\varepsilon$. Since x_1 is fixed, $f(x_1)$ and $g(x_1)$ are fixed, and so by (1) we may take x sufficiently close to x_0 such that $\left|\frac{f(x_1)-Ag(x_1)}{g(x)}\right|<\varepsilon$ and $\left|\frac{g(x_1)}{g(x)}\right|<\varepsilon$. Thus

$$\begin{aligned}\left|\frac{f(x)}{g(x)}-A\right|&\leqslant\left|\frac{f(x_1)-Ag(x_1)}{g(x)}\right|+\left|1-\frac{g(x_1)}{g(x)}\right|\left|\frac{f'(\xi)}{g'(\xi)}-A\right|\\&\leqslant\varepsilon+(1+\varepsilon)\varepsilon,\end{aligned}$$

which implies

$$\lim_{x\to x_0^+}\frac{f(x)}{g(x)}=A.\ \square$$

Note L'Hospital rules Ⅰ and Ⅱ remain valid for the cases $x\to x_0^-$, $x\to x_0$ or $x\to(\pm)\infty$, and can be used more than once if the conditions are satisfied.

§5.4.3 Other indeterminate types

The L'Hospital rule can be applied directly to the $\frac{0}{0}$ and $\frac{\pm\infty}{\pm\infty}$ types. The types $\infty-\infty$ and $0\cdot\infty$ can be changed to the $\frac{0}{0}$ type or $\frac{\pm\infty}{\pm\infty}$ type by using arithmetic operations. For the types 1^{∞}, 0^0 and ∞^0 we use the following procedure:

$$\lim f(x)^{g(x)} = \lim e^{g(x)\ln f(x)} = e^{\lim\frac{\ln f(x)}{(1/g(x))}}.$$

Examples (1) ($0\cdot\infty$ type)

$$\lim_{x\to0^+} x\ln x = \lim_{x\to0^+}\frac{\ln x}{1/x} = \lim_{x\to0^+}\frac{1/x}{-1/x^2} = \lim_{x\to0^+}(-x) = 0.$$

(2) ($\infty-\infty$ type)

$$\lim_{x\to\pi/2}(\sec x-\tan x) = \lim_{x\to\pi/2}\frac{1-\sin x}{\cos x} = \lim_{x\to\pi/2}\frac{-\cos x}{-\sin x} = 0.$$

(3) (0^0 type)

$$\lim_{x\to0^+} x^x = \lim_{x\to0^+} e^{x\ln x} = e^{\lim\limits_{x\to0^+} x\ln x} = e^0 = 1.$$

(4) (1^{∞} type)

$$\lim_{x\to0}(\cos x)^{1/x^2}:$$

Since

$$\lim_{x\to0}\ln(\cos x)^{1/x^2} = \lim_{x\to0}\frac{\ln(\cos x)}{x^2} = \lim_{x\to0}\frac{-\tan x}{2x} = -\frac{1}{2},$$

$$\lim_{x\to0}(\cos x)^{1/x^2} = \lim_{x\to0} e^{\ln(\cos x)^{1/x^2}} = e^{-1/2}.$$

(5) (∞^0 type)

$$\lim_{x\to 0^+}\left(1+\frac{1}{x}\right)^x:$$

Since

$$\lim_{x\to 0^+}\ln\left(1+\frac{1}{x}\right)^x = \lim_{x\to 0^+} x\ln\left(1+\frac{1}{x}\right) = \lim_{x\to 0^+}\frac{\ln(x+1)-\ln x}{1/x}$$

$$= \lim_{x\to 0^+}\frac{(x+1)^{-1}-x^{-1}}{-x^{-2}} = \lim_{x\to 0^+}\left(x-\frac{x^2}{x+1}\right) = 0,$$

$$\lim_{x\to 0^+}\left(1+\frac{1}{x}\right)^x = \lim_{x\to 0^+} e^{\ln\left(1+\frac{1}{x}\right)^x} = e^0 = 1.$$

§ 5.5 Newton-Raphson method

Let f be a differentiable function. Since the tangent to the graph of f at the point $(x_1, f(x_1))$ is the best linear approximation of f near $(x_1, f(x_1))$, it seems reasonable to hope that, if x_1 is close to a root c of the equation $f(x)=0$, then the number x_2 where the tangent at x_1 intersects the x-axis is also close to the root c, if not closer. The tangent at x_2 in turn produces the value x_3, which may be even closer to the root c. Iteration of this procedure is known as *Newton-Raphson Method* for approximating solutions of the equation $f(x)=0$. Now let's look into the methods a little further.

To identify the connection between successive approximations x_n and x_{n+1}, we examine the general case. We begin by writing the equation for the tangent at $(x_n, f(x_n))$:

$$y-f(x_n) = f'(x_n)(x-x_n).$$

The x-intercept x_{n+1} of this line can be found by setting $y=0$:

$$0-f(x_n) = f'(x_n)(x_{n+1}-x_n).$$

Solving this equation for x_{n+1}, we have Newton-Raphson Method as follows:

$$x_{n+1} = x_n - \frac{f(x_n)}{f'(x_n)} \quad (n = 1, 2, 3, \cdots).$$

Examples Using Newton-Raphson Method to approximate the real solutions of the following equations: ① $x^3 - x - 1 = 0$; ② $\cos x = x$.

Solution ① Let $f(x) = x^3 - x - 1$. So, $f'(x) = 3x^2 - 1$ and then

$$x_{n+1} = x_n - \frac{x_n^3 - x_n - 1}{3x_n^2 - 1}.$$

We will use $x_1 = 1.5$ as our first approximation ($x_1 = 1$ or $x_1 = 2$ would also be reasonable choices).

Letting $n = 1$ in the above formula and substituting $x_1 = 1.5$ yields

$$x_2 = 1.5 - \frac{(1.5)^3 - 1.5 - 1}{3 \times (1.5)^2 - 1} \approx 1.347\,826\,09.$$

Next, we let $n=2$ and substitute x_2 to obtain

$$x_3 = x_2 - \frac{x_2^3 - x_2 - 1}{3x_2^2 - 1} \approx 1.325\,200\,40.$$

If we continue this process until two identical approximations are generated in succession, we obtain

$$x_1 = 1.5, \quad x_2 \approx 1.347\,826\,09, \quad x_3 \approx 1.325\,200\,40,$$
$$x_4 \approx 1.324\,718\,17, \quad x_5 \approx 1.324\,717\,96, \quad x_6 \approx 1.324\,717\,96.$$

Suppose the calculator we used displays nine digits. Then at this stage there is no need to continue further because all subsequent approximations that are generated will likely be the same. Thus the solution is approximately $x \approx 1.324\,717\,96$.

② Rewrite the equation as $x - \cos x = 0$. Since $f'(x) = 1 + \sin x$,

$$x_{n+1} = x_n - \frac{x_n - \cos x_n}{1 + \sin x_n}.$$

We may use $x_1 = 1$ (radian) as our initial approximation. Letting $n = 1$ and substituting $x_1 = 1$ yields

$$x_2 = 1 - \frac{1 - \cos 1}{1 + \sin 1} \approx 0.750\ 363\ 868.$$

Next, letting $n = 2$ and substituting this value of x_2 yields

$$x_3 = x_2 - \frac{x_2 - \cos x_2}{1 + \sin x_2} \approx 0.739\ 112\ 891.$$

We continue this process until two identical approximations are generated in succession, obtaining

$$x_1 = 1,\ x_2 \approx 0.750\ 363\ 868,\ x_3 \approx 0.739\ 112\ 891,$$
$$x_4 \approx 0.739\ 085\ 133,\ x_5 \approx 0.739\ 085\ 133.$$

Thus the solution is approximately $x_5 \approx 0.739\ 085\ 133$.

Exercises

1. Consider whether the condition of Rolle theorem is satisfied in the following functions for the assigned interval:

(1) $f(x) = \frac{1}{x^2}$, $[-2, 2]$; (2) $f(x) = (x-4)^2$, $[0, 8]$;

(3) $f(x) = x^3$, $[-1, 3]$;

(4) $f(x) = \begin{cases} x^2 \sin \frac{1}{x} & (x \neq 1), \\ 0 & (x = 0), \end{cases}$ $[-1, 1]$;

(5) $f(x) = \frac{3}{2x^2 + 1}$, $[-1, 1]$;

(6) $f(x) = 1 - \sqrt[3]{x^2}$, $[-1, 1]$.

2. Prove the following inequalities by Lagrange theorem:

(1) $nb^{n-1}(a-b) < a^n - b^n < na^{n-1}(a-b)$ $(n > 1,\ a > b > 0)$;

(2) $\frac{a-b}{a} \leqslant \ln\frac{a}{b} \leqslant \frac{a-b}{b} \quad (a \geqslant b > 0)$;

(3) $e^x > ex \quad (x > 1)$;

(4) $|\sin x_1 - \sin x_2| \leqslant |x_1 - x_2|$.

3. Prove:

(1) Equation $x^3 - 3x + c = 0$ (c is a constant) can not have two different roots on $[0, 1]$;

(2) Equation $x^n + px + q = 0$ (p and q are constants and n is a positive integer) has at most two roots if n is an even number, and it has at most three roots if n is an odd number.

4. Show that $f(x) = kx + b$ if $f'(x) \equiv k$ (k is a constant).

5. Suppose $f''(x)$ is continuous at $x = a$. Show that

$$\lim_{h \to 0} \frac{f(a+h) + f(a-h) - 2f(a)}{h^2} = f''(a).$$

6. Explain why Cauchy theorem does not hold true for functions $f(x) = x^2$ and $g(x) = x^3$ on $[-1, 1]$.

7. Locate the monotone intervals of the following functions:

(1) $y = \frac{1}{3}x^3 - 4x$;

(2) $y = 2x^2 - \ln x$;

(3) $y = \arctan x - x$;

(4) $y = ax^2 + bx + c (a \neq 0)$;

(5) $y = (x-2)^2(2x+1)^4$;

(6) $y = x - 2\sin x$;

(7) $y = \sqrt{2x - x^2}$;

(8) $y = \frac{x^2 - 1}{x}$.

8. Find the local extremum and the local extremal points of the following functions:

(1) $y = 2x^2 - \ln x$;

(2) $y = x - 2\sin x$;

(3) $y = \ln(x^4 - 1)$;

(4) $y = \frac{(1-x)^3}{(3x-2)^2}$;

(5) $y = \frac{x^2}{1+x}$;

(6) $y = \frac{(\ln x)^2}{x}$;

(7) $y = 2x^3 - x^4$;

(8) $y = \arctan x - \frac{1}{2}\ln(1 + x^2)$;

(9) $y = |x(x^2-1)|$; (10) $y = \frac{x(x^2+1)}{x^4-x^2+1}$.

9. Determine a and b such that the function $y = a\ln x + bx^2 + x$ attains its extrema both at $x_1 = 1$ and $x_2 = 2$; In this case, is $f(x_i)$ $(i = 1, 2)$ its maximum or minimum respectively?

10. Find the absolute extremum of the following functions in the designated intervals respectively:

(1) $f(x) = x^4 - 2x^2 + 5$, $[-2, 2]$;

(2) $f(x) = x^{2/3} - (x^2-1)^{1/3}$, $(0, 2)$;

(3) $f(x) = (x-2)^6(2x+1)^4$, $(0, 1)$;

(4) $f(x) = x + 2\sqrt{x}$, $[0, 4]$.

11. Determine the convex intervals and the inflection points of the following functions:

(1) $y = 2x^3 - 3x^2 - 36x + 25$; (2) $y = x + \frac{1}{x}$;

(3) $y = \ln(x^2+1)$; (4) $y = x^2 + \frac{1}{x}$.

12. Sketch the graphs of the following functions:

(1) $y = x^3 + 6x^2 - 15x - 20$; (2) $y = \frac{x^3}{2(1+x)^2}$;

(3) $y = e^{-x^2}$; (4) $y = x - 2\arctan x$;

(5) $y = xe^{-x}$; (6) $\rho = a\sin 3\theta$.

13. Prove the following inequalities:

(1) $x > \arctan x > x - \frac{x^3}{3}$ $(x > 0)$;

(2) $x > \ln(1+x) > \frac{x}{1+x}$ $(-1 < x < 0)$;

(3) $e^x > 1 + x$ $(x \neq 0)$;

(4) $\sqrt[3]{abc} \leqslant \frac{a+b+c}{3} \leqslant \sqrt{\frac{a^2+b^2+c^2}{3}}$ $(a, b, c > 0)$.

14. Prove the following inequalities:

(1) $\tan x > x - \frac{x^3}{3}, x \in (0, \frac{\pi}{3})$;

(2) $\frac{2}{\pi}x < \sin x < x, x \in (0, \frac{\pi}{2})$;

(3) $x^a - 1 \leqslant a(x-1), x \in (0, \infty), a \in (0, 1)$.

15. Suppose $f(x)$ is differentiable on (a, b) and $f(x)$ is continuous at $x = b$. Prove that $f(x) \leqslant f(b)$ for any $x \in (a, b)$ if $f'(x) \geqslant 0$ and that $f(x) \geqslant f(b)$ for any $x \in (a, b)$ if $f'(x) \leqslant 0$.

16. Find the absolute maximum and absolute minimum of the following functions on the given intervals respectively:

(1) $y = x^5 - 5x^4 + 5x^3 + 1$, $[-1, 2]$;

(2) $y = 2\tan x - \tan^2 x$, $[0, \frac{\pi}{2})$;

(3) $y = \sqrt{x}\ln x$, $(0, \infty)$.

17. Find a positive number x_0 such that the sum of this number and its reciprocal attains the minimum.

18. Is the derivative of a monotone function still monotone?

19. Suppose $f(x)$ and $g(x)$ are both derivable on $[a, b]$ such that $f'(x) > g'(x)$ and $f(a) = g(a)$. Verify that $f(x) > g(x)$ on $[a, b]$.

20. Find the following limits:

(1) $\lim\limits_{x\to\pi} \frac{\sin 3x}{\tan 5x}$; (2) $\lim\limits_{x\to 0} \frac{1-\cos x}{x^2}$; (3) $\lim\limits_{x\to 0} \frac{e^x - 1}{xe^x + e^x - 1}$;

(4) $\lim\limits_{x\to e} \frac{\ln x - 1}{x - e}$; (5) $\lim\limits_{x\to 0} \frac{x}{\ln\cos x}$; (6) $\lim\limits_{x\to 0} \frac{\cos\alpha x - \cos\beta x}{x^2}$;

(7) $\lim\limits_{x\to 0} \frac{e^x - x - 1}{x(e^x - 1)}$; (8) $\lim\limits_{x\to 1} (\frac{1}{\ln x} - \frac{x}{\ln x})$;

(9) $\lim\limits_{x\to\frac{\pi}{2}^+} (\sec x - \tan x)$; (10) $\lim\limits_{x\to 0} (\cot x - \frac{1}{x})$;

(11) $\lim\limits_{x\to 0^+} \sin x \ln x$; (12) $\lim\limits_{x\to 1} (1-x)\tan(\frac{\pi}{2}x)$;

(13) $\lim\limits_{x\to0^+}\ln x\ln(1+x)$;　(14) $\lim\limits_{x\to a}\dfrac{a^x-x^a}{x-a}$　$(a>0,\ a\neq1)$;

(15) $\lim\limits_{x\to0^+}\dfrac{a^x-a^{\sin x}}{x^3}$　$(a>0,\ a\neq1)$;　(16) $\lim\limits_{x\to0^+}x^x$;

(17) $\lim\limits_{x\to0^+}\dfrac{\ln(\cos ax)}{\ln(\cos bx)}$　$(b\neq0)$; (18) $\lim\limits_{x\to0^+}x^n\ln x$　$(n>0)$;

(19) $\lim\limits_{x\to1}(2-x)^{\tan\frac{\pi x}{2}}$;

(20) $\lim\limits_{x\to\frac{\pi}{2}}(\cos x)^{\frac{\pi}{2}-x}$;

(21) $\lim\limits_{x\to\frac{\pi}{4}}(\tan x)^{\tan 2x}$;

(22) $\lim\limits_{x\to0^+}x^{\frac{a}{1+\ln x}}$;

(23) $\lim\limits_{x\to0^+}\left(\ln\dfrac{1}{x}\right)^x$;

(24) $\lim\limits_{x\to0}\dfrac{(1+x)^{\frac{1}{x}}-\mathrm{e}}{x}$.

(25) $\lim\limits_{x\to0}\dfrac{\mathrm{e}^x-1}{\sin x}$;

(26) $\lim\limits_{x\to\frac{\pi}{6}}\dfrac{1-2\sin x}{\cos 3x}$;

(27) $\lim\limits_{x\to0}\dfrac{\ln(1+x)-x}{\cos x-1}$;

(28) $\lim\limits_{x\to0}\dfrac{\tan x-x}{x-\sin x}$;

(29) $\lim\limits_{x\to\frac{\pi}{2}}\dfrac{\tan x-6}{\sec x+5}$;

(30) $\lim\limits_{x\to0}\left(\dfrac{1}{x}-\dfrac{1}{\mathrm{e}^x-1}\right)$;

(31) $\lim\limits_{x\to0^+}(\tan x)^{\sin x}$;

(32) $\lim\limits_{x\to1}x^{\frac{1}{1-x}}$;

(33) $\lim\limits_{x\to\infty}(1+x^2)^{1/x}$;

(34) $\lim\limits_{x\to0^+}\sin x\ln x$;

(35) $\lim\limits_{x\to1}\dfrac{\ln\cos(x-1)}{1-\sin\dfrac{\pi x}{2}}$;

(36) $\lim\limits_{x\to+\infty}(\pi-2\arctan x)\ln x$;

(37) $\lim\limits_{x\to\frac{\pi}{4}}(\tan x)^{\tan 2x}$;

(38) $\lim\limits_{x\to0}\left(\dfrac{\ln(1+x)^{(1+x)}}{x^2}-\dfrac{1}{x}\right)$;

(39) $\lim\limits_{x\to0}\left(\cot x-\dfrac{1}{x}\right)$.

21. Suppose $f(0)=0$, $f'(x)$ is continuous on some neighborhood of 0 and $f'(0)\neq0$. Prove that $\lim\limits_{x\to0}x^{f(x)}=1$.

22. Prove that $f(x)=x^3\mathrm{e}^{-x^2}$ is a bounded function.

23. Is the limit $\lim\limits_{x\to\infty}\dfrac{x+\sin x}{x-\sin x}$ existent? If yes, can we use

L'Hospital rule to find the limit?

24. Using Taylor formula to find the following limits:

(1) $\lim\limits_{x\to 0}\dfrac{\cos x-\mathrm{e}^{-\frac{x^2}{2}}}{x^4}$; (2) $\lim\limits_{x\to 0}\dfrac{\mathrm{e}^x\sin x-x(1+x)}{x^3}$;

(3) $\lim\limits_{x\to\infty}\left[x-x^2\ln\left(1+\dfrac{1}{x}\right)\right]$.

25. Let $p(x)$ be a polynomial. If $p'(x)$ has m real roots, prove that $p(x)$ has at most $m+1$ real roots.

26. Assume that $f(x)$, $g(x)$ and $h(x)$ are continuous on $[a, b]$ and differentiable on (a, b). Prove that there exists a constant $\xi\in(a, b)$ such that

$$\begin{vmatrix} f(a) & g(a) & h(a) \\ f(b) & g(b) & h(b) \\ f'(\xi) & g'(\xi) & h'(\xi) \end{vmatrix}=0,$$

and using this result to derive Lagrange theorem and Cauchy theorem.

27. Suppose $f(x)$ is differentiable on $(a, +\infty)$, $\lim\limits_{x\to+\infty}f(x)=k$, and $\lim\limits_{x\to+\infty}f'(x)$ exists. Prove that $\lim\limits_{x\to+\infty}f'(x)=0$.

28. (1) Let $f(x)=x^3+3x^2-2x+4$. Express $f(x)$ as a Taylor formula with remainder in Lagrange form at $x_0=1$ with order 1, 2 and 3 respectively.

(2) Let $f(x)=\cos(x^2)$. Express $f(x)$ as a Taylor formula with remainder in Peano form at $x_0=0$ with order $4n$.

(3) Let $f(x)=x^2+\mathrm{e}^{-x}$. Express $f(x)$ as a Taylor formula with remainder in Peano form at $x_0=0$ with order n.

(4) Let $f(x)=\mathrm{e}^x\cos x$. Express $f(x)$ as a Taylor formula with remainder in Peano form at $x_0=0$ with order 4.

(5) Let $P(x)=1+3x+5x^2-2x^3$. Express $P(x)$ as a polynomial of $(x+1)$ with positive integer power.

(6) Let $f(x) = \ln\cos x$. Express $f(x)$ as a 3-order Taylor formula with remainder in Peano form at $x_0 = 0$.

(7) Let $f(x) = \arctan x$. Express $f(x)$ as a 3-order Taylor formula with remainder in Lagrange form at $x_0 = 0$.

29. Express each of the following functions as a n-order Taylor formula with remainder in Lagrange form at the assigned point x_0.

(1) $f(x) = \mathrm{e}^{-x}$, $x_0 = 0$; (2) $f(x) = \dfrac{\mathrm{e}^x + \mathrm{e}^{-x}}{2}$, $x_0 = 0$;

(3) $f(x) = x\mathrm{e}^{-x}$, $x_0 = 0$; (4) $f(x) = \ln x$, $x_0 = 1$.

30. Using the Taylor formula with remainder in Peano form to find the following limits:

(1) $\lim\limits_{x\to 0}\dfrac{a^x - x\ln a - 1}{b^x - x\ln b - 1}$; (2) $\lim\limits_{x\to 0}\dfrac{\cos(x^2) - x^2\cos x - 1}{\sin(x^2)}$;

(3) $\lim\limits_{x\to 0}\left[\dfrac{(1+x)^{1/x}}{\mathrm{e}}\right]^{1/x}$.

31. Determine the order of the following infinitesimals compared with x as $x\to 0$:

(1) $\cos x - \mathrm{e}^{-\frac{x^2}{2}}$; (2) $\mathrm{e}^x\sin x - x(1+x)$; (3) $\dfrac{1}{x} - \dfrac{1}{\sin x}$;

(4) $x(\mathrm{e}^x + 1) - 2(\mathrm{e}^x - 1)$.

32. Let $a, b > 0$. Prove that the equation $x^3 + ax + b = 0$ has no positive root.

33. Let $k > 0$. Determine the value of k such that the equation $\arctan x - kx = 0$ has a positive root.

34. Judge whether the following statements are correct:

(1) If $f'(x) > g'(x)$, then $f(x) > g(x)$; conversely, if $f(x) > g(x)$, then $f'(x) > g'(x)$;

(2) Let $f(x) = \cos(x^{2/3})$, then $f'(x) = -\dfrac{3}{3\sqrt[3]{x}}\sin(x^{2/3})$. So, $f(x)$ is not differentiable at $x = 0$;

(3) If $\lim\limits_{x\to x_0} f'(x)$ is not existent, then $f(x)$ is not derivable at x_0.

Consider the example

$$f(x)=\begin{cases} x^2\sin\dfrac{1}{x} & (x\neq 0) \\ 0 & (x=0); \end{cases}$$

(4) Suppose $f(x)$ is continuous on a neighborhood of x_0, $U(x_0)$, and differentiable on $U(x_0)$ except x_0. If $\lim\limits_{x\to x_0} f'(x)$ exists, then $f(x)$ is differentiable at x_0 such that $\lim\limits_{x\to x_0} f'(x)=f'(x_0)$.

35. Prove: for any polynomial $P(x)$, there exist two points x_1 and x_2 such that $P(x)$ is strictly monotone on $(x_1, +\infty)$ and $(-\infty, x_2)$ respectively.

36. Prove the inequality

$$\frac{1}{2^{p-1}}\leqslant x^p+(1-x)^p\leqslant 1$$

for any $x\in[0, 1]$, where $p>1$ is a real number.

37. Suppose $f(x)$ is continuous on $[a, b]$ and differentiable on (a, b) and $\lim\limits_{x\to a^+} f'(x)=A$. Verify that the right derivative of $f(x)$ at a, $f'_+(a)$, exists and $f'_+(a)=A$.

38. Prove the following statements:

(1) Suppose $f(x)$ is differentiable on (a, b) and $\lim\limits_{x\to a^+} f(x)=\lim\limits_{x\to b^-} f(x)$. Then there exists at least one point $\xi\in(a, b)$ such that $f'(\xi)=0$;

(2) Suppose $f(x)$ is differentiable on (a, b) and $x_1, x_2\in(a, b)$. Let

$$g(x)=\begin{cases} \dfrac{f(x)-f(x_1)}{x-x_1} & (x\neq x_1), \\ f'(x_1) & (x=x_1). \end{cases}$$

Then for any number μ between $f'(x_1)$ and $g(x_2)$, there exists at least one point ξ between x_1 and x_2 such that $f'(\xi) = \mu$.

(3) Suppose $f''(x)$ exists on (a, b) and different points x_1, x_2, $\cdots$, $x_n \in (a, b)$. Let m_1, m_2, $\cdots$, m_n be positive constants. Then if $f''(x) > 0$ on (a, b), we have

$$f\left(\frac{m_1x_1 + m_2x_2 + \cdots + m_nx_n}{m_1 + m_2 + \cdots + m_n}\right) < \frac{m_1 f(x_1) + m_2 f(x_2) + \cdots + m_n f(x_n)}{m_1 + m_2 + \cdots + m_n};$$

and if $f''(x) < 0$ on (a, b), the above inequality also changes order.

39. (1) Let $f(x) = ax^2 + bx$ (a, b are constants and $b > 0$, $a \neq 0$). Find the absolute extremum of $f(x)$ on $[0, b/a]$;

(2) Let $f(x) = (x-c)^2 + b^2\left(1 - \frac{x^2}{a^2}\right)$ $(a, b, c > 0, a^2 - b^2 = c^2)$. Find the absolute extremum of $f(x)$ on $[-a, a]$.

40. (1) Suppose $f''(x) > 0$ for any $x \in [a, b]$. Show that for any x_1, $x_2 \in (a, b)$,

$$\frac{f(x_1) + f(x_2)}{2} \geqslant f\left(\frac{x_1 + x_2}{2}\right),$$

and the equality holds true only if $x_1 = x_2$.

(2) Under the condition of (1), show that the result still holds true for any n numbers x_1, x_2, $\cdots$, $x_n \in (a, b)$, i. e.

$$\frac{1}{n}\sum_{k=1}^{n} f(x_k) \geqslant f\left(\frac{\sum_{k=1}^{n} x_k}{n}\right),$$

and the equality holds true only if $x_1 = x_2 = \cdots = x_n$.

(3) Using (2) to prove that

(i) $$\frac{x_1 + x_2 + \cdots + x_n}{n} \leqslant \sqrt{\frac{x_1^2 + x_2^2 + \cdots + x_n^2}{n}};$$

(ii) $$\frac{x_1 + x_2 + \cdots + x_n}{n} \geqslant \sqrt[n]{x_1 x_2 \cdots x_n} \quad (x_1, x_2, \cdots, x_n > 0),$$

i. e. the arithmetic average number is no less than the geometric average number.

41. Let $a_1, a_2, \cdots, a_n > 0$ and

$$f(x) = \left(\frac{a_1^x + a_2^x + \cdots + a_n^x}{n}\right)^{1/x}.$$

Prove:

(1) $\lim\limits_{x \to 0} f(x) = \sqrt[n]{a_1 a_2 \cdots a_n}$;

(2) $\lim\limits_{x \to +\infty} f(x) = \max\{a_1, a_2, \cdots, a_n\}$.

42. Let

$$f(x) = \begin{cases} \dfrac{x}{2} + x^2 \sin \dfrac{1}{x} & (\text{if } x \neq 0), \\ 0 & (\text{if } x = 0). \end{cases}$$

(1) Is $f(x)$ derivable at $x = 0$?

(2) Is $f(x)$ monotone on a neighborhood of $x = 0$?

43. Suppose $f(x)$ has its n-order derivative and it takes on equal value at $n+1$ points $x_0, x_1, x_2, \cdots, x_n$ $(x_0 < x_1 < x_2 < \cdots < x_n)$. Prove that there exists $\xi \in (x_0, x_n)$ such that $f^{(n)}(\xi) = 0$.

44. Suppose function $f(x)$ is continuous on an interval I and $f'(x)$ is interiorly bounded on I, i. e. there exists $M > 0$ such that for any $(\alpha, \beta) \subset I$ and any $x \in (\alpha, \beta)$, $|f'(x)| \leqslant M$. Then, $f(x)$ is uniformly continuous on I.

45. Prove:

(1) $\arctan x$ is uniformly continuous on $(-\infty, +\infty)$;

(2) $\ln x$ is uniformly continuous on $[c, +\infty)$ $(c > 0)$.

Chapter 6 Indefinite integrals

Just as addition has its inverse operation subtraction and multiplication has its inverse operation division, differentiation has its inverse operation – integration. We know that differentiation is to study how to find the derivative function of a known function. The opposite problem is: how to find an unknown function such that its derivative function is exactly a given function? Solving this inverse problem is not only required by mathematics science itself, but also, more importantly, it arises in many practical problems. For example, given velocity $v(t)$, find displacement $s(t)$; given acceleration $a(t)$, find velocity $v(t)$; given slope of the tangent line at any point of a curve, find the equation of this curve, etc.

§ 6.1 Concept of indefinite integrals and fundamental formulas

§ 6.1.1 Antiderivatives and indefinite integrals

Definition Let $F(x)$ and $f(x)$ be defined on the interval I. If $F'(x) = f(x)$ on I, then $F(x)$ is called an *antiderivative* of $f(x)$ on I.

For example, $-(1/a)\cos ax\,(a \neq 0)$ is an antiderivative of $\sin ax$ on $(-\infty, +\infty)$, since $(-(1/a)\cos ax)' = \sin ax$; $(1/3)x^3$ is an antiderivative of x^2 on $(-\infty, +\infty)$, since $((1/3)x^3)' = x^2$.

Theorem 6.1.1 Suppose $F(x)$ is an antiderivative of $f(x)$ on I. Then

(1) $F(x) + C$ is also an antiderivative of $f(x)$, where C is an

arbitrary constant;

(2) The difference of any two antiderivatives of $f(x)$ is a constant.

Proof The first result is easily checked. We now show the second result. Let $F(x)$ and $G(x)$ be any two antiderivatives of $f(x)$. Then $(F(x) - G(x))' = f(x) - f(x) = 0$, which yields $F(x) - G(x)$ is a constant. □

This theorem indicates that if there is an antiderivative of $f(x)$, then there are infinitely many antiderivatives of $f(x)$, and if $F(x)$ is an antiderivative of $f(x)$, then the set of the antiderivatives of $f(x)$ is exactly the set $\{F(x) + C \mid C \in \mathbb{R}\}$. Then it is reasonable to introduce the following:

Definition The set of all the antiderivatives of $f(x)$ is called the *indefinite integral* of $f(x)$ on I, which is denoted by

$$\int f(x)\mathrm{d}x$$

where $\int$ is called *integral* sign, $f(x)$ *integrand*, x *integral variable*. The process of finding the indefinite integral is called *integration*. Suppose $F(x)$ is an antiderivative of $f(x)$, then

$$\int f(x)\mathrm{d}x = F(x) + C$$

where C is an arbitrary constant, called the *constant of integration*.

Two important questions concerning the indefinite integral are:

(1) Under which condition, the indefinite integral of a function is existent?

(2) If the indefinite integral of a function is existent, how to find it?

The following theorem, which will be verified in the proof of Theorem 7.3.2, answers the first question, and the later part of this chapter is mainly focused on the second question.

Theorem 6. 1. 2 If a function $f(x)$ is continuous on an interval I, then the indefinite integral of $f(x)$ is existent on I.

Remark

(1) Since any primary function is continuous on its domain interval, a primary function has its indefinite integral on its domain interval.

(2) By the definition of the indefinite integral, we see immediately the mutually inverse relationships between integration and differentiation as follows:

$$① \left(\int f(x)\mathrm{d}x\right)' = f(x);$$

$$② \mathrm{d}\left(\int f(x)\mathrm{d}x\right) = f(x)\mathrm{d}x;$$

$$③ \int F'(x)\mathrm{d}x = F(x) + C;$$

$$④ \int \mathrm{d}F(x) = F(x) + C.$$

(3) Not any function defined on an interval has its indefinite integral. In fact, any function which has discontinuities of the first class has no indefinite integral, e. g. the following function $f(x)$ has no indefinite integral on any interval including 0:

$$f(x) = \begin{cases} 0 & (x \neq 0), \\ 1 & (x = 0). \end{cases}$$

Proof We only need to show that the function has no antiderivative on any interval including 0. Suppose there is a function $F(x)$ such that $F'(x) = f(x)$. Then if $x > 0$, $F(x) \equiv C_1$ and if $x < 0$, $F(x) \equiv C_2$ for some constants C_1 and C_2. Since $F(x)$ has derivative at $x = 0$, $F(x)$ is continuous at $x = 0$ and so $C_1 = C_2$, i. e. $F(x) \equiv C_1$, but this is a contradiction to $F'(0) = f(0) = 1$. □

§ 6. 1. 2 Fundamental formulas for indefinite integrals

Integration is the inverse operation of differentiation. Hence fundamental formulas for indefinite integrals are just derived inversely from those for derivatives.

(1) $\int 0 \, dx = C$;

(2) $\int 1 \, dx = x + C$;

(3) $\int x^a \, dx = \dfrac{x^{a+1}}{a+1} + C \quad (a \neq -1, x > 0)$;

(4) $\int \dfrac{1}{x} dx = \ln |x| + C$;

(5) $\int e^x \, dx = e^x + C$;

(6) $\int a^x \, dx = \dfrac{a^x}{\ln a} + C \quad (a \neq -1, a > 0)$;

(7) $\int \sin x \, dx = -\cos x + C$;

(8) $\int \cos x \, dx = \sin x + C$;

(9) $\int \tan x \, dx = \ln |\sec x| + C$;

(10) $\int \cot x \, dx = \ln |\sin x| + C$;

(11) $\int \sec x \, dx = \ln |\sec x + \tan x| + C$;

(12) $\int \csc x \, dx = \ln |\csc x - \cot x| + C$;

(13) $\int \sec^2 x \, dx = \tan x + C$;

(14) $\int \csc^2 x \, dx = -\cot x + C$;

(15) $\int \sec x \tan x \, dx = \sec x + C$;

(16) $\int \csc x \tan x \, dx = -\csc x + C$;

(17) $\int \frac{dx}{a^2 + x^2} = \frac{1}{a} \arctan \frac{x}{a} + C$;

(18) $\int \frac{dx}{x^2 - a^2} = \frac{1}{2a} \ln \left| \frac{x - a}{x + a} \right| + C$;

(19) $\int \frac{dx}{(x + a)(x + b)} = \frac{1}{b - a} \ln \left| \frac{x + a}{x + b} \right| + C$;

(20) $\int \frac{dx}{\sqrt{a^2 - x^2}} = \arcsin \frac{x}{a} + C$;

(21) $\int \ln |x| \, dx = x \ln |x| - x + C$;

(22) $\int \sqrt{x^2 \pm a^2} \, dx = \frac{x}{2} \sqrt{x^2 \pm a^2} \pm \frac{a^2}{2} \ln(x + \sqrt{x^2 \pm a^2}) + C$;

(23) $\int \frac{dx}{\sqrt{x^2 \pm a^2}} = \ln(x + \sqrt{x^2 \pm a^2}) + C$;

§6.1.3 Operations of indefinite integrals

By the linear property of operations of differentiation, we may derive directly the following linear property of operations of indefinite integrals.

Theorem 6.1.3 If $f(x)$ and $g(x)$ both have antiderivatives on an interval I, then $f(x) \pm g(x)$ has its antiderivative on I too, and

$$\int (f(x) \pm g(x)) dx = \int f(x) dx \pm \int g(x) dx.$$

Proof Only need to note that

$$\begin{aligned}(\int f(x) dx \pm \int g(x) dx)' &= (\int f(x) dx)' \pm (\int g(x) dx)' \\ &= f(x) \pm g(x). \quad \square\end{aligned}$$

Theorem 6.1.4 If $f(x)$ has its antiderivative on an interval I, then for any constant $a \neq 0$, $af(x)$ has its antiderivative on I too,

and

$$\int (af(x))\mathrm{d}x = a\int f(x)\mathrm{d}x.$$

Proof Note

$$(a\int f(x)\mathrm{d}x)' = a(\int f(x)\mathrm{d}x)' = af(x). \ \square$$

Combining the preceding two theorems, evidently we have the following:

Corollary Suppose $f_i(x)(i = 1, 2, \cdots, n)$ have antiderivatives on an interval I, then

$$f(x) = \sum_{i=1}^{n} k_i f_i(x) \quad (k_i \in \mathbb{R}, i = 1, 2, \cdots, n)$$

also has its antiderivative on I, and

$$\int f(x)\mathrm{d}x = \sum_{i=1}^{n} (k_i \int f_i(x)\mathrm{d}x).$$

Examples (1) $\int (\frac{1}{x^2} - 3\cos x + \frac{2}{x})\mathrm{d}x$

$$= \int \frac{1}{x^2}\mathrm{d}x - 3\int \cos x \, \mathrm{d}x + 2\int \frac{1}{x}\mathrm{d}x$$

$$= -\frac{1}{x} - 3\sin x + 2\ln |x| + C;$$

(2) $$\int (x-\sqrt{x})^3 \mathrm{d}x = \int (x^3 - 3x^{5/2} + 3x^2 - x^{3/2})\mathrm{d}x$$

$$= \frac{1}{4}x^4 - \frac{6}{7}x^{7/2} + x^3 - \frac{2}{5}x^{5/2} + C;$$

(3) $$\int \frac{(x-\sqrt{x})(1+\sqrt{x})}{\sqrt[3]{x}}\mathrm{d}x = \int \frac{x\sqrt{x} - \sqrt{x}}{\sqrt[3]{x}}\mathrm{d}x$$

$$= \int (x^{7/6} - x^{1/6})\mathrm{d}x$$

$$= \frac{6}{13}x^{13/6} - \frac{6}{7}x^{7/6} + C;$$

(4) $\int \frac{2x^2}{1+x^2}\mathrm{d}x = \int \frac{2(x^2+1)-2}{1+x^2}\mathrm{d}x$

$= \int 2\,\mathrm{d}x - 2\int \frac{1}{1+x^2}\mathrm{d}x = 2x - 2\arctan x + C$;

(5) $\int \frac{1}{\sin^2 x \cos^2 x}\mathrm{d}x = \int \frac{\sin^2 x + \cos^2 x}{\sin^2 x \cos^2 x}\mathrm{d}x$

$= \int \frac{1}{\cos^2 x}\mathrm{d}x + \int \frac{1}{\sin^2 x}\mathrm{d}x$

$= \tan x - \cot x + C.$

§ 6.2 Techniques of integration

§ 6.2.1 Integration by substitution

We now study a technique, called substitution, that can often be used to transform complicated integration problem into simpler ones. From differentiation formulas we first introduce the following integration by substitution.

Theorem 6.2.1 (Integration by substitution Ⅰ) Suppose $u = \varphi(x)$ is differentiable on $[a, b]$, $\alpha \leqslant \varphi(x) \leqslant \beta$, $g(u)$ is defined on $[\alpha, \beta]$ and has an antiderivative $G(u)$. Then $\int g(\varphi(x))\varphi'(x)\mathrm{d}x$ exists and

$$\int g(\varphi(x))\varphi'(x)\mathrm{d}x = \int g(\varphi(x))\,\mathrm{d}(\varphi(x)) = G(\varphi(x)) + C.$$

Proof This is true because

$$\frac{\mathrm{d}}{\mathrm{d}x}(G(\varphi(x))) = (G'(u))_{u=\varphi(x)}\varphi'(x)$$

$$= (g(u))_{u=\varphi(x)}\varphi'(x) = (g(\varphi(x)))\varphi'(x). \quad \square$$

Examples (1) Find $\int \sin^3 x \cos x\,\mathrm{d}x$.

Solution Let $u = \sin x$, $g(u) = u^3$. Then $g(u)$ has an antiderivative $G(u) = u^4/4$. By the above theorem, we have

$$\int \sin^3 x \cos x \, dx = \int (\sin x)^3 (\sin x)' dx = \int (\sin x)^3 d(\sin x)$$
$$= \int u^3 du = \frac{1}{4}u^4 + C = \frac{1}{4}\sin^4 x + C.$$

You can use this method directly without the substitution variable u explicitly written if you get familiar with it.

(2) Find

$$I_1 = \int \frac{1-x}{\sqrt{9-4x^2}} dx; \qquad I_2 = \int \frac{dx}{2x^2-1};$$

$$I_3 = \int \frac{\ln x}{x\sqrt{1+\ln x}} dx; \qquad I_4 = \int \tan^4 x \, dx.$$

Solution

$$I_1 = \int \frac{dx}{\sqrt{9-4x^2}} - \int \frac{x\,dx}{\sqrt{9-4x^2}}$$
$$= \frac{1}{2}\int \frac{1}{\sqrt{1-(\frac{2}{3}x)^2}} d(\frac{2}{3}x) + \frac{1}{8}\int \frac{1}{\sqrt{9-4x^2}} d(9-4x^2)$$
$$= \frac{1}{2}\arcsin \frac{2}{3}x + \frac{1}{4}\sqrt{9-4x^2} + C.$$

$$I_2 = \int \frac{dx}{(\sqrt{2}x+1)(\sqrt{2}x-1)} = \int \left(\frac{1}{\sqrt{2}x-1} - \frac{1}{\sqrt{2}x+1}\right) dx$$
$$= \frac{1}{2\sqrt{2}}\int \frac{1}{\sqrt{2}x-1} d(\sqrt{2}x-1) - \frac{1}{2\sqrt{2}}\int \frac{1}{\sqrt{2}x+1} d(\sqrt{2}x+1)$$
$$= \frac{1}{2\sqrt{2}}\ln\left|\frac{\sqrt{2}x-1}{\sqrt{2}x+1}\right| + C.$$

$$I_3 = \int \frac{\ln x + 1 - 1}{x\sqrt{1+\ln x}} dx = \int \frac{\sqrt{\ln x + 1}}{x} dx - \int \frac{dx}{x\sqrt{1+\ln x}}$$
$$= \int (1+\ln x)^{1/2} d(1+\ln x) - \int (1+\ln x)^{-1/2} d(1+\ln x)$$
$$= \frac{2}{3}(1+\ln x)^{3/2} - 2\sqrt{1+\ln x} + C.$$

$$I_4 = \int [(\tan^4 x - 1) + 1] \mathrm{d}x = \int (\tan^2 x - 1)(\tan^2 x + 1) \mathrm{d}x + \int \mathrm{d}x$$

$$= \int (\tan^2 x - 1) \sec^2 x \, \mathrm{d}x + x$$

$$= \int \tan^2 x \cdot \sec^2 x \, \mathrm{d}x - \int \sec^2 x \, \mathrm{d}x + x$$

$$= \int \tan^2 x \, \mathrm{d}(\tan x) - \tan x + x = \frac{1}{3}\tan^3 x - \tan x + x + C.$$

(3) Find

$$I_1 = \int \frac{\mathrm{d}x}{\sin x \cos x}, \quad I_2 = \int \frac{\sin 2x}{\sqrt{a^2 \cos^2 x + b^2 \sin^2 x}} \mathrm{d}x \quad (a \neq b);$$

$$I_3 = \int \frac{\sin x \cos x}{\sqrt{9\sin^2 x + 4\cos^2 x}} \mathrm{d}x.$$

Solution

$$I_1 = \int \frac{\sin^2 x + \cos^2 x}{\sin x \cos x} \mathrm{d}x = \int \left(\frac{\sin x}{\cos x} + \frac{\cos x}{\sin x} \right) \mathrm{d}x$$

$$= -\int \frac{1}{\cos x} \mathrm{d}(\cos x) + \int \frac{1}{\sin x} \mathrm{d}(\sin x)$$

$$= -\ln |\cos x| + \ln |\sin x| + C$$

$$= \ln |\tan x| + C.$$

Since $(a^2 \cos^2 x + b^2 \sin^2 x)' = (b^2 - a^2) \sin 2x$,

$$I_2 = \frac{1}{b^2 - a^2} \int \frac{1}{\sqrt{a^2 \cos^2 x + b^2 \sin^2 x}} \mathrm{d}(a^2 \cos^2 x + b^2 \sin^2 x)$$

$$= \frac{2}{b^2 - a^2} \sqrt{a^2 \cos^2 x + b^2 \sin^2 x} + C.$$

$$I_3 = \frac{1}{10} \int \frac{18 \sin x \cos x}{\sqrt{9\sin^2 x + 4\cos^2 x}} \mathrm{d}x = \frac{1}{10} \int \frac{\mathrm{d}(9\sin^2 x + 4\cos^2 x)}{\sqrt{9\sin^2 x + 4\cos^2 x}}$$

$$= \frac{1}{5} \sqrt{9\sin^2 x + 4\cos^2 x} + C.$$

(4) Find

$$\int \frac{\sin x}{a \sin x + b \cos x} \mathrm{d}x.$$

Solution Let

$$A=\int\frac{\sin x}{a\sin x+b\cos x}\mathrm{d}x \text{ and } B=\int\frac{\cos x}{a\sin x+b\cos x}\mathrm{d}x.$$

Then $aA+bB=\int\mathrm{d}x=x+C_1$ and

$$\begin{aligned}aB-bA&=\int\frac{a\cos x-b\sin x}{a\sin x+b\cos x}\mathrm{d}x=\int\frac{\mathrm{d}(a\sin x+b\cos x)}{a\sin x+b\cos x}\\&=\ln|a\sin x+b\cos x|+C_2,\end{aligned}$$

from which it follows that

$$A=\frac{1}{a^2+b^2}(ax-b\ln|a\sin x+b\cos x|+C),$$

and also

$$B=\frac{1}{a^2+b^2}(bx+a\ln|a\sin x+b\cos x|+C).$$

(5) Find

$$\int\frac{\sin x}{\sqrt{2+\sin 2x}}\mathrm{d}x.$$

Solution Note $\sqrt{2+\sin 2x}=\sqrt{1+(\sin x+\cos x)^2}$

$$=\sqrt{3-(\sin x-\cos x)^2}.$$

Then

$$\begin{aligned}\int\frac{\sin x}{\sqrt{2+\sin 2x}}\mathrm{d}x&=-\frac{1}{2}\int\frac{(\cos x-\sin x)-(\cos x+\sin x)}{\sqrt{2+\sin 2x}}\mathrm{d}x\\&=-\frac{1}{2}\left[\int\frac{\cos x-\sin x}{\sqrt{1+(\sin x+\cos x)^2}}\mathrm{d}x\right.\\&\qquad\left.-\int\frac{\cos x+\sin x}{\sqrt{3-(\sin x-\cos x)^2}}\mathrm{d}x\right]\\&=-\frac{1}{2}\left[\int\frac{\mathrm{d}(\sin x+\cos x)}{\sqrt{1+(\sin x+\cos x)^2}}\right.\\&\qquad\left.-\int\frac{\mathrm{d}(\sin x-\cos x)}{\sqrt{3-(\sin x-\cos x)^2}}\right]\end{aligned}$$

$$=-\frac{1}{2}\{\ln[(\sin x+\cos x)+\sqrt{1+(\sin x+\cos x)^2}]$$

$$-\arcsin\frac{\sin x-\cos x}{\sqrt{3}}\}+C.$$

From above examples, we see that a key step in integration by substitution I is to convert $f(x)\mathrm{d}x$ into the form of $g(\varphi(x))\varphi'(x)\mathrm{d}x$, i. e. select $u=\varphi(x)$ properly such that the antiderivative of $g(u)$ becomes easier to obtain after transforming $f(x)\mathrm{d}x$ into $g(u)\mathrm{d}u$.

For some other indefinite integrals $\int f(x)\mathrm{d}x$, we may calculate $\int f(\varphi(t))\varphi'(t)\mathrm{d}t$ more conveniently if we choose a proper substitution $x=\varphi(t)$. This leads to the following.

Theorem 6.2.2 (Integration by substitution II) Suppose $x=\varphi(t)$ is differentiable on $[\alpha, \beta]$ with $a\leqslant\varphi(t)\leqslant b$ and $\varphi'(t)\neq 0$. Let $f(x)$ be defined on $[a, b]$ such that $f(\varphi(t))\varphi'(t)$ has an antiderivative $G(t)$. Then $\int f(x)\mathrm{d}x$ exists on $[a, b]$ and

$$\int f(x)\mathrm{d}x=\left(\int f(\varphi(t))\varphi'(t)\mathrm{d}t\right)_{t=\varphi^{-1}(x)}=G(\varphi^{-1}(x))+C.$$

Proof Note that the condition $\varphi'(t)\neq 0$ guarantees the existence of the inverse function $t=\varphi^{-1}(x)$ and $(\varphi^{-1}(x))'=\frac{1}{\varphi'(t)}$. By the differentiation laws of composite function and inverse function,

$$(G(\varphi^{-1}(x))+C)'=G'(t)(\varphi^{-1}(x))'=G'(t)\frac{1}{\varphi'(t)}$$

$$=f(\varphi(t))\varphi'(t)\frac{1}{\varphi'(t)}=f(x). \square$$

Note that, evidently, substitution I and II for indefinite integral are exactly the reversed process to each other.

Example Find

① $I_1 = \int \frac{1}{2^x(1+4^x)}\mathrm{d}x$;　② $I_2 = \int \sqrt{\frac{\mathrm{e}^x - 1}{\mathrm{e}^x + 1}}\,\mathrm{d}x$;

③ $I_3 = \int \frac{\sqrt{x+1}-1}{\sqrt{x+1}+1}\mathrm{d}x$.

Solution ① Let $2^x = t$, then $x = \ln t/\ln 2$ and $\mathrm{d}x = \mathrm{d}t/(t\ln 2)$. So

$$I_1 = \frac{1}{\ln 2}\int \frac{\mathrm{d}t}{t^2(1+t^2)}\mathrm{d}t = \frac{1}{\ln 2}\int\left(\frac{1}{t^2} - \frac{1}{1+t^2}\right)\mathrm{d}t$$

$$= \frac{-1}{\ln 2}\left(\frac{1}{t} + \arctan t\right) + C = -\frac{1}{\ln 2}(2^{-x} + \arctan 2^x) + C.$$

② Let

$$\frac{\sqrt{\mathrm{e}^x - 1}}{\sqrt{\mathrm{e}^x + 1}} = t,$$

then

$$x = \ln\frac{1+t^2}{1-t^2},\ \mathrm{d}x = \frac{4t}{(1+t^2)(1-t^2)}\mathrm{d}t.$$

Thus

$$I_2 = \int \frac{4t^2}{(1+t^2)(1-t^2)}\mathrm{d}t = 2\int\left(\frac{1}{1-t^2} - \frac{1}{1+t^2}\right)\mathrm{d}t$$

$$= \ln\left|\frac{1+t}{1-t}\right| - 2\arctan t + C$$

$$= \ln\left|\frac{1+\sqrt{(\mathrm{e}^x - 1)/(\mathrm{e}^x + 1)}}{1-\sqrt{(\mathrm{e}^x - 1)/(\mathrm{e}^x + 1)}}\right|$$

$$- 2\arctan\sqrt{(\mathrm{e}^x - 1)/(\mathrm{e}^x + 1)} + C$$

$$= \ln|\mathrm{e}^x + \sqrt{\mathrm{e}^{2x} - 1}| - 2\arctan\sqrt{(\mathrm{e}^x - 1)/(\mathrm{e}^x + 1)} + C.$$

③

$$I_3 = \int \frac{(\sqrt{x+1}-1)^2}{(\sqrt{x+1}+1)(\sqrt{x+1}-1)}\mathrm{d}x$$

$$= \int \frac{x+2-2\sqrt{x+1}}{x}\mathrm{d}x$$

$$= x + 2\ln|x| - 2\int\frac{\sqrt{x+1}}{x}\mathrm{d}x.$$

Let $t=\sqrt{x+1}$, then $x=t^2-1$, $\mathrm{d}x=2t\,\mathrm{d}t$. Thus

$$\int\frac{\sqrt{x+1}}{x}\mathrm{d}x=2\int\frac{t^2}{t^2-1}\mathrm{d}t=2\int\left(1+\frac{1}{t^2-1}\right)\mathrm{d}t$$

$$=2t+\ln\left|\frac{t-1}{t+1}\right|+C_1$$

$$=2\sqrt{x+1}+\ln\left|\frac{\sqrt{x+1}-1}{\sqrt{x+1}+1}\right|+C_1,$$

and so

$$I_3=x+2\ln|x|-4\sqrt{x+1}-2\ln\left|\frac{\sqrt{x+1}-1}{\sqrt{x+1}+1}\right|+C.$$

§ 6. 2. 2 Integration by parts

From the differentiation formula for product of two functions, it follows another fundamental technique of integration.

Theorem 6. 2. 3 (Integration by parts) If the functions $u(x)$ and $v(x)$ are both differentiable and $\int u'(x)v(x)\mathrm{d}x$ exists, then $\int u(x)v'(x)\mathrm{d}x$ also exists and

$$\int u(x)v'(x)\mathrm{d}x=u(x)v(x)-\int u'(x)v(x)\mathrm{d}x.$$

Proof Note $(u(x)v(x))'=u'(x)v(x)+u(x)v'(x)$ or $u(x)v'(x)=(u(x)v(x))'-u'(x)v(x)$. Take indefinite integral on both sides of the expression and the result follows. □

Remark The integral expression in the above theorem is called the *formula of integration by parts*, which can be stated in a brief way as

$$\int u\,\mathrm{d}v=uv-\int v\,\mathrm{d}u.$$

Examples (1) Find

$$I_1=\int \frac{\ln^3 x}{x^2}\mathrm{d}x;\qquad I_2=\int(\arcsin x)^2\mathrm{d}x;$$

$$I_3=\int \frac{x\mathrm{e}^{\arctan x}}{(1+x^2)^{3/2}}\mathrm{d}x;\quad I_4=\int \frac{x\mathrm{e}^{x}}{(1+x)^2}\mathrm{d}x.$$

Solution $I_1=-\int \ln^3 x\,\mathrm{d}(\frac{1}{x})=-\frac{1}{x}\ln^3 x+3\int \frac{\ln^2 x}{x^2}\mathrm{d}x$

$$=-\frac{1}{x}\ln^3 x-\frac{3}{x}\ln^2 x+6\int \frac{\ln x}{x^2}\mathrm{d}x$$

$$=-\frac{1}{x}\ln^3 x-\frac{3}{x}\ln^2 x-\frac{6}{x}\ln x+6\int \frac{1}{x^2}\mathrm{d}x$$

$$=-\frac{1}{x}\ln^3 x-\frac{3}{x}\ln^2 x-\frac{6}{x}\ln x-\frac{6}{x}+C.$$

$$I_2=x(\arcsin x)^2-\int x\,\mathrm{d}[(\arcsin x)^2]$$

$$=x(\arcsin x)^2-\int \frac{2x}{\sqrt{1-x^2}}\arcsin x\,\mathrm{d}x$$

$$=x(\arcsin x)^2+2\sqrt{1-x^2}\,\arcsin x-2\int \mathrm{d}x$$

$$=x(\arcsin x)^2+2\sqrt{1-x^2}\,\arcsin x-2x+C.$$

$$I_3=\int \frac{x}{\sqrt{1+x^2}}\mathrm{d}(\mathrm{e}^{\arctan x})$$

$$=\frac{x\mathrm{e}^{\arctan x}}{\sqrt{1+x^2}}-\int \frac{\mathrm{e}^{\arctan x}}{\sqrt{(1+x^2)^{3/2}}}\mathrm{d}x$$

$$=\frac{x\mathrm{e}^{\arctan x}}{\sqrt{1+x^2}}-\frac{\mathrm{e}^{\arctan x}}{\sqrt{1+x^2}}-\int \frac{x\mathrm{e}^{\arctan x}}{(1+x^2)^{3/2}}\mathrm{d}x$$

$$=\frac{x\mathrm{e}^{\arctan x}}{\sqrt{1+x^2}}-\frac{\mathrm{e}^{\arctan x}}{\sqrt{1+x^2}}-I_3,$$

and so

$$I_3=\frac{x-1}{2\sqrt{1+x^2}}\mathrm{e}^{\arctan x}+C.$$

$$I_4=-\int x\mathrm{e}^{x}\mathrm{d}\left(\frac{1}{1+x}\right)=-\frac{x\mathrm{e}^{x}}{1+x}+\int \frac{1}{1+x}\mathrm{d}(x\mathrm{e}^{x})$$

$$=-\frac{xe^x}{1+x}+\int \frac{1}{1+x}(e^x+xe^x)dx=-\frac{xe^x}{1+x}+\int e^x dx$$

$$=-\frac{xe^x}{1+x}+e^x+C=\frac{e^x}{1+x}+C.$$

(2) Find $I_n = \int \sin^n x \, dx$ and $J_n = \int \cos^n x \, dx$ (n is a natural number).

Solution $\int \sin^n x \, dx = \int \sin^{n-1} x \, d(-\cos x)$

$$=-\sin^{n-1} x \cos x + (n-1)\int \sin^{n-2} x \cos^2 x \, dx$$

$$=-\sin^{n-1} x \cos x+(n-1)\int \sin^{n-2} x \, dx-(n-1)\int \sin^n x \, dx,$$

and so

$$I_n = \int \sin^n x \, dx =-\frac{1}{n}\sin^{n-1} x \cos x+\frac{n-1}{n}\int \sin^{n-2} x \, dx$$

$$=-\frac{1}{n}\sin^{n-1} x \cos x+\frac{n-1}{n}I_{n-2}(x).$$

Similarly, we have

$$J_n = \frac{1}{n}\cos^{n-1} x \sin x+\frac{n-1}{n}J_{n-2}(x).$$

These formulas are called *recurrence formulas* with order n. Using recurrence formula repeatedly, we may reduce the integration of $I_n(J_n)$ to that of $I_0(J_0)$ or $I_1(J_1)$. For example,

$$\int \sin^4 x \, dx =-\frac{1}{4}\sin^3 x \cos x+\frac{3}{4}\int \sin^2 x \, dx$$

$$=-\frac{1}{4}\sin^3 x \cos x+\frac{3}{4}\left(-\frac{1}{2}\sin x \cos x+\frac{1}{2}\int 1dx\right)$$

$$=-\frac{1}{4}\sin^3 x \cos x-\frac{3}{8}\sin x \cos x+\frac{3}{8}x+C.$$

In an analogous manner, we can derive the recurrence formulas for the integration of

$$\int \sin^m x \cos^n x \, dx; \int \frac{1}{\cos^n x} dx \text{ and } \int \frac{1}{\sin^n x} dx,$$

where m and n are positive integers.

(3) Find

$$I_n = \int \frac{dt}{(t^2 + a^2)^n} (a > 0).$$

Solution Note

$$\begin{aligned} I_k &= \int \frac{dt}{(t^2 + a^2)^k} = \frac{t}{(t^2 + a^2)^k} - \int t d\left[\frac{1}{(t^2 + a^2)^k}\right] \\ &= \frac{t}{(t^2 + a^2)^k} + 2k \int \frac{t^2}{(t^2 + a^2)^{k+1}} dt \\ &= \frac{t}{(t^2 + a^2)^k} + 2k \int \frac{(t^2 + a^2) - a^2}{(t^2 + a^2)^{k+1}} dt \\ &= \frac{t}{(t^2 + a^2)^k} + 2kI_k - 2a^2 kI_{k+1}. \end{aligned}$$

Thus

$$I_{k+1} = \frac{1}{2a^2 k} \cdot \frac{t}{(t^2 + a^2)^k} + \frac{2k-1}{2a^2 k} I_k.$$

Let $k + 1 = n$. Then from the above expression it follows the recurrence formula

$$I_n = \frac{1}{a^2}\left[\frac{1}{2(n-1)} \cdot \frac{t}{(t^2 + a^2)^{n-1}} + \frac{2n-3}{2n-2} I_{n-1}\right] \quad (n > 1),$$

by which the calculation is reduced to the integration of

$$I_1 = \int \frac{1}{t^2 + a^2} dt = \frac{1}{a} \arctan \frac{t}{a} + C.$$

(4) Let $f^{-1}(x)$ be an inverse function of $f(x)$. Suppose

$$\int f(x) dx = F(x) + C.$$

Find $\int f^{-1}(x) dx$.

Solution $\int f^{-1}(x)\mathrm{d}x = xf^{-1}(x) - \int x\mathrm{d}(f^{-1}(x))$

$$= xf^{-1}(x) - \int f(f^{-1}(x))\mathrm{d}(f^{-1}(x))$$

$$= xf^{-1}(x) - F(f^{-1}(x)) + C.$$

(5) Find $I = \int \cos(\ln x)\mathrm{d}x$.

Solution $I = \int \cos(\ln x)\mathrm{d}x = x\cos(\ln x) - \int x\,\mathrm{d}(\cos(\ln x))$

$$= x\cos(\ln x) - \int x(-\sin(\ln x))\frac{1}{x}\mathrm{d}x$$

$$= x\cos(\ln x) + \int \sin(\ln x)\mathrm{d}x$$

$$= x\cos(\ln x) + x\sin(\ln x) - \int x\cos(\ln x)\frac{1}{x}\mathrm{d}x$$

$$= x\cos(\ln x) + x\sin(\ln x) - I,$$

and so $$I = \frac{1}{2}(x\cos(\ln x) + x\sin(\ln x)) + C.$$

It is sometimes efficient to find indefinite integral by substitution and by parts together.

Example Find

$$\int \frac{\arcsin \mathrm{e}^x}{\mathrm{e}^x}\mathrm{d}x \quad (x < 0).$$

Solution 1 (First by parts and then by substitution) Let $u = \arcsin \mathrm{e}^x$, $\mathrm{d}v = (1/\mathrm{e}^x)\mathrm{d}x = \mathrm{d}(-\mathrm{e}^{-x})$. Then

$$\int \frac{\arcsin \mathrm{e}^x}{\mathrm{e}^x}\mathrm{d}x = \int \arcsin \mathrm{e}^x\,\mathrm{d}(-\mathrm{e}^{-x})$$

$$= -\mathrm{e}^{-x}\arcsin \mathrm{e}^x + \int \mathrm{e}^{-x}\mathrm{d}(\arcsin \mathrm{e}^x)$$

$$= -\mathrm{e}^{-x}\arcsin \mathrm{e}^x + \int \mathrm{e}^{-x}\frac{\mathrm{e}^x}{\sqrt{1-\mathrm{e}^{2x}}}\mathrm{d}x$$

$$= -e^{-x}\arcsin e^x + \int \frac{dx}{\sqrt{1-e^{2x}}}.$$

Now let $e^x = \sin t$ $(0 < t < \pi/2)$. Then $\sqrt{1-e^{2x}} = \cos t$, and by $x = \ln(\sin t)$, $dx = (\cos t/\sin t)dt$. Thus

$$\int \frac{dx}{\sqrt{1-e^{2x}}} = \int \frac{1}{\sin t}dt = \ln|\csc t - \cot t| + C$$

$$= \ln\left(\frac{1}{e^x} - \frac{\sqrt{1-e^{2x}}}{e^x}\right) + C$$

$$= -x + \ln(1-\sqrt{1-e^{2x}}) + C.$$

So, finally we have

$$\int \frac{\arcsin e^x}{e^x}dx = -x - e^{-x}\arcsin e^x + \ln(1-\sqrt{1-e^{2x}}) + C \quad (x < 0).$$

Solution 2 (First by substitution and then by parts) Let $e^x = u$, then $dx = (1/u)du$, and so

$$\int \frac{\arcsin e^x}{e^x}dx = \int \frac{\arcsin u}{u^2}du = \int \arcsin u \, d\left(\frac{-1}{u}\right)$$

$$= -\frac{1}{u}\arcsin u + \int \frac{1}{u\sqrt{1-u^2}}du.$$

Let $u = \sin t$ $(0 < t < \pi/2)$, then

$$\int \frac{1}{u\sqrt{1-u^2}}du = \int \frac{1}{\sin t}dt = \ln|\csc t - \cot t| + C$$

$$= \ln\left(\frac{1}{u} - \frac{\sqrt{1-u^2}}{u}\right) + C.$$

Now put $u = e^x$ back in these two expressions, we can obtain the same result as in the first solution.

§ 6.3 Integration of some special kinds of functions

In this section, we consider the integration of functions which are rational functions or can be transformed into rational functions.

§ 6.3.1 Integration of rational functions

We first discuss the integration of the rational function, i. e. the function $P(x)/Q(x)$, where $P(x)$ and $Q(x)$ are polynomials. Without loss of generality, we may suppose that the function $P(x)/Q(x)$ is a *proper fraction*, i. e. the degree of $P(x)$ is less than the degree of $Q(x)$, because otherwise we will have

$$\frac{P(x)}{Q(x)} = G(x) + \frac{P_1(x)}{Q(x)}$$

where $G(x)$ is a polynomial and $P_1(x)/Q(x)$ is a proper fraction. Since it is easy to integrate $G(x)$, we need merely to consider the integration of the proper fraction. Here we present a consequence of the theory of algebra concerning the decomposition of the proper fraction without a proof:

Theorem 6.3.1 (Decomposition theorem)

(1) Any polynomial $Q_m(x)$ of degree m with real number coefficients can be uniquely decomposed as a product of one-degree factors and two-degree factors with real number coefficients, i. e.

$$\begin{aligned} Q_m(x) &= b_0 x^m + b_1 x^{m-1} + \cdots + b_m \\ &= b_0 (x-a)^k \cdots (x-b)^l (x^2+px+q)^\lambda \cdots (x^2+rx+s)^\mu, \end{aligned}$$

where $b_0 \neq 0$, $k, \cdots, l, \lambda, \cdots, \mu$ are positive integers, $k + \cdots + l + 2(\lambda + \cdots + \mu) = m$, $p^2 - 4q < 0, \cdots, r^2 - 4s < 0$.

(2) Suppose the polynomial $Q_m(x)$ is decomposed as in (1). Then the proper fraction $R(x) = P_n(x)/Q_m(x)$ can be uniquely decomposed as follows:

$$R(x) = \frac{P_n(x)}{Q_m(x)} = \frac{a_0 x^n + a_1 x^{n-1} + \cdots + a_n}{b_0 x^m + b_1 x^{m-1} + \cdots + b_m}$$

$$= \left[\frac{A_1}{(x-a)} + \frac{A_2}{(x-a)^2} + \cdots + \frac{A_k}{(x-a)^k}\right] + \cdots$$

$$+ \left[\frac{B_1}{(x-b)} + \frac{B_2}{(x-b)^2} + \cdots + \frac{B_l}{(x-b)^l}\right]$$

$$+ \left[\frac{C_1 x + D_1}{(x^2+px+q)} + \frac{C_2 x + D_2}{(x^2+px+q)^2} + \cdots + \frac{C_\lambda x + D_\lambda}{(x^2+px+q)^\lambda}\right] + \cdots$$

$$+ \left[\frac{E_1 x + F_1}{(x^2+rx+s)} + \frac{E_2 x + F_2}{(x^2+rx+s)^2} + \cdots + \frac{E_\mu x + F_\mu}{(x^2+rx+s)^\mu}\right],$$

where A_1, A_2, $\cdots$, A_k, $\cdots$, B_1, B_2, $\cdots$, B_l, C_1, $D_1 \cdots$, C_λ, D_λ, E_1, F_1, $\cdots$, E_μ, F_μ are all real numbers, $b_0 \neq 0$, k, $\cdots$, l, λ, $\cdots$, μ are positive integers, $k + \cdots + l + 2(\lambda + \cdots + \mu) = m$, $p^2 - 4q < 0$, $\cdots$, $r^2 - 4s < 0$.

Example By the theorem we have the decompositions of the following proper fractions:

$$\frac{x^4+1}{(x-1)^2(x+2)^3} = \frac{A_1}{x-1} + \frac{A_2}{(x-1)^2} + \frac{B_1}{x+2} + \frac{B_2}{(x+2)^2} + \frac{B_3}{(x+2)^3};$$

$$\frac{1}{(x+1)^3(x-1)(x+2)} = \frac{A_1}{x+1} + \frac{A_2}{(x+1)^2} + \frac{A_3}{(x+1)^3} + \frac{B_1}{x-1} + \frac{C_1}{x+2};$$

$$\frac{x^5}{(x^2+1)(x^2-4x+5)^2} = \frac{A_1 x + B_1}{x^2+1} + \frac{C_1 x + D_1}{x^2-4x+5} + \frac{C_2 x + D_2}{(x^2-4x+5)^2};$$

$$\frac{1-2x}{(x-1)^2(x^2+1)^2} = \frac{A_1}{x-1} + \frac{A_2}{(x-1)^2} + \frac{B_1 x + C_1}{x^2+1} + \frac{B_2 x + C_2}{(x^2+1)^2}.$$

The constants A_i, B_i and C_i can be determined by comparing coefficients of two sides, which will be illustrated later in the examples.

In the light of the theorem, any proper fraction can be decomposed as a sum of the following four kinds of simple fractions (also called *partial fractions*):

$$\frac{A}{x-a}, \frac{B}{(x-a)^k}, \frac{Ax+B}{x^2+px+q}, \frac{Ax+B}{(x^2+px+q)^k},$$

where $k \geqslant 2$ is a natural number, A, B, a, p, q are constants with $p^2-4q<0$. Hence, to integrate a proper fraction, we only need to attend to the integration of the partial fractions as follows:

(1) $\int \frac{A}{x-a} \mathrm{d}x = A\ln|x-a|+C$;

(2) $\int \frac{A}{(x-a)^k} \mathrm{d}x = \frac{A}{1-k} \cdot \frac{1}{(x-a)^{k-1}} + C \quad (k=2, 3, \cdots)$;

(3) Note $x^2+px+q=(x+\frac{p}{2})^2+\frac{4q-p^2}{4}$ where $4q-p^2>0$.

Let $a=p/2$, $b=(1/2)\sqrt{4q-p^2}$. Then $x^2+px+q=(x+a)^2+b^2$, and so

$$\begin{aligned}
\int \frac{Ax+B}{x^2+px+q} \mathrm{d}x &= \int \frac{\frac{A}{2}(x^2+px+q)'+B-\frac{1}{2}Ap}{x^2+px+q} \mathrm{d}x \\
&= \frac{A}{2}\ln(x^2+px+q)+(B-\frac{1}{2}Ap)\int \frac{1}{(x+a)^2+b^2} \mathrm{d}x \\
&= \frac{A}{2}\ln(x^2+px+q)+(B-\frac{1}{2}Ap) \cdot \frac{1}{b} \arctan \frac{x+a}{b} + C.
\end{aligned}$$

(4) $$\begin{aligned}
\int \frac{Ax+B}{(x^2+px+q)^k} \mathrm{d}x &= \int \frac{\frac{A}{2}(x^2+px+q)'+B-\frac{1}{2}Ap}{(x^2+px+q)^k} \mathrm{d}x \\
&= \frac{A}{2(1-k)} \cdot \frac{1}{(x^2+px+q)^{k-1}} \\
&\quad + (B-\frac{1}{2}Ap)\int \frac{1}{[(x+a)^2+b^2]^k} \mathrm{d}x.
\end{aligned}$$

$$\text{Let } I_k = \int \frac{1}{[(x+a)^2+b^2]^k}\mathrm{d}x.$$

Then using the recurrence formula obtained in an example of last subsection 6. 2. 2 (integration by parts), we have

$$I_k = \frac{1}{b^2}\left\{\frac{1}{2(k-1)} \cdot \frac{x+a}{[(x+a)^2+b^2]^{k-1}} + \frac{2k-3}{2k-2}I_{k-1}\right\} \quad (k = 2, 3, \cdots).$$

Thus, noting $I_1 = \int \frac{1}{(x+a)^2+b^2}\mathrm{d}x = \frac{1}{b}\arctan\frac{x+a}{b}+C$,

for any $k \geqslant 1$, I_k can be derived.

Examples (1) Find $\int \frac{x-1}{(x^2+2x+3)^2}\mathrm{d}x$.

Solution Since $x^2+2x+3=(x+1)^2+2$, so by setting $x+1=t$ we have

$$\begin{aligned}\int \frac{x-1}{(x^2+2x+3)^2}\mathrm{d}x &= \int \frac{t-2}{(t^2+2)^2}\mathrm{d}t = \int \frac{t}{(t^2+2)^2}\mathrm{d}t - 2\int \frac{1}{(t^2+2)^2}\mathrm{d}t \\ &= -\frac{1}{2}\frac{1}{t^2+2} - 2\times\frac{1}{2}\left(\frac{1}{2\times 1}\frac{t}{t^2+2}+\frac{1}{2}\int\frac{1}{t^2+2}\mathrm{d}t\right) \\ &= -\frac{t+1}{2(t^2+2)} - \frac{1}{2\sqrt{2}}\arctan\frac{t}{\sqrt{2}}+C \\ &= -\frac{x+2}{2(x^2+2x+3)} - \frac{1}{2\sqrt{2}}\arctan\frac{x+1}{\sqrt{2}}+C.\end{aligned}$$

(2) Find $\int \frac{2x+2}{(x-1)(x^2+1)^2}\mathrm{d}x$.

Solution By the above theorem,

$$\frac{2x+2}{(x-1)(x^2+1)^2} = \frac{A}{x-1} + \frac{B_1x+C_1}{x^2+1} + \frac{B_2x+C_2}{(x^2+1)^2},$$

which implies

$$\begin{aligned}2x+2 = {} & A(x^2+1)^2 + (B_1x+C_1)(x-1)(x^2+1) \\ & + (B_2x+C_2)(x-1).\end{aligned}$$

Then by comparing the coefficients of x's of the two sides of the equality, we have

$$\begin{cases} A+B_1=0, \\ C_1-B_1=0, \\ 2A+B_2+B_1-C_1=0, \\ C_2+C_1-B_2-B_1=2, \\ A-C_2-C_1=2. \end{cases}$$

Solving this group of equations, we obtain $A=1$, $B_1=C_1=-1$, $B_2=-2$, $C_2=0$. Thus

$$\begin{aligned} \int \frac{2x+2}{(x-1)(x^2+1)^2}\mathrm{d}x &= \int\left[\frac{1}{x-1}-\frac{x+1}{x^2+1}-\frac{2x}{(x^2+1)^2}\right]\mathrm{d}x \\ &= \ln|x-1|-\frac{1}{2}\ln(x^2+1)-\arctan x \\ &\quad +\frac{1}{x^2+1}+C \\ &= \ln\frac{|x-1|}{\sqrt{x^2+1}}-\arctan x+\frac{1}{x^2+1}+C. \end{aligned}$$

§ 6.3.2 Integration of a rational function of triangle functions

Now we discuss the integration of a *rational function of triangle functions* $R(\sin x, \cos x)$, which is a function obtained by a finite times of fundamental operations of $\sin x$ and $\cos x$.

Generally, take a substitution

$$t=\tan\frac{x}{2}.$$

Then

$$\sin x=\frac{2\sin\frac{x}{2}\cos\frac{x}{2}}{\cos^2\frac{x}{2}+\sin^2\frac{x}{2}}=\frac{2\tan\frac{x}{2}}{1+\tan^2\frac{x}{2}}=\frac{2t}{1+t^2},$$

$$\cos x = \frac{\cos^2 \frac{x}{2} - \sin^2 \frac{x}{2}}{\cos^2 \frac{x}{2} + \sin^2 \frac{x}{2}} = \frac{1 - 2\tan^2 \frac{x}{2}}{1 + \tan^2 \frac{x}{2}} = \frac{1-t^2}{1+t^2},$$

$$x = 2\arctan t, \ \mathrm{d}x = \frac{2\mathrm{d}t}{1+t^2}.$$

Thus the function $R(\sin x, \cos x)$ can be transformed into a rational function of t.

Example Find

$$I = \int \frac{1+\sin x}{\sin x(1+\cos x)}\mathrm{d}x.$$

Solution By the above substitution,

$$I = \int \frac{1+\frac{2t}{1+t^2}}{\frac{2t}{1+t^2}\left(1+\frac{1-t^2}{1+t^2}\right)} \cdot \frac{2}{1+t^2}\mathrm{d}t = \frac{1}{2}\int \frac{1+2t+t^2}{t}\mathrm{d}t$$

$$= \frac{1}{2}\int \left(t+2+\frac{1}{t}\right)\mathrm{d}t = \frac{1}{2}\left(\frac{t^2}{2}+2t+\ln|t|\right)+C$$

$$= \frac{1}{4}\tan^2 \frac{x}{2} + \tan \frac{x}{2} + \frac{1}{2}\ln\left|\tan \frac{x}{2}\right| + C.$$

However, the integration by the above substitution is not strongly recommended because of the amount of calculation. In particular, the following approaches may be more advisable in some cases.

I. Using $\mathrm{d}\sin x = \cos x \, \mathrm{d}x$ directly

Examples (1) $\int \tan x \, \mathrm{d}x = \int \frac{\sin x}{\cos x}\mathrm{d}x$

$$= -\int \frac{\mathrm{d}(\cos x)}{\cos x} = -\ln|\cos x| + C;$$

(2) $\int \cos^3 x \, \mathrm{d}x = \int \cos^2 \mathrm{d}(\sin x) = \int (1-\sin^2 x)\mathrm{d}(\sin x)$

$$= \sin x - \frac{1}{3}\sin^3 x + C;$$

$$(3) \int \frac{\mathrm{d}x}{\sin x} = -\int \frac{\mathrm{d}(\cos x)}{\sin^2 x} = \int \frac{\mathrm{d}(\cos x)}{\cos^2 x - 1}$$

$$= \frac{1}{2}\ln\left|\frac{\cos x - 1}{\cos x + 1}\right| + C = \frac{1}{2}\ln\left(\frac{1-\cos x}{1+\cos x}\right) + C;$$

$$(4) \int \sin^2 x \cos^3 x \,\mathrm{d}x = \int \sin^2 x \cos^2 x \,\mathrm{d}(\sin x)$$

$$= \int \sin^2 x (1 - \sin^2 x)\mathrm{d}(\sin x)$$

$$= \frac{1}{3}\sin^3 x - \frac{1}{5}\sin^5 x + C.$$

This approach is usually valid if the integrand contains $\sin^m x \cos^n x$ where at least one of m and n is an odd number.

II. Using $\mathrm{d}(\tan x) = \sec^2 x \,\mathrm{d}x$ or $\mathrm{d}(\cot x) = -\csc^2 x \,\mathrm{d}x$

Examples $(1) \int \frac{\mathrm{d}x}{\sin^2 x \cos^2 x} = \int \frac{\mathrm{d}(\tan x)}{\sin^2 x} = \int \csc^2 x \,\mathrm{d}(\tan x)$

$$= \int (1 + \cot^2 x)\mathrm{d}(\tan x)$$

$$= \tan x + \int \frac{\mathrm{d}(\tan x)}{\tan^2 x}$$

$$= \tan x - \frac{1}{\tan x} + C$$

$$= \tan x - \cot x + C;$$

$$(2) \int \frac{\mathrm{d}x}{\cos^4 x} = \int \frac{1}{\cos^2 x} \cdot \frac{1}{\cos^2 x}\mathrm{d}x = \int (\tan^2 x + 1)\frac{1}{\cos^2 x}\mathrm{d}x$$

$$= \int \tan^2 x \,\mathrm{d}(\tan x) + \int \mathrm{d}(\tan x)$$

$$= \frac{1}{3}\tan^3 x + \tan x + C.$$

III. Using trigonometric identities

Examples $(1) \int \sin 3x \cos 2x \,\mathrm{d}x = \int \frac{1}{2}(\sin 5x + \sin x)\mathrm{d}x$

$$= -\frac{1}{10}\cos 5x - \frac{1}{2}\cos x + C;$$

(2) $\int \sin^4 x \cos^2 x \, dx = \int \sin^2 x \sin^2 x \cos^2 dx$

$$= \int \frac{1-\cos 2x}{2} \cdot \frac{1}{4}\sin^2 2x \, dx$$

$$= \frac{1}{8}\int (\sin^2 2x - \cos 2x \sin^2 2x) dx$$

$$= \frac{1}{8}\int \sin^2 2x \, dx - \frac{1}{16}\int \sin^2 2x \, d(\sin 2x)$$

$$= \frac{x}{16} - \frac{\sin 4x}{64} - \frac{\sin^3 2x}{48} + C.$$

§6.3.3 Integration of some irrational functions

Now we consider some integrals in which the integrand contains irrational functions. The main idea is to convert irrational function into rational function by an appropriate transformation.

Ⅰ. The integrand contains $\sqrt[n]{ax+b}$. Let $u = \sqrt[n]{ax+b}$. Then

$$x = \frac{u^n - b}{a} \text{ and } dx = \frac{n}{a}u^{n-1} du.$$

Thus the integration can be converted into that of a rational function of u.

Example Find $I = \int x\sqrt{3+4x}\, dx$.

Solution Let $\sqrt{3+4x} = u$. Then $4x = u^2 - 3$, $4dx = 2u\, du$, and so

$$I = \int \left(\frac{u^2-3}{4}\right) \cdot u \cdot \frac{u\, du}{2} = \frac{1}{8}\int (u^4 - 3u^2) du$$

$$= \frac{1}{40}u^3(u^2-5) + C = \frac{1}{20}(2x-1)(4x+3)^{\frac{3}{2}} + C.$$

Ⅱ. The integrand contains $\sqrt{a^2-x^2}$, $\sqrt{a^2+x^2}$ or $\sqrt{x^2-a^2}$. Take substitutions $x = a\sin u$, $x = a\tan u$ or $x = a\sec u$ respectively. Then the square root can be removed in the expressions:

$$\sqrt{a^2 - a^2\sin^2 u} = \sqrt{a^2(1-\sin^2 u)} = a\cos u,$$

$$\sqrt{a^2 + a^2\tan^2 u} = \sqrt{a^2(1+\tan^2 u)} = a\sec u,$$

$$\sqrt{a^2\sec^2 u - a^2} = \sqrt{a^2(\sec^2 u - 1)} = a\tan u,$$

where absolute value symbols are omitted in understanding that this is merely true for some certain values of the variables.

Examples (1) Find $I = \int \frac{\mathrm{d}x}{(a^2 - x^2)^{3/2}}$.

Solution Let $x = a\sin u$. Then $\mathrm{d}x = a\cos u\,\mathrm{d}u$, and so

$$I = \int \frac{a\cos u\,\mathrm{d}u}{a^3\cos^3 u} = \frac{1}{a^2}\int \frac{\mathrm{d}u}{\cos^2 u} = \frac{\tan u}{a^2} + C$$

$$= \frac{1}{a^2}\frac{\frac{x}{a}}{\sqrt{1-\frac{x^2}{a^2}}} + C = \frac{x}{a^2\sqrt{a^2 - x^2}} + C.$$

(2) Find $I = \int \frac{\mathrm{d}x}{\sqrt{x^2 - a^2}}$.

Solution Let $x = a\sec u$.

Then $\mathrm{d}x = a\tan u\sec u\,\mathrm{d}u$, and so

$$I = \int \frac{1}{a\tan u}\cdot\frac{a\sin u}{\cos^2 u}\mathrm{d}u = \int \frac{\mathrm{d}u}{\cos u} = \int \frac{\cos u\,\mathrm{d}u}{\cos^2 u} = \int \frac{\mathrm{d}(\sin u)}{1-\sin^2 u}$$

$$= \frac{1}{2}\int\left(\frac{1}{1-\sin u} + \frac{1}{1+\sin u}\right)\mathrm{d}(\sin u) = \frac{1}{2}\ln\frac{1+\sin u}{1-\sin u} + C$$

$$= \frac{1}{2}\ln\frac{(1+\sin u)^2}{1-\sin^2 u} + C = \ln\left|\frac{1+\sin u}{\cos u}\right| + C.$$

Since

$$\cos u = \frac{a}{u},\ \sin u = \sqrt{1-\frac{a^2}{x^2}} = \frac{1}{x}\sqrt{x^2 - a^2},$$

$$\int \frac{\mathrm{d}x}{\sqrt{x^2 - a^2}} = \ln\left|\frac{x + \sqrt{x^2 - a^2}}{a}\right| + C = \ln|x + \sqrt{x^2 - a^2}| + C',$$

noting C represents an arbitrary constant, C' is also an arbitrary constant. In a similar way, we can find

$$\int \frac{\mathrm{d}x}{\sqrt{x^2+a^2}} = \ln|x+\sqrt{x^2+a^2}|+C.$$

Ⅲ. The integrand contains $\sqrt{ax^2+bx+c}$ where $a>0$ and $b^2-4ac\neq 0$ or $a<0$ and $b^2-4ac>0$, then

$$ax^2+bx+c = a\left[\left(x+\frac{b}{2a}\right)^2+\frac{4ac-b^2}{4a^2}\right].$$

$$\text{Let } u = x+\frac{b}{2a},\ k=\sqrt{\left|\frac{4ac-b^2}{4a^2}\right|}.$$

Then the integrand contains one of the following:

$$|a|(u^2+k^2),\ |a|(u^2-k^2),\ |a|(k^2+u^2).$$

This is the case discussed in Ⅱ.

Examples (1) Find $I=\int \frac{\mathrm{d}x}{\sqrt{x^2+2x+3}}$.

Solution By the above integral,

$$I=\int\frac{\mathrm{d}x}{\sqrt{(x+1)^2+2}}=\int\frac{\mathrm{d}(x+1)}{\sqrt{(x+1)^2+2}}$$

$$=\ln|x+1+\sqrt{x^2+2x+3}|+C.$$

(2) Find $I=\int\frac{(x+4)\mathrm{d}x}{(x^2+2x+4)\sqrt{x^2+2x+5}}$.

Solution Since $x^2+2x+5=(x+1)^2+4$, let $y=x+1$, then

$$I=\int\frac{y+3}{(y^2+3)\sqrt{y^2+4}}\mathrm{d}y.$$

Now let $y=2\tan t\ (|t|<\pi/2)$, then $\mathrm{d}y=2\sec^2 t\,\mathrm{d}t$. Thus

$$I=\int\frac{2\tan t+3}{(4\tan^2 t+3)2\sec t}2\sec^2 t\,\mathrm{d}t=\int\frac{2\sin t+3\cos t}{4\sin^2 t+3\cos^2 t}\mathrm{d}t$$

$$=2\int\frac{\sin t}{4\sin^2 t+3\cos^2 t}\mathrm{d}t+3\int\frac{\cos t}{4\sin^2 t+3\cos^2 t}\mathrm{d}t$$

$$= 2\int \frac{\sin t}{4-\cos^2 t}dt + 3\int \frac{\cos t}{4\sin^2 t + 3}dt$$

$$= \frac{1}{2}\ln\frac{2-\cos t}{2+\cos t} + \sqrt{3}\arctan\left(\frac{\sin t}{\sqrt{3}}\right) + C.$$

Since $y = 2\tan t$, $\tan t = y/2$, $\sin t = y/\sqrt{4+y^2}$, $\cos t = 2/\sqrt{4+y^2}$, and

$$I = \frac{1}{2}\ln\left(\frac{2-\dfrac{2}{\sqrt{4+y^2}}}{2+\dfrac{2}{\sqrt{4+y^2}}}\right) + \sqrt{3}\arctan\frac{y}{\sqrt{3(4+y^2)}} + C$$

$$= \frac{1}{2}\ln\frac{\sqrt{4+y^2}-1}{\sqrt{4+y^2}+1} + \sqrt{3}\arctan\frac{y}{\sqrt{3(4+y^2)}} + C.$$

Since $x+1 = y$, $y^2+4 = x^2+2x+5$, and so

$$I = \frac{1}{2}\ln\frac{\sqrt{x^2+2x+5}-1}{\sqrt{x^2+2x+5}+1} + \sqrt{3}\arctan\frac{x+1}{\sqrt{3(x^2+2x+5)}} + C.$$

Ⅳ. The integrand contains

$$\sqrt[n]{\frac{ax+b}{cx+d}} \quad (n > 1,\ ad-bc \neq 0).$$

Take the substitutions

$$u = \sqrt[n]{\frac{ax+b}{cx+d}}.$$

Then the integral can be transformed into that of a rational function.

Examples (1) Find $I = \int \frac{1}{x}\sqrt{\frac{x+2}{x-2}}\,dx$.

Solution Let

$$t = \sqrt{\frac{x+2}{x-2}},\ \text{then } x = \frac{2(t^2+1)}{t^2-1},\ dx = -\frac{8t}{(t^2-1)^2}.$$

Thus, $I=\int \frac{4t^2}{(1-t^2)(1+t^2)}\mathrm{d}t = 2\int\left(\frac{1}{1-t^2}-\frac{1}{1+t^2}\right)\mathrm{d}t$

$$= \ln\left|\frac{1+t}{1-t}\right| - 2\arctan t + C$$

$$= \ln\left|\frac{1+\sqrt{(x+2)/(x-2)}}{1-\sqrt{(x+2)/(x-2)}}\right| - 2\arctan\sqrt{\frac{x+2}{x-2}} + C.$$

(2) Find $I=\int \frac{\mathrm{d}x}{\sqrt[3]{(x-1)^2(x+2)}}$.

Solution Note

$$\sqrt[3]{(x-1)^2(x+2)} = \sqrt[3]{\frac{(x-1)^2}{(x+2)^2}(x+2)^3} = (x+2)\sqrt[3]{(\frac{x-1}{x+2})^2}.$$

Let $\frac{x-1}{x+2}=t^3$, then

$$I=\int \frac{\mathrm{d}x}{(x+2)\sqrt[3]{\left(\frac{x-1}{x+2}\right)^2}} = \int\frac{3\mathrm{d}t}{1-t^3} = 3\int\frac{1}{3}\left(\frac{1}{1-t}+\frac{t+2}{1+t+t^2}\right)\mathrm{d}t$$

$$=\int\left[\frac{1}{1-t}+\frac{1}{2}\frac{1+2t}{1+t+t^2}+\frac{3}{2}\frac{1}{\left(\frac{1}{2}+t\right)^2+\frac{3}{4}}\right]\mathrm{d}t$$

$$=-\ln|1-t|+\frac{1}{2}\ln|1+t+t^2|+\sqrt{3}\arctan\frac{1+2t}{\sqrt{3}}+C$$

$$=-\frac{3}{2}\ln|\sqrt[3]{x+2}-\sqrt[3]{x-1}|+\sqrt{3}\arctan\frac{2\sqrt[3]{x-1}+\sqrt[3]{x+2}}{\sqrt{3}\sqrt[3]{x+2}}+C'.$$

Remark The antiderivative of a primary function is not necessarily a primary function, i. e. it can not be expressed as a primary function, though it is existent. If this is the case, we also say "it can not be integrated out". For example, the following integrals are such cases:

$$\int \mathrm{e}^{-x^2}\mathrm{d}x,\quad \int\frac{\mathrm{e}^x}{x}\mathrm{d}x,\quad \int\sin(x^2)\mathrm{d}x,$$

$$\int \cos(x^2)\,dx,\quad \int \frac{\sin x}{x}dx,\quad \int \frac{\cos x}{x}dx,\quad \int \frac{dx}{\ln x}.$$

Exercises

1. Which of the following functions are antiderivatives of a same function?

(1) $\frac{1}{2}\sin^2 x$, $-\frac{1}{4}\cos 2x$, $-\frac{1}{2}\cos^2 x$;

(2) $\ln x$, $\ln ax\ (a>0)$, $\ln|x|$, $\ln x+c$;

(3) $\tan\frac{x}{2}$, $-\cot x+\frac{1}{\sin x}$, $-\cos x+\cot x+5$.

2. Find antiderivatives of the following functions on the given intervals:

(1) $e^{-|x|}\quad (-\infty<x<+\infty)$;

(2) $|\ln x|\quad (0<x<+\infty)$;

(3) $\max\{1, x^2\}\quad (-\infty<x<+\infty)$;

(4) $f'(2x)\quad (a<x<b)$.

3. Suppose $f(x)$ satisfies the given conditions. Find $f(x)$:

(1) $f'(\sin^2 x)=\cos^2 x$, $f(0)=0$;

(2) $f'(x)=\frac{1}{4}x$, $f(2)=\frac{5}{2}$;

(3) $f(x)f'(x)=1\quad (x>0)$;

(4) $f(0)=0$, $f'(\ln x)=\begin{cases}1 & (0\leqslant x\leqslant 1),\\ x & (1<x<+\infty).\end{cases}$

4. Find the following indefinite integrals:

(1) $\int \sqrt[n]{x^m}\,dx$; (2) $\int x\sqrt{x}\,dx$; (3) $\int (2^x+3^x)^2 dx$;

(4) $\int\left(\frac{3}{\sqrt{4-4x^2}}+\sin x\right)dx$; (5) $\int \frac{10x^3+3}{x^4}dx$;

(6) $\int \frac{x^2}{1+x^2}dx$; (7) $\int a^x e^x dx$;

(8) $\int e^x\left(1-\frac{e^{-x}}{x^2}\right)dx$;　(9) $\int \frac{2^t-3^t}{5^t}dt$;

(10) $\int\left(3\sin t+\frac{1}{\sin^2 t}\right)dt$;　(11) $\int\left(\frac{2}{\sqrt{1-x^2}}-\frac{3}{1+x^2}\right)dx$;

(12) $\int(a\sinh x+b\cosh x)dx$;　(13) $\int(1-2x)^5dx$;

(14) $\int\frac{1}{(2x-5)^5}dx$;　(15) $\int\sqrt{8-2x}\,dx$;

(16) $\int\frac{1}{4+9x^2}dx$;　(17) $\int e^{1-3x}dx$;

(18) $\int\frac{1}{1-4x^2}dx$;　(19) $\int\frac{1}{(x+1)(2x+3)}dx$;

(20) $\int\frac{1}{2x^2+x-1}dx$;　(21) $\int\frac{1}{x^2+2x}dx$;

(22) $\int\frac{1}{\sqrt{1-(2x-3)^2}}dx$;　(23) $\int a^{-2x}dx$;

(24) $\int\frac{1}{x^2+2x+3}dx$;　(25) $\int(e^x+1)^3dx$;

(26) $\int\sin(\omega t+\varphi)dt\quad(\varphi\neq 0)$;　(27) $\int\csc\frac{\varphi}{2}d\varphi$;

(28) $\int(\sinh(2x+1)+\cosh(2x-1))dx$;

(29) $\int\frac{1}{\sinh^5 x}dx$;　(30) $\int\frac{1}{\cosh^2\frac{x}{2}}dx$;

(31) $\int\frac{1}{\cos^2\left(2x-\frac{\pi}{4}\right)}dx$.

5. Find the following indefinite integrals:

(1) $\int\frac{2x-3}{x^2-3x+8}dx$;　(2) $\int\frac{\sin x}{1+\cos x}dx$;　(3) $\int\frac{\sec^2 x}{\tan x}dx$;

(4) $\int\frac{1}{x(\ln x)\ln(\ln x)}dx$;　(5) $\int\frac{1}{\ln(\sin x)}\cdot\frac{\cos x}{\sin x}dx$.

6. Prove that $y=\frac{x^2}{2}\operatorname{sgn} x$ is an antiderivative of $|x|$ on $(-\infty, +\infty)$.

7. Find the following indefinite integrals:

(1) $\int \frac{x}{\sqrt{1+x^2}}\mathrm{d}x$; (2) $\int x\mathrm{e}^{-x^2}\,\mathrm{d}x$; (3) $\int \frac{x^2}{4+x^6}\mathrm{d}x$;

(4) $\int \frac{x^3}{\sqrt{1-x^8}}\mathrm{d}x$; (5) $\int \frac{1}{x^2}\sin\frac{1}{x}\mathrm{d}x$; (6) $\int \frac{\cos\sqrt{x}}{\sqrt{x}}\mathrm{d}x$.

8. Find the following indefinite integrals:

(1) $\int \frac{\cos x}{\sin^2 x}\mathrm{d}x$; (2) $\int \frac{1}{\cos^2 x\sqrt{1+\tan x}}\mathrm{d}x$; (3) $\int \frac{\arctan^2 x}{1+x^2}\mathrm{d}x$;

(4) $\int \frac{1}{\sqrt{x+1}+\sqrt{x-1}}\mathrm{d}x$; (5) $\int \frac{2^x}{\sqrt{1-4^x}}\mathrm{d}x$;

(6) $\int \frac{\sqrt{1+\ln x}}{x}\mathrm{d}x$; (7) $\int \frac{1}{1+\sin x}\mathrm{d}x$;

(8) $\int \frac{1}{1-\cos x}\mathrm{d}x$; (9) $\int \frac{1}{\sin x}\mathrm{d}x$;

(10) $\int \sqrt{\frac{1-x}{1+x}}\,\mathrm{d}x$; (11) $\int \frac{\mathrm{e}^x}{1+\mathrm{e}^x}\mathrm{d}x$;

(12) $\int \frac{x^2+7x+12}{x+4}\mathrm{d}x$; (13) $\int \frac{1}{\sqrt{1-2x-x^2}}\mathrm{d}x$;

(14) $\int \frac{x^4}{1-x}\mathrm{d}x$; (15) $\int \frac{\mathrm{e}^x}{\cosh x+\sinh x}\mathrm{d}x$;

(16) $\int \frac{\sin x\cos x}{1+\cos^2 x}\mathrm{d}x$; (17) $\int \frac{\ln(\tan x)}{\sin x\cos x}\mathrm{d}x$;

(18) $\int \mathrm{e}^{\sin x}\cos x\,\mathrm{d}x$; (19) $\int \sin x\cos 3x\,\mathrm{d}x$;

(20) $\int \cos 3x\cos 5x\,\mathrm{d}x$; (21) $\int \mathrm{e}^{x+\mathrm{e}^x}\,\mathrm{d}x$;

(22) $\int \mathrm{e}^{2x^2+\ln x}\mathrm{d}x$; (23) $\int \cos^4 x\sin^3 x\,\mathrm{d}x$;

(24) $\int \tan^3 x\,\mathrm{d}x$.

9. Consider the idea of finding the following integrals (not need to find the final results)

(1) $\int \cos^n x \, dx$, $\int \tan^n x \, dx \quad (n = 1, 2, 3, 4, 5, 6)$;

(2) $\int \sec^n x \, dx \quad (n = 1, 2, 4)$, $\int \sec^n x \tan x \, dx \quad (n = 1, 2, 3, 4)$;

(3) $\int \frac{dx}{1+\cos x}$, $\int \frac{dx}{1-\cos x}$, $\int \frac{dx}{1+\sin x}$, $\int \frac{dx}{1-\sin x}$;

(4) $\int \frac{dx}{\sin x + \cos x}$, $\int \frac{dx}{\cos^3 x \sin x}$, $\int \frac{dx}{\sin^5 x \cos^3 x}$;

(5) $\int \frac{x^n}{x^2 - 2x - 3} dx$, $\int \frac{x^n}{4x^2 + 4x + 10} dx \quad (n = 0, 1, 2, 3)$;

(6) $\int \frac{x+1}{\sqrt{3+4x-4x^2}} dx$.

10. Find the following indefinite integrals by substitution:

(1) $\int \frac{1}{1+\sqrt{1+x}} dx$; (2) $\int \frac{x^3}{(x-1)^{10}} dx$; (3) $\int \frac{\sqrt{x}}{\sqrt{x} - \sqrt[3]{x}} dx$;

(4) $\int \frac{x^3}{\sqrt{1+x^2}} dx$; (5) $\int \frac{1}{(x^2 - a^2)^{3/2}} dx$; (6) $\int \frac{x^2}{\sqrt{a^2 - x^2}} dx$;

(7) $\int \frac{\sqrt{x^2 + a^2}}{x^2} dx$; (8) $\int \frac{1}{x\sqrt{1-x^2}} dx$;

(9) $\int x \sqrt{\frac{x}{2a - x}} dx \quad (a > 0)$; (10) $\int \frac{e^x \sqrt{1+e^x}}{\sqrt{1-e^{2x}}} dx$;

(11) $\int x^5 (2 - 5x^3)^{2/3} dx$.

11. Using the substitutions (1) $x = \sec t$; (2) $x = \csc t$; (3) $\sqrt{x^2 - 1} = t$; (4) $x = 1/t$ respectively to find the integral

$$\int \frac{1}{x\sqrt{x^2 - 1}} dx \quad (x > 1).$$

12. Find the following indefinite integrals by parts:

(1) $\int x \cos mx \, dx$; (2) $\int x e^{-3x} dx$; (3) $\int x^3 e^{x^2} dx$;

(4) $\int \ln(1+x^2)\mathrm{d}x$;　(5) $\int \arccos x\,\mathrm{d}x$;　(6) $\int \frac{1}{\sqrt{x}}\arcsin\sqrt{x}\,\mathrm{d}x$;

(7) $\int x\sec^2 x\,\mathrm{d}x$;　(8) $\int \mathrm{e}^{-3x}\sin 2x\,\mathrm{d}x$;　(9) $\int x^2\sin 2x\,\mathrm{d}x$;

(10) $\int \mathrm{e}^{2y}(y^2-2y+5)\mathrm{d}y$;　(11) $\int \frac{\ln(x+1)}{\sqrt{x+1}}\mathrm{d}x$;

(12) $\int \sin(\ln x)\mathrm{d}x$;　(13) $\int x^2\sqrt{x^2+a^2}\,\mathrm{d}x$;

(14) $\int x^2\sqrt{4-x^2}\,\mathrm{d}x$;　(15) $\int \ln(x+\sqrt{1+x^2})\mathrm{d}x$;

(16) $\int \frac{\ln(\cos x)}{\cos^2 x}\mathrm{d}x$;　(17) $\int \frac{\arcsin\sqrt{x}}{\sqrt{1-x}}\mathrm{d}x$;

(18) $\int \ln x\,\mathrm{d}x$;　(19) $\int x^2\cos x\,\mathrm{d}x$;

(20) $\int \frac{\ln x}{x^3}\mathrm{d}x$;　(21) $\int (\ln x)^2\mathrm{d}x$;

(22) $\int x\arctan x\,\mathrm{d}x$;　(23) $\int \left(\ln(\ln x)+\frac{1}{\ln x}\right)\mathrm{d}x$;

(24) $\int \sec^3 x\,\mathrm{d}x$.

13. Find the following indefinite integrals:

(1) $\int \frac{x+1}{(x-1)(x-2)}\mathrm{d}x$;　(2) $\int \frac{5x^2-10x-4}{(2x+1)(x-1)^2}\mathrm{d}x$;

(3) $\int \frac{1}{x^4(2x^2-1)}\mathrm{d}x$;　(4) $\int \frac{\mathrm{d}x}{x^4+1}$;

(5) $\int \frac{2x^4+x^3-2x^2-8x-3}{2x^3+x^2-3x}\mathrm{d}x$;　(6) $\int \frac{4x}{(x+1)(x^2+1)^2}\mathrm{d}x$;

(7) $\int \frac{\mathrm{d}x}{x^4-1}$;　(8) $\int \frac{x^{11}}{(6+x^6)^3}\mathrm{d}x$;

(9) $\int \frac{x^5}{x^{12}-1}\mathrm{d}x$;　(10) $\int \frac{x^7}{(1-x^2)^5}\mathrm{d}x$;

(11) $\int \frac{2x^2-x-2}{(x-1)^2}\mathrm{d}x$;　(12) $\int \frac{1}{x(1+x^2)}\mathrm{d}x$;

(13) $\int \frac{1+x^2}{(1+x)^2(x-1)}\mathrm{d}x$; (14) $\int \frac{x^3+x+6}{(x^2+2x+2)(x^2-4)}\mathrm{d}x$;

(15) $\int \frac{x^3}{(x-1)^{100}}\mathrm{d}x$; (16) $\int \frac{1}{x(x^{10}-2)}\mathrm{d}x$;

(17) $\int \frac{x(1-x^2)}{1+x^4}\mathrm{d}x$; (18) $\int \frac{x^2}{1-x^4}\mathrm{d}x$;

(19) $\int \frac{1}{8}\left(\frac{x-1}{x+1}\right)^4\mathrm{d}x$; (20) $\int \frac{1}{1+x^3}\mathrm{d}x$.

14. Find the following indefinite integrals of triangle functions:

(1) $\int \sin^3 x\cos^2 x\,\mathrm{d}x$; (2) $\int \cos^2 x\sin^4 x\,\mathrm{d}x$;

(3) $\int \frac{1}{(\sin x+\cos x)^2}\mathrm{d}x$; (4) $\int \frac{1}{5+4\sin x}\mathrm{d}x$;

(5) $\int \frac{1}{5-4\sin x+3\cos x}\mathrm{d}x$; (6) $\int \frac{1}{1+\tan x}\mathrm{d}x$;

(7) $\int \frac{1+\tan x}{\sin 2x}\mathrm{d}x$; (8) $\int \frac{\sin x\cos^3 x}{1+\cos^2 x}\mathrm{d}x$;

(9) $\int \frac{1}{\sin^4 x\cos^4 x}\mathrm{d}x$; (10) $\int \frac{\sin x}{\sin x+\cos x}\mathrm{d}x$;

(11) $\int \frac{1-\tan x}{1+\tan x}\mathrm{d}x$; (12) $\int \frac{\sin^5 x}{\cos^4 x}\mathrm{d}x$;

(13) $\int \frac{\cos x-\sin x}{\cos x+\sin x}\mathrm{d}x$; (14) $\int \frac{1}{\sin x+\cos x}\mathrm{d}x$;

(15) $\int \frac{\sin x\cos x}{1+\sin^4 x}\mathrm{d}x$; (16) $\int \frac{\sin x}{\sin^2 x+4\cos^2 x}\mathrm{d}x$;

(17) $\int \cos^6 x\,\mathrm{d}x$; (18) $\int \frac{\mathrm{d}x}{\sin^3 x}$;

(19) $\int \tan^5 x\,\mathrm{d}x$; (20) $\int \frac{\mathrm{d}x}{\sqrt{\tan x}}$;

(21) $\int \frac{\mathrm{d}x}{\sqrt{\sin^3 x\cos^5 x}}$; (22) $\int \cos x\cos 2x\cos 3x\,\mathrm{d}x$;

(23) $\int \frac{\mathrm{d}x}{\sin x-\sin a}\quad(\cos a\neq 0)$;

(24) $\int \frac{dx}{\sin(x+a)\sin(x+b)}$ $[\sin(a-b) \neq 0]$;

(25) $\int \frac{dx}{(2+\cos x)\sin x}$; (26) $\int \frac{dx}{2\sin x - \cos x + 5}$;

(27) $\int \frac{dx}{1+\varepsilon \cos x}$ (i) $0 < \varepsilon < 1$; (ii) $\varepsilon > 1$.

15. Find the following indefinite integrals of irrational functions:

(1) $\int \frac{1}{\sqrt[3]{(x+1)^2(x-1)^4}} dx$; (2) $\int x\sqrt{3x+2}\, dx$;

(3) $\int y^2\sqrt{1-y^2}\, dy$; (4) $\int \frac{x}{1+\sqrt{1+x^2}} dx$;

(5) $\int \frac{1}{\sqrt{5-4x+4x^2}} dx$; (6) $\int \sqrt{4x^2+4x+5}\, dx$;

(7) $\int \sqrt{x^3+2x-3}\, dx$; (8) $\int \frac{1-x+x^2}{\sqrt{1+x-x^2}} dx$;

(9) $\int \frac{\sqrt{x(x+1)}}{\sqrt{x}+\sqrt{x+1}} dx$; (10) $\int \frac{x^3}{\sqrt{1+2x^2}} dx$;

(11) $\int \sqrt{\frac{a-x}{x-b}}\, dx$; (12) $\int x\sqrt{x+x^2}\, dx$;

(13) $\int \frac{\sqrt[3]{1+\sqrt[4]{x}}}{\sqrt{x}} dx$; (14) $\int \frac{1}{\sqrt[4]{1+x^4}} dx$;

(15) $\int \frac{x}{\sqrt{1+\sqrt[3]{x^2}}} dx$.

16. Find the following indefinite integrals:

(1) $\int \frac{\sqrt{1+\cos x}}{\sin x} dx$; (2) $\int \frac{x\cos x}{\sin^3 x} dx$; (3) $\int \frac{1}{\sin x + 5} dx$;

(4) $\int e^{\sin^2 x} \sin 2x\, dx$; (5) $\int \sqrt{1+\csc x}\, dx$; (6) $\int \frac{x}{1-\cos x} dx$;

(7) $\int \frac{1+\cos x}{(x+\sin x)^{3/2}} dx$; (8) $\int \frac{\cos\sqrt{x}-1}{\sqrt{x}\sin^2\sqrt{x}} dx$; (9) $\int \frac{\ln^3 x}{x^2} dx$;

(10) $\int \sqrt{\frac{x-1}{x+1}} \cdot \frac{1}{x^2} \mathrm{d}x \quad (x>1)$; (11) $\int x^5 \mathrm{e}^{-x^2} \mathrm{d}x$;

(12) $\int \frac{x^2+1}{x^4+1} \mathrm{d}x$; (13) $\int \frac{\ln x - 1}{\ln^2 x} \mathrm{d}x$;

(14) $\int \frac{\arctan \sqrt{x}}{\sqrt{x}(x+1)} \mathrm{d}x$; (15) $\int \frac{1}{\sqrt{x+1}+\sqrt[3]{x+1}} \mathrm{d}x$;

(16) $\int \frac{x^2-1}{x^4+1} \mathrm{d}x$; (17) $\int \frac{2-\sin x}{2+\cos x} \mathrm{d}x$;

(18) $\int \frac{\arctan x}{x^2(1+x^2)} \mathrm{d}x$; (19) $\int \frac{1}{\sin x \sqrt{1+\cos x}} \mathrm{d}x$;

(20) $\int \frac{x \ln x}{(1+x^2)^2} \mathrm{d}x$; (21) $\int \frac{1+\sin x}{1+\cos x} \mathrm{e}^x \mathrm{d}x$.

(22) $\int \frac{\sqrt{x} - 2\sqrt[8]{x} - 1}{\sqrt[4]{x}} \mathrm{d}x$; (23) $\int x \arcsin x \, \mathrm{d}x$;

(24) $\int \frac{1}{1+\sqrt{x}} \mathrm{d}x$; (25) $\int \frac{\arcsin x}{x^2} \mathrm{d}x$;

(26) $\int \frac{1}{x\sqrt{x^2-1}} \mathrm{d}x$; (27) $\int x \ln\left(\frac{1+x}{1-x}\right) \mathrm{d}x$;

(28) $\int \frac{x+\sin x}{1+\cos x} \mathrm{d}x$; (29) $\int (\ln x)^n \mathrm{d}x$; (30) $\int (\arcsin x)^n \mathrm{d}x$.

17. Find the following indefinite integrals:

(1) $\int \frac{f'(x)}{1+(f(x))^2} \mathrm{d}x$; (2) $\int \frac{f'(x)}{f(x)} \mathrm{d}x$;

(3) $\int \mathrm{e}^{f(x)} f'(x) \mathrm{d}x$; (4) $\int (f(x))^\alpha f'(x) \mathrm{d}x$.

18. Let $P(x)$ be a polynomial of degree n, $a \neq 0$. Prove:

(1) $\int \mathrm{e}^{ax} P(x) \mathrm{d}x = \mathrm{e}^{ax}\left(\frac{P(x)}{a} - \frac{P'(x)}{a^2} + \cdots + (-1)^n \frac{P^{(n)}(x)}{a^{n+1}}\right) + C$;

(2) $$\int P(x) \sin ax \, \mathrm{d}x = -\frac{\cos ax}{a}\left(P(x) - \frac{P''(x)}{a^2} + \frac{P^{(4)}(x)}{a^4} - \cdots\right) + \frac{\sin ax}{a^2}\left(P'(x) - \frac{P'''(x)}{a^2} + \frac{P^{(5)}(x)}{a^4} - \cdots\right) + C.$$

Chapter 7 Definite integrals

Calculus deals principally with two geometric problems: finding the tangent line to a curve, and finding the area of a region under a curve. The first is studied by a limit process known as differentiation; the second by another limit process-definite integration-to which we turn now. We will see in this chapter that the approaches of the two problems are so closed related essentially that the distinction between them is often hard to discern.

§ 7.1 Concept of definite integrals

In this section we will begin by attempting to solve the area problem-we will discuss what the term "area" means, and we will outline the approach to defining and calculating areas, which will lead to the concept of the definite integral naturally.

§ 7.1.1 Raising of the problem

If $f(x)$ is a continuous non-negative function on the interval $[a, b]$. A plane region bounded by the curve $y = f(x)(\geqslant 0)$, the straight lines $x = a$, $x = b(a < b)$ and the x-axis is called a *curved trapezoid*.

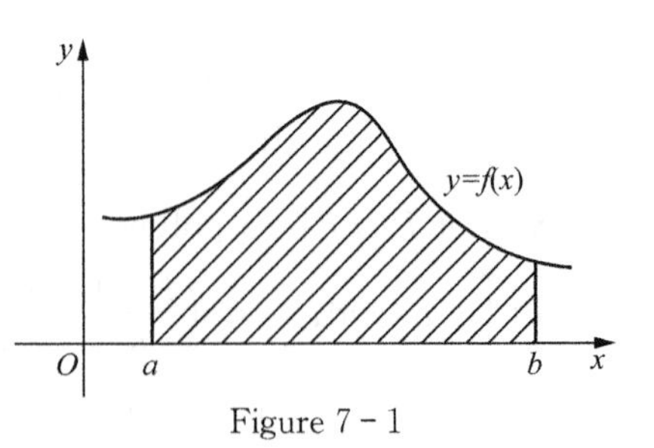

Figure 7 - 1

We refer to the area of the curved trapezoid as the area under the graph of $f(x)$ from a to b (cf. Figure 7 - 1).

The computation of the area is not a trivial matter when the top boundary of the region is curved. However, we can estimate the area to any desired degree of accuracy. The basic idea is to construct rectangles whose total area is approximately the same as the area to be computed. The area of each rectangle, of course, is easy to compute.

Figure 7 - 2 shows three rectangular approximations to the area under a graph. When the rectangles are thin, the mismatch between the rectangles and the region under the graph is quite small. In general, a rectangular approximation can be made as closed as desired to the exact area simply by making the width of the rectangles sufficiently small.

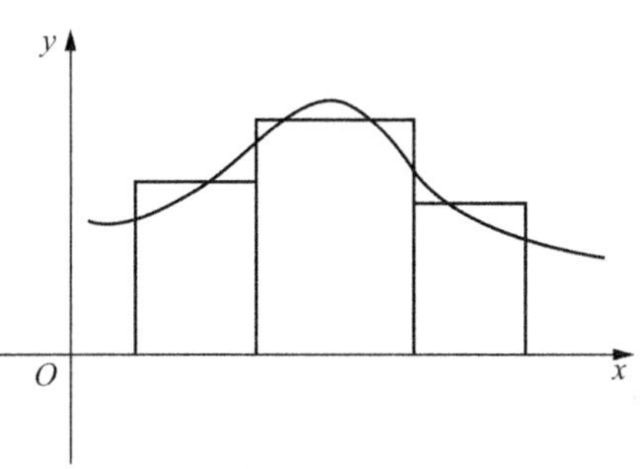

Figure 7 - 2

Now divide the interval $[a, b]$ into n equal subintervals, where n is some positive integer. Since the entire interval is of width $b-a$, the width of each of the n subintervals is $(b-a)/n$. For brevity, denote the width by Δx. That is

$$\Delta x = \frac{b-a}{n} \quad \text{(width of one interval)}.$$

In each subinterval, select a point arbitrarily. Let x_i be the point selected from the ith interval. These points are used to form rectangles that approximate the region under the graph of $f(x)$. Construct the ith rectangle with height $f(x_i)$ and the ith subinterval as base. Notice that

Area of the ith rectangule=height · width$=f(x_i)\Delta x$.

Thus, the combined area A of the n rectangles is

$$A = f(x_1)\Delta x + f(x_2)\Delta x + \cdots + f(x_n)\Delta x.$$

This sum provides an approximation to the area under the graph of $f(x)$ when $f(x)$ is non-negative and continuous. In fact, as the number of subintervals increases indefinitely, the sums approach a limiting value, the area under the graph. This idea motivates the notion of the definite integral (cf. Figure 7 - 3).

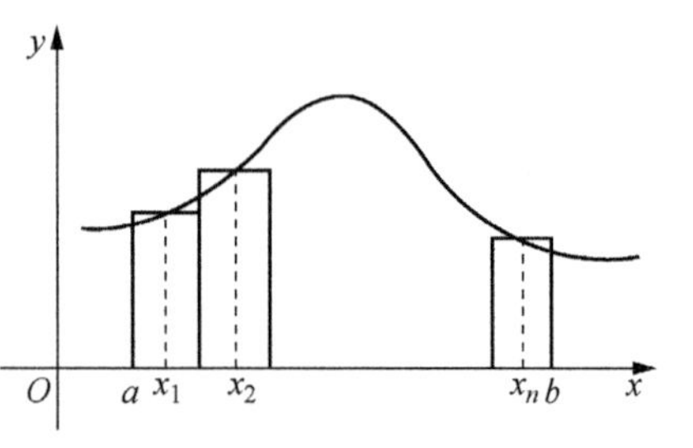

Figure 7 - 3

§7.1.2 Definition of the definite integral

Now we introduce the concept of the definite integral, which will link the notion of area to other important notions such as length, volume, density and work.

Definition Suppose the function $f(x)$ is defined on $[a, b]$, Insert arbitrarily $n-1$ points between a and b:

$$a = x_0 < x_1 < x_2 < \cdots < x_{i-1} < x_i < \cdots < x_n = b,$$

and divide $[a, b]$ into n subintervals $\Delta_i = [x_{i-1}, x_i](i = 1, 2, \cdots n, x_0 = a, x_n = b)$ (called a *partition* of the interval $[a, b]$, denoted by T, while the points $x_0, x_1, \cdots, x_n$ are called *partition points*). Denote $\Delta x_i = x_i - x_{i-1}$ and take arbitrarily $\xi_i \in [x_{i-1}, x_i]$, called *intermediate point*. Then the sum

$$\sigma = \sum_{i=1}^{n} f(\xi_i)\Delta x_i$$

is called *Riemann sum* or just *integral sum*. The *intermediate point set*, $\{\xi_i \mid i = 1, 2, \cdots, n\}$ is composed of all intermediate points ξ_i. If the limit of the Riemann sum exists as $\| T \| = \max\{\Delta x_i; i = 1, 2, \cdots, n\} \to 0$ ($\| T \|$ is called the *module* of the partition T), then this limit is called the *definite integral* of $f(x)$ on the interval $[a, b]$, which is denoted by

$$\int_a^b f(x)\mathrm{d}x = \lim_{\|T\|\to 0}\sum_{i=1}^{n} f(\xi_i)\Delta x_i,$$

where x is called *integral variable*, $f(x)$ *integrand function* or *just integrand*, $f(x)\mathrm{d}x$ *integrand expression*, $[a, b]$ *integral interval*, b and a *upper limit* and *lower limit* of the integral respectively, and $\int$ *integral sign*. Also in this case, we say the function $f(x)$ is *integrable* on $[a, b]$, or *the integral of* $f(x)$ *on* $[a, b]$ *is existent*. Otherwise, we say $f(x)$ is *unintegrable* on $[a, b]$, or *the integral of* $f(x)$ *on* $[a, b]$ *is non-existent*.

As soon as $f(x)$ and the partition T are fixed, different choices of the intermediate point set derive different integral sums. But, we denote all these different integral sums as $\sum_f(T)$ indifferently, or just $\sum(T)$ if no obscurity may be incurred as for which function is referred to.

Remark

(1) The definite integral is essentially a number, which depends on the integrand $f(x)$ and the integral interval $[a, b]$, and so it is irrelevant to the integral variable x, i. e.

$$\int_a^b f(x)\mathrm{d}x = \int_a^b f(t)\mathrm{d}t = \int_a^b f(w)\mathrm{d}w.$$

Notice that the indefinite integral $\int f(x)\,\mathrm{d}x$ is a family of antiderivatives of the function $f(x)$. So, the definite integral and the indefinite integral are two different concepts.

(2) The geometric interpretation of the definite integral:

As we know, if $f(x)\geqslant 0$, $\int_a^b f(x)\,\mathrm{d}x$ represents the area of the region between the graph of the function $f(x)$ and the x-axis over

$[a, b]$, as illustrated in the following figures. In particular, it is clear that $\int_a^b dx = b - a$.

(3) For convenience, we define

$$\int_a^a f(x)dx = 0; \int_a^b f(x)dx = -\int_b^a f(x)dx.$$

(4) A partition T on $[a, b]$:

$$a = x_0 < x_1 < x_2 < \cdots < x_{i-1} < x_i < \cdots < x_n = b,$$

such that $\Delta x_i = \Delta x_j$ for any $i, j \in \{1, 2, \cdots, n\}$ is called a *uniform partition* on $[a, b]$. In this case, $x_i = a + i \cdot \frac{b-a}{n}$ and $\Delta x_i = \frac{b-a}{n}$ for any i, and so $\|T\| \to 0$ if and only if $n \to +\infty$. Thus, we have

$$\int_a^b f(x)dx = \lim_{\|T\| \to 0} \sum_{i=1}^n f(\xi_i)\Delta x_i = \lim_{n \to \infty} \frac{b-a}{n} \sum_{i=1}^n f(\xi_i).$$

Furthermore, if we take $\xi_i = x_i (= a + i \cdot \frac{b-a}{n})$,

$$\int_a^b f(x)dx = \lim_{n \to \infty} \frac{b-a}{n} \sum_{i=1}^n f\left(a + i \cdot \frac{b-a}{n}\right).$$

In particular, if $a = 0$ and $b = 1$,

$$\int_0^1 f(x)dx = \lim_{n \to \infty} \frac{1}{n} \sum_{i=1}^n f\left(\frac{i}{n}\right).$$

Examples (1) Find $\int_0^1 e^x dx$ by definition.

Solution $$\int_0^1 e^x dx = \lim_{n \to \infty} \frac{1}{n} \sum_{i=1}^n f\left(\frac{i}{n}\right) = \lim_{n \to \infty} \frac{1}{n} \sum_{i=1}^n e^{\frac{i}{n}}$$

$$= \lim_{n \to \infty} \left\{ \frac{1}{n} \cdot \frac{e^{\frac{1}{n}}[1 - (e^{\frac{1}{n}})^n]}{1 - e^{\frac{1}{n}}} \right\}$$

$$= (e - 1) \lim_{n \to \infty} \left(\frac{1/n}{e^{1/n} - 1} e^{1/n} \right) = e - 1.$$

(2) Find $\int_a^b \frac{1}{x^2}\, dx$ by definition.

Solution Let T be a partition on $[a, b]$:

$$a = x_0 < x_1 < x_2 < \cdots < x_{i-1} < x_i < \cdots < x_n = b.$$

Take $\xi_i = \sqrt{x_i x_{i+1}} \in [x_i, x_{i+1}] (i = 0, 1, \cdots, n-1)$. Then we have the Riemann sum

$$\sigma = \sum_{i=0}^{n-1} \frac{1}{\xi_i^2} \Delta x_i = \sum_{i=0}^{n-1} \frac{1}{x_i x_{i+1}} (x_{i+1} - x_i)$$
$$= \sum_{i=0}^{n-1} \left(\frac{1}{x_i} - \frac{1}{x_{i+1}} \right) = \frac{1}{x_0} - \frac{1}{x_n} = \frac{1}{a} - \frac{1}{b}.$$

Thus, $\int_a^b \frac{1}{x^2} dx = \lim \sigma = \frac{1}{a} - \frac{1}{b}$.

§7.1.3 Condition of the existence of the definite integral

The definite integral is a limit of the Riemann sum $\sigma = \sum_{i=1}^{n} f(\xi_i) \Delta x_i$. However, this limit is not only relevant to the partition T of the interval $[a, b]$ and $\| T \|$, the length of the longest subinterval of T, but also the choice of ξ_i. Thereupon the problem concerning the existence of the limit is quite complicated. Here we will discuss some preliminary results in this regard.

Theorem 7.1.1 (Necessity for integrability) If $f(x)$ is integrable on $[a, b]$, $f(x)$ is bounded on $[a, b]$.

Proof Suppose $f(x)$ is unbounded on $[a, b]$. Then for any partition T of $[a, b]$, we can always find a subinterval $\Delta_i = [x_{i-1}, x_i]$ of T such that $f(x)$ is unbounded on Δ_i, and so there is a choice of $\xi_i \in \Delta_i$ with $| f(\xi_i) \Delta x_i |$ being greater than any given number M, which can further make the absolute value of the Riemann sum

$\sum_{i=1}^{n} f(\xi_i)\Delta x_i$ arbitrarily great. Thus the limit of Riemann sum is not existent, i. e. $f(x)$ is unintegrable on $[a, b]$ by the definition. □

Note, however, that the boundedness of $f(x)$ is not a sufficiency for $f(x)$ to be integrable.

Example Let $f(x)$ be Dirichlet function, i. e.

$$f(x) = \begin{cases} 1 & (\text{if } x \text{ is rational}), \\ 0 & (\text{if } x \text{ is irrational}). \end{cases}$$

Then, $f(x)$ is bounded but unintegrable on $[0, 1]$.

Proof For any subinterval $\Delta_i = [x_{i-1}, x_i]$ of any partition T of $[0, 1]$, we can always take a rational number $\xi_i \in \Delta_i$ and also an irrational number $\zeta_i \in \Delta_i$. Then

$$\sum_{i=1}^{n} f(\xi_i)\Delta x_i = 1; \ \sum_{i=1}^{n} f(\zeta_i)\Delta x_i = 0.$$

Thus the limits of these two sum expressions are 1 and 0 respectively as $\| T \| \to 0$, which shows $f(x)$ is unintegrable on $[0, 1]$. □

In order to study the sufficiency of integrability for a function on $[a, b]$, we should discuss the evaluation of the Riemann sum as $\| T \| \to 0$.

Suppose $f(x)$ is bounded on $[a, b]$. Then $f(x)$ has its supremum and infimum on $[a, b]$. Denote

$$M = \sup_{x \in [a, b]} \{f(x)\}; m = \inf_{x \in [a, b]} \{f(x)\}.$$

Let T be a partition on $[a, b]$ and let $\Delta_i (i = 1, 2, \cdots, n)$ be the subintervals belonging to T. Denote

$$M_i = \sup_{x \in \Delta_i} \{f(x)\}; \ m_i = \inf_{x \in \Delta_i} \{f(x)\} \quad (i = 1, 2, \cdots, n).$$

Obviously, $m \leqslant m_i \leqslant f(\xi_i) \leqslant M_i \leqslant M$, $\xi_i \in \Delta_i \quad (i = 1, 2, \cdots, n)$,

(1)

Denote $S_f(T) = \sum_{i=1}^{n} M_i \Delta x_i$; $s_f(T) = \sum_{i=1}^{n} m_i \Delta x_i$, called (*Darboux*) *upper sum* and (*Darboux*) *lower sum* of $f(x)$ about the partition T respectively (or briefly, denoted by $S(T)$ and $s(T)$). Note that the value of upper sum and lower sum are determined if the partition is determined, different from that of Riemann sum.

Now, assuming $f(x)$ is a bounded function defined on $[a, b]$, we consider the properties of upper sum and lower sum.

Property 1 The lower sum $s(T)$ is the infimum of all the integral sum $\sum(T)$ and the upper sum $S(T)$ is the supremum of all the integral sum $\sum(T)$, and

$$m(b-a) \leqslant s(T) \leqslant \sum(T) \leqslant S(T) \leqslant M(b-a).$$

Proof By the inequality (1),

$$m(b-a) \leqslant \sum_{i=1}^{n} m_i \Delta x_i \leqslant \sum_{i=1}^{n} f(\xi_i) \Delta x_i \leqslant \sum_{i=1}^{n} M_i \Delta x_i \leqslant M(b-a),$$

which is the required inequality. Now we proceed to show that $s(T)$ and $S(T)$ are not only the lower bound and the upper bound of all the $\sum(T)$, but also the infimum and the supremum. For any $\varepsilon > 0$ and any Δ_i, since $M_i = \sup\limits_{x \in \Delta_i} f(x)$, we may select an intermediate ξ_i such that $f(\xi_i) > M_i - \varepsilon$. Then

$$\begin{aligned} \sum(T) &= \sum_{i=1}^{n} f(\xi_i) \Delta x_i > \sum_{i=1}^{n} (M_i - \varepsilon) \Delta x_i \\ &= \sum_{i=1}^{n} M_i \Delta x_i - \varepsilon \sum_{i=1}^{n} \Delta x_i = S(T) - \varepsilon(b-a), \end{aligned}$$

which implies that $S(T)$ is the supremum of all the $\sum(T)$. Similarly, we may prove $s(T)$ is the infimum of all the $\sum(T)$. □

Property 2 Let T' be a new partition obtained by adding p new partition points to the partition T. Then

$$s(T) \leqslant s(T') \leqslant s(T) + p(M-m)\,\|T\|\,,$$
$$S(T) \geqslant S(T') \geqslant S(T) - p(M-m)\,\|T\|\,.$$

Proof Adding p new partition points to the partition T can be manipulated in "one by one" manner, i. e. adding one new partition point to T to obtain T_1, and then adding one partition point to T_1 to obtain T_2 and so on. Therefore, we first consider the case $p = 1$.

Adding one new partition point to T, this point must falls into a subinterval Δ_k, which is thereby divided into two small intervals, say, Δ_k' and Δ_k'', and the other subintervals in T keep unchanged in T_1. Hence, comparing the terms in $S(T)$ and $S(T_1)$, we may see that the only difference between them is that the term $M_k\Delta x_k$ in $S(T)$ is now substituted with two terms $M_k'\Delta x_k'$ and $M_k''\Delta x_k''$ in $S(T_1)$, where M'_k and M''_k represent the supremums of $f(x)$ in Δ'_k and Δ''_k respectively. Thus

$$\begin{aligned} S(T) - S(T_1) &= M_k\Delta x_k - (M_k'\Delta x_k' + M_k''\Delta x_k'') \\ &= M_k(\Delta x_k' + \Delta x_k'') - (M_k'\Delta x_k' + M_k''\Delta x_k'') \\ &= (M_k - M_k')\Delta x_k' + (M_k - M_k'')\Delta x_k''. \end{aligned}$$

Since $m \leqslant M_k' \leqslant M_k \leqslant M$ and $m \leqslant M_k'' \leqslant M_k \leqslant M$,

$$\begin{aligned} 0 &\leqslant S(T) - S(T_1) \leqslant (M-m)\Delta x_k' + (M-m)\Delta x_k'' \\ &= (M-m)\Delta x_k \leqslant (M-m)\,\|T\|\,. \end{aligned}$$

Generally, adding one new partition point to T_i to obtain T_{i+1}, we have

$$0 \leqslant S(T_i) - S(T_{i+1}) \leqslant (M-m)\,\|T_i\| \quad (i = 0,\ 1,\ 2,\ \cdots,\ p-1).$$

By summing all these inequalities, we have

$$0 \leqslant S(T_0) - S(T_p) \leqslant (M - m)\sum_{i=1}^{p} \| T_i \| \leqslant (M - m)p \| T_0 \|, \text{ i.e.}$$

$$S(T) \geqslant S(T') \geqslant S(T) - p(M - m) \| T \|.$$

Another inequality can be proved in an analogous manner. □

The above property can be interpreted as the fact that the lower sum is not decreasing and the upper sum is not increasing as the partition becomes finer.

Property 3 Let T and T' be two partitions and let T'' be a partition which is constituted by the union of the partition points of T and the partition points of T', denoted by $T'' = T + T'$. Then

$$s(T'') \geqslant s(T), \ S(T'') \leqslant S(T); \ s(T'') \geqslant s(T'), \ S(T'') \leqslant S(T').$$

Proof Only need to notice that T'' can be regarded as a partition obtained by adding new partition points to T, as well as a partition obtained by adding new partition points to T'. Then, the result follows directly from the property 2. □

Property 4 For any two partitions T and T', $s(T) \leqslant S(T')$.

Proof Let $T'' = T + T'$. By the properties 1 and 3,

$$s(T) \leqslant s(T'') \leqslant S(T'') \leqslant S(T'). \ \square$$

This property indicates that for any two partitions, the lower sum of one partition is never greater than the upper sum of the other. Hence, considering all partitions, the set of all the lower sums has an upper bound and the set of all the upper sums has a lower bound, and furthermore, they have supremum and infimum respectively, denoted by s and S, i. e.

$$s = \sup_{T} \{ s(T) \},$$

which is called *the lower integral* of $f(x)$ on $[a, b]$, and

$$S = \inf_{T} \{ S(T) \},$$

which is called *the upper integral* of $f(x)$ on $[a, b]$.

Property 5 $m(b-a) \leqslant s \leqslant S \leqslant M(b-a)$.

This property can be derived by the property 4 and the definitions of infimum and supermum.

Property 6 (Darboux theorem) $\lim_{\|T\| \to 0} S(T) = S$ and $\lim_{\|T\| \to 0} s(T) = s$, i. e. the upper and lower integrals are respectively the limits of the upper and lower sums as $\|T\| \to 0$.

Proof We only prove the first limit. The proof of the second limit will be left as an exercise for readers.

Let ε be a positive number. By the definition of S, there exists a partition T' such that

$$S(T') < S + \frac{\varepsilon}{2}. \tag{2}$$

Assume that T' is a partition which is composed of p partition points. Thus, for any other partition T, $T + T'$ contains at most p more partition points than T. Then, by the properties 2 and 3, we have

$$S(T) - p(M-m)\|T\| \leqslant S(T+T') \leqslant S(T'),$$

and so

$$S(T) \leqslant S(T') + p(M-m)\|T\|.$$

Therefore, so long as $\|T\| < \varepsilon/[2p(M-m)]$, $S(T) \leqslant S(T') + \frac{\varepsilon}{2}$, and combining (2), it follows that $S(T) < S + \varepsilon$. Noting $S \leqslant S(T)$ is always true, we have $\lim_{\|T\| \to 0} S(T) = S$. □

By Theorem 7. 1. 1, boundedness is a necessity for integrability. Now applying properties of upper sum and lower sum, we may derive sufficient and necessary conditions for functions to be integrable.

Theorem 7. 1. 2 The sufficient and necessary condition for a bounded function $f(x)$ to be integrable is: The upper integral of $f(x)$ on $[a, b]$ is equal to its lower integral on $[a, b]$, i. e. $S = s$.

Proof Necessity. Assume that $f(x)$ is integrable on $[a, b]$ with $\int_a^b f(x)\,dx = J$, i. e. for any $\varepsilon > 0$, there exists $\delta > 0$ such that $|\sum(T) - J| < \varepsilon$ whenever $\|T\| < \delta$. Since $S(T)$ and $s(T)$ are supremum and infimum of $\sum(T)$ respectively, if only $\|T\| < \delta$, we also have

$$|S(T) - J| \leqslant \varepsilon, \ |s(T) - J| \leqslant \varepsilon,$$

which implies that $\lim\limits_{\|T\| \to 0} S(T) = J$ and $\lim\limits_{\|T\| \to 0} s(T) = J$. So, by Darboux theorem, $S = s(= J)$.

Sufficiency. Let $S = s = J$ for some number J. By Darboux theorem,

$$\lim_{\|T\| \to 0} S(T) = \lim_{\|T\| \to 0} s(T) = J. \tag{3}$$

Applying the inequality $s(T) \leqslant \sum(T) \leqslant S(T)$ in the property 1 and the squeezing law of the limit, we have $\lim\limits_{\|T\| \to 0} \sum(T) = J$. □

The above mentioned unintegrability of Dirichlet function on $[0, 1]$ is exactly due to the inequality of its upper integral ($S=1$) and its lower integral ($s=0$).

The result of Theorem 7. 1. 2 is quite concise, but the next theorem is more convenient in application.

Theorem 7. 1. 3 Let $f(x)$ be a bounded function defined on $[a, b]$. Then the sufficient and necessary condition for $f(x)$ to be integrable is: For any $\varepsilon > 0$, there exists a partition T of $[a, b]$ such that $S(T) - s(T) < \varepsilon$.

Proof Necessity. Suppose $f(x)$ is integrable on $[a, b]$. By (3) in the proof of Theorem 7. 1. 2 we have,

$$\lim_{\|T\| \to 0} (S(T) - s(T)) = 0,$$

and thereby, for any $\varepsilon > 0$, there exists a partition T of $[a, b]$ such that $S(T) - s(T) < \varepsilon$.

Sufficiency. Assume the condition of the theorem is satisfied. Noticing $s(T) \leqslant s \leqslant S \leqslant S(T)$, we have

$$0 \leqslant S - s \leqslant S(T) - s(T) \leqslant \varepsilon,$$

which implies $S = s$. Then $f(x)$ is integrable on $[a, b]$ by Theorem 7.1.2. □

Let $f(x)$ be a bounded function defined on $[a, b]$. Denote $M = \sup\{f(x)\}$, $m = \inf\{f(x)\}$ and $\omega = M - m$, which is called the *amplitude of vibration* of $f(x)$ on $[a, b]$. For a partition $T = \{\Delta_i \mid i = 1, 2, \cdots, n\}$ of $[a, b]$, denote $\omega_i (i = 1, 2, \cdots, n)$ as amplitude of vibration of $f(x)$ on Δ_i. Then

$$S(T) - s(T) = \sum_{i=1}^{n} (M_i - m_i) \Delta x_i = \sum_{i=1}^{n} \omega_i \Delta x_i,$$

where M_i and m_i are the supremum and infimum of $f(x)$ on Δ_i respectively. Now, we can restate Theorem 7.1.3 as follows:

Theorem 7.1.4 Let $f(x)$ be a bounded function defined on $[a, b]$. Then the sufficient and necessary condition for $f(x)$ to be integrable is: For any $\varepsilon > 0$, there exists a partition T of $[a, b]$ such that

$$\sum_{i=1}^{n} \omega_i \Delta x_i < \varepsilon.$$

According to the above theorems, we can give some classes of integrable functions on $[a, b]$.

Theorem 7.1.5 If $f(x)$ is continuous on $[a, b]$, then $f(x)$ is integrable on $[a, b]$.

Proof By Theorem 3.8.1 (Cantor theorem), $f(x)$ must be

uniformly continuous on $[a, b]$. So, for any $\varepsilon > 0$, there exists $\delta > 0$ such that

$$| f(x') - f(x'') | < \frac{\varepsilon}{b-a}$$

for any x', $x'' \in [a, b]$ with $| x' - x'' | < \delta$. Thus, for any partition $T = \{\Delta_i \mid i = 1, 2, \cdots, n\}$ of $[a, b]$, if only $\| T \| < \delta$, the amplitude of vibration of $f(x)$ on Δ_i: $\omega_i = M_i - m_i < \frac{\varepsilon}{b-a}$ and so

$$\sum_{i=1}^{n} \omega_i \Delta x_i < \frac{\varepsilon}{b-a} \sum_{i=1}^{n} \Delta x_i = \frac{\varepsilon}{b-a}(b-a) = \varepsilon,$$

from which and Theorem 7. 1. 4 it follows that $f(x)$ is integrable on $[a, b]$. □

Theorem 7. 1. 6 If $f(x)$ is bounded and it has only a finite number of discontinuities on $[a, b]$, then $f(x)$ is integrable on $[a, b]$.

Proof Let M and m be respectively the supremum and infimum of $f(x)$ on $[a, b]$. Let ε be any positive number. Suppose $f(x)$ has k discontinuities on $[a, b]$ sequentially as $x_1, x_2, \cdots, x_k$. Now take a partition T' on $[a, b]$ with partition points $x_i - \delta$, $x_i + \delta (i = 1, 2, \cdots, k)$ (if $x_i - \delta \leqslant a$, take a instead, and if $x_i + \delta \geqslant b$, take b instead) where $\delta < \frac{\varepsilon}{4k(M-m)}$. Thus, there are two classes of subintervals in T': one class is composed of subintervals Δ' which contains discontinuities. The total length of all such Δ' is less than or equal to $\frac{\varepsilon}{2(M-m)}$; The other class is composed of subintervals Δ'' which does not contain discontinuities, i. e. $f(x)$ is continuous on each Δ''. By Theorems 7. 1. 5 and 7. 1. 4, for any $\varepsilon > 0$, there exists an overall partition T'' of all these Δ''s such that $\sum_{\Delta''} \omega_i \Delta x_i < \frac{\varepsilon}{2}$. Let $T = T' + T''$. Then T is a partition of $[a, b]$, for which

$$\sum \omega_i \Delta x_i = \sum_{\Delta'} \omega_i \Delta x_i + \sum_{\Delta''} \omega_i \Delta x_i < (M-m) \sum_{\Delta'} \Delta x_i + \frac{\varepsilon}{2}$$

$$< (M-m) \frac{\varepsilon}{2(M-m)} + \frac{\varepsilon}{2} = \varepsilon.$$

Therefore, by Theorem 7.1.4 $f(x)$ is integrable on $[a, b]$. □

From Theorem 7.1.6, it follows immediately the next:

Corollary If $f(x)$ is piecewise continuous on $[a, b]$, then $f(x)$ is integrable on $[a, b]$.

Theorem 7.1.7 If $f(x)$ is a monotone function on $[a, b]$, then $f(x)$ is integrable on $[a, b]$.

Proof Suppose $f(x)$ is an increasing function on $[a, b]$ (without loss of generality we may suppose $f(a) \neq f(b)$). Let $T = \{\Delta_i \mid i = 1, 2, \cdots, n\}$ be a partition of $[a, b]$. Because of the monotony of $f(x)$, $M_i = \sup\limits_{\Delta_i} f(x) = f(x_i)$, $m_i = \inf\limits_{\Delta_i} f(x) = f(x_{i-1})$, and so

$$\sum_T \omega_i \Delta x_i = \sum_T (f(x_i) - f(x_{i-1})) \Delta x_i \leqslant \sum_T (f(x_i) - f(x_{i-1})) \| T \|$$

$$= \| T \| (f(b) - f(a)).$$

Hence, for any $\varepsilon > 0$, take a partition T on $[a, b]$ such that

$$\| T \| < \frac{\varepsilon}{f(b) - f(a)},$$

then $\sum \omega_i \Delta x_i < \varepsilon$, which implies $f(x)$ is integrable on $[a, b]$ by Theorem 7.1.4(Similar proof for a decreasing function). □

Remark A monotone function on a closed interval may possess infinitely many discontinuities and is yet integrable. We give examples as follows:

Examples (1) $f(x) = \begin{cases} 0 & (\text{if } x = 0), \\ \dfrac{1}{n} & \left(\text{if } \dfrac{1}{n+1} < x \leqslant \dfrac{1}{n}\right). \end{cases}$

This function is increasing on $[0, 1]$. It is discontinuous at $x=\frac{1}{n}(n=2, 3, \cdots)$, but by Theorem 7.1.7, it is integrable on $[0, 1]$.

(2) Riemann function (cf. Chapter 1)

$$f(x)=\begin{cases}\frac{1}{q} & \left(\text{if } x=\frac{p}{q}\in[0, 1] \text{ where } p, q\in\mathbb{N} \text{ and}\right.\\ & p/q \text{ is irreducible fraction}),\\ 0 & (\text{if } x\in[0, 1] \text{ is irrational, or } x=0, 1)\end{cases}$$

is integrable.

Proof For any $\varepsilon>0$, in all rationals $\frac{p}{q}\in[0, 1]$, there are merely a finite number of them satisfying the inequality $\frac{1}{q}>\frac{\varepsilon}{2}$ (denoting 0 and 1 as $\frac{0}{1}$ and $\frac{1}{1}$ respectively), which may be denoted as $r_1, r_2, \cdots, r_{N(\varepsilon)}$. Now divide $[0, 1]$ into subintervals such that each $r_i(1\leqslant i\leqslant N(\varepsilon))$ belongs to exactly one of these subintervals and the total length of those subintervals containing r_i does not exceed $\frac{\varepsilon}{2}$, and on other subintervals Δ_i, $M_i := \sup\limits_{\Delta_i}\leqslant\frac{\varepsilon}{2}$. This process constructs a partition T of $[a, b]$. Hence, it is easy to check that

$$\begin{aligned}S(T)-s(T)&=\sum_{i=1}^{n}(M_i-m_i)\Delta x_i<\frac{\varepsilon}{2}\sum{}'\Delta x_i+1\cdot\frac{\varepsilon}{2}\\&<\frac{\varepsilon}{2}+\frac{\varepsilon}{2}=\varepsilon,\end{aligned}$$

where the sum $\sum{}'$ is for those subintervals each of which does not contain $r_i(1\leqslant i\leqslant N(\varepsilon))$. Then, by Theorem 7.1.3, $f(x)$ is integrable on $[0, 1]$. □

At the end of this section, we prove some commonly used

statements about the integrable functions.

Theorem 7. 1. 8 If a bounded function $f(x)$ is integrable on $[a, b]$, then $|f(x)|$ is also integrable on $[a, b]$.

Proof Denote by ω_i and ω_i^* $(i=1, 2, \cdots, n)$ the amplitude of vibration of $f(x)$ and $|f(x)|$ on $\Delta_i = [x_{i-1}, x_i]$ respectively. It is easy to see: if the signs of the supremum M_i and the infimum m_i of $f(x)$ on Δ_i are the same, then

$$\omega_i^* = M_i - m_i = \omega_i;$$

if the signs of M_i and m_i of $f(x)$ on Δ_i are different (note $m_i < 0 < M_i$ in this case), then

$$\omega_i^* < |M_i| + |m_i| = M_i - m_i = \omega_i.$$

That is we have $\omega_i^* \leqslant \omega_i$ in any case. Hence, for any partition T of $[a, b]$,

$$\sum_i \omega_i^* \Delta x_i \leqslant \sum_i \omega_i \Delta x_i.$$

Thus, for any $\varepsilon > 0$, since $f(x)$ is integrable on $[a, b]$, there is a partition T of $[a, b]$ such that the right expression of the above inequality is less than ε, and so the left expression of the above inequality is also less than ε. Therefore, by Theorem 7. 1. 4 $|f(x)|$ is also integrable on $[a, b]$. □

Theorem 7. 1. 9 If a function $f(x)$ is integrable on $[a, b]$ and $[c, d] \subseteq [a, b]$, then $f(x)$ is also integrable on $[c, d]$.

Proof Since $f(x)$ is integrable on $[a, b]$, by Theorem 7. 1. 3, for any $\varepsilon > 0$ there is a partition T of $[a, b]$ such that $S(T) - s(T) < \varepsilon$. By adding the points c and d to the set of partition points of T, we have a new partition T' of $[a, b]$. By the property (2) of the upper sum and the lower sum, $s(T) \leqslant s(T') \leqslant S(T') \leqslant S(T)$, and so $S(T') - s(T') \leqslant S(T) - s(T) < \varepsilon$, which implies

$$\sum_{i=1}^{n} \omega_i \Delta x_i < \varepsilon$$

for the partition T' of $[a, b]$. Now, from the left expression of the above inequality we remove all such terms which contain the small subintervals outside $[c, d]$, and we denote the sum of the remainder terms as $\sum' \omega_i \Delta x_i$, which is exactly the sum expression over $[c, d]$. As the terms removed are all non-negative,

$$\sum' \omega_i \Delta x_i < \varepsilon.$$

This demonstrates that for any $\varepsilon > 0$, we have constructed a partition T^* of $[c, d]$, which is obtained by removing those points outside $[c, d]$ from T', and for this T^*, $\sum' \omega_i \Delta x_i < \varepsilon$. Then, by Theorem 7.1.4, $f(x)$ is integrable on $[c, d]$. □

Theorem 7.1.10 Suppose $f(x)$ and $g(x)$ are both integrable on $[a, b]$. Then $f(x)g(x)$ is also integrable on $[a, b]$.

Proof Since $f(x)$ and $g(x)$ are integrable on $[a, b]$, they are bounded on $[a, b]$, and so we may let $|f(x)| \leqslant A$ and $|g(x)| \leqslant B$ for some $A, B \in \mathbb{R}$. Let T be a partition of $[a, b]$ and let x' and x'' be any two points in the small interval $[x_{i-1}, x_i]$. We have the identity

$$f(x'')g(x'') - f(x')g(x') = (f(x'') - f(x'))g(x'') + (g(x'') - g(x'))f(x') \tag{1}$$

Denote by ω_i, ω_i' and ω_i'' the amplitudes of vibration of $f(x)g(x)$, $f(x)$ and $g(x)$ on $[x_{i-1}, x_i]$ respectively. Since

$$|f(x'')g(x'') - f(x')g(x')| \leqslant \omega_i, \ |f(x'') - f(x')| \leqslant \omega_i' \text{ and } |g(x'') - g(x')| \leqslant \omega_i'',$$

by (1) we have

$$\omega_i \leqslant B\omega_i' + A\omega_i'',$$

and so,

$$\sum_{i=1}^{n} \omega_i \Delta x_i \leqslant B \sum_{i=1}^{n} \omega_i' \Delta x_i + A \sum_{i=1}^{n} \omega_i'' \Delta x_i.$$

Due to the integrability of $f(x)$ and $g(x)$ on $[a, b]$, for any $\varepsilon > 0$, there is a partition T of $[a, b]$ such that

$$\sum_{i=1}^{n} \omega_i' \Delta x_i < \frac{\varepsilon}{2B}, \quad \sum_{i=1}^{n} \omega_i'' \Delta x_i < \frac{\varepsilon}{2A}.$$

Thus, for this partition T, we have

$$\sum_{i=1}^{n} \omega_i \Delta x_i \leqslant B \frac{\varepsilon}{2B} + A \frac{\varepsilon}{2A} = \varepsilon,$$

i. e. $f(x)g(x)$ is integrable on $[a, b]$. □

§ 7.2 Properties of definite integrals

Because definite integrals are defined as limits, they inherit many of the properties of limits. For example, we know that constants can be moved through limit sign and that the limit of a sum or difference is the sum or difference of the limits. Thus, we should not be surprised by the properties listed below.

Theorem 7.2.1 (Linearity) Suppose $f(x)$ and $g(x)$ are both integrable on $[a, b]$, then $f(x) \pm g(x)$ and $kf(x)$ are also integrable on $[a, b]$ where k is a constant, and furthermore

$$\int_a^b (f(x) \pm g(x))\,dx = \int_a^b f(x)\,dx \pm \int_a^b g(x)\,dx;$$

$$\int_a^b kf(x)\,dx = k\int_a^b f(x)\,dx.$$

Proof Note that

$$\lim_{\|T\|\to 0}\sum_{i=1}^{n}(f(\xi_i)\pm g(\xi_i))\Delta x_i = \lim_{\|T\|\to 0}\sum_{i=1}^{n}f(\xi_i)\Delta x_i \pm \lim_{\|T\|\to 0}\sum_{i=1}^{n}g(\xi_i)\Delta x_i,$$

and

$$\lim_{\|T\|\to 0}\sum_{i=1}^{n}kf(\xi_i)\Delta x_i = k\lim_{\|T\|\to 0}\sum_{i=1}^{n}f(\xi_i)\Delta x_i.$$

The results follow immediately. □

Theorem 7. 2. 2 (Path Property) $\int_a^b f(x)\mathrm{d}x = \int_a^c f(x)\mathrm{d}x + \int_c^b f(x)\mathrm{d}x$.

Proof ⓘ Let $a<c<b$. Notice the limit of the Riemann sum is irrelevant to the formation of the partition. Thus it is allowed to keep taking c as an endpoint, say x_{n_1}, of a subinterval for the Riemann sum, and so

$$\sum_{i=1}^{n}f(\xi_i)\Delta x_i = \sum_{i=1}^{n_1}f(\xi_i)\Delta x_i + \sum_{n=n_1+1}^{n}f(\xi_i)\Delta x_i,$$

where the two latter terms are the Riemann sum of $f(x)$ on intervals $[a, c]$ and $[c, b]$ respectively. Then by taking the limit on both sides of the equality we have the result immediately.

ⓘⓘ Suppose c is not between a and b, e. g. $a<b<c$. Then by ⓘ we see

$$\int_a^c f(x)\mathrm{d}x = \int_a^b f(x)\mathrm{d}x + \int_b^c f(x)\mathrm{d}x = \int_a^b f(x)\mathrm{d}x - \int_c^b f(x)\mathrm{d}x,$$

which clearly implies the result expected. □

Theorem 7. 2. 3 (Order-Preserving Property) (1) If $f(x)\geqslant 0$ for any $x\in[a, b]$, then

$$\int_a^b f(x)\mathrm{d}x \geqslant 0;$$

(2) If $f(x) \geqslant g(x)$ for any $x \in [a, b]$, then

$$\int_a^b f(x)\mathrm{d}x \geqslant \int_a^b g(x)\mathrm{d}x.$$

Proof (1) By the condition, for any partition of $[a, b]$ and any choices of ξ_i, we have

$$\sum_{i=1}^{n} f(\xi_i)\Delta x_i \geqslant 0,$$

from which the result follows directly.

(2) Using (1) to the function $F(x) = f(x) - g(x)$, we can obtain the result immediately. □

Examples (1) Suppose $f(x)$ is continuous and decreasing on $[0, 1]$ and $\lambda \in (0, 1)$. Show that

$$\int_0^\lambda f(x)\mathrm{d}x \geqslant \lambda \int_0^1 f(x)\mathrm{d}x.$$

Proof Since f is decreasing on $[0, 1]$, if $x \in (\lambda, 1)$, $f(x) \leqslant f(\lambda)$ and so $\int_\lambda^1 f(x)\mathrm{d}x \leqslant \int_\lambda^1 f(\lambda)\mathrm{d}x = (1-\lambda)f(\lambda)$, i. e.

$$f(\lambda) \geqslant \frac{1}{1-\lambda}\int_\lambda^1 f(x)\mathrm{d}x.$$

Now let $x \in (0, \lambda)$. Then $f(x) \geqslant f(\lambda)$ and so

$$\int_0^\lambda f(x)\mathrm{d}x \geqslant \int_0^\lambda f(\lambda)\mathrm{d}x = \lambda f(\lambda) \geqslant \frac{\lambda}{1-\lambda}\int_\lambda^1 f(x)\mathrm{d}x.$$

Thus $(1-\lambda)\int_0^\lambda f(x)\mathrm{d}x \geqslant \lambda\int_\lambda^1 f(x)\mathrm{d}x$, i. e.

$$\int_0^\lambda f(x)\mathrm{d}x \geqslant \lambda\int_0^\lambda f(x)\mathrm{d}x + \lambda\int_\lambda^1 f(x)\mathrm{d}x = \lambda\int_0^1 f(x)\mathrm{d}x. \quad \square$$

(2) Let $f(x)$ be a continuous function on $[a, b]$ such that $\int_a^b f(x)\varphi(x)\mathrm{d}x = 0$ for any continuous function $\varphi(x)$. Show that $f(x) \equiv 0$ on $[a, b]$.

Proof Suppose $f(x) \not\equiv 0$ on $[a, b]$. Then there exists $x_0 \in [a, b]$ with $f(x_0) \neq 0$. W. l. o. g., we may let $x_0 \in (a, b)$ with $f(x_0) > 0$. Since $f(x)$ is continuous at x_0, there exists $\delta > 0$ such that $(x_0 - \delta, x_0 + \delta) \subseteq (a, b)$ and $|f(x) - f(x_0)| < \frac{1}{2} f(x_0)$ whenever $|x - x_0| < \delta$. Then, $f(x) > \frac{1}{2} f(x_0)$ on $(x_0 - \delta, x_0 + \delta)$, and so

$$\int_a^b f^2(x)\mathrm{d}x \geqslant \int_{x_0-\delta}^{x_0+\delta} f^2(x)\mathrm{d}x \geqslant \int_{x_0-\delta}^{x_0+\delta} \left(\frac{1}{2} f(x_0)\right)^2 \mathrm{d}x = \frac{1}{2}\delta f^2(x_0) > 0.$$

However, by taking $\varphi(x) = f(x)$, $\int_a^b f^2(x)\mathrm{d}x = 0$, which yields a contradiction. □

(3) Suppose integrable functions $f(x)$ and $g(x)$ are both increasing on $[0, 1]$. Verify that

$$\int_0^1 f(x)g(x)\mathrm{d}x \geqslant \int_0^1 f(x)\mathrm{d}x \cdot \int_0^1 g(x)\mathrm{d}x.$$

Proof Notice $(f(x) - f(y)) \cdot (g(x) - g(y)) \geqslant 0$ for any x, $y \in [0, 1]$, i. e.

$$f(x)g(x) + f(y)g(y) \geqslant f(x)g(y) + f(y)g(x).$$

Fixing y and integrating both sides of the above inequality about x from 0 to 1, we have

$$\int_0^1 f(x)g(x)\mathrm{d}x + f(y)g(y)\int_0^1 \mathrm{d}x \geqslant g(y)\int_0^1 f(x)\mathrm{d}x + f(y)\int_0^1 g(x)\mathrm{d}x.$$

Now, integrating both sides of this inequality about y from 0 to 1, we have

$$\int_0^1 f(x)g(x)\mathrm{d}x \cdot \int_0^1 \mathrm{d}y + \int_0^1 f(y)g(y)\mathrm{d}y \geqslant$$

$$\int_0^1 g(y)\mathrm{d}y \cdot \int_0^1 f(x)\mathrm{d}x + \int_0^1 g(y)\mathrm{d}y \cdot \int_0^1 g(x)\mathrm{d}x,$$

which yields

$$\int_0^1 f(x)g(x)\mathrm{d}x \geqslant \int_0^1 f(x)\mathrm{d}x \cdot \int_0^1 g(x)\mathrm{d}x. \ \square$$

(4) Suppose $f(x)$ and $g(x)$ are integrable on $[a, b]$. Show that

① Schwarz inequality: $\left(\int_a^b fg\ \mathrm{d}x\right)^2 \leqslant \int_a^b f^2\mathrm{d}x \int_a^b g^2\mathrm{d}x$;

② Minkowski inequality : $\left[\int_a^b (f+g)^2\mathrm{d}x\right]^{1/2} \leqslant \left(\int_a^b f^2\mathrm{d}x\right)^{1/2} + \left(\int_a^b g^2\mathrm{d}x\right)^{1/2}$.

③ if $f(x) > 0$, $\int_a^b f\mathrm{d}x \int_a^b \frac{1}{f}\mathrm{d}x \geqslant (b-a)^2$;

④ $\left(\int_a^b f\ \mathrm{d}x\right)^2 \leqslant (b-a)\int_a^b f^2\mathrm{d}x$.

Proof Let $\int_a^b g^2\mathrm{d}x = A$, $\int_a^b fg\ \mathrm{d}x = B$ and $\int_a^b f^2\mathrm{d}x = C$.

① As $(f+tg)^2 \geqslant 0$ for any t,

$$At^2 + 2Bt + C = \int_a^b (f+tg)^2\mathrm{d}x \geqslant 0,$$

which implies $(2B)^2 - 4AC \leqslant 0$, i. e. $B^2 \leqslant AC$.

② As A, $C \geqslant 0$, by (1) we see $B \leqslant A^{1/2}C^{1/2}$. Then

$$\int_a^b (f+g)^2\mathrm{d}x = A + 2B + C \leqslant A + 2A^{1/2}C^{1/2} + C$$

$$= (A^{1/2} + C^{1/2})^2 = \left[\left(\int_a^b f^2\mathrm{d}x\right)^{1/2} + \left(\int_a^b g^2\mathrm{d}x\right)^{1/2}\right]^2.$$

③ By (1),

$$\int_a^b f\,dx\int_a^b \frac{1}{f}dx = \int_a^b (\sqrt{f})^2 dx\int_a^b \left(\frac{1}{\sqrt{f}}\right)^2 dx \geqslant \left[\int_a^b \left(\sqrt{f}\cdot\frac{1}{\sqrt{f}}\right)dx\right]^2$$
$$= (b-a)^2.$$

④ Set $g\equiv 1$ in (1), then

$$\left(\int_a^b f\,dx\right)^2 = \left(\int_a^b fg\,dx\right)^2 \leqslant \int_a^b f^2 dx\cdot\int_a^b g^2 dx$$
$$= \int_a^b f^2 dx\cdot\int_a^b 1^2 dx = (b-a)\int_a^b f^2 dx. \square$$

(5) Prove:

① $\lim\limits_{n\to+\infty}\int_0^{\frac{\pi}{2}} \sin^n x\,dx = 0$;

② $\lim\limits_{n\to+\infty}\left\{\int_0^{\frac{\pi}{2}} \sin^n x\,dx\right\}^{\frac{1}{n}} = 1.$

Proof ① For any $\varepsilon > 0$, we have

$$0\leqslant\int_0^{\frac{\pi}{2}} \sin^n x\,dx = \int_{\frac{\pi}{2}-\frac{\varepsilon}{2}}^{\frac{\pi}{2}} \sin^n x\,dx + \int_0^{\frac{\pi}{2}-\frac{\varepsilon}{2}} \sin^n x\,dx$$
$$\leqslant\frac{\varepsilon}{2}+\frac{\pi}{2}\sin^n\frac{\pi-\varepsilon}{2} < \frac{\varepsilon}{2}+2\sin^n\frac{\pi-\varepsilon}{2}.$$

As $\sin\frac{\pi-\varepsilon}{2} < 1$, $\lim\limits_{n\to+\infty}\sin^n\frac{\pi-\varepsilon}{2} = 0$, and so by squeezing law,

$$0\leqslant\lim_{n\to+\infty}\int_0^{\frac{\pi}{2}} \sin^n x\,dx \leqslant\frac{\varepsilon}{2}.$$

Consequently, from the arbitrariness of ε the result follows.

② Note that $0\leqslant\sin x\leqslant 1$ as $x\in[0, \frac{\pi}{2}]$. Then,

$$\left\{\int_0^{\frac{\pi}{2}} \sin^n x\,dx\right\}^{\frac{1}{n}} \leqslant \left\{\int_0^{\frac{\pi}{2}} \sin x\,dx\right\}^{\frac{1}{n}} = 1 \qquad (1)$$

Since $\sin x$ is continuous, for any $\varepsilon > 0$ there is a $\delta > 0$ such that

$1-\varepsilon \leqslant \sin x \leqslant 1$ if $\frac{\pi}{2}-\delta \leqslant x \leqslant \frac{\pi}{2}$. Thus

$$\left\{\int_0^{\frac{\pi}{2}} \sin^n x \,\mathrm{d}x\right\}^{\frac{1}{n}} \geqslant \left\{\int_{\frac{\pi}{2}-\delta}^{\frac{\pi}{2}} \sin^n x \,\mathrm{d}x\right\}^{\frac{1}{n}} > (1-\varepsilon)\delta^{\frac{1}{n}} \qquad (2)$$

Taking the upper limit and the lower limit respectively in the inequalities (1) and (2), we obtain

$$1-\varepsilon \leqslant \varliminf_{n\to+\infty} \left\{\int_0^{\frac{\pi}{2}} \sin^n x \,\mathrm{d}x\right\}^{\frac{1}{n}} \leqslant \varlimsup_{n\to+\infty} \left\{\int_0^{\frac{\pi}{2}} \sin^n x \,\mathrm{d}x\right\}^{\frac{1}{n}} \leqslant 1.$$

Due to the arbitrariness of ε, the upper limit is equal to the lower limit, and so the limit exists and is equal to 1. □

Theorem 7.2.4 $\left|\int_a^b f(x)\mathrm{d}x\right| \leqslant \int_a^b |f(x)| \,\mathrm{d}x \quad (a<b)$

Proof Note that for any $x \in [a, b]$, $-|f(x)| \leqslant f(x) \leqslant |f(x)|$. Then by the above theorem,

$$-\int_a^b |f(x)|\,\mathrm{d}x \leqslant \int_a^b f(x)\mathrm{d}x \leqslant \int_a^b |f(x)|\,\mathrm{d}x,$$

from which the result follows directly. □

Theorem 7.2.5 (Estimation theorem) Suppose $m \leqslant f(x) \leqslant M$ for any $x \in [a, b]$, where m and M are constants. Then

$$m(b-a) \leqslant \int_a^b f(x)\mathrm{d}x \leqslant M(b-a).$$

Proof By the above theorem,

$$\int_a^b m \,\mathrm{d}x \leqslant \int_a^b f(x)\mathrm{d}x \leqslant \int_a^b M \,\mathrm{d}x,$$

which clearly implies the consequence required.

In order to introduce the mean value theorem of integral, we first prove the following:

Lemma Suppose $f(x)$ is continuous on $[a, b]$ such that $f(x) \geqslant$

0 on $[a, b]$ or $f(x) \leqslant 0$ on $[a, b]$. If $\int_a^b f(x)\mathrm{d}x = 0$, then $f(x) \equiv 0$ on $[a, b]$ (i. e. $f(x) = 0$ for any $x \in [a, b]$).

Proof Suppose $f(x) \geqslant 0$ on $[a, b]$. If $f(x) \not\equiv 0$ on $[a, b]$, assume that there is $x_0 \in (a, b)$ with $f(x_0) > 0$. By continuity of $f(x)$ there exists a neighborhood, say $U(x_0) = (x_0 - \delta, x_0 + \delta) \subset (a, b)$ such that $f(x) > 0$ for any $x \in U(x_0)$, and so

$$\int_a^b f(x)\mathrm{d}x \geqslant \int_{x_0-\delta}^{x_0+\delta} f(x)\mathrm{d}x > 0,$$

which yields a contradiction. If $f(a) > 0$ or $f(b) > 0$, a contradiction will be similarly derived.

Suppose $f(x) \leqslant 0$ on $[a, b]$. Then $-f(x) \geqslant 0$ on $[a, b]$ such that $\int_a^b (-f(x))\mathrm{d}x = -\int_a^b f(x)\mathrm{d}x = 0$. Thus $-f(x) \equiv 0$ on $[a, b]$, i. e. $f(x) \equiv 0$ on $[a, b]$.

Theorem 7. 2. 6 (The first mean-value theorem of integral) Suppose $f(x)$, $g(x)$ are functions with the following properties on $[a, b]$: (i) $f(x)$, $g(x)$ are both integrable; (ii) there exist constants m and M such that $m \leqslant f(x) \leqslant M$; (iii) $g(x)$ is continuous and keeps its sign on $[a, b]$ (i. e. either $g(x) \geqslant 0$ or $g(x) \leqslant 0$ on $[a, b]$). Then

(1) there is μ with $m \leqslant \mu \leqslant M$ such that

$$\int_a^b f(x)g(x)\mathrm{d}x = \mu\int_a^b g(x)\mathrm{d}x;$$

(2) there is $\xi \in [a, b]$ such that

$$\int_a^b f(x)g(x)\mathrm{d}x = f(\xi)\int_a^b g(x)\mathrm{d}x,$$

if $f(x)$ is continuous on $[a, b]$.

Proof (1) Suppose $g(x) \geqslant 0$. If $\int_a^b g(x)\mathrm{d}x = 0$, by Lemma

$g(x) \equiv 0$, and so the result holds true evidently. Now let $\int_a^b (x)\mathrm{d}x >$ 0. Since $mg(x) \leqslant f(x)g(x) \leqslant Mg(x)$,

$$m\int_a^b g(x)\mathrm{d}x \leqslant \int_a^b f(x)g(x)\mathrm{d}x \leqslant M\int_a^b g(x)\mathrm{d}x,$$

and so

$$m \leqslant \frac{\int_a^b f(x)g(x)\mathrm{d}x}{\int_a^b g(x)\mathrm{d}x} \leqslant M,$$

where the fraction of two integrals may be taken as the μ required. For the case $g(x) \leqslant 0$, the result can be similarly proved.

(2) This result follows immediately from (1) and the intermediate-value theorem of the continuous function. □

Now by taking $g(x) \equiv 1$ in the preceding theorem, we have the following:

Corollary Suppose $f(x)$ is integrable on $[a, b]$ and $m \leqslant f(x) \leqslant M$ for some $m, M \in \mathbb{R}$. Then

(1) there is $\mu \in [m, M]$ such that

$$\int_a^b f(x)\mathrm{d}x = \mu(b-a);$$

(2) there is $\xi \in [a, b]$ such that

$$\int_a^b f(x)\mathrm{d}x = f(\xi)(b-a),$$

if $f(x)$ is continuous on $[a, b]$.

Examples (1) Suppose $f(x)$ is continuous on $[0, 1]$, differentiable on $(0, 1)$ and $3\int_{2/3}^1 f(x)\mathrm{d}x = f(0)$. Show that there exists $\xi \in (0, 1)$ such that $f'(\xi) = 0$.

Proof Because there exists $x_0 \in (0, 1)$ with

$$\int_{2/3}^{1} f(x)\mathrm{d}x = f(x_0)(1-\frac{2}{3}) = \frac{1}{3}f(x_0),$$

we have $f(0) = f(x_0)$. Then by Rolle theorem there exists $\xi \in (0, x_0) \subseteq (0, 1)$ such that $f'(\xi) = 0$. □

(2) (cf. Example of Theorem 7.2.3) Suppose $f(x)$ is continuous and decreasing on $[0, 1]$ and $\lambda \in (0, 1)$. Show that

$$\int_0^{\lambda} f(x)\mathrm{d}x \geqslant \lambda \int_0^1 f(x)\mathrm{d}x.$$

Proof

$$\begin{aligned}
&\int_0^{\lambda} f(x)\mathrm{d}x - \lambda \int_0^1 f(x)\mathrm{d}x \\
&= \int_0^{\lambda} f(x)\mathrm{d}x - \lambda \int_0^{\lambda} f(x)\mathrm{d}x - \lambda \int_{\lambda}^1 f(x)\mathrm{d}x \\
&= (1-\lambda)\int_0^{\lambda} f(x)\mathrm{d}x - \lambda \int_{\lambda}^1 f(x)\mathrm{d}x \\
&= (1-\lambda)f(\xi_1)(\lambda - 0) - \lambda f(\xi_2)(1-\lambda) \\
&= \lambda(1-\lambda)f(\xi_1) - \lambda(1-\lambda)f(\xi_2) \\
&= \lambda(1-\lambda)(f(\xi_1) - f(\xi_2)) \geqslant 0,
\end{aligned}$$

where $\xi_1 \in (0, \lambda)$ and $\xi_2 \in (\lambda, 1)$. □

§ 7.3 The fundamental theorems of calculus

The essential connection between differentiation and integration is provided by the theorem of the title of this section. This theorem goes back to a simple intuitive idea: That the total change in a quantity ought to be exactly the sum of the successive small instantaneous change-an idea which in other forms arises early in geometry, with the rule that the whole is the sum of its parts. Specifically, if the quantity in question is a function $F(t)$ of time t, the total change from time $t = a$ to a later time $t = b$ is just the difference $F(b) - F(a)$. Now suppose the function $F(t)$ has at each *time* t a derivative $F'(t) = f(t)$. The instantaneous change in an infinitesimal interval $\mathrm{d}t$ of time is then the product of this interval $\mathrm{d}t$ by the instantaneous rate of change $F'(t)$, and so the sum of all the successive instantaneous change is the definite integral $\int_a^b f(t)\mathrm{d}t = F(b) - F(a)$.

Theorem 7.3.1 Let $f(x)$ be integrable on $[a, b]$. Then the function

$$\varphi(x) = \int_0^x f(t)\mathrm{d}t \quad (x \in [a, b])$$

is continuous on $[a, b]$.

Proof W.l.o.g., we may suppose $\Delta x > 0$ and x, $x + \Delta x \in [a, b]$. Then

$$\varphi(x + \Delta x) - \varphi(x) = \int_a^{x+\Delta x} f(t)\mathrm{d}t - \int_a^x f(t)\mathrm{d}t = \int_x^{x+\Delta x} f(t)\mathrm{d}t.$$

Since f is integrable on $[a, b]$, f is bounded on $[a, b]$. Thus there exist constants m and M with $m \leqslant f \leqslant M$ on $[x, x + \Delta x]$. So by mean-value theorem of integral, there exists $\mu \in [m, M]$ such that

$$\int_x^{x+\Delta x} f(t)\mathrm{d}t = \mu\Delta x.$$

i.e. $\varphi(x + \Delta x) - \varphi(x) = \mu\Delta x$, which implies

$$\lim_{\Delta x \to 0}\varphi(x + \Delta x) = \varphi(x). \ \square$$

Remark In general, let

$$\varphi_1(x) = \int_a^{\beta(x)} f(t)\mathrm{d}t,\ \varphi_2(x) = \int_{\alpha(x)}^b f(t)\mathrm{d}t,\ \varphi_3(x) = \int_{\alpha(x)}^{\beta(x)} f(t)\mathrm{d}t,$$

where $f(x)$ is integrable on $[a, b]$ and $\alpha(x)$, $\beta(x)$ are continuous on $[a, b]$. Then φ_1, φ_2 and φ_3 are all continuous on $[a, b]$.

Theorem 7. 3. 2 (The first fundamental theorem of calculus) Let $f(x)$ be continuous on $[a, b]$. Then the function

$$\varphi(x) = \int_a^x f(t)\,\mathrm{d}t \quad (x \in [a, b])$$

is differentiable on $[a, b]$ and

$$\frac{\mathrm{d}}{\mathrm{d}x}\varphi(x) = f(x).$$

Proof Note

$$\begin{aligned}\Delta\varphi(x) &= \varphi(x+\Delta x) - \varphi(x)\\ &= \int_a^{x+\Delta x} f(t)\,\mathrm{d}t - \int_a^x f(t)\,\mathrm{d}t = \int_x^{x+\Delta x} f(t)\,\mathrm{d}t.\end{aligned}$$

Thus

$$\frac{\mathrm{d}}{\mathrm{d}x}\varphi(x) = \lim_{\Delta x\to 0}\frac{\Delta\varphi}{\Delta x} = \lim_{\Delta x\to 0}\frac{f(\xi)\Delta x}{\Delta x} = \lim_{\Delta x\to 0} f(\xi) = \lim_{\xi\to x} f(\xi) = f(x),$$

where the second equality is due to the mean-value theorem and the fourth equality is by the fact $x \leqslant \xi \leqslant x+\Delta x$. □

The following result is a direct consequence of Theorem 7. 3. 2.

Corollary Let $f(x)$ be a continuous function on $[a, b]$. Then

$$\varphi(x) = \int_a^x f(t)\,\mathrm{d}t \quad (x \in [a, b])$$

is an antiderivative of $f(x)$ on $[a, b]$.

This corollary tells us that any continuous function has its antiderivative, and so has its indefinite integral, which was already mentioned in Theorem 6. 1. 2.

Examples (1) Find

$$L = \lim_{x\to 0}\frac{\int_0^x \cos(t^2)\,\mathrm{d}t}{x}.$$

Solution 1 $L = \lim_{x\to 0} \frac{\frac{d}{dx}\int_0^x \cos(t^2)\,dt}{\frac{d}{dx}x} = \lim_{x\to 0} \cos(x^2) = 1.$

Solution 2 $L = \lim_{x\to 0} \frac{x\cos(\xi^2)}{x}$,

where ξ is between 0 and x by mean-value theorem, and so

$$L = \lim_{\xi\to 0} \cos(\xi^2) = 1.$$

(2) Let $f(x)$ be continuous on $[0, 1]$ with $f(x) < 1$. Verify that the equation

$$2x - \int_0^x f(t)\,dt = 1$$

has a unique root $r \in [0, 1]$.

Proof Let

$$F(x) = 2x - \int_0^x f(t)\,dt - 1.$$

By mean-value theorem, $F(1) = 1 - \int_0^1 f(t)\,dt = 1 - f(\xi)$ for some $\xi \in (0, 1)$, and so $F(1) > 0$. As $F(0) = -1 < 0$, there exists $r \in (0, 1)$ with $F(r) = 0$. Noticing $F'(x) = 2 - f(x) > 0$ on $[0, 1]$, $F(x)$ is strictly increasing on $[0, 1]$. Hence the root r is unique. □

(3) Determine a and b such that

$$L = \lim_{x\to 0} \frac{1}{bx - \sin x}\int_0^x \frac{t^2}{\sqrt{a+t^2}}\,dt = 1.$$

Solution By L' Hospital rule and the above theorem,

$$L = \lim_{x\to 0} \frac{\frac{x^2}{\sqrt{a+x^2}}}{b - \cos x} = 1,$$

which, as

$$\lim_{x\to 0}\frac{x^2}{\sqrt{a+x^2}}=0,$$

implies that $\lim\limits_{x\to 0}(b-\cos x)=0$. Thus $b=1$ and so

$$\lim_{x\to 0}\frac{\dfrac{x^2}{\sqrt{a+x^2}}}{1-\cos x}=\lim_{x\to 0}\frac{\dfrac{x^2}{\sqrt{a+x^2}}}{\dfrac{1}{2}x^2}=1,$$

i. e.

$$\lim_{x\to 0}\frac{2}{\sqrt{a+x^2}}=\frac{2}{\sqrt{a}}=1.$$

Then $a=4$.

(4) Suppose $f(x)$ is derivable on $[0,1]$ with $0\leqslant f'(x)\leqslant 1$ and $f(0)=0$. Prove that

$$\left(\int_0^1 f(x)\mathrm{d}x\right)^2\geqslant\int_0^1 f^3(x)\mathrm{d}x. \tag{1}$$

Proof Let

$$F(t)=\left(\int_0^t f(x)\mathrm{d}x\right)^2-\int_0^t f^3(x)\mathrm{d}x. \tag{2}$$

Thus, $F(0)=0$, $F(x)$ is secondly derivable and

$$F'(t)=2f(t)\int_0^t f(x)\mathrm{d}x-f^3(t)=f(t)G(t) \tag{3}$$

where $G(t)=2\int_0^t f(x)\mathrm{d}x-f^2(t)$ is defined on $[0,1]$ such that $G(0)=0$ and

$$G'(t)=2f(t)-2f(t)f'(t)=2f(t)(1-f'(t))\geqslant 0,$$

which implies that $G(t)$ is an increasing function on $[0,1]$ and so

$G(t) \geqslant 0$.

Since $f(0) = 0$ and $f'(t) \geqslant 0$, $f(t) \geqslant 0$ on $[0, 1]$. By(3), $F'(t) \geqslant 0$, and as $F(0) = 0$, $F(t) \geqslant 0$ on $[0, 1]$. In particular, $F(1) \geqslant 0$, and in consequence, the inequality (1) holds true. □

(5) Suppose $f(x)$ is defined on $[0, +\infty)$ with $\lim\limits_{x \to +\infty} f(x) = a$.

① If $f(x)$ is continuous on $[0, +\infty)$, prove

$$\lim_{x \to +\infty} \frac{1}{x}\int_0^x f(t)\mathrm{d}t = a; \tag{1}$$

② If $f(x)$ is integrable on $[0, x]$ for any $x > 0$, prove the limit (1) still holds true.

Proof ① Since $f(x)$ is continuous on $[0, +\infty)$, the function $\int_0^x f(t)\mathrm{d}t$ is derivable on $[0, +\infty)$. Then, by L'Hospital rule,

$$\lim_{x \to +\infty} \frac{1}{x}\int_0^x f(t)\mathrm{d}t = \lim_{x \to +\infty} f(x) = a.$$

② In this case, we can not use L'Hospital rule. Without loss of generality, let $a = 0$, since otherwise, we may just consider $f(x) - a$. So, for any $\varepsilon > 0$, there exists $A > 0$ such that

$$|f(x)| < \frac{\varepsilon}{2} \tag{2}$$

whenever $x \geqslant A$. Clearly,

$$\frac{1}{x}\int_0^x f(t)\mathrm{d}t = \frac{1}{x}\int_0^A f(t)\mathrm{d}t + \frac{1}{x}\int_A^x f(t)\mathrm{d}t. \tag{3}$$

For the first term on the right side of (3), there is $B > 0$ such that for any $x > B$,

$$\left|\frac{1}{x}\int_0^A f(t)\mathrm{d}t\right| < \frac{\varepsilon}{2}. \tag{4}$$

For the second term on the right side of (3), by formula(2),

$$\left|\frac{1}{x}\int_A^x f(t)\,\mathrm{d}t\right| < \frac{\varepsilon}{2}. \tag{5}$$

Take $M = \max\{A, B\}$. Then, by formula (3), (4) and formula (5), we have

$$\left|\frac{1}{x}\int_0^x f(t)\,\mathrm{d}t\right| < \varepsilon$$

whenever $x > M$, from which the result follows immediately. □

Theorem 7.3.3 Let

$$\varphi(x) = \int_a^{\beta(x)} f(t)\,\mathrm{d}t, \quad \psi(x) = \int_{\alpha(x)}^b f(t)\,\mathrm{d}t,$$

where $f(x)$ is continuous, $\alpha(x)$ and $\beta(x)$ are differentiable, and a, b are constants. Then

$$\varphi'(x) = f(\beta(x))\beta'(x), \psi'(x) = -f(\alpha(x))\alpha'(x).$$

Proof Let $u = \beta(x)$ and $h(u) = \int_a^u f(t)\,\mathrm{d}t$. Then $\varphi(x) = h(\beta(x))$, and so by the above theorem

$$\frac{\mathrm{d}\varphi}{\mathrm{d}x} = \frac{\mathrm{d}h}{\mathrm{d}u}\frac{\mathrm{d}u}{\mathrm{d}x} = f(u)\beta'(x) = f(\beta(x))\beta'(x).$$

Since $\psi(x) = -\int_b^{\alpha(x)} f(t)\,\mathrm{d}t$, $\psi'(x) = -f(\alpha(x))\alpha'(x)$. □

Now we have a more general result as follows:

Theorem 7.3.4 (Leibniz rule) Suppose $f(x)$ is continuous, $\alpha(x)$ and $\beta(x)$ are differentiable. Then

$$\frac{\mathrm{d}}{\mathrm{d}x}\left(\int_{\alpha(x)}^{\beta(x)} f(t)\,\mathrm{d}t\right) = f(\beta(x))\beta'(x) - f(\alpha(x))\alpha'(x).$$

Proof Only need to notice that

$$\int_{\alpha(x)}^{\beta(x)} f(t)\,\mathrm{d}t = \int_a^{\beta(x)} f(t)\,\mathrm{d}t + \int_{\alpha(x)}^a f(t)\,\mathrm{d}t,$$

for any constant a. The consequence follows directly from the above theorem. □

Examples (1) Suppose $f(x)$ is continuous on $[0, 1]$ and $f(x) > 0$. Let

$$F(x) = \int_{x/2}^{1} f(t)\,\mathrm{d}t - \int_{0}^{x/2} \frac{1}{f(t)}\,\mathrm{d}t.$$

Prove the equation $F(x) = 0$ has exactly one root $r \in (0, 2)$.

Proof Since $F'(x) = f(x/2)(-1/2) - \dfrac{1}{f(x/2)}(1/2) = -(1/2)\left(f(x/2) + \dfrac{1}{f(x/2)}\right) < 0$, $F(x)$ is strictly decreasing on $[0, 2]$. Furthermore, $F(0) = \int_0^1 f(t)\,\mathrm{d}t > 0$ and $F(2) = -\int_0^1 \dfrac{1}{f(t)}\,\mathrm{d}t < 0$. Hence, the result follows immediately. □

(2) Find

$$L = \lim_{x\to 0} \frac{x^2 - \int_0^{x^2} \cos(t^2)\,\mathrm{d}t}{\sin^{10} x}.$$

Solution By L' Hospital rule and the above theorem

$$L = \lim_{x\to 0} \frac{2x - 2x\cos(x^4)}{10\sin^9 x \cos x} = \lim_{x\to 0} \frac{2x[1 - \cos(x^4)]}{10x^9}$$

$$= \lim_{x\to 0} \frac{2\left[1 - \left(1 - \frac{1}{2}(x^4)^2 + o((x^4)^{2+1})\right)\right]}{10x^8}$$

$$= \lim_{x\to 0} \frac{2\left(\frac{1}{2}x^8 + o(x^{12})\right)}{10x^8} = \frac{1}{10}.$$

The following theorem establishes the basic relationship between definite and indefinite integrals. In words, the definite integral can be evaluated by finding any antiderivative of the integrand and then subtracting the value of this antiderivative at the lower limit of the integral from its value at the upper limit of the integral.

Theorem 7.3.5 (The second fundamental theorem of calculus) Suppose $f(x)$ is continuous on interval $[a, b]$ and $F(x)$ is an antiderivative of $f(x)$. Then

$$\int_a^b f(x)\mathrm{d}x = F(b) - F(a) = F(x)\Big|_a^b,$$

which is called *Newton-Leibniz formula*.

Proof Let $\varphi(x) = \int_a^x f(t)\mathrm{d}t$. Clearly, $\varphi'(x) = f(x)$ and so $\varphi(x)$ is an antiderivative of $f(x)$. Note $F(x)$ is also an antiderivative of $f(x)$. Then, $F(x) - \varphi(x) = C$ for some constant C. Setting $x = a$ we see $C = F(a)$ and thus

$$\int_a^x f(t)\mathrm{d}t = \varphi(x) = F(x) - F(a),$$

which, as we set $x = b$, implies

$$\int_a^b f(x)\mathrm{d}x = F(b) - F(a). \square$$

Examples (1) Suppose $f(x)$ is continuous. Show

$$\lim_{h\to 0}\int_a^b \frac{f(x+h) - f(x)}{h}\mathrm{d}x = f(b) - f(a).$$

Proof Let $F(x)$ be an antiderivative of $f(x)$. Then

$$\frac{\mathrm{d}}{\mathrm{d}x}\left(\frac{F(x+h) - F(x)}{h}\right) = \frac{f(x+h) - f(x)}{h},$$

and so by the above theorem

$$\begin{aligned}\lim_{h\to 0}\int_a^b \frac{f(x+h) - f(x)}{h}\mathrm{d}x &= \lim_{h\to 0}\left(\frac{F(x+h) - F(x)}{h}\right)\Big|_a^b \\ &= \lim_{h\to 0}\frac{F(b+h) - F(b)}{h} - \lim_{h\to 0}\frac{F(a+h) - F(a)}{h}\end{aligned}$$

$$= F'(b) - F'(a) = f(b) - f(a). \square$$

(2) Find

$$L = \lim_{n \to +\infty} \frac{1}{n}\left(\sqrt{1+\frac{1}{n}} + \sqrt{1+\frac{2}{n}} + \cdots + \sqrt{1+\frac{n}{n}}\right).$$

Solution $L = \int_0^1 \sqrt{1+x}\,\mathrm{d}x = \frac{2}{3}(1+x)^{3/2}\Big|_0^1 = \frac{2}{3}(2\sqrt{2}-1)$,

where the first equality is due to the definition of definite integral and the second follows from Newton-Leibniz formula.

(3) Suppose $f(x)$ is differentiable on $[a, b]$ with $f'(x) \leqslant M$ (M is a constant) and $f(a) = 0$. Prove

$$\int_a^b f(x)\,\mathrm{d}x \leqslant \frac{M}{2}(b-a)^2.$$

Proof By Lagrange theorem there exists $\xi \in (a, x)$ with $f(x) = f(x) - f(a) = f'(\xi)(x-a)$, and then

$$\int_a^b f(x)\,\mathrm{d}x = \int_a^b f'(\xi)(x-a)\,\mathrm{d}x \leqslant M\int_a^b (x-a)\,\mathrm{d}x$$

$$= M\frac{(x-a)^2}{2}\Big|_a^b = \frac{M}{2}(b-a)^2. \square$$

Note if $f(x)$ is discontinuous on $[a, b]$, the Newton-Leibniz formula may be invalid. e. g.

$$\int_{-1}^1 \frac{1}{x^2}\,\mathrm{d}x \neq -\frac{1}{x}\Big|_{-1}^1 = -2,$$

since clearly

$$\int_{-1}^1 \frac{1}{x^2}\,\mathrm{d}x \geqslant \int_{-1}^1 0\,\mathrm{d}x = 0.$$

In fact this is a so-called improper integral and we will discuss it later.

Theorem 7.3.5 can be generalized as follows:

Theorem 7.3.6 Let $F(x)$ be a continuous function on $[a, b]$ such that except for a finite number of points, it has the derivative: $F'(x) = f(x)$, and $f(x)$ is integrable on $[a, b]$. Then

$$\int_a^b f(x)\mathrm{d}x = F(b) - F(a).$$

Proof Let S be a point set such that for any $x \in S$, $F(x)$ has no derivative at x. Construct a partition T of $[a, b]$: $a = x_0 < x_1 < x_2 < \cdots < x_n = b$ with $S \subset \{x_1, x_2, \cdots, x_n\}$. Then, clearly

$$\sum_{i=1}^{n}(F(x_i) - F(x_{i-1})) = F(b) - F(a).$$

By the corollary of Theorem 7.2.6, for any $i = 1, 2, \cdots, n$,

$$F(x_i) - F(x_i - 1) = f(\xi_i)\Delta x_i,$$

where $\Delta x_i = x_i - x_{i-1}$ and $x_{i-1} < \xi_i < x_i$, and so we have

$$F(b) - F(a) = \sum_{i=1}^{n} f(\xi_i)\Delta x_i.$$

Now, let $\| T \| \to 0$, and the result follows immediately.

§ 7.4 Integration techniques of definite integrals

In this section, we will discuss the techniques for calculating definite integrals, the idea of which are similar to that for indefinite integrals.

§ 7.4.1 Definite integral by substitution

Theorem 7.4.1 (Substitution I) Let $u = g(x)$. If $g'(x)$ is continuous on $[a, b]$ and $f(u)$ is continuous on interval I, the range of $g(x)$, then

$$\int_a^b f(g(x))g'(x)\mathrm{d}x = \int_a^b f(g(x))\mathrm{d}g(x) = \int_{g(a)}^{g(b)} f(u)\mathrm{d}u.$$

Proof Let $F(u)$ be an antiderivative of $f(u)$. Then $F'(u) = f(u)$, and so

$$(F(g(x)))' = F'(g(x))g'(x) = f(g(x))g'(x),$$

i. e. $F(g(x))$ is an antiderivative of $f(g(x))g'(x)$. Thus by the second fundamental theorem,

$$\int_a^b f(g(x))g'(x)\mathrm{d}x = F(g(b)) - F(g(a)).$$

On the other hand, since $F(u)$ is an antiderivative of $f(u)$,

$$\int_{g(a)}^{g(b)} f(u)\mathrm{d}u = F(g(b)) - F(g(a)). \ \square$$

Example $\int_{e^{1/4}}^{e^4} \frac{\sqrt{\ln x}}{x}\mathrm{d}x = \int_{e^{1/4}}^{e^4} \sqrt{\ln x}(\ln x)'\mathrm{d}x = \int_{e^{1/4}}^{e^4} \sqrt{\ln x}\,\mathrm{d}(\ln x) = \left(\frac{2}{3}(\ln x)^{3/2}\right)\Big|_{e^{1/4}}^{e^4} = \frac{21}{4}.$

Theorem 7. 4. 2 (Substitution Ⅱ) Suppose $f(x)$ is continuous on $[a, b]$, and $x = g(t)$ satisfies the following conditions:

(1) $g(t)$ is continuous on $[\alpha, \beta]$ with $a \leqslant g(t) \leqslant b(\alpha \leqslant t \leqslant \beta)$;

(2) $g(\alpha) = a$, $g(\beta) = b$;

(3) $g'(t)$ is continuous on $[\alpha, \beta]$. Then

$$\int_a^b f(x)\mathrm{d}x = \int_\alpha^\beta f(g(t))g'(t)\mathrm{d}t.$$

Proof By Substitution Ⅰ,

$$\int_\alpha^\beta f(g(t))g'(t)\mathrm{d}t = \int_\alpha^\beta f(g(t))\mathrm{d}g(t) = \int_{g(\alpha)}^{g(\beta)} f(x)\mathrm{d}x = \int_a^b f(x)\mathrm{d}x. \ \square$$

Remark Substitution Ⅰ and Ⅱ for definite integral, just as for indefinite integral, are the reversed process to each other.

Proposition (1) Let $f(x)$ be continuous on $[-a, a]$. If $f(x)$ is an odd function, then

$$\int_{-a}^{a} f(x)\mathrm{d}x = 0;$$

if $f(x)$ is an even function, then

$$\int_{-a}^{a} f(x)\mathrm{d}x = 2\int_{0}^{a} f(x)\mathrm{d}x.$$

(2) If $f(x)$ is a periodical function with period T, i. e. $f(x) = f(x+T)$, then

$$\int_{a}^{a+T} f(x)\mathrm{d}x = \int_{0}^{T} f(x)\mathrm{d}x.$$

Proof (1) Let

$$I = \int_{-a}^{a} f(x)\mathrm{d}x = \int_{-a}^{0} f(x)\mathrm{d}x + \int_{0}^{a} f(x)\mathrm{d}x = I_1 + I_2.$$

For I_1, let $x = -t$, then

$$I_1 = \int_{a}^{0} (-f(-t))\mathrm{d}t = \int_{0}^{a} f(-t)\mathrm{d}t = \int_{0}^{a} f(-x)\mathrm{d}x.$$

If $f(x)$ is odd on $[-a, a]$, $I_1 = \int_{0}^{a} (-f(x))\mathrm{d}x = -I_2$ and so $I = 0$; if $f(x)$ is even, $I_1 = \int_{0}^{a} f(x)\mathrm{d}x = I_2$ and so $I = 2I_2$.

(2) Let

$$\begin{aligned} I &= \int_{a}^{a+T} f(x)\mathrm{d}x = \int_{a}^{0} f(x)\mathrm{d}x + \int_{0}^{T} f(x)\mathrm{d}x + \int_{T}^{a+T} f(x)\mathrm{d}x \\ &= I_1 + I_2 + I_3. \end{aligned}$$

For I_3, let $x = t+T$, then $t: 0 \to a$ as $x: T \to a+T$, and so $I_3 = \int_{0}^{a} f(t+T)\mathrm{d}t = \int_{0}^{a} f(t)\mathrm{d}t = -I_1$. Thus, $I = I_2$. □

Examples (1) Find

$$I=\int_0^2 \frac{x^2}{\sqrt{2x-x^2}}\mathrm{d}x.$$

Solution Let $x-1=\sin t$. Then $t:-\pi/2 \to \pi/2$ as $x:0\to 2$ and $\mathrm{d}x=\cos t\,\mathrm{d}t$. Thus

$$I=\int_0^2 \frac{x^2}{\sqrt{1-(x-1)^2}}\mathrm{d}x=\int_{-\pi/2}^{\pi/2}\frac{(1+\sin t)^2}{\cos t}\cos t\,\mathrm{d}t$$

$$=2\int_0^{\pi/2}(1+\sin^2 t)\mathrm{d}t=\frac{3}{2}\pi.$$

(2) Find

$$I=\int_0^{\pi}\frac{x\sin x}{1+\cos^2 x}\mathrm{d}x.$$

Solution (1) Let $x=\frac{\pi}{2}-t$. Then $\sin x=\cos t$, $\cos x=\sin t$ and $t:\pi/2\to -\pi/2$ as $x:0\to\pi$. Thus

$$I=\int_{\frac{\pi}{2}}^{-\frac{\pi}{2}}\frac{\left(\frac{\pi}{2}-t\right)\cos t}{1+\sin^2 t}\mathrm{d}\left(\frac{\pi}{2}-t\right)=\frac{\pi}{2}\int_{-\frac{\pi}{2}}^{\frac{\pi}{2}}\frac{\cos t}{1+\sin^2 t}\mathrm{d}t-\int_{-\frac{\pi}{2}}^{\frac{\pi}{2}}\frac{t\cos t}{1+\sin^2 t}\mathrm{d}t$$

$$=\pi\int_0^{\frac{\pi}{2}}\frac{\mathrm{d}(\sin t)}{1+\sin^2 t}=\pi(\arctan(\sin t))\Big|_0^{\frac{\pi}{2}}=\frac{\pi^2}{4}.$$

Solution (2) Let $x=\pi-t$. Then $\sin x=\sin t$, $\cos x=-\cos t$ and $t:\pi\to 0$ as $x:0\to\pi$. Thus

$$I=\int_{\pi}^{0}\frac{(\pi-t)\sin t}{1+\cos^2 t}\mathrm{d}(\pi-t)=\pi\int_0^{\pi}\frac{\sin t}{1+\cos^2 t}\mathrm{d}t-\int_0^{\pi}\frac{t\sin t}{1+\cos^2 t}\mathrm{d}t$$

$$=\pi\int_0^{\pi}\frac{\sin t}{1+\cos^2 t}\mathrm{d}t-I,\text{ and so,}$$

$$I=\frac{\pi}{2}\int_0^{\pi}\frac{\sin t}{1+\cos^2 t}\mathrm{d}t=\frac{\pi}{2}\int_0^{\pi}\frac{-\mathrm{d}(\cos t)}{1+\cos^2 t}\mathrm{d}t=\frac{\pi^2}{4}.$$

(3) Find

$$I=\int_0^1 \frac{\ln(1+x)}{1+x^2}\mathrm{d}x.$$

Solution Let $x=\tan t$. Then $t:0\to\pi/4$ as $x:0\to 1$. Thus

$$I=\int_0^{\pi/4}\frac{\ln(1+\tan t)}{1+\tan^2 t}\mathrm{d}(\tan t)=\int_0^{\pi/4}\frac{\ln(1+\tan t)}{\sec^2 t}\cdot\sec^2 t\,\mathrm{d}t$$

$$=\int_0^{\pi/4}\ln\left(\frac{\cos t+\sin t}{\cos t}\right)\mathrm{d}t=\int_0^{\pi/4}\ln\frac{\cos t+\cos\left(\frac{\pi}{2}-t\right)}{\cos t}\mathrm{d}t$$

$$=\int_0^{\pi/4}\ln\frac{2\cos\frac{\pi}{4}\cdot\cos\left(\frac{\pi}{4}-t\right)}{\cos t}\mathrm{d}t$$

$$=\int_0^{\pi/4}\ln\sqrt{2}\,\mathrm{d}t+\int_0^{\pi/4}\ln\left[\cos\left(\frac{\pi}{4}-t\right)\right]\mathrm{d}t-\int_0^{\pi/4}\ln(\cos t)\mathrm{d}t;$$

$$=I_1+I_2-I_3.$$

For I_2, let $u=\frac{\pi}{4}-t$, then $u:\frac{\pi}{4}\to 0$ as $t:0\to\frac{\pi}{4}$, and so $I_2=\int_{\pi/4}^{0}\ln(\cos u)(-\mathrm{d}u)=I_3$. Thus

$$I=I_1=\frac{\pi}{4}\ln\sqrt{2}=\frac{\pi}{8}\ln 2.$$

(4) Prove

$$2-\frac{\pi}{2}\leqslant\int_{-1}^{1}\frac{x^2+x\cos x}{1+\sin^2 x}\mathrm{d}x\leqslant\frac{2}{3}.$$

Proof First noticing $x\cos x/(1+\sin^2 x)$ is odd, we have

$$\int_{-1}^{1}\frac{x^2+x\cos x}{1+\sin^2 x}\mathrm{d}x=\int_{-1}^{1}\frac{x^2}{1+\sin^2 x}\mathrm{d}x.$$

Since

$$\frac{x^2}{1+x^2}\leqslant\frac{x^2}{1+\sin^2 x}\leqslant x^2\quad(x\in[-1,1]),$$

whereas

$$\int_{-1}^{1} \frac{x^2}{1+x^2} dx = \int_{-1}^{1} \left(1 - \frac{1}{1+x^2}\right) dx = (x - \arctan x)\Big|_{-1}^{1} = 2 - \frac{\pi}{2},$$

and

$$\int_{-1}^{1} x^2 dx = \frac{x^3}{3}\Big|_{-1}^{1} = \frac{2}{3},$$

the result follows immediately. □

§ 7.4.2 Definite integral by parts

Theorem 7.4.3 (Definite integral by parts) Suppose $f(x)$ and $g(x)$ have continuous derivatives on $[a, b]$. Then

$$\int_a^b f(x) g'(x) dx = ((f(x) g(x))\Big|_a^b - \int_a^b g(x) f'(x) dx,$$

i.e.

$$\int_a^b f(x) dg(x) = (f(x) g(x))\Big|_a^b - \int_a^b g(x) df(x).$$

Proof Since $(f(x)g(x))' = f(x)g'(x) + f'(x)g(x)(a \leqslant x \leqslant b)$, $f(x)g(x)$ is an antiderivative of $f(x)g'(x) + f'(x)g(x)$. Then, by Newton-Leibniz formula

$$(f(x)g(x))\Big|_a^b = \int_a^b f(x) g'(x) dx + \int_a^b g(x) f'(x) dx. \ \square$$

Examples (1) $\int_0^1 x e^x dx = \int_0^1 x \, de^x$

$$= x e^x \Big|_0^1 - \int_0^1 e^x dx = x e^x \Big|_0^1 - e^x \Big|_0^1 = 1.$$

(2) $\int_1^e x \ln x \, dx = \frac{1}{2}\int_1^e \ln x \, d(x^2) = \frac{1}{2} x^2 \ln x \Big|_1^e - \frac{1}{2}\int_1^e x \, dx$

$$= \frac{1}{4}(e^2 + 1).$$

(3) Prove $\int_0^{\frac{\pi}{2}} \sin^n x \, dx = \int_0^{\frac{\pi}{2}} \cos^n x \, dx$

$$= \begin{cases} \dfrac{(n-1)!!}{n!!} \dfrac{\pi}{2} & (\text{if } n \text{ is even}), \\ \dfrac{(n-1)!!}{n!!} & (\text{if } n \text{ is odd}). \end{cases}$$

Proof Let $x = \frac{\pi}{2} - t$. Then

$$\int_0^{\frac{\pi}{2}} \cos^n x \, dx = \int_{\frac{\pi}{2}}^0 \left(-\cos^n\left(\frac{\pi}{2} - t\right)\right) dt = \int_0^{\frac{\pi}{2}} \sin^n t \, dt = \int_0^{\frac{\pi}{2}} \sin^n x \, dx.$$

Let

$$J_n = \int_0^{\frac{\pi}{2}} \sin^n x \, dx = \int_0^{\frac{\pi}{2}} \cos^n x \, dx \quad (n \geqslant 2).$$

Then

$$\begin{aligned} J_n &= \int_0^{\frac{\pi}{2}} \sin^n x \, dx = \int_0^{\frac{\pi}{2}} (-\sin^{n-1} x) d(\cos x) \\ &= (-(\sin x)^{n-1} \cos x) \Big|_0^{\frac{\pi}{2}} + (n-1) \int_0^{\frac{\pi}{2}} \sin^{n-2} x \cos^2 x \, dx \\ &= (n-1) \int_0^{\frac{\pi}{2}} \sin^{n-2} x (1 - \sin^2 x) dx \\ &= (n-1) \int_0^{\frac{\pi}{2}} \sin^{n-2} x \, dx - (n-1) \int_0^{\frac{\pi}{2}} \sin^n x \, dx \\ &= (n-1) J_{n-2} - (n-1) J_n. \end{aligned}$$

Thus

$$J_n = \frac{n-1}{n} J_{n-2} \quad (n \geqslant 2).$$

Notice that

$$J_0 = \int_0^{\frac{\pi}{2}} \mathrm{d}x = \frac{\pi}{2},\ J_1 = \int_0^{\frac{\pi}{2}} \sin x\, \mathrm{d}x = 1.$$

The result follows directly. □

Note By a similar proof, we can get a generalization of the above example as follows:

$$\begin{aligned} J_{mn} &= \int_0^{\frac{\pi}{2}} \sin^m x \cdot \cos^n x\, \mathrm{d}x \\ &= \frac{(m-1)!! \cdot (n-1)!!}{(m+n)!!} k, \end{aligned}$$

where $k = \pi/2$ if m and n are both even and $k = 1$ otherwise.

(4) Let $f(x)$ be a continuous function. Prove that

$$\int_0^x (x-u) f(u) \mathrm{d}u = \int_0^x \left(\int_0^u f(x) \mathrm{d}x \right) \mathrm{d}u.$$

Proof Let $g(u) = \int_0^u f(x) \mathrm{d}x$. Then $g'(u) = f(u)$, i. e. $g'(x) = f(x)$. Thus

$$\begin{aligned} \int_0^x \left(\int_0^u f(x) \mathrm{d}x \right) \mathrm{d}u &= \int_0^x g(u) \mathrm{d}u = (ug(u)) \Big|_0^x - \int_0^x ug'(u) \mathrm{d}u \\ &= xg(x) - \int_0^x uf(u) \mathrm{d}u \\ &= x\int_0^x f(u) \mathrm{d}u - \int_0^x uf(u) \mathrm{d}u \\ &= \int_0^x (x-u) f(u) \mathrm{d}u. \ \square \end{aligned}$$

(5) Let

$$a_n = \frac{\sqrt[n]{(n+1)(n+2)\cdots(n+n)}}{n}.$$

Find $\lim\limits_{n \to +\infty} a_n$.

Solution $$\lim_{n\to+\infty}\ln a_n = \lim_{n\to+\infty}\ln\left[\frac{(n+1)(n+2)\cdots(n+n)}{n^n}\right]^{1/n}$$

$$= \lim_{n\to+\infty}\frac{1}{n}\sum_{i=1}^{n}\ln\left(1+\frac{i}{n}\right)=\int_0^1\ln(1+x)\mathrm{d}x$$

$$= (x\ln(1+x))\Big|_0^1-\int_0^1\left(1-\frac{1}{1+x}\right)\mathrm{d}x$$

$$= \ln\frac{4}{\mathrm{e}},\ \text{and so}\ \lim_{n\to+\infty}a_n=\frac{4}{\mathrm{e}}.$$

§ 7.5 Improper integrals

The definite integrals discussed so far are restricted in two categories, namely, the integral interval is limited and the integrand is bounded on the integral interval, which are called *proper* integrals. However, in practical problems it tends to be needed to generalize the definite integral in two cases, i. e. the case that the integral interval is unlimited or the case that the integrand is unbounded on the integral interval. Such generalized definite integrals are called *improper* integrals.

§ 7.5.1 Integrals over unbounded intervals

We now first extend the concept of a definite integral to allow for integrals over unbounded intervals. Here are some examples:

$$\int_1^{+\infty}\frac{\mathrm{d}x}{x^2},\ \int_{-\infty}^{0}\mathrm{e}^x\mathrm{d}x,\ \int_{-\infty}^{+\infty}\frac{\mathrm{d}x}{1+x^2}.$$

Definition Suppose that a function f is defined on $[a,+\infty)$ and f is integrable on any interval $[a,A]$ $(A\geqslant a)$. Then we define the following *improper integrals over unbounded intervals* $[a,+\infty)$ when the limit exists:

$$\int_a^{+\infty}f(x)\mathrm{d}x=\lim_{b\to+\infty}\int_a^b f(x)\mathrm{d}x.$$

If the limit exists, we also say that the improper integral is *convergent and the limit is called the value of the improper integral* $\int_a^{+\infty} f(x)\mathrm{d}x$; otherwise it is said to be *divergent*. Similarly, we define *improper integral over unbounded interval* $(-\infty, b]$ or $(-\infty, +\infty)$:

$$\int_{-\infty}^{b} f(x)\mathrm{d}x = \lim_{a\to-\infty}\int_a^b f(x)\mathrm{d}x;$$

$$\int_{-\infty}^{+\infty} f(x)\mathrm{d}x = \int_{-\infty}^{c} f(x)\mathrm{d}x + \int_{c}^{+\infty} f(x)\mathrm{d}x,$$

provided both of the integrals on the right-hand side exist for some c.

Remark Let $F(x)$ be an antiderivative of $f(x)$ and denote $\lim\limits_{x\to+\infty} F(x) = F(+\infty)$ and $\lim\limits_{x\to-\infty} F(x) = F(-\infty)$. Then, if the above improper integrals are convergent, by definition we have

$$\int_a^{+\infty} f(x)\mathrm{d}x = F(x)\Big|_a^{+\infty} = F(+\infty) - F(a);$$

$$\int_{-\infty}^{b} f(x)\mathrm{d}x = F(x)\Big|_{-\infty}^{b} = F(b) - F(-\infty);$$

$$\int_{-\infty}^{+\infty} f(x)\mathrm{d}x = F(x)\Big|_{-\infty}^{+\infty} = F(+\infty) - F(-\infty).$$

In short, Newton-Leibniz formula is still valid for improper integrals over unbounded intervals.

Examples (1) Determine the convergence of the following improper integrals, and find the value if it is convergent:

① $\int_1^{+\infty} x\mathrm{e}^{-x^2}\mathrm{d}x$; ② $\int_1^{+\infty} \frac{1}{x^p}\mathrm{d}x$; ③ $\int_2^{+\infty} \frac{\mathrm{d}x}{x(\ln x)^p}$;

④ $\int_{-\infty}^{+\infty} \frac{1}{1+x^2}\mathrm{d}x$; ⑤ $\int_{-\infty}^{a} \cos x\,\mathrm{d}x$.

Solution (1) ① $\lim\limits_{b\to+\infty}\int_1^b x\mathrm{e}^{-x^2}\mathrm{d}x = \lim\limits_{b\to+\infty}\int_1^b\left(-\frac{1}{2}\right)\mathrm{e}^{-x^2}\mathrm{d}(-x^2)$

$$= \lim_{b\to+\infty}\left(-\frac{1}{2}\mathrm{e}^{-x^2}\right)\Big|_1^b.$$

$$= \left(-\frac{1}{2}\right)\lim_{b\to+\infty}\left(e^{-b^2}-\frac{1}{e}\right)=\frac{1}{2e}.$$

So the improper integral is convergent with value $1/2e$.

② Noticing that

$$\int_1^b \frac{1}{x^p}dx = \begin{cases} \left.\dfrac{x^{-p+1}}{-p+1}\right|_1^b = \dfrac{b^{1-p}}{1-p}-\dfrac{1}{1-p} & (\text{if } p\neq 1), \\ \ln x\,|_1^b = \ln b & (\text{if } p=1). \end{cases}$$

Then

$$\int_1^{+\infty}\frac{1}{x^p}dx = \lim_{b\to+\infty}\int_1^b \frac{1}{x^p}dx = \frac{1}{p-1} \quad (\text{if } p>1);$$

$$\int_1^{+\infty}\frac{1}{x^p}dx = \lim_{b\to+\infty}\int_1^b \frac{1}{x^p}dx = +\infty \quad (\text{if } p\leqslant 1).$$

Hence, if $p>1$, the improper integral is convergent with value $1/(p-1)$; otherwise it is divergent.

③ Let $u=\ln x$. Then

$$\int_2^{+\infty}\frac{dx}{x(\ln x)^p} = \int_2^{+\infty}\frac{d(\ln x)}{(\ln x)^p} = \int_{\ln 2}^{+\infty}\frac{du}{u^p}.$$

Thus, by the above example, we conclude that if $p>1$ the improper integral is convergent with value $(\ln 2)^{1-p}/(p-1)$; otherwise it is divergent.

④ Let c be any real number. Since

$$\int_c^{+\infty}\frac{1}{1+x^2}dx = \lim_{b\to+\infty}\int_c^b\frac{1}{1+x^2}dx = \lim_{b\to+\infty}(\arctan x)\Big|_c^b$$

$$= \lim_{b\to+\infty}(\arctan b-\arctan c) = \frac{\pi}{2}-\arctan c;$$

similarly, we have

$$\int_{-\infty}^c\frac{1}{1+x^2}dx = \arctan c+\frac{\pi}{2}.$$

Thus, we see that the improper integral is convergent with value π.

⑤ Since

$$\lim_{a\to-\infty}\int_a^b \cos x\,\mathrm{d}x = \lim_{a\to-\infty}(\sin b - \sin a)$$

does not exist, the integral is divergent.

(2) Find the following improper integrals

① $I=\int_0^{+\infty}\frac{\arctan x}{(1+x^2)^{3/2}}\mathrm{d}x$; ② $I_n=\int_0^{+\infty}x^n\mathrm{e}^{-x}\mathrm{d}x$.

Solution ① Let $x=\tan t$. Then I can be converted into a proper integral:

$$I=\int_0^{\pi/2}\frac{t\sec^2 t}{\sec^3 t}\mathrm{d}t=\int_0^{\pi/2}\frac{t}{\sec t}\mathrm{d}t=\int_0^{\pi/2}t\,\mathrm{d}(\sin t)$$

$$=(t\sin t+\cos t)\Big|_0^{\pi/2}=\frac{\pi}{2}-1;$$

② $$I_n=\int_0^{+\infty}x^n\mathrm{d}(-\mathrm{e}^{-x})=-x^n\mathrm{e}^{-x}\Big|_0^{+\infty}+n\int_0^{+\infty}x^{n-1}\mathrm{e}^{-x}\mathrm{d}x$$

$$=n\int_0^{+\infty}x^{n-1}\mathrm{e}^{-x}\mathrm{d}x$$

$$=nI_{n-1}=n(n-1)I_{n-2}=n(n-1)(n-2)I_{n-3}$$

$$=\cdots=n!\cdot I_0=n!\cdot\int_0^{+\infty}\mathrm{e}^{-x}\mathrm{d}x=n!.$$

(3) Find the following improper integrals

① $I=\int_0^{+\infty}\frac{\mathrm{d}x}{(1+x^2)(1+x^\beta)}\quad(\beta\geqslant 0)$;

② $I=\int_0^{+\infty}\frac{\ln x}{1+x^2}\mathrm{d}x$.

Solution ① Let $x=1/t$. Then

$$I=\int_{+\infty}^0\frac{-\frac{1}{t^2}\mathrm{d}t}{\left(1+\frac{1}{t^2}\right)\left(1+\frac{1}{t^\beta}\right)}=\int_0^{+\infty}\frac{t^\beta\mathrm{d}t}{(1+t^2)(1+t^\beta)}$$

$$= \int_0^{+\infty} \frac{x^\beta \mathrm{d}x}{(1+x^2)(1+x^\beta)},$$

and so

$$I = \frac{1}{2}\left[\int_0^{+\infty} \frac{\mathrm{d}x}{(1+x^2)(1+x^\beta)} + \int_0^{+\infty} \frac{x^\beta \mathrm{d}x}{(1+x^2)(1+x^\beta)}\right]$$

$$= \frac{1}{2}\int_0^{+\infty} \frac{\mathrm{d}x}{1+x^2} = \left(\frac{1}{2}\arctan x\right)\Big|_0^{+\infty} = \frac{\pi}{4}.$$

②

$$I = \int_0^{+\infty} \frac{\ln x}{1+x^2}\mathrm{d}x = \int_0^1 \frac{\ln x}{1+x^2}\mathrm{d}x + \int_1^{+\infty} \frac{\ln x}{1+x^2}\mathrm{d}x.$$

For the second term, let $x=1/t$. Then

$$\int_0^{+\infty} \frac{\ln x}{1+x^2}\mathrm{d}x = \int_1^0 \frac{\ln t}{1+t^2}\mathrm{d}t = -\int_0^1 \frac{\ln t}{1+t^2}\mathrm{d}t = -\int_0^1 \frac{\ln x}{1+x^2}\mathrm{d}x.$$

Thus $I = 0$.

(4) Prove that

$$\int_0^{+\infty} f\left(ax+\frac{b}{x}\right)\mathrm{d}x = \frac{1}{a}\int_0^{+\infty} f(\sqrt{x^2+4ab})\mathrm{d}x \quad (a,\ b>0).$$

Proof Let $ax - \frac{b}{x} = t$.

Then $t: -\infty \to +\infty$ as $x: 0 \to +\infty$, and

$$ax + \frac{b}{x} = \sqrt{t^2+4ab}.$$

Moreover

$$x = \frac{1}{2a}(t+\sqrt{t^2+4ab}),\ \mathrm{d}x = \frac{t+\sqrt{t^2+4ab}}{2a\sqrt{t^2+4ab}}\mathrm{d}t.$$

Thus

$$\int_0^{+\infty} f\left(ax+\frac{b}{x}\right)\mathrm{d}x = \frac{1}{2a}\int_{-\infty}^{+\infty} f(\sqrt{t^2+4ab})\frac{t+\sqrt{t^2+4ab}}{\sqrt{t^2+4ab}}\mathrm{d}t$$

$$= \frac{1}{2a}\int_{-\infty}^{0} + \frac{1}{2a}\int_0^{+\infty} = I_1 + I_2.$$

Let $t=-s$ in I_1. Then

$$I_1 = \frac{1}{2a}\int_{+\infty}^{0} f(\sqrt{s^2+4ab})\frac{-s+\sqrt{s^2+4ab}}{\sqrt{s^2+4ab}}(-1)\mathrm{d}s$$

$$= \frac{1}{2a}\int_0^{+\infty} f(\sqrt{s^2+4ab})\frac{\sqrt{s^2+4ab}-s}{\sqrt{s^2+4ab}}\mathrm{d}s$$

$$= \frac{1}{2a}\int_0^{+\infty} f(\sqrt{t^2+4ab})\frac{\sqrt{t^2+4ab}-t}{\sqrt{t^2+4ab}}\mathrm{d}t.$$

Thus

$$I_1 + I_2 = \frac{1}{2a}\int_0^{+\infty} f(\sqrt{t^2+4ab})\frac{(\sqrt{t^2+4ab}-t)+(\sqrt{t^2+4ab}+t)}{\sqrt{t^2+4ab}}\mathrm{d}t$$

$$= \frac{1}{a}\int_0^{+\infty} f(\sqrt{t^2+4ab})\mathrm{d}t. \ \square$$

The following theorem can be proved just by the definition of the improper integral over unbounded interval and the properties of limit operations.

Theorem 7.5.1

(1) Suppose $\int_a^{+\infty} f_1(x)\mathrm{d}x$, $\int_a^{+\infty} f_2(x)\mathrm{d}x$

are both convergent and k_1 and k_2 are constants. Then

$\int_a^{+\infty}(k_1 f_1(x)+k_2 f_2(x))\mathrm{d}x$ is convergent and

$$\int_a^{+\infty}(k_1 f_1(x)+k_2 f_2(x))\mathrm{d}x = k_1\int_a^{+\infty} f_1(x)\mathrm{d}x + k_2\int_a^{+\infty} f_2(x)\mathrm{d}x.$$

(2) Suppose $f(x)$ is integrable on any interval $[a, A]$ $(a<A)$. Then

$\int_{a}^{+\infty} f(x)\mathrm{d}x$ is convergent, if and only if $\int_{b}^{+\infty} f(x)\,\mathrm{d}x$ is convergent, where $a < b$.

(3) Suppose $f(x)$, $g(x)$, $f'(x)$, $g'(x)$ are continuous on $[a, +\infty)$.
If two terms in the following expression are existent, then the third term is also existent and

$$\int_{a}^{+\infty} f(x)\mathrm{d}g(x) = f(x)g(x)\Big|_{a}^{+\infty} - \int_{a}^{+\infty} g(x)\mathrm{d}f(x).$$

If the antiderivative of the integrand can not be found or it is difficult to be found, the convergence of the improper integral can only be determined by the property of the integral. By the definition, the convergence of the integral $\int_{a}^{+\infty} f(x)\mathrm{d}x$ depends on whether the limit $\lim\limits_{b\to+\infty} F(b) = \lim\limits_{b\to+\infty}\int_{a}^{b} f(x)\mathrm{d}x$ exists. Therefore the following Cauchy criterion for convergence of the improper integral can be derived from that for the convergence of function limit.

Theorem 7.5.2 (Cauchy criterion) $\int_{a}^{+\infty} f(x)\mathrm{d}x$ is convergent if and only if for any $\varepsilon > 0$, there exists $M > 0$, such that

$$\left|\int_{A_1}^{A_2} f(x)\mathrm{d}x\right| < \varepsilon \text{ for any } A_1,\ A_2 > M.$$

Example Show that the following two improper integrals are convergent:

(1) $\int_{1}^{+\infty} \frac{\sin x}{x}\mathrm{d}x$; (2) $\int_{1}^{+\infty} \frac{\cos x}{x}\mathrm{d}x$.

Proof Let $A_2 > A_1 > 1$.

$$\int_{A_1}^{A_2} \frac{\sin x}{x}\mathrm{d}x = -\int_{A_1}^{A_2} \frac{1}{x}\mathrm{d}(\cos x)$$

$$= -\frac{1}{x}\cos x\Big|_{A_1}^{A_2} + \int_{A_1}^{A_2} \cos x \cdot \left(-\frac{1}{x^2}\right)\mathrm{d}x.$$

Then

$$\left|\int_{A_1}^{A_2} \frac{\sin x}{x}\mathrm{d}x\right| \leqslant \left|\frac{\cos A_1}{A_1}\right| + \left|\frac{\cos A_2}{A_2}\right| + \left|\int_{A_1}^{A_2} \frac{1}{x^2}\mathrm{d}x\right|$$

$$\leqslant \frac{1}{A_1} + \frac{1}{A_2} + \left|\left(\frac{x^{-2+1}}{-2+1}\right)\Big|_{A_1}^{A_2}\right| \leqslant \frac{2}{A_1} + \frac{2}{A_2} \leqslant \frac{4}{A_1} < \varepsilon.$$

Thus $A_1 > 4/\varepsilon$. Hence, for any $\varepsilon > 0$, take $M = 4/\varepsilon$, and then

$$\left|\int_{A_1}^{A_2} \frac{\sin x}{x}\mathrm{d}x\right| < \varepsilon \text{ whenever } A_2 > A_1 > M,$$

i. e. $\int_1^{+\infty} \frac{\sin x}{x}\mathrm{d}x$ is convergent. Similarly, we can prove $\int_1^{+\infty} \frac{\cos x}{x}\mathrm{d}x$ is convergent. □

We introduce the concept of absolute convergence and conditional convergence as follows:

Definition Assume $f(x)$ is defined on $[a, +\infty)$ such that it is integrable on any finite interval $[a, A]$. If $\int_a^{+\infty} |f(x)|\mathrm{d}x$ is convergent, then we say $\int_a^{+\infty} f(x)\mathrm{d}x$ is *absolutely convergent*; If $\int_a^{+\infty} f(x)\mathrm{d}x$ is convergent, but $\int_a^{+\infty} |f(x)|\mathrm{d}x$ is divergent, then we say $\int_a^{+\infty} f(x)\mathrm{d}x$ is *conditionally convergent*.

Theorem 7.5.3 If $\int_a^{+\infty} f(x)\mathrm{d}x$ is absolutely convergent, then it is convergent, i. e. the convergence of $\int_a^{+\infty} |f(x)|\mathrm{d}x$ implies the convergence of $\int_a^{+\infty} f(x)\mathrm{d}x$, and in this case $\left|\int_a^{+\infty} f(x)\mathrm{d}x\right| \leqslant \int_a^{+\infty} |f(x)|\mathrm{d}x$.

Proof As $\int_a^{+\infty} |f(x)| \mathrm{d}x$ is convergent, by Cauchy criterion, for any $\varepsilon > 0$, there exists $M > 0$, such that

$$\left| \int_{A_1}^{A_2} \right| f(x) \mid \mathrm{d}x \mid < \varepsilon \text{ for any } A_1, A_2 > M.$$

Since

$$\left| \int_{A_1}^{A_2} f(x) \mathrm{d}x \right| < \left| \int_{A_1}^{A_2} \right| f(x) \mid \mathrm{d}x \mid,$$

we have

$$\left| \int_{A_1}^{A_2} f(x) \mathrm{d}x \right| < \varepsilon.$$

So, by Cauchy criterion, $\int_a^{+\infty} f(x) \mathrm{d}x$ is convergent. The inequality just follows from the properties of the limit and the definition of the improper definite integral over unbounded intervals.

In practical use, we sometimes only need to know whether an improper integral is convergent or not, not need to know what its value is. Hence, we introduce some tests here just for determining the convergence. It is clear that the convergence of an improper integral such as $\int_a^{\infty} f(x) \mathrm{d}x$ depends upon the behavior of $f(x)$ when x is large. The following simple test is often used.

Theorem 7.5.4 (The comparison test) Suppose $f(x)(\geqslant 0)$ and $g(x)$ are defined on $[a, +\infty)$ and integrable on any finite interval $[a, A]$ $(A > a)$.

(1) If $f(x) \leqslant g(x)$ and $\int_a^{+\infty} g(x) \mathrm{d}x$ is convergent, then $\int_a^{+\infty} f(x) \mathrm{d}x$ is convergent;

(2) If there exists $p > 1$ such that

$$f(x) \leqslant \frac{1}{x^p},$$

then $\int_a^{+\infty} f(x)\mathrm{d}x$ is convergent; If there exists $p \leqslant 1$ such that

$$f(x) \geqslant \frac{1}{x^p},$$

then $\int_a^{+\infty} f(x)\mathrm{d}x$ is divergent.

Proof The result (1) follows from Cauchy criterion directly. Taking $g(x) = 1/x^p$ in (1), we have the result (2) immediately. □

Examples (1) Discuss the convergence of the following improper integrals.

① $\int_0^{+\infty} \mathrm{e}^{-ax}\sin bx\,\mathrm{d}x (a > 0)$; ② $\int_1^{+\infty} \frac{1+\mathrm{e}^{-x}}{x}\mathrm{d}x$;

③ $\int_1^{+\infty} \frac{\sin x}{x\sqrt{1+x^2}}\mathrm{d}x$.

Solution ① Note $|\mathrm{e}^{-ax}\sin bx| \leqslant \mathrm{e}^{-ax}$ and

$$\int_0^{+\infty} \mathrm{e}^{-ax}\mathrm{d}x = \lim_{b\to+\infty}\int_0^b \mathrm{e}^{-ax}\mathrm{d}x = \lim_{b\to+\infty}\left(-\frac{1}{a}\mathrm{e}^{-ax}\right)\Bigg|_0^b$$
$$= -\frac{1}{a}\lim_{b\to+\infty}(\mathrm{e}^{-ab} - 1) = \frac{1}{a}.$$

So, by the comparison test, $\int_0^{+\infty} \mathrm{e}^{-ax}\sin bx\,\mathrm{d}x$ is absolutely convergent, and is also convergent.

② Since

$$\frac{1+\mathrm{e}^{-x}}{x} > \frac{1}{x} \text{ on } [1, +\infty) \text{ and } \int_1^{+\infty} \frac{1}{x}\mathrm{d}x \text{ is divergent},$$

so,

$$\int_1^{+\infty} \frac{1+\mathrm{e}^{-x}}{x}\mathrm{d}x \text{ is divergent.}$$

③ Since

$$\int_1^{+\infty} \frac{1}{x^2} \mathrm{d}x \text{ is convergent, and } \left| \frac{\sin x}{x\sqrt{1+x^2}} \right| \leqslant \frac{1}{x^2},$$

so,

$$\int_1^{+\infty} \frac{\sin x}{x\sqrt{1+x^2}} \mathrm{d}x \text{ is absolutely convergent, and also convergent.}$$

(2) Suppose $f(x)$ is differentiable on $[a, +\infty)$ and

$$\int_a^{+\infty} f(x)\mathrm{d}x \text{ and } \int_a^{+\infty} f'(x)\mathrm{d}x$$

are both convergent. Prove that $\lim_{x \to +\infty} f(x) = 0$.

Proof Since

$$\lim_{A \to +\infty} \int_a^A f'(x)\mathrm{d}x = \lim_{A \to +\infty} (f(A) - f(a)) \text{exists,}$$

$\lim_{A \to +\infty} f(A)$ exists, i. e. $\lim_{x \to +\infty} f(x) = B$ exists.

Suppose $B > 0$. Then for $\varepsilon \in (0, B)$, there exists $M > 0$ such that $|f(x) - B| < \varepsilon$ whenever $x \geqslant M$, and so $0 < B - \varepsilon < f(x) < B + x$ for any $x \in [M, +\infty)$. Noticing

$$\int_M^{+\infty} (B-\varepsilon)\mathrm{d}x = (B-\varepsilon)\int_M^{+\infty} \mathrm{d}x$$

is divergent, by the comparison test $\int_M^{+\infty} f(x)\mathrm{d}x$, and also $\int_a^{+\infty} f(x)\mathrm{d}x$, is divergent, which is a contradiction. Similarly, we can prove $B < 0$ is also impossible. Then it follows that $B = 0$. □

(3) Suppose $\int_a^{+\infty} f(x)\mathrm{d}x$ is convergent and $\lim_{x \to +\infty} f(x)$ exists. Show that $\lim_{x \to +\infty} f(x) = 0$.

Proof Let $\lim_{x \to +\infty} f(x) = A$. If $A \neq 0$, say $A > 0$, then there exists

$N > 0$ such that $f(x) > A/2$ whenever $x > N$, and so

$$\int_N^{+\infty} f(x)\,\mathrm{d}x \geqslant \int_N^{+\infty} \frac{A}{2}\,\mathrm{d}x.$$

Clearly $\int_N^{+\infty} \frac{A}{2}\,\mathrm{d}x$ is divergent, and so is $\int_N^{+\infty} f(x)\,\mathrm{d}x$ by the comparison test, which yields a contradiction. □

The next example shows that the convergence does not imply the absolute convergence.

(4) Show that

$$\int_1^{+\infty} \frac{\sin x}{x}\,\mathrm{d}x \text{ and } \int_1^{+\infty} \frac{\cos x}{x}\,\mathrm{d}x$$

are both conditionally convergent.

Proof We only show $\int_1^{+\infty} \frac{\sin x}{x}\,\mathrm{d}x$ is conditionally convergent, and the latter statement can be similarly proved.

Recalling that $\int_1^{+\infty} \frac{\sin x}{x}\,\mathrm{d}x$ is convergent by Cauchy criterion, we now prove it is not absolutely convergent. Noticing that

$$\int_1^{+\infty} \frac{\sin^2 x}{x}\,\mathrm{d}x = \int_1^{+\infty} \frac{1-\cos 2x}{2x}\,\mathrm{d}x = \int_1^{+\infty} \frac{1}{2x}\,\mathrm{d}x - \int_1^{+\infty} \frac{\cos 2x}{2x}\,\mathrm{d}x,$$

where the first integral on the right side is divergent and the second is convergent, we see that

$$\int_1^{+\infty} \frac{\sin^2 x}{x}\,\mathrm{d}x$$

is divergent. Furthermore, since

$$\left|\frac{\sin x}{x}\right| \geqslant \left|\frac{\sin^2 x}{x}\right| = \frac{\sin^2 x}{x} \quad (1 \leqslant x < +\infty),$$

by the comparison test,

$$\int_1^{+\infty}\left|\frac{\sin x}{x}\right|dx \text{ is divergent.} \square$$

Theorem 7.5.5 (The limit test) Suppose $f(x)(\geqslant 0)$ and $g(x)$ are defined on $[a, +\infty)$ and integrable on any finite interval $[a, A]$ $(A > a)$.

(1) Suppose

$$\lim_{x\to+\infty}\frac{f(x)}{g(x)} = r.$$

If $0 \leqslant r < +\infty$, then the convergence of $\int_a^{+\infty} g(x)dx$ implies the convergence of $\int_a^{+\infty} f(x)dx$; if $0 < r \leqslant +\infty$, then the divergence of $\int_a^{+\infty} g(x)dx$ implies the divergence of $\int_a^{+\infty} f(x)dx$.

(2) Suppose

$$\lim_{x\to+\infty}\frac{f(x)}{\frac{1}{x^p}} = \lim_{x\to+\infty} x^p f(x) = r.$$

If $0 \leqslant r < +\infty$ and $p > 1$, then $\int_a^{+\infty} f(x)dx$ is convergent; if $0 < r \leqslant +\infty$ and $p \leqslant 1$, then $\int_a^{+\infty} f(x)dx$ is divergent.

Proof (1) Assume

$$\lim_{x\to+\infty}\frac{f(x)}{g(x)} = r \in (0, +\infty).$$

Take $\varepsilon = r/2$. Then there exists $M > 0$ such that if $x > M$,

$$\frac{r}{2} = r - \varepsilon < \frac{f(x)}{g(x)} < r + \varepsilon = \frac{3}{2}r,$$

i. e.

$$\frac{r}{2}g(x) < f(x) < \frac{3r}{2}g(x).$$

Thus, by the comparison test (1), $\int_a^{+\infty} g(x)\mathrm{d}x$ is convergent if and only if $\int_a^{+\infty} f(x)\mathrm{d}x$ is convergent.

If $r=0$, take $\varepsilon=1/2$. Then there exists $M>0$ such that if $x>M$,

$$\frac{f(x)}{g(x)}<\varepsilon=\frac{1}{2}, \text{ i. e. } f(x)<\frac{1}{2}g(x).$$

So, by the comparison test (1), the convergence of $\int_a^{+\infty} g(x)\mathrm{d}x$ implies the convergence of $\int_a^{+\infty} f(x)\mathrm{d}x$.

For the case $r=+\infty$, it can be similarly proved that the divergence of $\int_a^{+\infty} g(x)\mathrm{d}x$ implies the divergence of $\int_a^{+\infty} f(x)\mathrm{d}x$.

(2) Taking $g(x)=1/x^p$ in (1), we have the result (2) immediately. □

Examples 1. Determine the convergence of the following improper integrals

(1) $\int_0^{+\infty} \frac{x^{3/2}}{1+x^2}\mathrm{d}x$; (2) $\int_1^{+\infty} \frac{1}{x\sqrt{1+x^2}}\mathrm{d}x$;

(3) $\int_1^{+\infty} \frac{\arctan x}{x}\mathrm{d}x$; (4) $\int_0^{+\infty} \frac{x^2}{x^4+x^2+1}\mathrm{d}x$;

(5) $\int_1^{+\infty} \frac{\mathrm{d}x}{x\cdot\sqrt[3]{x^2+1}}$.

Solution (1) Since $\lim\limits_{x\to+\infty} x\cdot\frac{x^{3/2}}{1+x^2}=+\infty$ $(\lambda=1)$, it is divergent;

(2) Since $\lim\limits_{x\to+\infty} x^2\cdot\frac{1}{x\sqrt{1+x^2}}=1$ $(\lambda=2)$, it is convergent;

(3) Since $\lim\limits_{x\to+\infty} x\cdot\frac{\arctan x}{x}=\pi/2$ $(\lambda=1)$, it is divergent;

(4) Since $\lim\limits_{x\to+\infty} x^2 \cdot \dfrac{x^2}{x^4+x^2+1} = 1 \quad (\lambda = 2)$, it is convergent;

(5) Since $\lim\limits_{x\to+\infty} x^{5/3} \cdot \dfrac{1}{x \cdot \sqrt[3]{x^2+1}} = 1 \quad (\lambda = 5/3)$, it is convergent.

2. Prove $\int_1^{+\infty} x^a e^{-x} dx$ is convergent for any constant a.

Proof It is easy to check that

$$\lim_{x\to+\infty} \frac{x^a e^{-x}}{x^{-2}} = \lim_{x\to+\infty} x^a x^2 e^{-x} = 0.$$

Since $\int_1^{+\infty} x^{-2} dx$ is convergent, so $\int_1^{+\infty} x^a e^{-x} dx$ is convergent.

As we have noticed that the comparison test and the limit test are valid for $f(x) \geqslant 0$. Now we consider the tests for more general cases, i. e. $f(x)$ may not keep its sign on $[a, +\infty)$. First, we introduce an important identity, which is called *Abel Transformation* (Abel, 1802-1829), by which we further present the second mean value theorem of integral.

Abel tranformation Let α_i, $\beta_i (i=1, 2, \cdots, m)$ be two groups of numbers:

$$A_p = \sum_{i=1}^{p} \alpha_i \quad (p = 1, 2, \cdots, m).$$

Then

$$\sum_{i=1}^{m} \alpha_i \beta_i = A_m \beta_m + \sum_{i=1}^{m-1} A_i (\beta_i - \beta_{i+1}).$$

Proof As $\alpha_1 = A_1$, $\alpha_2 = A_2 - A_1$, $\alpha_3 = A_3 - A_2$, $\cdots$, $\alpha_m = A_m - A_{m-1}$,

$$\begin{aligned}\sum_{i=1}^{m} \alpha_i \beta_i &= A_1\beta_1 + (A_2 - A_1)\beta_2 + (A_3 - A_2)\beta_3 + \cdots + (A_m - A_{m-1})\beta_m \\ &= A_1(\beta_1 - \beta_2) + A_2(\beta_2 - \beta_3) + \cdots + A_{m-1}(\beta_{m-1} - \beta_m) + A_m \beta_m. \quad \square\end{aligned}$$

Theorem 7.5.6 (The second mean value theorem of integral)

Suppose $g(x)$ is integrable on $[a, b]$.

(1) If $f(x)$ is decreasing and non-negative on $[a, b]$. Then, there exists $\xi \in [a, b]$ such that

$$\int_a^b f(x)g(x)\mathrm{d}x = f(a)\int_a^\xi g(x)\mathrm{d}x;$$

(2) if $f(x)$ is increasing and non-negative on $[a, b]$. Then, there exists $\xi \in [a, b]$ such that

$$\int_a^b f(x)g(x)\mathrm{d}x = f(b)\int_\xi^b g(x)\mathrm{d}x.$$

Proof We only prove the first case and the second case can be proved similarly. Let

$$G(x) = \int_a^x g(t)\mathrm{d}t.$$

Then $G(x)$ is continuous on $[a, b]$. Let $M = \sup\limits_{x\in[a, b]} G(x)$ and $m = \inf\limits_{x\in[a, b]} G(x)$, let $T: a = x_0 < x_1 < \cdots < x_n = b$ be a partition of $[a, b]$ and let $M_i = \sup\limits_{x\in[x_{i-1}, x_i]} f(x)$ and $m_i = \inf\limits_{x\in[x_{i-1}, x_i]} f(x)$. As $f(x)$ is integrable on $[a, b]$, $\lim\limits_{\|T\|\to 0} \sum\limits_{i=1}^n (M_i - m_i)\Delta x_i = 0$. So, by the inequality

$$\sum_{i=1}^n \int_{x_{i-1}}^{x_i} |f(x) - f(x_{i-1})| |g(x)| \mathrm{d}x \leqslant \sup |g(x)| \sum_{i=1}^n (M_i - m_i)\Delta x_i,$$

we have

$$\lim_{\|T\|\to 0} \sum_{i=1}^n \int_{x_{i-1}}^{x_i} |f(x) - f(x_{i-1})| |g(x)| \mathrm{d}x = 0.$$

Furthermore,

$$I = \int_a^b f(x)g(x)\mathrm{d}x = \lim_{\|T\|\to 0} \sum_{i=1}^n \int_{x_{i-1}}^{x_i} f(x_{i-1})g(x)\mathrm{d}x$$

$$= \lim_{\|T\| \to 0} \sum_{i=1}^{n} f(x_{i-1})(G(x_i) - G(x_{i-1})). \qquad (*)$$

Then, by Abel Transformation,

$$\sum_{i=1}^{n} f(x_{i-1})(G(x_i) - G(x_{i-1})) = \sum_{i=1}^{n-1} (f(x_i) - f(x_{i-1}))G(x_i) + f(x_{n-1})G(b).$$

Since $f(x)$ is decreasing and non-negative,

$$mf(a) \leqslant \sum_{i=1}^{n-1} (f(x_i) - f(x_{i-1}))G(x_i) + f(x_{n-1})G(b) \leqslant Mf(a).$$

Combining (*), we have

$$mf(a) \leqslant \int_a^b f(x)g(x)\mathrm{d}x \leqslant Mf(a).$$

If $f(a) = 0$, $\int_a^b f(x)g(x)\mathrm{d}x = 0$, and so ξ may take on any number; if $f(a) > 0$,

$$m \leqslant \frac{1}{f(a)} \int_a^b f(x)g(x)\mathrm{d}x \leqslant M.$$

By the continuity of $G(x)$, there exists $\xi \in [a, b]$ such that

$$\frac{1}{f(a)} \int_a^b f(x)g(x)\mathrm{d}x = G(\xi) = \int_a^{\xi} g(x)\mathrm{d}x,$$

i. e.

$$\int_a^b f(x)g(x)\mathrm{d}x = f(a)\int_a^{\xi} g(x)\mathrm{d}x. \ \square$$

Corollary Suppose $g(x)$ is integrable on $[a, b]$, and $f(x)$ is monotone on $[a, b]$. Then, there exists $\xi \in [a, b]$ such that

$$\int_a^b f(x)g(x)\mathrm{d}x = f(a)\int_a^{\xi} g(x)\mathrm{d}x + f(b)\int_{\xi}^b g(x)\mathrm{d}x.$$

Proof Without loss of generality, suppose $f(x)$ is increasing on

$[a, b]$. Let $F(x) = f(x) - f(a)$. Then, $F(x)$ is increasing and also non-negative. Thus, by Theorem 7.5.6, there exists $\xi \in [a, b]$ such that

$$\int_a^b F(x)g(x)\mathrm{d}x = F(b)\int_\xi^b g(x)\mathrm{d}x = (f(b) - f(a))\int_\xi^b g(x)\mathrm{d}x,$$

while

$$\int_a^b F(x)g(x)\mathrm{d}x = \int_a^b f(x)g(x)\mathrm{d}x - f(a)\int_a^b g(x)\mathrm{d}x.$$

Consequently,

$$\int_a^b f(x)g(x)\mathrm{d}x = f(a)\int_a^\xi g(x)\mathrm{d}x + f(b)\int_\xi^b g(x)\mathrm{d}x.$$

Theorem 7.5.7 (Dirichlet test) Suppose $f(x)$ is defined on $[a, +\infty)$ and integrable on any finite interval $[a, A]$. Also suppose the following two conditions are satisfied:

(1) There exists $M \geqslant 0$ such that for any $A \geqslant a$, $\left|\int_b^A f(x)\mathrm{d}x\right| \leqslant M$;

(2) $g(x)$ is decreasing and as $x \to +\infty$, $g(x)$ approaches 0.

Then the improper integral $\int_a^{+\infty} f(x)g(x)\mathrm{d}x$ is convergent.

Proof By Condition (2), for any $\varepsilon > 0$, there exists A_0 such that if $A \geqslant A_0$, $|g(A)| < \varepsilon$. Hence, if $A_1, A_2 \geqslant A_0$, by the corollary of Theorem 7.5.6,

$$\begin{aligned}\left|\int_{A_1}^{A_2} f(x)g(x)\mathrm{d}x\right| &= \left|g(A_1)\int_{A_1}^{\xi} f(x)\mathrm{d}x + g(A_2)\int_\xi^{A_2} f(x)\mathrm{d}x\right| \\ &\leqslant \left|g(A_1)\right|\left|\int_{A_1}^{\xi} f(x)\mathrm{d}x\right| + \left|g(A_2)\right|\left|\int_\xi^{A_2} f(x)\mathrm{d}x\right| \\ &\leqslant \varepsilon \cdot \left|\int_a^\xi f(x)\mathrm{d}x - \int_a^{A_1} f(x)\mathrm{d}x\right| \\ &\quad + \varepsilon \cdot \left|\int_a^{A_2} f(x)\mathrm{d}x - \int_a^\xi f(x)\mathrm{d}x\right| \leqslant 4\varepsilon M.\end{aligned}$$

Thus, by Cauchy criterion, $\int_a^{+\infty} f(x)g(x)\mathrm{d}x$ is convergent.

As an example we use Dirichlet test to show again $\int_1^{+\infty} \frac{\sin x}{x}\mathrm{d}x$ is convergent.

Example Prove the improper integral $\int_1^{+\infty} \frac{\sin x}{x}\mathrm{d}x$ is convergent.

Proof Clearly, for any $A \geqslant 1$,

$$\left|\int_1^A \frac{\sin x}{x}\mathrm{d}x\right| = |\cos A - \cos 1| \leqslant 2$$

and the function $\frac{1}{x}$ is decreasing on $[1, +\infty)$ with limit 0 as $x \to +\infty$. So by Dirichlet test, $\int_1^{+\infty} \frac{\sin x}{x}\mathrm{d}x$ is convergent. □

Theorem 7. 5. 8 (Abel test) Suppose $f(x)$ is defined on $[a, +\infty)$ and the improper integrable $\int_a^{+\infty} f(x)\mathrm{d}x$ is convergent. If $g(x)$ is monotone and bounded on $[a, +\infty)$, then the improper integral $\int_a^{+\infty} f(x)g(x)\mathrm{d}x$ is convergent.

Proof The convergence of $\int_a^{+\infty} f(x)\mathrm{d}x$ implies that $f(x)$ satisfies Condition (1) of Dirichlet test. Besides, since $g(x)$ is monotone and bounded on $[a, +\infty)$, the limit of $g(x)$ is existent as $x \to +\infty$. W. l. o. g., we may suppose that $g(x)$ has limit p as $x \to +\infty$, and so $g(x) - p$ is decreasing with limit 0. By Dirichlet test,

$$\int_a^{+\infty} f(x)(g(x) - p)\mathrm{d}x$$

is convergent. Moreover, as $\int_a^{+\infty} f(x)\mathrm{d}x$ is convergent, $\int_a^{+\infty} f(x)g(x)\mathrm{d}x$ is convergent. □

§7.5.2 Integrals of unbounded functions

At present our objective is to extend the concept of a definite integral to allow for integrands being unbounded on the interval of integration, i. e. integrands with vertical asymptotes within the interval of integration. Here are some examples:

$$\int_{-3}^{3}\frac{dx}{x^2},\ \int_{1}^{2}\frac{dx}{x-1},\ \int_{0}^{\pi}\tan x\,dx$$

Definition Let a be a real number. If for any $\delta > 0$, $f(x)$ is unbounded on $(a, a+\delta)$ or $(a-\delta, a)$ or $(a-\delta, a+\delta)$, then a is said to be a *singular point* (or *singularity*) of $f(x)$.

Definition (1) Suppose $f(x)$ is defined on $[a, b)$ such that $f(x)$ is integrable on any $[a, b-\varepsilon] \subset [a, b)$ $(\varepsilon > 0)$, where b is a singular point of $f(x)$. We denote

$$\int_{a}^{b} f(x)\,dx = \lim_{t\to b^-}\int_{a}^{t} f(x)\,dx.$$

If the limit exists, then we say the improper integral $\int_{a}^{b} f(x)\,dx$ is *convergent*, or the improper integral is existent; otherwise we say it is *divergent*.

(2) Suppose $f(x)$ is defined on $(a, b]$ such that $f(x)$ is integrable on any $[a+\varepsilon, b] \subset (a, b]$ $(\varepsilon > 0)$, where a is a singular point of $f(x)$. We denote

$$\int_{a}^{b} f(x)\,dx = \lim_{t\to a^+}\int_{t}^{b} f(x)\,dx.$$

If the limit exists, then we say the improper integral $\int_{a}^{b} f(x)\,dx$ is *convergent*, or the improper integral is existent; otherwise we say it is *divergent*.

(3) Suppose $f(x)$ is defined on $[a, c) \cup (c, b]$ with c being a singular point of $f(x)$. We denote

$$\int_a^b f(x)\mathrm{d}x = \lim_{s \to c^-}\int_a^s f(x)\mathrm{d}x + \lim_{t \to c^+}\int_t^b f(x)\mathrm{d}x.$$

If the two limits both exist, then we say the improper integral $\int_a^b f(x)\mathrm{d}x$ is *convergent*, or the improper integral is existent; otherwise we say it is *divergent*.

Each of above integrals is called *improper integral of unbounded function*. If it is convergent, then the corresponding limit value is called the *value of the improper integral*.

Remark (1) Let $F(x)$ is an antiderivative of $f(x)$. Denote

$$F(b-0) = \lim_{x \to b^-} F(x) \text{ and } F(a+0) = \lim_{x \to a^+} F(x).$$

Then the above definitions can be rewritten respectively as follows, which may be regarded as the generalization of Newton-Leibniz formula in improper integrals of unbounded function:

$$① \int_a^b f(x)\mathrm{d}x = F(b-0) - F(a) = F(x)\Big|_a^{b^-};$$

$$② \int_a^b f(x)\mathrm{d}x = F(b) - F(a+0) = F(x)\Big|_{a^+}^{b};$$

$$③ \int_a^b f(x)\mathrm{d}x = (F(c-0) - F(a)) + (F(b) - F(c+0))$$

$$= F(x)\Big|_a^{c^-} + F(x)\Big|_{c^+}^{b}.$$

(2) The above definition expressions can also be rewritten respectively in terms of $\varepsilon(>0)$:

$$① \int_a^b f(x)\mathrm{d}x = \lim_{\varepsilon \to 0}\int_a^{b-\varepsilon} f(x)\mathrm{d}x;$$

$$② \int_a^b f(x)\mathrm{d}x = \lim_{\varepsilon \to 0}\int_{a+\varepsilon}^{b} f(x)\mathrm{d}x;$$

③ $\int_a^b f(x)\mathrm{d}x = \lim_{\varepsilon_1 \to 0}\int_a^{c-\varepsilon_1} f(x)\mathrm{d}x$

$+ \lim_{\varepsilon_2 \to 0}\int_{c+\varepsilon_2}^{b} f(x)\mathrm{d}x \quad (\varepsilon_1, \varepsilon_2 > 0).$

Example Discuss the convergence of the following improper integrals

① $\int_0^1 \frac{\mathrm{d}x}{\sqrt{1-x^2}}$; ② $\int_a^b \frac{\mathrm{d}x}{(x-a)^q} \quad (q>0,\ b>a)$;

③ $\int_0^1 \frac{\mathrm{d}x}{x^q} \quad (q>0)$; ④ $\int_0^3 \frac{\mathrm{d}x}{(x-1)^{2/3}}$.

Solution ① Note $x=1$ is a singularity, and so

$$\int_0^1 \frac{\mathrm{d}x}{\sqrt{1-x^2}} = \lim_{\varepsilon \to 0}\int_0^{1-\varepsilon} \frac{\mathrm{d}x}{\sqrt{1-x^2}} = \lim_{\varepsilon \to 0} \arcsin(1-\varepsilon)$$

$$= \arcsin 1 = \frac{\pi}{2},$$

i. e. it is convergent with value $\frac{\pi}{2}$.

② Note $x=a$ is a singularity. Let $t \in (a, b)$.

$$\int_t^b f(x)\mathrm{d}x = \begin{cases} \frac{1}{1-q}[(b-a)^{1-q} - (t-b)^{1-q}] & (q \neq 1), \\ \ln(b-a) - \ln(t-a) & (q = 1). \end{cases}$$

Then

$$\lim_{t \to a^+}\int_t^b f(x)\mathrm{d}x = \begin{cases} \frac{(b-a)^{1-q}}{1-q} & (q<1), \\ +\infty & (q \geqslant 1). \end{cases}$$

So, if $0<q<1$, the integral is convergent with value $\frac{(b-a)^{1-q}}{1-q}$; if $q \geqslant 1$, the integral is divergent. Note that if $q \leqslant 0$, $\int_a^b \frac{\mathrm{d}x}{(x-a)^q}$ is a proper integral.

③ just by ②, we see that if $0 < q < 1$, it is convergent with value $\frac{1}{1-q}$; if $q \geqslant 1$, it is divergent.

④ Note $x = 1$ is a singularity. Hence

$$\begin{aligned}\int_0^3 f(x)\mathrm{d}x &= \int_0^1 f(x)\mathrm{d}x + \int_1^3 f(x)\mathrm{d}x \\ &= \lim_{\varepsilon_1 \to 0}\int_0^{1-\varepsilon_1} f(x)\mathrm{d}x + \lim_{\varepsilon_2 \to 0}\int_{1+\varepsilon_2}^3 f(x)\mathrm{d}x \\ &= \lim_{\varepsilon_1 \to 0}[3(x-1)^{1/3}]\Big|_0^{1-\varepsilon_1} + \lim_{\varepsilon_2 \to 0}[(3(x-1)^{1/3})]\Big|_{1+\varepsilon_2}^3 \\ &= 3 + 3\sqrt[3]{2},\end{aligned}$$

i. e. it is convergent with value $3 + 3\sqrt[3]{2}$.

Just as the tests for the improper integral over unbounded interval, we have the similar results for that of unbounded functions, and we will present them without proof as follows.

Theorem 7. 5. 9 (Cauchy criterion) The improper integral $\int_a^b f(x)\mathrm{d}x$ is convergent ($x = a$ is a singularity) if and only if for any $\varepsilon > 0$, there exists $\delta > 0$ such that

$$\left|\int_{a+\varepsilon_1}^{a+\varepsilon_2} f(x)\mathrm{d}x\right| < \varepsilon \text{ whenever } 0 < \varepsilon_1 < \delta \text{ and } 0 < \varepsilon_2 < \delta.$$

Theorem 7. 5. 10 (The comparison test) Suppose $f(x)(\geqslant 0)$ and $g(x)$ are defined on $(a, b]$ and integrable on any $[a + \varepsilon, b]$ $(\varepsilon > 0)$.

(1) If $f(x) \leqslant g(x)$ and $\int_a^b g(x)\mathrm{d}x$ is convergent, then $\int_a^b f(x)\mathrm{d}x$ is convergent;

(2) If

$$0 \leqslant f(x) \leqslant \frac{c}{(x-a)^p} \quad (0 < p < 1,\ c \text{ is some constant}),$$

then $\int_a^b f(x)\mathrm{d}x$ is convergent; if

$$f(x) \geqslant \frac{c}{(x-a)^p} \quad (p \geqslant 1, \ c \text{ is some constant}),$$

then $\int_a^b f(x)\mathrm{d}x$ is divergent.

Theorem 7. 5. 11 (The limit test) Suppose $f(x)(\geqslant 0)$ is defined on $(a, b]$ and integrable on any $[a+\varepsilon, b](\varepsilon > 0)$.

Suppose

$$\lim_{x \to a^+} \frac{f(x)}{\dfrac{1}{(x-a)^p}} = \lim_{x \to a^+} (x-a)^p f(x) = r.$$

If $0 \leqslant r < +\infty$ and $0 < p < 1$, then $\int_a^b f(x)\mathrm{d}x$ is convergent; if $0 < r \leqslant +\infty$ and $p \geqslant 1$, then $\int_a^b f(x)\mathrm{d}x$ is divergent.

Examples (1) Discuss the convergence of the following improper integrals

① $\int_0^1 \frac{\ln x}{\sqrt{x}}\mathrm{d}x$; ② $\int_1^2 \frac{\sqrt{x}}{\ln x}\mathrm{d}x$; ③ $\int_0^{\pi/2} \frac{\ln(\sin x)}{\sqrt{x}}\mathrm{d}x$.

Solution ① Note $x = 0$ is the singularity and

$$\lim_{x \to 0^+} x^{3/4} \frac{\ln x}{\sqrt{x}} = \lim_{x \to 0^+} \frac{\ln x}{x^{-1/4}} = \lim_{x \to 0^+} \frac{1/x}{-\frac{1}{4}x^{-\frac{1}{4}-1}} = 0.$$ So, it is convergent.

② Note $x = 1$ is the singularity and

$$\lim_{x \to 1^+} (x-1) \frac{\sqrt{x}}{\ln x} = \lim_{x \to 1^+} \sqrt{x} \cdot \lim_{x \to 1^+} \frac{x-1}{\ln[1+(x-1)]} = 1.$$

So, it is divergent.

③ Note $x = 0$ is the singularity and

$$\lim_{x\to 0^+} x^{5/6} \cdot \frac{\ln(\sin x)}{\sqrt{x}} = \lim_{x\to 0^+} x^{1/3}\ln(\sin x) = \lim_{x\to 0^+} \frac{\ln(\sin x)}{x^{-1/3}}$$

$$= \lim_{x\to 0^+} \frac{\cos x/\sin x}{-\dfrac{1}{3}x^{-4/3}}$$

$$= -3\lim_{x\to 0^+}\left(\frac{x}{\sin x}\cdot x^{1/3}\cdot \cos x\right)$$

$= 0$. So, it is convergent.

(2) Discuss the convergence of the improper integral: $\int_0^1 x^{a-1}(1-x)^{b-1}\mathrm{d}x \quad (a<1,\ b<1)$.

Solution Let's study the following two improper integrals:

$$I_1 = \int_0^{1/2} x^{a-1}(1-x)^{b-1}\mathrm{d}x,\ I_2 = \int_{1/2}^1 x^{a-1}(1-x)^{b-1}\mathrm{d}x.$$

The function $x^{a-1}(1-x)^{b-1}$ is continuous, but unbounded on $(0, \frac{1}{2}]$, and

$$\lim_{x\to 0^+} x^{1-a}x^{a-1}(1-x)^{b-1} = 1.$$

So, by Theorem 7.5.11, I_1 is convergent if $0 < 1-a < 1$, i.e. $0<a<1$, and divergent if $1-a\geqslant 1$, i.e. $a\leqslant 0$.

Similarly, $x^{a-1}(1-x)^{b-1}$ is continuous, but unbounded on $\left[\frac{1}{2}, 1\right)$, and

$$\lim_{x\to 1^-}(1-x)^{1-b}x^{a-1}(1-x)^{b-1} = 1.$$

So, by Theorem 7.5.11, I_2 is convergent if $0<b<1$, and divergent if $b\leqslant 0$.

In summary, the improper integral $\int_0^1 x^{a-1}(1-x)^{b-1}\mathrm{d}x$ $(a<1,\ b<1)$ is convergent if and only if $a>0$ and $b>0$.

In the above improper integral, if we regard a and b as independent variables (a and b are not necessarily < 1), then $\beta(a, b) = \int_0^1 x^{a-1}(1-x)^{b-1}\mathrm{d}x$ defines a function of a and b, called *β-function* (pronounced as Beta function), which will have many applications in practice.

An improper integral over unbounded interval is also called *improper integral of type* I. An improper integral of unbounded function is also called *improper integral of type* II. An integral of unbounded function may be also over unbounded interval. Such integral is called *improper integral of mixed type*. It is clear that an improper integral of mixed type is convergent if and only if each type in it is convergent. The following problems are the examples for the improper integral of mixed type.

(3) Discuss the convergence of the improper integral

$$\int_0^{+\infty} \frac{\ln(1+x)}{x^b}\mathrm{d}x.$$

Solution Notice $\int_0^{+\infty} = \int_0^1 + \int_1^{+\infty} = I_1 + I_2$, where I_1 is of unbounded function and I_2 is over unbounded interval.

I_1: Since

$$\lim_{x\to 0^+} x^{b-1} \cdot \frac{\ln(1+x)}{x^b} = \lim_{x\to 0^+} \frac{\ln(1+x)}{x} = 1,$$

we see that I_1 is convergent only if $b-1 < 1$, i. e. $b < 2$;

I_2: If $b > 1$, there exists $a > 0$ such that $b - a > 1$, and

$\lim\limits_{x\to+\infty} x^{b-a} \cdot \dfrac{\ln(1+x)}{x^b} = \lim\limits_{x\to+\infty} \dfrac{\ln(1+x)}{x^a} = 0$, and so I_2 is convergent;

if $b \leqslant 1$, $\lim\limits_{x\to+\infty} x^b \cdot \dfrac{\ln(1+x)}{x^b} = +\infty$, and so I_2 is divergent.

In summary, the integral is convergent if and only if $1 < b < 2$.

(4) Discuss the convergence of the improper integral

$$\int_0^{+\infty} \frac{dx}{x^p + x^q}.$$

Solution 1 Notice $\int_0^{+\infty} = \int_0^1 + \int_1^{+\infty} = I_1 + I_2$, where I_1 is an improper integral of unbounded function and I_2 is that over unbounded interval.

I_1: Since

$$\lim_{x \to 0^+} x^{\min\{p, q\}} \cdot \frac{1}{x^p + x^q} = 1,$$

I_1 is convergent if and only if $\min\{p, q\} < 1$;

I_2: Since

$$\lim_{x \to +\infty} x^{\max\{p, q\}} \cdot \frac{1}{x^p + x^q} = 1,$$

I_2 is convergent if and only if $\max\{p, q\} > 1$.

In summary, the integral is convergent if and only if $\min\{p, q\} < 1 < \max\{p, q\}$.

Solution 2 Without loss of generality, let $p \leqslant q$. Also $\int_0^{+\infty} = \int_0^1 + \int_1^{+\infty}$. For

$$\int_0^1 = \int_0^1 \frac{dx}{x^p(1 + x^{q-p})},$$

we see that $\int_0^1$ is convergent if $p < 1$, and $\int_0^1$ is divergent if $p \geqslant 1$; for

$$\int_1^{+\infty} = \int_1^{+\infty} \frac{dx}{x^q(1 + x^{p-q})},$$

we see that $\int_1^{+\infty}$ is convergent if $q > 1$, and $\int_1^{+\infty}$ is divergent if $q \leqslant 1$. So, the integral is convergent if and only if $\min\{p, q\} < 1 < \max\{p, q\}$.

(5) Let $f(x)$ be continuous on $(0, +\infty)$ with $\lim_{x\to+\infty} f(x) = l$ for some real number l. Suppose the integral

$$\int_0^A \frac{f(x)}{x}\mathrm{d}x$$

is convergent for any $A > 0$. Verify that for any $a > b > 0$,

$$\int_0^{+\infty} \frac{f(ax) - f(bx)}{x}\mathrm{d}x = l \ln \frac{a}{b}.$$

Proof Note that

$$\begin{aligned}
\int_0^A \frac{f(ax)-f(bx)}{x}\mathrm{d}x &= \int_0^A \frac{f(ax)}{x}\mathrm{d}x - \int_0^A \frac{f(bx)}{x}\mathrm{d}x \\
&= \int_0^{aA} \frac{f(x)}{x}\mathrm{d}x - \int_0^{bA} \frac{f(x)}{x}\mathrm{d}x \\
&= \int_0^{aA} \frac{f(x)}{x}\mathrm{d}x + \int_{bA}^{0} \frac{f(x)}{x}\mathrm{d}x = \int_{bA}^{aA} \frac{f(x)}{x}\mathrm{d}x \\
&= \int_{bA}^{aA} \left(\frac{f(x)}{x} - \frac{l}{x}\right)\mathrm{d}x + \int_{bA}^{aA} \frac{l}{x}\mathrm{d}x \\
&= \int_{bA}^{aA} \frac{f(x)-l}{x}\mathrm{d}x + l \cdot \ln \frac{a}{b}.
\end{aligned}$$

Since $\lim_{x\to+\infty} f(x) = l$, for any $\varepsilon > 0$, there exists $M > 0$ such that $|f(x) - l| < \varepsilon$ whenever $x > M$. Now take $M_1 = M/b$. Then if $A > M_1$, $aA > bA > M$ and so

$$\begin{aligned}
\left|\int_0^A \frac{f(ax)-f(bx)}{x}\mathrm{d}x - l \cdot \ln \frac{a}{b}\right| &= \left|\int_{bA}^{aA} \frac{f(x)-l}{x}\mathrm{d}x\right| \\
&\leqslant \int_{bA}^{aA} \frac{|f(x)-l|}{x}\mathrm{d}x \\
&< \int_{bA}^{aA} \frac{\varepsilon}{x}\mathrm{d}x = \varepsilon \cdot \int_{bA}^{aA} \frac{1}{x}\mathrm{d}x \\
&= \varepsilon \cdot \ln \frac{a}{b}. \ \square
\end{aligned}$$

There is a close connection between the improper integral over unbounded interval and that of unbounded function. Suppose the function $f(x)$ in $\int_a^b f(x)\mathrm{d}x$ has a singularity a. By a transformation $y=\frac{1}{x-a}$, we have

$$\int_a^b f(x)\mathrm{d}x=\int_{\frac{1}{b-a}}^{+\infty}\frac{f\left(a+\frac{1}{y}\right)}{y^2}\mathrm{d}y,$$

the latter of which is the improper integral over unbounded interval. Through such a transformation, we may yield Dirichlet test and Abel test for the improper integral of unbounded function as follows:

Theorem 7.5.12 (Dirichlet test) Suppose $f(x)$ is defined on $(a, b]$ and integrable on $[a+\varepsilon, b]$ for any $\varepsilon>0$. Also suppose the following two conditions are satisfied:

(1) There exists $M\geqslant 0$ such that $\left|\int_{a_1}^b f(x)\mathrm{d}x\right|\leqslant M$ for any $a_1\in(a, b)$;

(2) $g(x)$ is monotone on $(a, b]$ and as $x\to a^+$, $g(x)$ approaches 0.

Then the improper integral $\int_a^b f(x)g(x)\mathrm{d}x$ is convergent.

Theorem 7.5.13 (Abel test) Suppose $f(x)$ is defined on $(a, b]$ and integrable on $[a+\varepsilon, b]$ for any $\varepsilon>0$. Also suppose the following two conditions are satisfied:

(1) $\int_a^b f(x)\mathrm{d}x$ is convergent;

(2) $g(x)$ is monotone and also bounded on $(a, b]$.

Then the improper integral $\int_a^b f(x)g(x)\mathrm{d}x$ is convergent.

Similar to the improper integral over unbounded interval, we may also introduce the concept of absolute convergence and conditional convergence for the improper integral of unbounded function.

Example Discuss the convergence of the integral

$$I=\int_0^1 \frac{\sin\frac{1}{x}}{x^r}\mathrm{d}x,$$

where $0<r\leqslant 2$.

Solution If $0<r<1$,

$$\left|\frac{\sin\frac{1}{x}}{x^r}\right|\leqslant\frac{1}{x^r}, \text{ and so } I \text{ is absolutely convergent.}$$

If $1\leqslant r<2$, for any $\eta>0$,

$$\left|\int_\eta^1 \frac{\sin\frac{1}{x}}{x^2}\right|\leqslant\left|\left(\cos\frac{1}{x}\right)\Big|_\eta^1\right|=\left|\cos 1-\cos\frac{1}{\eta}\right|\leqslant 2,$$

and

$$\int_0^1 \frac{\sin\frac{1}{x}}{x^r}\mathrm{d}x=\int_0^1 \frac{1}{x^{r-2}}\cdot\frac{\sin\frac{1}{x}}{x^2}\mathrm{d}x.$$

Let $g(x)=\frac{1}{x^{r-2}}$. Then $g(x)$ is decreasing and approaches 0, and so by Dirichlet test we conclude I is convergent; we can show similarly that

$$\int_0^1 \frac{\cos\frac{2}{x}}{2x^r}\mathrm{d}x$$

is convergent. Noting

$$\left|\frac{\sin\frac{1}{x}}{x^r}\right|\geqslant\frac{\sin^2\frac{1}{x}}{x^r}=\frac{1}{2x^r}-\frac{\cos\frac{2}{x}}{2x^r},$$

and

$$\int_0^1 \frac{1}{2x^r}\mathrm{d}x$$

is divergent, we see that

$$\int_0^1 \left|\frac{\sin\frac{1}{x}}{x^r}\right| \mathrm{d}x$$

is divergent, i. e. if $1 \leqslant r < 2$, I is conditionally convergent.

If $r = 2$,

$$\int_0^1 \frac{\sin\frac{1}{x}}{x^2}\mathrm{d}x = \lim_{\eta\to 0^+}\int_\eta^1 \frac{\sin\frac{1}{x}}{x^2}\mathrm{d}x = \lim_{\eta\to 0^+}\left(\cos 1 - \cos\frac{1}{\eta}\right)$$

is not existent. So,

$$\int_0^1 \frac{\sin\frac{1}{x}}{x^2}\mathrm{d}x$$

is divergent.

In summary, if $0 < r < 1$, I is absolutely convergent; if $1 \leqslant r < 2$, I is conditionally convergent; if $r = 2$, I is divergent.

§ 7.6 Numerical integration

As we mentioned at the end of Chapter 6, not all integrals can be computed in the closed form in terms of the elementary functions. Thus, it becomes necessary to use approximation methods to calculate the value of the definite integral. Recall that the definite integral is defined as a limit of Riemann sums, so any Riemann sum could be used as an approximation to the integral. Some of the simplest numerical methods of integration are introduced below.

Let us take a partition of $[a, b]$ into n subintervals of equal length $\Delta x = (b - a)/n$. Then we have

$$\int_a^b f(x)\mathrm{d}x \approx \sum_{i=1}^{n} f(x_i^*)\Delta x,$$

where x_i^* is any point in the ith subinterval $[x_{i-1}, x_i]$ of the partition. if x_i^* is chosen to be the left endpoint of the interval, then $x_i^* = x_{i-1}$ and in this case we have the *left endpoint approximation*:

$$\int_a^b f(x)\mathrm{d}x \approx L_n = \sum_{i=1}^{n} f(x_{i-1})\Delta x \tag{1}$$

if x_i^* is chosen to be the right endpoint of the interval, then $x_i^* = x_i$ and thus we have the *right endpoint approximation*:

$$\int_a^b f(x)\mathrm{d}x \approx R_n = \sum_{i=1}^{n} f(x_i)\Delta x. \tag{2}$$

Another approximation, called *Trapezoidal rule*, results from averaging the left endpoint approximation and the right endpoint approximation:

$$\begin{aligned}\int_a^b f(x)\mathrm{d}x \approx T_n &= \frac{1}{2}\left(\sum_{i=1}^{n} f(x_{i-1})\Delta x + \sum_{i=1}^{n} f(x_i)\Delta x\right) \\ &= \frac{\Delta x}{2}(f(x_0) + 2f(x_1) + 2f(x_2) + \cdots \\ &\quad + 2f(x_{n-1}) + f(x_n))\end{aligned} \tag{3}$$

The reason for the name Trapezoidal rule can be seen from the fact that if $f(x) \geqslant 0$, the area of the trapezoid that lies above the ith interval $[x_{i-1}, x_i]$ is

$$\Delta x\left(\frac{f(x_{i-1}) + f(x_i)}{2}\right) = \frac{\Delta x}{2}(f(x_{i-1}) + f(x_i)).$$

If we choose x_i^* to be the midpoint $\overline{x_i} = \frac{1}{2}(x_{i-1} + x_i)$ of the interval $[x_{i-1}, x_i]$, then we have the *midpoint approximation*:

$$\int_a^b f(x)\,dx \approx M_n = \sum_{i=1}^{n} f(\overline{x_i})\Delta x. \tag{4}$$

Example Using (1) the Trapezoidal rule and (2) the midpoint approximation with $n = 5$ to approximate the integral $\int_1^2 \frac{1}{x}dx$.

Solution (1) With $n = 5$, $a = 1$ and $b = 2$, we have $\Delta x = \frac{1}{5}$, and so the Trapezoidal rule gives

$$\begin{aligned}\int_1^2 \frac{1}{x}dx \approx T_5 &= \frac{0.2}{2}(f(1) + 2f(1.2) + 2f(1.4) \\ &\quad + 2f(1.6) + 2f(1.8) + f(2)) \\ &= 0.1\left(\frac{1}{1} + \frac{2}{1.2} + \frac{2}{1.4} + \frac{2}{1.6} + \frac{2}{1.8} + \frac{1}{2}\right) \\ &\approx 0.695\,635.\end{aligned}$$

(2) The midpoints of the five intervals are 1.1, 1.3, 1.5, 1.7 and 1.9, and so the midpoint approximation gives

$$\begin{aligned}\int_1^2 \frac{1}{x}dx \approx M_5 &= \frac{1}{5}(f(1.1) + f(1.3) + f(1.5) + f(1.7) + f(1.9)) \\ &= \frac{1}{5}\left(\frac{1}{1.1} + \frac{1}{1.3} + \frac{1}{1.5} + \frac{1}{1.7} + \frac{1}{1.9}\right) \approx 0.691\,908.\end{aligned}$$

Another rule for approximate integration, called *Simpson rule*, results from using parabolas instead of straight line segments to approximate a curve. Without further explanations we now present Simpson rule briefly as follows:

$$\begin{aligned}\int_a^b f(x)\,dx \approx S_n = \frac{\Delta x}{3}(f(x_0) + 4f(x_1) + 2f(x_2) + 4f(x_3) + \cdots \\ + 2f(x_{n-2}) + 4f(x_{n-1}) + f(x_n))\end{aligned} \tag{5}$$

where n is an even positive integer and $\Delta x = (b - a)/n$.

Example Using Simpson rule with $n = 10$ to approximate

$\int_1^2 \frac{1}{x} dx.$

Solution With $f(x) = \frac{1}{x}$, $n = 10$ and $\Delta x = 0.1$ in (5), we obtain that

$$\int_1^2 \frac{1}{x} dx \approx S_{10} = \frac{\Delta x}{3}(f(1) + 4f(1.1) + 2f(1.2) + 4f(1.3) + \cdots$$
$$+ 2f(1.8) + 4f(1.9) + f(2))$$
$$= \frac{0.1}{3}\left(\frac{1}{1} + \frac{4}{1.1} + \frac{2}{1.2} + \frac{4}{1.3} + \frac{2}{1.4} + \frac{4}{1.5} + \frac{2}{1.6}\right.$$
$$\left. + \frac{4}{1.7} + \frac{2}{1.8} + \frac{4}{1.9} + \frac{1}{2}\right)$$
$$\approx 0.693\,150.$$

Exercises

1. Evaluate $\int_0^\beta (ax + b) dx$ by definition.

2. Prove $\int_a^b k\, dx = k(b - a)$ by definition.

3. Answer the following questions and give your reasons:

(1) Suppose $|f(x)|$ is integrable on $[a, b]$. Is $f(x)$ also integrable on $[a, b]$?

(2) Suppose $f^2(x)$ is integrable on $[a, b]$. Is $f(x)$ also integrable on $[a, b]$?

(3) Suppose $f^3(x)$ is integrable on $[a, b]$. Is $f(x)$ also integrable on $[a, b]$?

4. Suppose $f(x)$ is continuous on $[a, b]$ such that $\int_a^b f^2(x) dx = 0$. Prove: $f(x) \equiv 0$ on $[a, b]$.

5. Suppose $f(x)$ is integrable on $[a, b]$ and $\int_a^b f(x) dx \neq 0$.

Prove that there exists $\xi \in (a, b)$ such that $\int_a^{\xi} f(x)\mathrm{d}x = \int_{\xi}^{b} f(x)\mathrm{d}x$. What if $\int_a^b f(x)\mathrm{d}x = 0$?

6. Let $f(x) (\geqslant 0)$ be an integrable function on $[a, b]$ and $\int_a^b f(x)\mathrm{d}x = 1$. Prove:

$$\left(\int_a^b f(x)\cos kx \,\mathrm{d}x\right)^2 + \left(\int_a^b f(x)\sin kx \,\mathrm{d}x\right)^2 \leqslant 1.$$

7. Let $f(x) (\geqslant 0)$ be a continuous function on $[a, b]$. Prove:

$$\lim_{n\to+\infty}\left(\int_a^b f^n(x)\mathrm{d}x\right)^{\frac{1}{n}} = \max_{x\in[a, b]} f(x).$$

8. Let $f(x)$ be an integrable function on $[a, b]$. Prove:

$$\lim_{n\to+\infty}\int_0^{\frac{\pi}{2}} f(x) \mid \sin nx \mid \mathrm{d}x = \frac{2}{\pi}\int_0^{\frac{\pi}{2}} f(x)\mathrm{d}x.$$

9. Suppose $f(y)$ and $y = \varphi(x)$ are both integrable. Is it always true that the composite function $f(\varphi(x))$ is integrable?

10. Prove: If $f(x)$ is integrable on $[a, b]$, then $f(x)$ has infinitely many continuous points on $[a, b]$.

11. Prove: If $f(x)$ is integrable on $[a, b]$ and $\int_a^b f(x)\mathrm{d}x > 0$, then there exists an subinterval $[\alpha, \beta] \subseteq [a, b]$ such that $f(x) > 0$ on $[\alpha, \beta]$.

12. Suppose $f(x)$ is integrable on $[a, b]$. Prove: $\int_a^b f^2(x)\mathrm{d}x = 0$ if and only if $f(x) = 0$ at any continuous point x of $f(x)$ on $[a, b]$.

13. Express the following limits of sums as definite integrals:

(1) $\lim\limits_{n\to+\infty}\left(\frac{1}{n+1} + \frac{1}{n+2} + \cdots + \frac{1}{n+n}\right)$;

(2) $\lim\limits_{n\to+\infty}\left(\frac{1}{\sqrt{4n^2-1}} + \frac{1}{\sqrt{4n^2-2^2}} + \cdots + \frac{1}{\sqrt{4n^2-n^2}}\right)$;

(3) $\lim\limits_{n\to+\infty} \frac{1}{n}\left(f\left(a+\frac{h}{n}\right)+f\left(a+\frac{2h}{n}\right)\cdots+f\left(a+\frac{nh}{n}\right)\right)$.

where $h=b-a$ and $f(x)$ is integrable on $[a, b]$.

14. Express the following definite integrals as the limits of the Riemann sums:

(1) $\int_a^b 2x\,dx$;　　(2) $\int_0^1 \frac{1}{1+x^2}dx$;

(3) $\int_0^{\pi} \sin x\,dx$;　　(4) $\int_a^b (f(x)+g(x))dx$.

15. Using the definition of the definite integral to find the following limits:

(1) $\lim\limits_{n\to+\infty} n\left(\frac{1}{n^2+1^2}+\frac{1}{n^2+2^2}+\cdots+\frac{1}{n^2+n^2}\right)$;

(2) $\lim\limits_{n\to+\infty} \frac{1}{n}\left(\sin\frac{\pi}{n}+\sin\frac{2\pi}{n}+\cdots+\sin\frac{n-1}{n}\pi\right)$;

(3) $\lim\limits_{n\to+\infty} \left(\frac{1}{\sqrt{n^2}}+\frac{1}{\sqrt{n(n+1)}}+\cdots+\frac{1}{\sqrt{n(2n-1)}}\right)$.

16. Prove:

(1) $\ln(n+1)<1+\frac{1}{2}+\cdots+\frac{1}{n}<1+\ln n$;

(2) $\lim\limits_{n\to+\infty} \frac{1+\frac{1}{2}+\cdots+\frac{1}{n}}{\ln n}=1$.

17. Suppose $f(x)$ is a continuous function. Prove:

(1) $\int_0^{\frac{\pi}{2}} f(\sin x)dx=\int_0^{\frac{\pi}{2}} f(\cos x)dx$;

(2) $\int_0^{\pi} xf(\sin x)dx=\frac{\pi}{2}\int_0^{\pi} f(\sin x)dx$;

(3) $\int_1^a f\left(x^2+\frac{a^2}{x^2}\right)\frac{dx}{x}=\int_1^a f\left(x+\frac{a^2}{x}\right)\frac{dx}{x}$.

18. Assume $f'(x)$ is continuous. Find $\frac{d}{dx}\int_a^x (x-t)f'(t)dt$, and using this result to find $\frac{d}{dx}\int_0^x (x-t)\cos t\,dt$.

19. Identify the larger in the following pairs:

(1) $\int_3^4 (\ln x)^2 dx$ and $\int_3^4 (\ln x)^3 dx$;

(2) $\int_{-2}^{-1} \left(\frac{1}{3}\right)^x dx$ and $\int_0^1 3^x dx$;

(3) $\int_0^1 x\,dx$ and $\int_0^1 x^2 dx$;

(4) $\int_0^{\frac{\pi}{2}} x\,dx$ and $\int_0^{\frac{\pi}{2}} \sin x\,dx$.

20. Prove the following inequalities:

(1) $1 < \int_0^{\frac{\pi}{2}} \frac{\sin x}{x} dx < \frac{\pi}{2}$;

(2) $\frac{3}{e^4} < \int_{-1}^{2} e^{-x^2} dx < 3$;

(3) $1 < \int_0^1 e^{x^2} dx < e$;

(4) $\frac{\pi}{2} < \int_0^{\frac{\pi}{2}} \frac{1}{\sqrt{1-\frac{1}{2}\sin^2 x}} dx < \frac{\pi}{\sqrt{2}}$.

21. Prove:

(1) $\lim\limits_{n\to+\infty}\int_0^1 \frac{x^n}{1+x} dx = 0$;   (2) $\lim\limits_{n\to+\infty}\int_0^{\frac{\pi}{2}} \sin^n x\,dx = 0$.

22. Suppose $f(x)$ is continuous on $[a, b]$ and $F(x) = \int_0^x f(t)(x-t)dt$. Show that $F''(x) = f(x)$.

23. Suppose $f(x)$ and $g(x)$ are both integrable on $[a, b]$. Prove that $M(x) = \max\limits_{x\in[a,b]}\{f(x), g(x)\}$ and $m(x) = \min\limits_{x\in[a,b]}\{f(x), g(x)\}$ are also integrable on $[a, b]$.

24. Suppose $f(x)$ is integrable on $[a, b]$ and $|f(x)| \geqslant m > 0$

for any $x \in [a, b]$. Prove $\dfrac{1}{f(x)}$ is also integrable on $[a, b]$.

25. Suppose $f''(x)$ is continuous on $[a, b]$. Prove

$$\int_a^b x f''(x)\mathrm{d}x = (bf'(b) - f(b)) - (af'(a) - f(a)).$$

26. Find the following derivatives:

(1) $\dfrac{\mathrm{d}}{\mathrm{d}x}\left(\int_1^x \dfrac{\sin t}{t}\mathrm{d}t\right) \quad (x > 0)$; (2) $\dfrac{\mathrm{d}}{\mathrm{d}y}\left(\int_y^0 \sqrt{1+x^4}\,\mathrm{d}x\right)$;

(3) $\dfrac{\mathrm{d}}{\mathrm{d}x}\left(\int_0^{x^2} \dfrac{x \sin x}{1+\cos^2 x}\mathrm{d}x\right)$; (4) $\dfrac{\mathrm{d}}{\mathrm{d}x}\left(\int_x^{x^2} \mathrm{e}^{-t^2}\,\mathrm{d}t\right)$.

27. Find the following definite integrals:

(1) $\int_0^1 \mathrm{e}^{-x}\,\mathrm{d}x$; (2) $\int_0^{\pi/2} \sin x\,\mathrm{d}x$; (3) $\int_{\frac{1}{\sqrt{3}}}^{\sqrt{3}} \dfrac{1}{1+x^2}\mathrm{d}x$;

(4) $\int_{-1/2}^{1/2} \dfrac{1}{\sqrt{1-x^2}}\mathrm{d}x$; (5) $\int_{\pi/6}^{\pi/2} \cos^2 x\,\mathrm{d}x$; (6) $\int_4^9 \sqrt{x}(1+\sqrt{x})\mathrm{d}x$;

(7) $\int_0^2 |1-x|\sqrt{(x-4)^2}\,\mathrm{d}x$; (8) $\int_0^{\pi/4} \dfrac{1+\sin^2 x}{\cos^2 x}\mathrm{d}x$;

(9) $\int_0^1 \dfrac{x}{(x^2+1)^2}\mathrm{d}x$; (10) $\int_1^{\mathrm{e}} \dfrac{1+\ln x}{x}\mathrm{d}x$;

(11) $\int_0^{\pi/2} \sqrt{1-\sin 2x}\,\mathrm{d}x$; (12) $\int_2^3 \dfrac{1}{2x^2+3x-2}\mathrm{d}x$;

(13) $\int_1^2 \dfrac{1}{x+x^3}\mathrm{d}x$. (14) $\int_1^2 \dfrac{(x+1)(x^2-3)}{3x^2}\mathrm{d}x$;

(15) $\int_0^1 \left(\dfrac{x-1}{x+1}\right)^4 \mathrm{d}x$; (16) $\int_0^1 \dfrac{1+x^2}{1+x^4}\mathrm{d}x$;

(17) $\int_0^1 x^2(2-3x^2)^2\,\mathrm{d}x$; (18) $\int_0^{2\pi} x^2 \cos x\,\mathrm{d}x$;

(19) $\int_0^{\ln 2} x\mathrm{e}^{-x}\,\mathrm{d}x$; (20) $\int_0^1 \arccos x\,\mathrm{d}x$;

(21) $\int_0^1 \dfrac{x}{1+\sqrt{1+x}}\mathrm{d}x$; (22) $\int_0^{\pi} (x \sin x)^2\,\mathrm{d}x$;

(23) $\int_1^{\mathrm{e}} (x \ln x)^3\,\mathrm{d}x$; (24) $\int_0^3 \arcsin \dfrac{x}{1+x}\mathrm{d}x$;

(25) $\int_0^{2\pi}\frac{1}{(2+\cos x)(3+\cos x)}dx$; (26) $\int_0^{\pi}e^x\cos^2 x\,dx$;

(27) $I_n=\int_0^1(1-x^2)^n dx \quad (n\in\mathbb{N})$.

28. (1) Suppose that the function $y=y(x)$ is defined by the following equation:

$$\int_0^y e^{-t^2}dt+\int_0^x\cos(t^2)dt=0. \quad \text{Find } \frac{dy}{dx}.$$

(2) Suppose $f(x)$ is continuous at any x and $\int_0^x f(x)dx = x^2(1+x)$. Find $f(2)$.

(3) Let $f(x)=\int_0^x xe^{-x^2}dx$. Find extremum points of $f(x)$.

(4) Find

$$\lim_{x\to+\infty}\frac{(\int_0^x e^{t^2}dt)^2}{\int_0^x e^{2t^2}dt}.$$

29. Find the following definite integrals:

(1) $\int_4^9\frac{\sqrt{x}}{\sqrt{x}-1}dx$; (2) $\int_{\sqrt{2}/2}^1\frac{\sqrt{1-x^2}}{x^2}dx$;

(3) $\int_0^{3/4}\frac{1}{(x+1)\sqrt{x^2+1}}dx$; (4) $\int_0^{-a}\sqrt{x^2+a^2}\,dx \quad (a>0)$;

(5) $\int_0^1\sqrt{2x+x^2}\,dx$; (6) $\int_{-\sqrt{2}}^{-2}\frac{1}{x\sqrt{x^2-1}}dx$;

(7) $\int_0^1\sqrt{(1-x^2)^3}\,dx$; (8) $\int_0^{\ln 2}\sqrt{e^x-1}\,dx$;

(9) $\int_0^{\pi/4}\frac{1}{1+\sin^2 x}dx$; (10) $\int_0^1\frac{1-x^2}{1+x^2}dx$;

(11) $\int_0^1\frac{e^x-e^{-x}}{2}dx$; (12) $\int_{\frac{1}{e}}^{e}\frac{(\ln x)^2}{x}dx$;

(13) $\int_0^{\frac{\pi}{3}} \tan^2 x \, dx$;   (14) $\int_0^{\frac{\pi}{2}} \cos^5 x \sin 2x \, dx$;

(15) $\int_0^{\pi/2} \frac{\cos x}{\sin x + \cos x} dx$.

30. Find the following definite integrals by parts:

(1) $\int_0^1 x e^x \, dx$;   (2) $\int_0^{\pi} x \sin x \, dx$;   (3) $\int_0^{\pi/2} x \sin^2 x \, dx$;

(4) $\int_0^{\pi/2} e^{2t} \cos t \, dt$;   (5) $\int_{1/e}^{e} |\ln x| \, dx$;   (6) $\int_0^{\sqrt{3}} x \arctan x \, dx$.

31. Prove the following statements:

(1) Let $f(x)$ be a continuous function. Then

$$\int_0^x f(u)(x-u)du = \int_0^x \left(\int_0^u f(x)dx\right)du. \quad \text{(Hint: by parts)}$$

(2) Let $f(x)$ be a continuous function. Then

$$\int_0^a x^3 f(x^2)dx = \frac{1}{2}\int_0^{a^2} xf(x)dx \quad (a > 0).$$

(3) Let $f(x)$ be a continuous function such that $f(x)$ is symmetric about $x = T(a < T < b)$. Then

$$\int_a^b f(x)dx = 2\int_T^b f(x)dx + \int_a^{2T-b} f(x)dx.$$

(4) Suppose $f(x)$ is continuous and odd. Then $\int_0^x f(t) \, dt$ is an even function; conversely, if $f(x)$ is continuous and even, then $\int_0^x f(t)dt$ is an odd function; Also, all the antiderivatives of an odd function are even functions, while only one of the antiderivatives of an even function is an odd functions.

32. Prove the following statements:

(1) Let $f(x)$ be a continuous increasing function on $[a, b]$. Then the function

$$F(x) = \frac{1}{x-a}\int_a^x f(t)\mathrm{d}t$$

is an increasing function on (a, b).

(2) Let $f(x)$ be a continuous function on $[0, +\infty)$ and $f(x) > 0$. Then the function

$$\varphi(x) = \frac{\int_0^x tf(t)\mathrm{d}t}{\int_0^x f(t)\mathrm{d}t}$$

is a strictly increasing function on $(0, +\infty)$.

33. Suppose $f(x)$ is a continuous function on $[0, +\infty)$ and $\lim\limits_{x\to+\infty} f(x) = A$. Prove that

$$\lim_{T\to+\infty} \frac{1}{T}\int_0^T f(x)\mathrm{d}x = A.$$

34. Let $f(x)$ be a continuous function on $[a, b]$ and $f(x) > 0$. Prove that

$$\ln\left(\frac{1}{b-a}\int_a^b f(x)\mathrm{d}x\right) \geqslant \frac{1}{b-a}\int_a^b \ln(f(x))\mathrm{d}x.$$

35. Let $f(x)$ be a continuous decreasing function on $(0, +\infty)$ and $f(x) > 0$. Prove the sequence $\{a_n\}$ is convergent where

$$a_n = \sum_{k=1}^n f(k) - \int_1^n f(x)\mathrm{d}x.$$

36. Approximate the value of each of the following integrals for a given value of n and using (a) left endpoint approximation, (b) right endpoint approximation, (c) midpoint approximation and (d) Trapezoidal rule respectively:

(1) $\int_1^3 \frac{1}{x}\mathrm{d}x$, $n = 10$; (2) $\int_2^4 \frac{1}{\sqrt{x}}\mathrm{d}x$ $(n = 10)$;

(3) $\int_0^1 \frac{1}{1+\sqrt{x}} dx \quad (n=10)$; (4) $\int_1^2 \frac{1}{1+x^2} dx \quad (n=10)$;

(5) $\int_0^1 \frac{1+\sqrt{x}}{1+x} dx \quad (n=10)$; (6) $\int_0^2 x^3 dx \quad (n=10)$;

(7) $\int_0^2 (x^2-2x) dx \quad (n=10)$; (8) $\int_0^1 \sqrt{1+x^2}\, dx \quad (n=10)$;

(9) $\int_0^1 \sqrt{1+x^3}\, dx \quad (n=8)$; (10) $\int_0^1 \sqrt{1+x^4}\, dx \quad (n=8)$.

37. Find the following improper integrals of unbounded functions:

(1) $\int_0^1 \ln x\, dx$; (2) $\int_{-1}^1 \frac{dx}{\sqrt{1-x^2}}$; (3) $\int_0^1 \frac{dx}{(2-x)\sqrt{1-x}}$.

38. Identify the convergence of the following improper integrals of unbounded functions:

(1) $\int_0^1 \frac{e^x}{\sqrt{1-x^2}} dx$; (2) $\int_1^2 \frac{dx}{(x-1)^{3/2}}$; (3) $\int_1^2 \frac{dx}{(\ln x)^2}$;

(4) $\int_0^1 \frac{dx}{\sqrt[4]{1-x^4}}$; (5) $\int_0^1 \frac{\ln x}{1-x^2} dx$; (6) $\int_0^{\pi/2} \sqrt[3]{\tan x}\, dx$;

(7) $\int_0^{\pi/3} \frac{dx}{\sqrt{2\cos x - 1}}$; (8) $\int_0^1 \frac{dx}{\sqrt{(1-x^2)(1-k^2x^2)}} \quad (k^2 < 1)$.

39. Find the following improper integrals over unbounded intervals:

(1) $\int_{-\infty}^{+\infty} \frac{dx}{1+x^2}$; (2) $I_1 = \int_0^{+\infty} x e^{-x}\, dx$;

(3) $I_n = \int_0^{+\infty} x^n e^{-x} dx \quad (n=1, 2, \cdots)$.

40. Prove: the improper integral over unbounded interval

$$\int_3^{+\infty} \frac{dx}{x(\ln x)^{\lambda}}$$

is convergent if and only if $\lambda > 1$.

41. Discuss the convergence of the following improper integrals over unbounded intervals:

(1) $\int_0^{+\infty} \frac{x^{3/2}}{1+x^2}\mathrm{d}x$; (2) $\int_1^{+\infty} \frac{\mathrm{d}x}{x\sqrt[3]{x^2+1}}$; (3) $\int_0^{+\infty} \frac{\cos ax}{k^2+x^2}\mathrm{d}x$;

(4) $\int_1^{+\infty} \frac{\ln x}{x}\mathrm{d}x$; (5) $\int_1^{+\infty} \frac{\sin x\,\mathrm{d}x}{x\sqrt{2+3x^2}}$; (6) $\int_1^{+\infty} \frac{x\ln x\,\mathrm{d}x}{(1+x)^3}$;

(7) $\int_1^{+\infty} \frac{\sin x\cos x}{x^4}\mathrm{d}x$; (8) $\int_{\mathrm{e}}^{+\infty} \frac{\ln x\sin x}{x}\mathrm{d}x$;

(9) $\int_0^{+\infty} \cos(x^2)\mathrm{d}x$; (10) $\int_0^{+\infty} \frac{\sin^2 x}{x}\mathrm{d}x$;

(11) $\int_0^{+\infty} \frac{\ln(1+x)}{\sqrt[3]{1+x^4}}\mathrm{d}x$; (12) $\int_1^{+\infty} \frac{\mathrm{d}x}{\sqrt{1+x^2}\ln^2(1+x)}$;

(13) $\int_0^{+\infty} \frac{\mathrm{d}x}{1+x\,|\sin x|}$; (14) $\int_1^{+\infty} \frac{\mathrm{d}x}{\sqrt{x}\ln^2 x}$;

(15) $\int_1^{+\infty} \frac{\frac{\pi}{2}-\arctan x}{x}\mathrm{d}x$; (16) $\int_{-\infty}^{+\infty} x^2\mathrm{e}^{-x^2}\mathrm{d}x$;

(17) $\int_1^{+\infty} \left(\ln\left(1+\frac{1}{x}\right)-\frac{1}{x+1}\right)\mathrm{d}x$;

(18) $\int_{-\infty}^{+\infty} \frac{x}{\mathrm{e}^x+x^4}\mathrm{d}x$;

(19) $\int_1^{+\infty} \ln\left(\cos\frac{1}{x}+\sin\frac{1}{x}\right)\mathrm{d}x$;

(20) $\int_2^{+\infty} \left(\cos\left(\frac{\mathrm{e}^{-\frac{1}{x}}}{x}\right)-\cos\left(\frac{\mathrm{e}^{\frac{1}{x}}}{x}\right)\right)\mathrm{d}x$.

42. Calculate the following improper integrals over unbounded intervals:

(1) $\int_0^{+\infty} \frac{\mathrm{d}x}{1+x^3}$; (2) $\int_{-\infty}^{+\infty} \frac{\mathrm{d}x}{x^2+4}$; (3) $\int_{\mathrm{e}^2}^{+\infty} \frac{\mathrm{d}x}{x(\ln x)\ln^2(\ln x)}$;

(4) $\int_0^{+\infty} \mathrm{e}^{-ax}\sin x\,\mathrm{d}x$; (5) $\int_0^{+\infty} \frac{1+x^2}{1+x^4}\mathrm{d}x$; (6) $\int_0^{+\infty} \frac{\sqrt{x}}{1+x^2}\mathrm{d}x$.

43. Discuss the convergence of the following improper integrals over unbounded intervals:

(1) $\int_1^{+\infty} x^{-p} \ln x \, dx \quad (p > 0)$;

(2) $\int_0^{+\infty} \frac{x^p}{1+x^q} dx \quad (p > 0, q > 0)$;

(3) $\int_0^{+\infty} \frac{dx}{1+x^p \mid \sin x \mid^q} \quad (p > 0, q > 0)$.

44. Determine the absolute convergence and the conditional convergence of the following improper integrals:

(1) $\int_0^{+\infty} \frac{x \cos x}{1+x^2} dx$;

(2) $\int_1^{+\infty} \frac{\sin x}{\ln x} dx$;

(3) $\int_0^{+\infty} e^{\sin x} \sin(\sin x) \frac{1}{x} dx$;

(4) $\int_0^{+\infty} \frac{\sin(\sin x) \cos x}{x} dx$;

(5) $\int_1^{+\infty} \sin \frac{1}{x^2} dx$;

(6) $\int_e^{+\infty} \frac{\ln \ln x}{\ln x} \sin x \, dx$;

(7) $\int_0^{+\infty} e^{\cos x} \sin(\sin x) \frac{1}{x} dx$;

(8) $\int_1^{+\infty} \frac{\sin \frac{1}{x}}{x} dx$;

(9) $\int_0^{+\infty} \frac{\sin x}{x + \frac{\pi}{2} \sin x} dx$.

45. Suppose $f(x)$ is uniformly continuous on $[a, +\infty)$ and the improper integral $\int_a^{+\infty} f(x) \, dx$ is convergent. Prove that $\lim_{x \to +\infty} f(x) = 0$.

46. Suppose $f(x)$ is continuous on $[a, +\infty)$, and the improper integrals $\int_a^{+\infty} f(x) dx$ is convergent. Is it necessary that $\lim_{x \to +\infty} f(x) = 0$? What if there is another condition $f(x) > 0$?

47. Suppose $f(x)$ is monotonic on $[a, +\infty)$ and the improper integral $\int_0^{+\infty} f(x) dx$ is convergent. Prove that $\lim_{x \to +\infty} x f(x) = 0$.

48. Suppose $f(x)$ is continuous on $[0, +\infty)$ and $\lim_{x \to +\infty} f(x) = A \in \mathbb{R}$. Prove:

$$\lim_{\alpha\to 0^+}\left(\alpha\int_0^{+\infty} e^{-\alpha x} f(x)\,dx\right)=A.$$

49. Suppose $f(x)$ and $g(x)$ are both continuous on $[a, +\infty)$ and $\lim\limits_{x\to+\infty}\frac{g(x)}{f(x)}=1$. If the improper integral $\int_a^{+\infty} f(x)dx$ is convergent, is $\int_a^{+\infty} g(x)dx$ also convergent? Why?

50. Suppose $f(x)$ is continuous on $[a, +\infty)$ and the improper integral $\int_a^{+\infty} f(x)dx$ is convergent. Is $\int_a^{+\infty} f^3(x)dx$ also convergent? Why?

51. Suppose $f(x)\geqslant 0$ on $(-\infty, +\infty)$, integrable on any finite interval and

$$\int_{-\infty}^{+\infty} f(x)dx=\int_{-\infty}^{+\infty} x^2 f(x)dx=1,\ \int_{-\infty}^{+\infty} xf(x)dx=0.$$

Prove: for any $\alpha\leqslant 0$,

$$\int_{-\infty}^{\alpha} f(x)dx\leqslant\frac{1}{1+\alpha^2}.$$

52. Calculate the following improper integrals of unbounded functions:

(1) $\int_{-1}^{1}\frac{dx}{\sqrt{1-x^2}}$;　　(2) $\int_0^1\frac{dx}{(2-x)\sqrt{1-x}}$;

(3) $\int_0^1 x\ln^2 x\,dx$;　　(4) $\int_0^1\frac{\arcsin x}{x}dx$.

53. Determine the convergence of the following improper integrals:

(1) $\int_0^1\frac{dx}{\ln x}$;　　(2) $\int_0^1\frac{\ln x}{1-x}dx$;

(3) $\int_0^1\ln x\ln(1-x)dx$;　　(4) $\int_0^{\pi}\frac{dx}{\sqrt{\sin x}}$;

(5) $\int_0^{+\infty}\frac{\mathrm{d}x}{\sqrt{x(x-1)(x-2)}}$;　(6) $\int_0^{\frac{\pi}{2}}\frac{\ln\sin x}{\sqrt{x}}\mathrm{d}x$.

54. Discuss the convergence of the following improper integrals:

(1) $\int_0^{\frac{\pi}{2}}\frac{x^p}{\sin^q x}\mathrm{d}x\quad(p>0,\ q>0)$;

(2) $\int_0^1\frac{x^{\alpha-1}+x^{-\alpha}}{1+x}\mathrm{d}x\quad(\alpha\in\mathbb{R})$;

(3) $\int_0^1\frac{x^p}{\ln^q(1+x)}\mathrm{d}x\quad(p>0,\ q>0)$;

(4) $\int_0^{+\infty}\frac{\mathrm{e}^x-1}{x^p}\mathrm{d}x\quad(p>0)$;

(5) $\int_0^{+\infty}\frac{x^p}{1+x+x^2}\mathrm{d}x\quad(p\in\mathbb{R})$;

(6) $\int_0^{+\infty}\frac{\mathrm{d}x}{x^p(1+x^q)}\quad(p>0,\ q>0)$.

Chapter 8 Applications of definite integrals

As we see in the last chapter, the usual formal definition of the definite integral is presented as the calculation of an area – specially the area below a curve $y = f(x)$, above the x-axis, and bounded by the ordinates $x=a$ at the left and $x=b$ at the right. The basic idea of doing this is that the area is broken up into the usual thin vertical and rectangular strips of widths dx running from $x=a$ to $x=b$. The area of such a strip is altitude $f(x)$ times base dx, hence is written $f(x)dx$ and the sum of them all – and hence the total area desired – is the definite integral $\int_a^b f(x)\,dx$. It is remarkable that so many different processes of approximating total measured quantities by adding together little bits of these quantities can all be subsumed under one process, that of integration. Area, volume, length, work, pressure, moment of inertia, and the like all can be managed by such sum.

§ 8.1 Applications in geometry

§ 8.1.1 Area of plane region

At the beginning of Chapter 7 we learned that the area of a plane region bounded by a continuous curve $y=f(x)(\geqslant 0)$, the straight lines $x = a$, $x = b$ $(a < b)$ and the x-axis (a curved trapezoid) was given by

$$A=\int_a^b f(x)\mathrm{d}x.$$

If $y=f(x)$ is not always non-negative on $[a, b]$, then the area of the enclosed region is given by

$$A=\int_a^b |f(x)|\mathrm{d}x.$$

Generally, for a region which is surrounded by two continuous curves $y=f_1(x)$, $y=f_2(x)$ and the lines $x=a$, $x=b$ $(a<b)$, its area can be calculated by the formula

$$A=\int_a^b |f_2(x)-f_1(x)|\mathrm{d}x.$$

Example Suppose a plane region is bounded by a parabola $y^2=x$ and a line $x-2y-3=0$. Find its area A.

Solution The parabola is $x=y^2=g_1(y)$, the line is $x=2y+3=g_2(y)$ and their two intersection points are $(1, -1)$ and $(9, 3)$. So,

$$A=\int_{-1}^3 |g_2(y)-g_1(y)|\mathrm{d}y=\int_{-1}^3 [(2y+3)-y^2]\mathrm{d}y=10\frac{2}{3}.$$

Suppose a plane curve is described by a *parametric equation*:

$$C:\begin{cases} x=x(t) \\ y=y(t) \end{cases}, \quad (\alpha\leqslant t\leqslant\beta),$$

and also $x(t)$ and $y(t)$ are continuous on $[\alpha, \beta]$, $x'(t)>0$ (for $x'(t)<0$ or $y'(t)\neq 0$, similarly discuss). Denote $a=x(\alpha)$ and $b=y(\beta)$. Then the area of the plane region bounded by C, $x=a$, $x=b$ and the x-axis is given by:

$$A=\int_a^b |f(x)|\mathrm{d}x=\int_\alpha^\beta |y(t)|x'(t)\mathrm{d}t.$$

Example Suppose a plane region is bounded by an arch of the pendulum curve $x=a(t-\sin t)$, $y=a(1-\cos t)$ $(a>0)$ and the x-axis. Find its area A.

Solution If $t=0$ or $t=2\pi$, $y=0$. So, as t changes from 0 to 2π, the curve takes exactly one arch. Hence, we obtain

$$A=\int_0^{2\pi} a(1-\cos t)[a(t-\sin t)]'\mathrm{d}t=\int_0^{2\pi} a^2(1-\cos t)^2\mathrm{d}t=3\pi a^2.$$

Now we suppose a curve C is a *polar coordinates curve*:

$$C: r=r(\theta) \quad (\alpha\leqslant\theta\leqslant\beta),$$

where $r(\theta)$ is continuous on $[\alpha, \beta]$ and $\beta-\alpha\leqslant 2\pi$. The plane region surrounded by C and two rays $\theta=\alpha$, $\theta=\beta$ is called a *fan*, denoted by S. We now proceed to find the area of the fan. (cf. Figure 8-1)

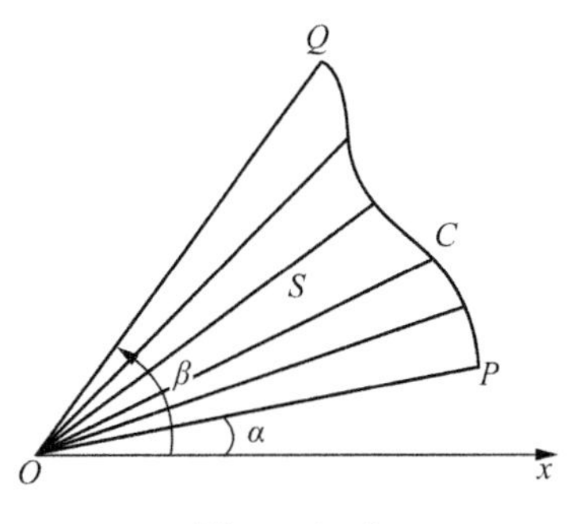

Figure 8-1

Take $n-1$ points on $[\alpha, \beta]$: θ_1, θ_2, $\cdots$, θ_{n-1} and denote $\theta_0=\alpha$, $\theta_n=\beta$. Thus we have a partition T on $[\alpha, \beta]$: $\alpha=\theta_0<\theta_1<\theta_2<\cdots<\theta_{n-1}<\theta_n=\beta$. Since $r(\theta)$ is continuous on $[\alpha, \beta]$, each small fan is approximately shaped like a small circular sector, e. g, the ith small fan S_i is approximately a small circular sector which has radius $r(\theta_i)$ and the circle-center-angle $\Delta\theta_i=\theta_i-\theta_{i-1}$, and so the area of S_i:

$$A(S_i)\approx\frac{1}{2}r^2(\theta_i)\Delta\theta_i.$$

Furthermore, we obtain the area of the fan

$$A(S)=\sum_{i=1}^{n}A(S_i)\approx\sum_{i=1}^{n}\frac{1}{2}r^2(\theta_i)\Delta\theta_i. \tag{1}$$

Let $||T||$ be the module of the partition T, i. e. $||T||=\max_i\{\Delta\theta_i\}$.

Then, by the definition of the definite integral and the integrability of a continuous function, as $||T||\to o$, the limit of the right side of the expression (1) is existent, and

$$A=\frac{1}{2}\int_{\alpha}^{\beta}r^{2}(\theta)\mathrm{d}\theta.$$

Example Let $C: r^2 = a^2\cos 2\theta$. Find the area A of the region enclosed by C.

Solution Since $r^2\geqslant 0$, θ takes on its values on $[-\frac{\pi}{4},\frac{\pi}{4}]$ and $[\frac{3\pi}{4},\frac{5\pi}{4}]$. Considering the symmetry of the region, by the above formula we have

$$A=2\times\frac{1}{2}\int_{-\frac{\pi}{4}}^{\frac{\pi}{4}}a^{2}\cos 2\theta\,\mathrm{d}\theta=a^{2}.$$

§8.1.2 Volume of solid

Suppose D is a solid with volume $V(D)$ shown in Figure 8-2. Now we try to find this $V(D)$.

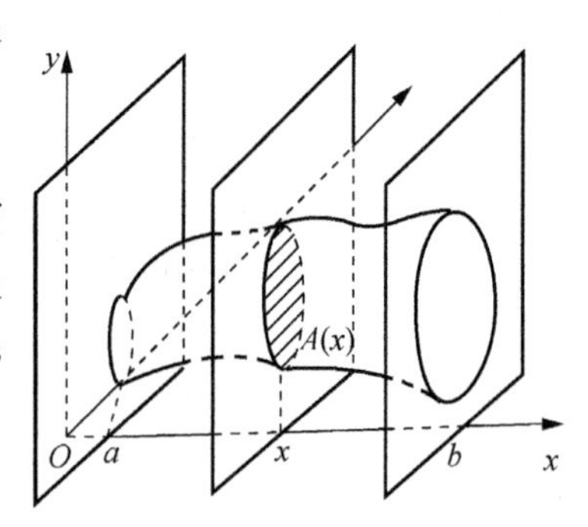

Figure 8-2

Assume at the point x of x-axis, the area of the cross section which is perpendicular to x-axis is A. Clearly, A is a function of x, denoted by $A(x)$. Let $P=\{x_0=a, x_1, x_2, \cdots, x_{n-1}, x_n=b\}$ be a partition of the interval $[a, b]$. For any $k\in\{1, 2, \cdots, n\}$, let ξ_k be an arbitrary point belonging to the subinterval $[x_{k-1}, x_k]$, and let ΔV be the volume of the part of D which is located between x_{k-1} and x_k. Then, ΔV should approximately be equal to $A(\xi_k)\Delta x_k$, where $\Delta x_k = x_k - x_{k-1}$. Since $V(D)$ is the sum of V_1, V_2, $\cdots$, V_n,

$$V(D) \approx \sum_{k=1}^{n} A(\xi_k)\Delta x_k,$$

which is exactly the Riemann sum of $A(x)$ over $[a, b]$. Denote $\lambda = \max\{\Delta x_1, \Delta x_2 \cdots, \Delta x_n\}$. Thus, by the definition of the definite integral, we have

$$\begin{aligned} V(D) &= \lim_{\lambda \to 0} \sum_{k=1}^{n} A(\xi_k)\Delta x_k \\ &= \int_a^b A(x)\mathrm{d}x. \end{aligned}$$

We may summarize the above approach to the calculation formula as: first take a little piece (here ΔV) and find its approximate expression (here $\Delta V \approx A(\xi_k)\Delta x_k$), then add them all as a Riemann sum, finally take limit to yield a definite integral.

The approach we used here is called the *approach of infinitesimal element*, which may be briefed as: first take an infinitesimal element (here the volume element $\mathrm{d}V$) and find its expression (here $\mathrm{d}V = A(x)\mathrm{d}x$), then integrate this expression on both sides to get

$$V = \int_a^b A(x)\mathrm{d}x.$$

We will see that this is also an effective way to derive calculation formulas in other applications of definite integral.

Examples Suppose a circle with changeable radius moves in this way: one point of the circle is kept on the x-axis, while the center of the circle moves one round along the circle $x^2 + y^2 = r^2 (r > 0)$. Assume the plane on which the moving circle lies is perpendicular to the x-axis. Find the volume of the solid obtained in this way.

Solution Let $x \in [-r, r]$. Then the cross section of the solid at x is composed of two disks which have a same radius $y = \sqrt{r^2 - x^2}$, and so the area of the cross section is

$$A(x) = 2 \cdot \pi y^2(x) = 2\pi(r^2 - x^2).$$

Thus, the volume of the solid

$$V = \int_{-r}^{r} A(x)\mathrm{d}x = \int_{-r}^{r} 2\pi(r^2 - x^2)\mathrm{d}x = \frac{8}{3}\pi r^3.$$

Volume of a solid of revolution

A special way to create a solid having known cross section is to revolve a region in the plane around some line in space, which is called a *solid of revolution*.

Suppose a plane region bounded by $y = f(x) > 0$, $x = a$, $x = b$ and the x-axis sweeps out a solid of revolution. To calculate the volume of this solid, we first approximate it by a finite sum of thin circular cylinders. We divide the interval $[a, b]$ into equal subintervals, each of length Δx. The radius of each cylinder is $f(x_i)$, where x_i is the right hand endpoint of the subinterval that determines that cylinder. Since the volume of a circular cylinder is given by

$$V = \pi r^2 h,$$

each of the approximating cylinders has volume

$$\pi f(x_i)^2 \Delta x.$$

The volume of the solid of revolution is approximated by the sum of the volumes of all the cylinders:

$$V \approx \sum_{i=1}^{n} \pi(f(x_i))^2 \Delta x.$$

The actual volume is the limit as the thickness of the cylinders approaches zero or the number of them approaches infinity:

$$V = \lim_{n \to +\infty} \sum_{i=1}^{n} \pi (f(x_i))^2 \Delta x = \int_a^b \pi (f(x))^2 \mathrm{d}x.$$

This process for finding the volume is called *flat-cylinder-approach*. We present this result in the following theorem:

Theorem 8. 1. 1 For a continuous function $f(x)$ defined on an interval $[a, b]$, the volume of the solid of revolution obtained by rotating about the x-axis the area under the graph of f over $[a, b]$ is given by

$$V = \int_a^b \pi (f(x))^2 \mathrm{d}x.$$

Examples (1) Suppose a disk D is defined by $x^2 + (y-b)^2 \leqslant a^2$ $(b > a > 0)$ and a solid S is obtained by rotating D one round about the x-axis. Find the volume of S.

Solution The boundary of the disk is the circle: $x^2 + (y-b)^2 = a^2$. Then the equations of the upper half circle and the lower half circle are respectively:

$$y_u = b + \sqrt{a^2 - x^2}\,;\ y_l = b - \sqrt{a^2 - x^2}.$$

Thus, the volume of S is

$$V(S) = \int_{-a}^{a} \pi (y_u^2 - y_l^2) \mathrm{d}x = 8b\pi \int_0^a \sqrt{a^2 - x^2}\, \mathrm{d}x = 2a^2 b \pi^2.$$

(2) Let C be a curve described by $y = ax^2 + bx + c$. Suppose $y > 0$ as $0 \leqslant x \leqslant 1$ and the point $(0, 0)$ is on C. Let R be a plane region surrounded by the curve C, the line $x = 1$ and the x-axis with the area $A(R) = 1/3$, and let S be a solid formed by rotating R one round about the x-axis. Determine a, b and c such that the volume of S, $V(S)$, attains its minimum and find this $V(S)$.

Solution As $(0, 0)$ is on C, the equation of C is $y = ax^2 + bx$,

and so

$$A(R) = \int_0^1 (ax^2 + bx)\mathrm{d}x = \frac{a}{3} + \frac{b}{2} = \frac{1}{3}.$$

Thus, $b = (2/3)(1-a)$, $y = ax^2 + (2/3)(1-a)x$, and

$$V(S) = \int_0^1 \pi y^2(x)\mathrm{d}x = \pi\int_0^1 [ax^2 + \frac{2}{3}(1-a)x]^2\mathrm{d}x$$
$$= \pi\left(\frac{2}{135}a^2 + \frac{1}{27}a + \frac{4}{27}\right).$$

Letting

$$\frac{\mathrm{d}V(S)}{\mathrm{d}a} = \pi\left(\frac{4}{135}a + \frac{1}{27}\right) = 0,$$

we have $a = -5/4$ and so $b = (2/3)[1 + (5/4)] = 3/2$. Since

$$\frac{\mathrm{d}^2V(S)}{\mathrm{d}a^2} = \frac{4}{135}\pi > 0,$$

$$V(S)\big|_{a=-\frac{5}{4}} = \pi\left(\frac{2}{135}a^2 + \frac{1}{27}a + \frac{4}{27}\right)\Bigg|_{a=-\frac{5}{4}} = \frac{\pi}{8} = \min.$$

In summary, $V(S)$ attains its minimum value $\pi/8$ if $a = -5/4$, $b = 3/2$ and $c = 0$.

The flat-cylinder-approach is not quite effective in finding the volume of some kinds of solid of revolution, for example, a solid of revolution which is obtained by rotating about the y-axis the region between $y = x(x-1)^2$ and $y = 0$. Now we introduce a method called *cylindrical shell approach* to solve such problems efficiently.

Theorem 8.1.2 Let R be a plane region bounded by two curves $y = y_1(x)$, $y = y_2(x)(\geqslant y_1(x))$, and two vertical lines $x = a$ and $x = b(>a)$. Let S be a solid obtained by rotating R about y-axis one round. Then the volume of S is given as follows:

$$V(S) = \int_a^b 2\pi x(y_2(x) - y_1(x))\mathrm{d}x.$$

Proof We still use the approach of infinitesimal element to derive this formula. Let $x \in [a, b]$. We consider the little piece of R which is bounded by $y_1(x)$, $y_2(x)$, $x = x$ and $x = x + \mathrm{d}x$, denoted as R_x. Then the little piece of S obtained by rotating R_x about the y-axis, denotedby S_x, is a "cylindrical shell", and the volume of the cylindrical shell is the volume element. Noting that the circumference of S_x is $2\pi x$, the height of S_x is $y_2(x) - y_1(x)$, and the thickness of S_x is $\mathrm{d}x$, we have the volume element

$$\mathrm{d}V = 2\pi x(y_2(x) - y_1(x))\mathrm{d}x,$$

which implies that

$$V(S) = \int_a^b 2\pi x(y_2(x) - y_1(x))\mathrm{d}x. \quad \square$$

Remark The cylindrical shell approach can also be used to compute the volume of a solid of revolution which is obtained by rotating about the x-axis the region between $x = x_1(y)$ and $x = x_2(y)(\geqslant x_1(y))$ from $y = a$ to $y = b$, i. e. merely by interchanging the roles of x and y in the above theorem, we have

$$V = \int_a^b 2\pi y(x_2(y) - x_1(y))\mathrm{d}y.$$

Examples (1) Find the volume of the solid obtained by rotating about the y-axis the region between $y = x$ and $y = x^2$.

Solution It is easy to check that the intersection points of $y = x$ and $y = x^2$ are $(0, 0)$ and $(1, 1)$, and $x \geqslant x^2$ on the interval $[0, 1]$. So the volume required is

$$V = \int_0^1 2\pi x(x - x^2)\mathrm{d}x = \frac{\pi}{6}.$$

(2) Find the volume of the solid obtained by rotating about the y-axis the region between $y = x(x-1)^2$ and $y = 0$.

Solution By the above theorem, we have

$$V = \int_0^1 2\pi x[x(x-1)^2]\mathrm{d}x = \frac{\pi}{15}.$$

§ 8.1.3 Arc length of plane curve

We first introduce the concept of arc length of a plane curve. The idea is to approximate the curve by inscribed polygons, a technique learned from ancient geometers. Since a straight line is the shortest path between two points, the length of any inscribed polygon should not exceed that of the curve, so the length of a curve should be an upper bound to the lengths of all inscribed polygons. Therefore, it seems natural to define the length of a curve to be the least upper bound of the lengths of all possible inscribed polygons. The definite integral can also be used to compute the length of a plane curve. Suppose we wish to compute the *arc length*, as it is usually called, of a curve $y = f(x)$ from $x = a$ to $x = b$, where $a < b$. As usual, we divide the curve into many little pieces; each one is approximately straight.

Figure 8 - 3 shows that a small change Δx in the x coordinate produces a corresponding small change in the y coordinate of approximately $\Delta y \approx f'(x)\Delta x$. The length of the little piece of the curve is then approximately

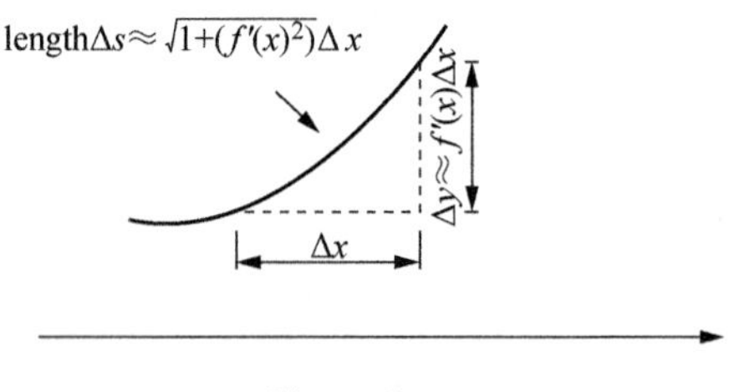

Figure 8 - 3

$$\begin{aligned}\Delta S &\approx \sqrt{(\Delta x)^2 + (\Delta y)^2} \approx \sqrt{(\Delta x)^2 + (f'(x)\Delta x)^2} \\ &= \sqrt{1 + (f'(x))^2}\,\Delta x.\end{aligned}$$

Thus, for $a<b$, the arc length of the curve $y=f(x)$ from $x=a$ to $x=b$ is approximated by a Riemann sum:

$$\text{Arc length} = S \approx \sum \sqrt{1+(f'(x))^2}\,\Delta x.$$

As we let Δx tend to zero, the sum becomes a definite integral. Since x varies between a and b, we have the following expression for the arc length:

$$\text{Arc length} = S = \int_a^b \sqrt{1+(f'(x))^2}\,\mathrm{d}x,$$

which also implies the differential of S is

$$\mathrm{d}S = \sqrt{1+(f'(x))^2}\,\mathrm{d}x = \sqrt{1+\left(\frac{\mathrm{d}y}{\mathrm{d}x}\right)^2}\,\mathrm{d}x = \sqrt{(\mathrm{d}x)^2+(\mathrm{d}y)^2},$$

which is usually called the *differential of arc*.

Example Prove the circumference S of a circle C with radius r is $2\pi r$.

Proof Note C: $x^2+y^2=r^2$. Then $2x+2yy'=0$, $y'=-x/y$. So,

$$\begin{aligned} S &= 4\int_0^r \sqrt{1+(f'(x))^2}\,\mathrm{d}x = 4\int_0^r \sqrt{1+\left(-\frac{x}{y}\right)^2}\,\mathrm{d}x \\ &= 4\int_0^r \frac{r}{\sqrt{r^2-x^2}}\,\mathrm{d}x = 4r\left(\arcsin\frac{x}{r}\right)\Big|_0^r = 4r\cdot\frac{\pi}{2} = 2\pi r. \quad \square \end{aligned}$$

Remark (1) Suppose a plane curve C is given by a parameter equation, i. e.

$$C: \begin{cases} x=\varphi(t), \\ y=\psi(t) \end{cases} \quad (\alpha \leqslant t \leqslant \beta).$$

Then, the length of the curve C is:

$$S = \int_\alpha^\beta \sqrt{\varphi'^2+\psi'^2}\,\mathrm{d}t.$$

Proof Just note that

$$dS=\sqrt{1+\left(\frac{dy}{dx}\right)^2}\,dx=\sqrt{1+\left(\frac{\psi'(t)}{\varphi'(t)}\right)^2}\,d\varphi(t)=\sqrt{\varphi'^2+\psi'^2}\,dt.\ \square$$

(2) Suppose a plane curve C is given by a polar equation, i. e.

$$C: r=r(\theta)\quad (\theta_1\leqslant\theta\leqslant\theta_2).$$

Then, the length of the curve C is:

$$S=\int_{\theta_1}^{\theta_2}\sqrt{r'^2(\theta)+r^2(\theta)}\,d\theta.$$

Proof Regarding θ as a parameter, we have

$$C:\begin{cases}x=r(\theta)\cos\theta,\\ y=r(\theta)\sin\theta\end{cases}\quad (\theta_1\leqslant\theta\leqslant\theta_2).$$

Then

$$dx=(r'(\theta)\cos\theta-r(\theta)\sin\theta)\,d\theta;$$
$$dy=(r'(\theta)\sin\theta+r(\theta)\cos\theta)\,d\theta.$$

Hence

$$dS=\sqrt{(dx)^2+(dy)^2}=\sqrt{r'^2(\theta)+r^2(\theta)}\,d\theta.$$

Examples (1) Let C be a pendulum curve defined by

$$\begin{cases}x=a\ (t-\sin t)\ ,\\ y=a\ (1-\cos t),\end{cases}$$

where $a(>0)$ is a constant. Find the arc length of its first arch, i. e. $t: 0\to 2\pi$.

Solution Since $x'(t)=a(1-\cos t)$ and $y'(t)=a\sin t$,

$$dS=\sqrt{x'^2+y'^2}\,dt=a\sqrt{(1-\cos t)^2+(\sin t)^2}\,dt$$
$$=a\sqrt{2-2\cos t}\,dt=2a\left|\sin\frac{t}{2}\right|dt,$$

and so

$$S = 2a\int_0^{2\pi}\left|\sin\frac{t}{2}\right|dt = 2a\int_0^{2\pi}\sin\frac{t}{2}dt = -4a\left[\cos\frac{t}{2}\right]\Big|_0^{2\pi} = 8a.$$

(2) Let C be a heart curve defined by $r = a(1+\cos\theta)$ $(0 \leqslant \theta \leqslant 2\pi)$, where $a(>0)$ is a constant. Find its circumference.

Solution As $r'(\theta) = -a\sin\theta$,

$$\begin{aligned} dS &= a\sqrt{(1+\cos\theta)^2 + (-\sin\theta)^2}\,d\theta \\ &= a\sqrt{2+2\cos\theta}\,d\theta = 2a\left|\cos\frac{\theta}{2}\right|d\theta, \end{aligned}$$

and so by symmetry we have

$$S = 2\int_0^{\pi} 2a\left|\cos\frac{\theta}{2}\right|d\theta = 4a\int_0^{\pi}\cos\frac{\theta}{2}d\theta = 8a.$$

§8.1.4 Curvature of plane curve

Now we consider the curvature of a plane curve. We can see intuitively that a curve changes direction slowly when it is fairly straight, but it changes direction more quickly when it bends or twists more sharply. As shown in Figure 8-4, the length of arc $\overparen{PQ}$ is approximately equal to that of arc $\overparen{P_1Q_1}$, but the degrees of their bending are quite different. Actually, as a point moves from point P to point Q along the curve, the angle formed by the corresponding tangent line is $\Delta\alpha$, which is greater than $\Delta\beta$. The curvature of a smooth curve at a given point is a measure of how quickly the curve changes direction at that point. Specifically, we will define it to be the magnitude of the rate of change of the increment of the direction angle with respective to the increment of the arc length.

Assume a plane curve C is given by a parametric equation: $x = x(t)$, $y = y(t)$. Suppose $\alpha(t)$ represents the slope angle of the tangent line of a smooth curve C at the point $P(x(t), y(t))$, and $\Delta\alpha = \alpha(t+\Delta t) - \alpha(t)$ *represents the increment of the slope angle of the tangent line when a point moves from point* P to point Q. Denote ΔS as the arc length of the arc $\overset{\frown}{PQ}$, and

$$\overline{K} = \frac{\Delta\alpha}{\Delta S},$$

which is called the *average curvature* of the arc $\overset{\frown}{PQ}$.

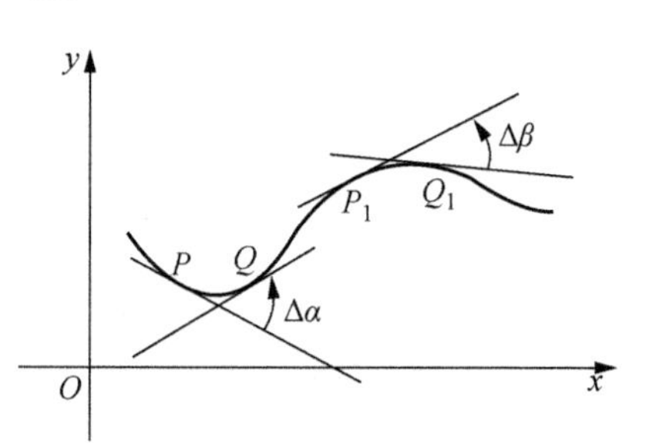

Figure 8 - 4

If $\lim\limits_{\Delta t \to 0} \frac{\Delta\alpha}{\Delta S}$ exists, then the absolute value of the limit is called the *curvature* of the curve C at point Q, which is denoted as

$$K = \left| \frac{\mathrm{d}\alpha}{\mathrm{d}S} \right|.$$

Since $x'(t)$ and $y'(t)$ are not both zeroes, we have

$$\alpha(t) = \arctan \frac{y'(t)}{x'(t)} \text{ or } \alpha(t) = \operatorname{arccot} \frac{x'(t)}{y'(t)}.$$

If $x''(t)$ and $y''(t)$ exist, then

$$\lim_{\Delta t \to 0} \frac{\Delta\alpha}{\Delta S} = \lim_{\Delta t \to 0} \frac{\Delta\alpha/\Delta t}{\Delta S/\Delta t} = \frac{\alpha'(t)}{S'(t)} = \frac{x'y'' - y'x''}{(x'^2 + y'^2)^{3/2}}.$$

Thus, we obtain the formula for the curvature as follows

$$K = \frac{|x'y'' - y'x''|}{(x'^2 + y'^2)^{3/2}}.$$

If the curve is described as $y = f(x)$, then

$$K = \frac{|y''|}{(1 + y'^2)^{3/2}}.$$

Example Find the points on the ellipse C: $x = a\cos t$, $y = b\sin t$ $(0 \leqslant t \leqslant 2\pi,\ a > b > 0)$ at which the curvature of C attains its maximum and minimum respectively.

Solution Since $x' = -a\sin t$, $x'' = -a\cos t$, $y' = b\cos t$, $y'' = -b\sin t$,

$$K = \frac{|x'y'' - y'x''|}{(x'^2 + y'^2)^{3/2}} = \frac{ab}{(a^2\sin^2 t + b^2\cos^2 t)^{3/2}}$$

$$= \frac{ab}{[(a^2 - b^2)\sin^2 t + b^2]^{3/2}}.$$

Since $\sin^2 t$ attains its minimum 0 as $t = 0, \pi$ and its maximum 1 as $t = \pi/2, 3\pi/2$,

$$K(0) = K(\pi) = \max = \frac{a}{b^2};\ K\left(\frac{\pi}{2}\right) = K\left(\frac{3\pi}{2}\right) = \min = \frac{b}{a^2}.$$

In the preceding example, if $a=b=R$, then the ellipse becomes a circle with radius R, and $K \equiv \frac{1}{R}$, i. e. at any point of a circle the curvature is identically equal to the reciprocal of its radius.

Let C be a smooth curve and let P be a point on C. Suppose the curvature of C at P is non-zero, i. e. $K \neq 0$. Construct a circle W with radius $\frac{1}{K}$ such that the following conditions are satisfied: (1) W passes through P; (2) W and C share a same tangent line l at P; (3) W and C are on a same side of l. Such a circle W is called the *curvature circle* of C at the point P, its radius $R = \frac{1}{K}$ is called the *curvature radius* of C at P and the center of the circle W is called the *curvature center* of C at P.

Proposition Suppose C: $y = f(x)$ is a smooth curve, $P(a, f(a))$ isa point on C, and W: $y = g(x)$ is the curvature circle of C at the point P. Then $f(a) = g(a)$, $f'(a) = g'(a)$ and $f''(a) = g''(a)$.

Proof Since the point P is on C and also on W, $f(a)=g(a)$. As C and W share a same tangent line at P, $f'(a)=g'(a)$. Noticing that C and W has a same curvature at the point P, we have

$$\frac{|f''(a)|}{(1+f'^2(a))^{3/2}}=\frac{|g''(a)|}{(1+g'^2(a))^{3/2}},$$

and so $|f''(a)|=|g''(a)|$. Furthermore, since W and C are on a same side of the tangent line, they have the same convexity at the point P. Thus, $f''(a)$ and $g''(a)$ must have a same sign, and therefore $f''(a)=g''(a)$. □

§8.1.5 Area of surface of revolution

Consider the graph of $y=f(x)$. If the upper half-plane is rotated about the x-axis, then each point on the graph has a circular path, and the whole graph sweeps out a certain surface, called a *surface of revolution*. The area of a surface of revolution can be calculated by the definite integral.

Suppose a surface of revolution is obtained by rotating the curve $\overset{\frown}{AB}$ one round about the x-axis. The area of the surface is defined as follows:

Construct an inscribed broken line $M_0M_1M_2\cdots M_{n-1}M_n$ of the curve $\overset{\frown}{AB}$ (cf. Figure 8-5), when $\lambda=\max\{|M_{i-1}M_i|: i=1, 2, \cdots,$

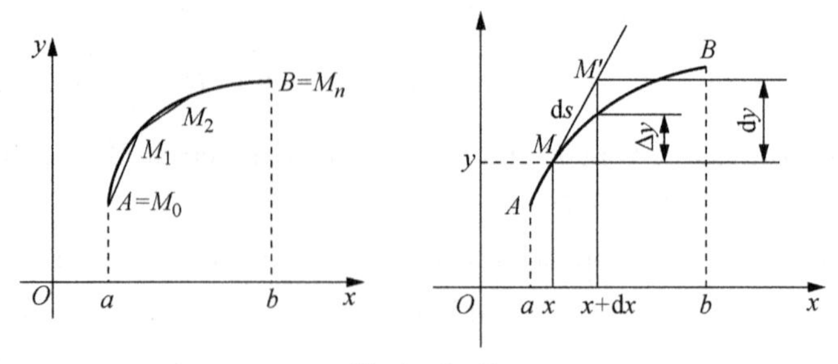

Figure 8-5

$n\} \to 0$, the limit value of the area of the surface of revolution obtained by rotating the broken line one round about the x-axis is defined as *the area of the surface of revolution obtained by rotating* $\overset{\frown}{AB}$ *about the x-axis*.

Let the equation of the curve $\overset{\frown}{AB}$ is $y = f(x) > 0 (a \leqslant x \leqslant b)$. Take arbitrarily a subinterval $[x, x+dx]$ in $[a, b]$. Let MM' be the tangent segment over $[x, x+dx]$ with length dS (the differential of arc), let $A(x)$ be the lateral area of the little circular truncated cone, and let ΔF be the area of the surface of revolution obtained by rotating the arc $\overset{\frown}{MM'}$ about the x-axis. Then, by the approach of infinitesimal element,

$$\Delta F \approx A(x) = \pi[y + (y + \mathrm{d}y)]\mathrm{d}S.$$

Neglecting the higher order infinitesimal term dydS, we have

$$\mathrm{d}F = 2\pi y \,\mathrm{d}S = 2\pi y \sqrt{1 + (y'(x))^2}\,\mathrm{d}x.$$

Therefore, the area of the surface of revolution obtained by rotating the curve $\overset{\frown}{AB}$about the x-axis is

$$F = \int_a^b 2\pi y \sqrt{1 + (y'(x))^2}\,\mathrm{d}x.$$

Remark (1) If the curve is described as $x = g(y)$ $(a \leqslant y \leqslant b)$, and the surface of revolution is obtained by rotating the curve one round about the y-axis, then the formula for surface area becomes

$$F = \int_a^b 2\pi x \sqrt{1 + (x'(y))^2}\,\mathrm{d}y = \int_a^b 2\pi g(y) \sqrt{1 + \left(\frac{\mathrm{d}x}{\mathrm{d}y}\right)^2}\,\mathrm{d}y.$$

(2) If the curve $\overset{\frown}{AB}$ is given by a parameter equation $x = x(t)$, $y = y(t)$ $(\alpha \leqslant t \leqslant \beta)$, and the surface of revolution is obtained by

rotating the curve one round about the x-axis, then the formula for surface area is

$$F = \int_{\alpha}^{\beta} 2\pi y(t) \sqrt{x'^2(t) + y'^2(t)}\, dt.$$

Readers may derive a corresponding calculation formula without much difficulty if the curve $\overset{\frown}{AB}$ is given by a polar coordinate equation.

Examples (1) Prove that the area of a sphere with radius r is $4\pi r^2$.

Proof Clearly, the sphere is exactly the surface of revolution obtained by rotating the upper half circle $y = \sqrt{r^2 - x^2}$ one round about the x-axis. So, the area of the sphere

$$\begin{aligned} F &= 2\pi \int_{-r}^{r} y \sqrt{1 + (y'(x))^2}\, dx \\ &= 2\pi \int_{-r}^{r} \sqrt{r^2 - x^2} \sqrt{1 + \frac{x^2}{r^2 - x^2}}\, dx \\ &= 2\pi \int_{-r}^{r} r\, dx = 4\pi r^2. \square \end{aligned}$$

(2) The arc of the parabola $y = x^2$ from (1, 1) to (2, 4) is rotated about the y-axis. Find the area of the resulting surface.

Solution Noting $x = \sqrt{y}$ and $\dfrac{dx}{dy} = \dfrac{1}{2\sqrt{y}}$ we have the area of the resulting surface

$$\begin{aligned} F &= \int_{1}^{4} 2\pi g(y) \sqrt{1 + \left(\frac{dx}{dy}\right)^2}\, dy = \int_{1}^{4} 2\pi \sqrt{y} \sqrt{1 + \frac{1}{4y}}\, dy \\ &= \pi \int_{1}^{4} \sqrt{4y + 1}\, dy = \frac{\pi}{6}(17\sqrt{17} - 5\sqrt{5}). \end{aligned}$$

(3) Let C be a curve given by $x = r\cos^3 t$, $y = r\sin^3 t$ (a star curve) (cf. Figure 8-6), and let S be the surface of revolution obtained by rotating C one round about the x-axis. Find the area of S.

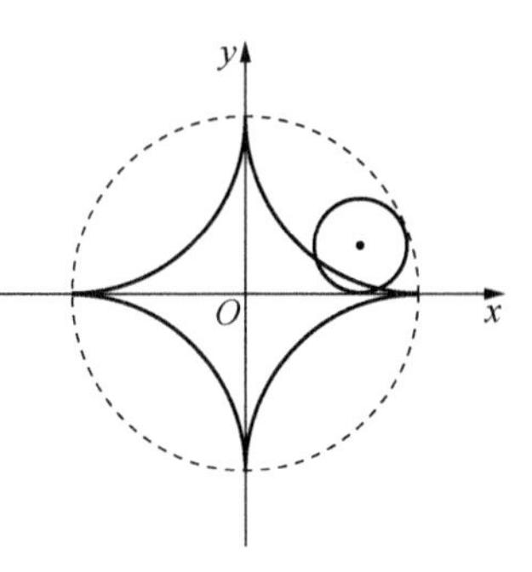

Figure 8-6

Solution Let $A(S)$ be the area of S. By the symmetry of S,

$$A(S) = \int_{\alpha}^{\beta} 2\pi y(t) \sqrt{x'^2(t) + y'^2(t)}\, dt$$

$$= 2 \times 2\pi \int_0^{\frac{\pi}{2}} r \sin^3 t \sqrt{(3r\cos^2 t \sin t)^2 + (3r \sin^2 t \cos t)^2}\, dt$$

$$= 12\pi r^2 \int_0^{\frac{\pi}{2}} \sin^4 t \cos t\, dt = \frac{12}{5}\pi r^2.$$

§ 8.2 Applications in physics

§ 8.2.1 Work

As we know, provided the force f exerted on an object is a constant, i. e. $f=c$, and the object moves from a point a to another point b along a straight line under the exertion of the force, the work done by the force is $W = c(b-a)$, i. e. work=force × distance. If the force is measured in newtons and the distance in meters, then the unit for the work is a newton-meter, which is called a joule (J); if the force is measured in pounds and the distance in feet, then the unit for the work is a foot-pound (ft-lb), which is about 1.36 J.

Now we suppose the force is not necessarily constant. Let

$$P = \{x_0, x_1, \cdots, x_{n-1}, x_n\} \ (x_0 = a, x_n = b)$$

be a partition of $[a, b]$, and let ξ_k be any point in the subinterval $[x_{k-1}, x_k]$ for any $k \in \{1, 2, \cdots, n\}$. Denote $x_k - x_{k-1} = \Delta x_k$ and $||P|| = \max\{\Delta x_k\}$, the module of the partition. As the object moves from x_{k-1} to x_k, the work done by the force can be approximated by $f(\xi_k)(x_k - x_{k-1}) = f(\xi_k)\Delta x_k$, i. e. $W_k \approx f(\xi_k)\Delta x_k$. Thus, it is reasonable to think that the work done by the force as the object moves from a to b can be approximated by the sum of all these terms, i. e.

$$W \approx \sum_{k=1}^{n} f(\xi_k)\Delta x_k,$$

which is a Riemann sum. Therefore the work W should be defined as

$$W = \lim_{||P|| \to 0} \sum_{k=1}^{n} f(\xi_k)\Delta x_k = \int_a^b f(x)\mathrm{d}x.$$

Examples (1) Suppose that the work needed to stretch a spring by a length of 1 foot beyond its natural length is 10 foot-pound. How much work W is required to stretch it by a length of 2 feet beyond its natural length?

Solution Since the force exerted on the spring is described as $f(x) = kx$ by Hooke law, where x represents the length stretched beyond its natural length, by the assumption, we have $\int_0^1 kx\,\mathrm{d}x = 10$, and so $k=20$. Thus

$$W = \int_0^2 kx\,\mathrm{d}x = \int_0^2 20x\,\mathrm{d}x = 40(\text{foot - pound})^*.$$

(2) A tank has the shape of an inverted circular cone with height 10 meters and base radius 4 meters. It is filled with water to a height of 8 meters. Find the work required to empty the tank by pumping all

---

* 1 ft · pd=1 ft · 1bf=1. 355 82 J≈1. 36 J

of the water to the top of the tank.

Solution Set the vertical coordinate with the y-axis downward with the origin (0, 0) at the center of base disk of the tank. Thus the water extends from a depth of 2 m to a depth of 10 m, i. e. from $y=2$ to $y=10$. Take a thin layer of the tank at $y\in[2, 10]$ with thickness dy. Note that the layer approximated by circular cylinder with radius $r=(4/10)(10-y)$. Then its volume is

$$\pi r^2 \mathrm{d}y=\pi\left[\frac{4}{10}(10-y)\right]^2 \mathrm{d}y.$$

Noting the density of water is 1 000 kg/m^3, the work element

$$\begin{aligned}\mathrm{d}W &= g\cdot 1\,000\cdot\pi\left[\frac{4}{10}(10-y)\right]^2 \mathrm{d}y\cdot y\\ &= g\cdot 160\pi y(10-y)^2\mathrm{d}y\approx 1\,570\pi y(10-y)^2\mathrm{d}y,\end{aligned}$$

where $g\approx 9.8$ is the acceleration due to gravity, and so

$$W=\int_2^{10} 1\,570\pi y(10-y)^2\mathrm{d}y\approx 3.4\times 10^6\,(\mathrm{J}).$$

(3) Suppose a 15 - meter uniform cable with a mass of 45 kg is dangling over the edge of a tall building 10 meters high. How much work is needed to pull it up to the top of the building?

Solution We consider the cable in two portions: the first portion, which is exactly 10 meters long, is from the top of the building to the bottom of the building; the second portion, the remainder lying on the ground, is exactly 5 meters long.

Let W_1 be the work needed to pull the first portion up to the top of the building, and W_2 be for the second. Set the y-axis downward with $y=0$ for the top of the building. By the approach of infinitesimal element,

$$\mathrm{d}W_1=\left(\frac{45}{15}\cdot\mathrm{d}y\right)\cdot y=3y\,\mathrm{d}y,$$

and so

$$W_1 = \int_0^{10} 3y\,dy = 150(\text{kg}-\text{meter}).$$

Clearly, $W_2 = 5(45/15) \times 10 = 150(\text{kg}-\text{meter})$. Then, the total work needed is

$$W = W_1 + W_2 = 300(\text{kg}-\text{meter}) \approx 300 \times 9.8(\text{J}) = 2\,940(\text{J}).$$

Now we consider the total work done to move an object infinitely far from the earth. If you throw an object into the air, you expect gravity to bring it back to earth. However, the harder you throw it, the longer you expect to wait before it returns. You may wonder if it is possible to throw something hard enough so that it never returns to earth. We will calculate the total amount of work done to move an object infinitely far away from the earth. Amazingly enough, this total amount of work is finite, and it is possible to launch something hard enough that it never returns to earth.

Since we are thinking of a body moving into outer space, we have to take into account the fact that, as we move away from the earth, its gravitational force gets weaker. Newton's law of Gravitation says that the force, F, exerted on a mass m at a distance, r, from the center of the earth is given by

$$F = \frac{GMm}{r^2},$$

where $r > R$, the earth's radius, M is the mass of the earth, and G is called the gravitational constant, whose value is about 6.67×10^{-11} if mass is measured in kilograms, distance in meters, and the force in newtons.

Example Find the total work done to move an object of mass m infinitely far from the earth, and find the minimum vertical velocity that must be given to the object so that it is never drawn back to earth by gravity (which is called the *escape velocity*).

Solution By Newton's law of Gravitation, the work needed to move an object a distance dr further away is

$$\mathrm{d}W = \left(\frac{GMm}{r^2}\right)\mathrm{d}r.$$

Noting that the object should move from the surface of the earth to infinitely far away, i. e. the value r runs from R to $+\infty$, we have the total work required

$$W = \int_R^{+\infty} \frac{GMm}{r^2}\mathrm{d}r = \frac{GMm}{R}.$$

Let v be the escape velocity. Then by setting the initial kinetic energy (The kinetic energy is defined as the energy of the object by virtue of its motion and is equal to $\frac{1}{2}mv^2$) equal to the work done to escape infinitely far from the earth, we get

$$\frac{1}{2}mv^2 = \frac{GMm}{R}.$$

Solving for v gives

$$v = \sqrt{\frac{2GM}{R}} \approx \sqrt{\frac{2(6.67\times10^{-11})\times(6\times10^{24})}{6.4\times10^6}}$$
$$\approx 11\times10^3\,(\text{meters/sec}),$$

since the mass of the earth is $M\approx 6\times10^{24}$ (kg) and its radius is $R\approx 6.4\times10^6$ (meters).

§8.2.2 Moment and mass center

The *density* of a substance, e. g. air, wood, or metal, is the mass of a unit volume of the substance and is measured in, say, grams per cubic centimeter. A population density is measured in, say, people per mile along the edge of a road, or people per unit area in some region. Suppose we want to calculate the total mass or total population, but the density is not constant over a region. Then we may also use the approach of infinitesimal element to find the total:

Divide the region into small pieces in such a way that the density is approximately constant on each piece, and add up the contributions of all the pieces. That is to take the quantity element first with a differential expression, and then take definite integral over the region to find the total quantity.

Examples (1) The air density (in $\mathrm{kg/m^3}$) h meters above the earth's surface is $P = f(h)$. Find the mass of a cylindrical column of air 2 meters in diameter and 25 kilometers high.

Solution Since the air density varies with altitude but remains constant horizontally, we take horizontal slice of air for mass element, which may be obtained by multiplying its volume and its density together. Note the thickness of the slice is $\mathrm{d}h$, its volume is $\pi r^2 \mathrm{d}h = \pi 1^2 \mathrm{d}h = \pi \mathrm{d}h (\mathrm{m^3})$, and the density of the slice is roughly $f(h)$ $(\mathrm{kg/m^3})$. Thus, the mass element

$$\mathrm{d}M = \pi \mathrm{d}h \cdot f(h) = \pi f(h) \mathrm{d}h,$$

and so the total mass

$$M = \int_0^{25\,000} \pi f(h) \mathrm{d}h \quad (\mathrm{kg}).$$

(2) Suppose the population density of some circle region with a radius of 5 miles is a function from the city center. At r miles from the center, the density is $P = f(r)$ people per square mile. Write a definite integral that expresses the total population of the region.

Solution We want to slice the region up and estimate the population on each slice. If we were to take straight line slice, the population density would vary on each slice, since it depends on the distance from the city center. Therefore we take slices that are thin rings around the center, clearly on each of which the population density is pretty close to constant. Noting the circumference of the slice is $2\pi r$ and the width is dr, the area of the slice is $2\pi r dr$, and so the population element

$$\mathrm{d}N = f(r) \cdot 2\pi r \,\mathrm{d}r = 2\pi r f(r)\mathrm{d}r,$$

which implies that the total population

$$N = \int_0^5 2\pi r f(r)\mathrm{d}r.$$

The *center mass* of a mechanical system can be thought of as the balancing point. For a system consisting of n discrete masses m_i along the x-axis, each located at coordinate x_i, the center of the mass is defined by the point whose x-coordinate is $\overline{x}$ where

$$\overline{x} = \frac{\sum_{i=1}^{n} x_i m_i}{\sum_{i=1}^{n} m_i}.$$

The terms $x_i m_i$ in the numerator are called the *moments of the masses* m_i with respect to the origin, and $\sum_{i=1}^{n} x_i m_i$ is called the *moment of the system* with respect to the origin; the denominator is the total mass of the system. The moment measures the tendency of the mass or mass system to rotate about the origin. If the mechanical system is not composed of discrete masses, but is instead an object lying on the x-axis between $x=a$ and $x=b$, with mass density $\rho(x)$, in units of mass/length, then the center of mass is given by

$$\overline{x} = \frac{\int_a^b x\rho(x)\mathrm{d}x}{\int_a^b \rho(x)\mathrm{d}x}.$$

Again, the denominator is the total mass of the system.

We can see how this last formula relates to the formula for discrete masses using Riemann sums. If we divide the object into n small pieces, each of length Δx, then the ith piece has mass $m_i \approx \rho(x_i)\Delta x$ (where x_i is a point in the ith piece). Then the numerator of the formula for $\overline{x}$ is $\sum_{i=1}^{n} x_i\rho(x_i)\Delta x$, which approaches $\int_a^b x\rho(x)\mathrm{d}x$ as $n\to+\infty$. Similarly, the denominator $\sum_{i=1}^{n} m_i \approx \sum_{i=1}^{n} \rho(x_i)\Delta x$ approaches $\int_a^b \rho(x)\mathrm{d}x$ as $n\to+\infty$.

Example Find the center of mass of a 2-meter rod lying on the x-axis with its left end at the origin if:

① The density is constant and the total mass is 5 kg;

② The density is $\rho(x)=15x^2$ kg/m.

Solution ① In this case, $\rho(x)=\frac{5}{2}$ kg/m, and so

$$\overline{x} = \frac{\int_0^2 x \cdot \frac{5}{2}\mathrm{d}x}{5} = 1 \quad \text{(meter)}.$$

② Total mass $= \int_0^2 15x^2\ \mathrm{d}x = 40$ kg. Thus

$$\overline{x} = \frac{\int_0^2 x \cdot 15x^2\mathrm{d}x}{40} = \frac{3}{2} \quad \text{(meter)}.$$

Now we consider the two-dimensional case, i. e. a mass system in plane. Let a system of n masses $m_1, m_2, \cdots, m_n$ be located at the points $\{(x_i, y_i) \mid i=1, 2, \cdots, n\}$ in the xy-plane. By analogy with the one-dimensional case, we define the *moment of the system about the y-axis* to be

$$M_y = \sum_{i=1}^{n} x_i m_i$$

and the *moment of the system about the x-axis* to be

$$M_x = \sum_{i=1}^{n} y_i m_i.$$

Then, M_y measures the tendency of the system to rotate about the y-axis and M_x measures the tendency of the system to rotate about the x-axis.

As in the one dimensional case, the coordinate $(\overline{x}, \overline{y})$ of the center of the mass are given in terms of the moments by the formulas

$$\overline{x} = \frac{M_y}{m}, \overline{y} = \frac{M_x}{m},$$

where $m = \sum m_i$ is the total mass of the system. Since $m\overline{x} = M_y$ and $m\overline{y} = M_x$, the center of the mass $(\overline{x}, \overline{y})$ is the point where a single particle of mass m would have the same moments as the system.

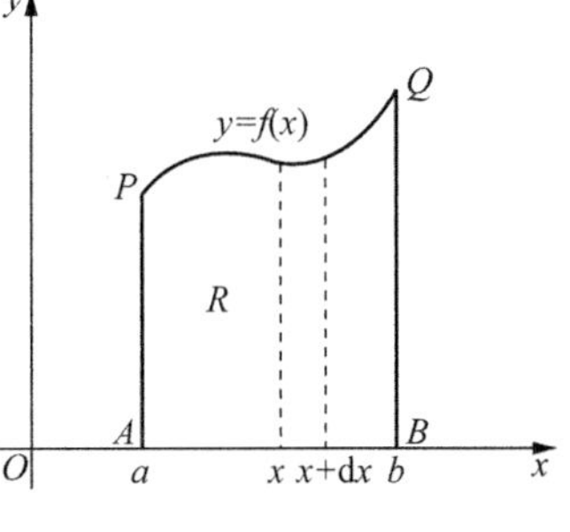

Figure 8 - 7

We next discuss the mass center of flat plate, which can be abstracted as that of a plane region. Let $R = APQB$ be a plane region (cf. Figure 8 - 7) (a curved trapezoid), bounded by the curve $\widehat{PQ}: y = f(x) (>0)$, $x = a$, $x = b (\geqslant a)$ and the x-axis. We suppose its plane density ρ is constant. (Note: the mass center of a flat plate with constant plane density is also called the *centroid*.) Then the total mass of the curved trapezoid is

$$M = \rho \cdot (\text{area of } APQB) = \rho \int_a^b f(x)\mathrm{d}x.$$

To find its total moment about coordinate axes, we may divide the region into several slim slices using a line parallel to the y-axis. Any slim slice over $[x, x+\Delta x]$ can approximately be regarded as a rectangle R_x, and so the moment elements

$$\mathrm{d}M_x = \rho \cdot \frac{y}{2} \cdot y\mathrm{d}x = \frac{1}{2}\rho y^2 \mathrm{d}x; \ \mathrm{d}M_y = \rho \cdot x \cdot y\mathrm{d}x = \rho x y \mathrm{d}x.$$

Thus, the total moments of the region with respective to x-axis and y-axis are respectively

$$M_x = \rho \int_a^b \frac{1}{2} y^2 \mathrm{d}x; \ M_y = \rho \int_a^b xy\, \mathrm{d}x.$$

Furthermore, the coordinate $(\overline{x}, \overline{y})$ of the mass center of the region can be described as

$$\begin{cases} \overline{x} = \dfrac{M_y}{m} = \dfrac{\rho \int_a^b xy\,\mathrm{d}x}{\rho \int_a^b y\,\mathrm{d}x} = \dfrac{\int_a^b xy\,\mathrm{d}x}{\int_a^b y\,\mathrm{d}x}, \\ \overline{y} = \dfrac{M_x}{m} = \dfrac{\rho \int_a^b \frac{1}{2} y^2\,\mathrm{d}x}{\rho \int_a^b y\,\mathrm{d}x} = \dfrac{\frac{1}{2}\int_a^b y^2\,\mathrm{d}x}{\int_a^b y\,\mathrm{d}x}. \end{cases}$$

In summary, the mass center of the plate (or the *centroid* of the curved trapezoid) is located at the point $(\overline{x}, \overline{y})$:

$$\begin{cases} \overline{x} = \dfrac{1}{A}\int_a^b x f(x)\mathrm{d}x, \\ \overline{y} = \dfrac{1}{A}\int_a^b \dfrac{1}{2}(f(x))^2\mathrm{d}x, \end{cases}$$

where A is the area of the region.

If the region R lies between two curves $y=f(x)$ and $y=g(x)$

$(f(x) \geqslant g(x))$ $(a \leqslant x \leqslant b)$, then in a similar manner we can derive that the centroid $(\overline{x}, \overline{y})$ of the region R is given by

$$\begin{cases} \overline{x} = \dfrac{1}{A}\displaystyle\int_a^b x(f(x)-g(x))\mathrm{d}x, \\ \overline{y} = \dfrac{1}{A}\displaystyle\int_a^b \dfrac{1}{2}(f^2(x) - g^2(x))\mathrm{d}x, \end{cases}$$

where A is the area of the region.

Examples (1) Find the centroid of a semicircular plate with radius r.

Proof Let R be the semicircular plate given by $x^2 + y^2 = r^2$ $(y \geqslant 0)$. By the symmetry, we have $\overline{x} = 0$. Since the area of the region $A(R) = \dfrac{1}{2}\pi r^2$ and

$$\frac{1}{2}\int_{-r}^{r} y^2 \mathrm{d}x = \frac{1}{2}\int_{-r}^{r}(r^2 - x^2)\mathrm{d}x = \int_0^r (r^2 - x^2)\mathrm{d}x = \frac{2}{3}r^3,$$

$$\overline{y} = \frac{\frac{2}{3}r^3}{\frac{1}{2}\pi r^2} = \frac{4r}{3\pi}.$$

Thus, we have the centroid of the semicircular plate

$$(\overline{x}, \overline{y}) = \left(0, \frac{4r}{3\pi}\right). \square$$

(2) Find the centroid of the region bounded by the curves $y = \cos x, x = \pi/2$, the x-axis and the y-axis.

Solution The area of the region is

$$A = \int_0^{\frac{\pi}{2}} \cos x \,\mathrm{d}x = 1,$$

and so

$$\overline{x} = \frac{1}{A}\int_0^{\frac{\pi}{2}} xf(x)\mathrm{d}x = \int_0^{\frac{\pi}{2}} x \cos x \,\mathrm{d}x = \frac{\pi}{2} - 1$$

and

$$\overline{y} = \frac{1}{A}\int_0^{\frac{\pi}{2}} \frac{1}{2}(f(x))^2 \mathrm{d}x = \int_0^{\frac{\pi}{2}} \frac{1}{2}\cos^2 x \,\mathrm{d}x = \frac{\pi}{8},$$

i. e. we have the centroid

$$(\overline{x}, \overline{y}) = \left(\frac{\pi}{2} - 1, \frac{\pi}{8}\right). \square$$

The next theorem, which is named after the Greek mathematician Pappus of Alexandria, shows that the centroid can be used to find the volume of a solid of revolution.

Theorem 8.2.1 (Theorem of Pappus for volume) Let R be a plane region with area A that lies entirely on one side of a line l in the plane. Let $P(\overline{x}, \overline{y})$ be the centroid of R and let r be the distance between P and l. If R is rotated about l one round, then the volume of the resulting solid is $V = 2\pi r A$.

Proof We only consider the special case in which the region lies between the curves $y = f(x)$ and $y = g(x)$ $(f(x) \geqslant g(x))$ $(a \leqslant x \leqslant b)$ and the line l is the y-axis. Other cases can be treated similarly. By the approach of cylindrical shell,

$$\begin{aligned} V &= \int_a^b 2\pi x(f(x) - g(x))\mathrm{d}x = 2\pi\int_a^b x(f(x) - g(x))\mathrm{d}x \\ &= 2\pi(\overline{x}A) = 2\pi\overline{x}A = 2\pi r A. \square \end{aligned}$$

Example Let S be a solid of revolution formed by rotating a circle of radius r with center P about a line l in the plane of the circle (Such a solid is called a *torus*). Let the distance of P and l be $R (R > r)$. Prove that the volume of S is

$$V(S) = 2\pi^2 r^2 R.$$

Proof Clearly, the area of the circle is $A = \pi r^2$. So by Theorem of Pappus, we have

$$V(S) = 2\pi R \cdot A = 2\pi^2 r^2 R. \square$$

We now discuss the moments and the centroid of a plane curve.

Theorem 8. 2. 2 Suppose a plane curve C with constant linear density ρ is described by the parameter equation

$$\begin{cases} x = x(t), \\ y = y(t) \end{cases} \quad (\alpha \leqslant t \leqslant \beta).$$

Let $P(\overline{x}, \overline{y})$ be the centroid of the curve. Then

(1) the moments

$$\begin{cases} M_x = \rho \int_\alpha^\beta y(t) \sqrt{(x'(t))^2 + (y'(t))^2}\, dt, \\ M_y = \rho \int_\alpha^\beta x(t) \sqrt{(x'(t))^2 + (y'(t))^2}\, dt; \end{cases}$$

(2)

$$\begin{cases} \overline{x} = \frac{1}{L} \int_\alpha^\beta x(t) \sqrt{(x'(t))^2 + (y'(t))^2}\, dt, \\ \overline{y} = \frac{1}{L} \int_\alpha^\beta y(t) \sqrt{(x'(t))^2 + (y'(t))^2}\, dt, \end{cases}$$

where L is the length of the curve.

Proof The idea of the proof is still the approach of infinitesimal element.

(1) Since the arc length element $ds = \sqrt{(x'(t))^2 + (y'(t))^2}\, dt$ and the mass element $dm = \rho\, ds$, we have the moment element

$$\begin{cases} dM_x = y\, dm = \rho y\, ds = \rho y \sqrt{(x'(t))^2 + (y'(t))^2}\, dt, \\ dM_y = x\, dm = \rho x\, ds = \rho x \sqrt{(x'(t))^2 + (y'(t))^2}\, dt, \end{cases}$$

from which the expression of M_x and M_y follow directly;

(2) Only note that

$$\begin{cases} \overline{x} = \dfrac{M_y}{m} = \dfrac{M_y}{\rho \cdot L}, \\ \overline{y} = \dfrac{M_x}{m} = \dfrac{M_x}{\rho \cdot L}, \end{cases}$$

where m is the total mass of the curve. Then the result just follows from the result (1). □

The following theorem shows how the centroid of a curve can be used in finding the area of a surface of revolution.

Theorem 8. 2. 3 Suppose a plane curve C with constant linear density and centroid $(\overline{x}, \overline{y})$ is described by the parameter equation

$$\begin{cases} x = x(t), \\ y = y(t) \end{cases} \quad (\alpha \leqslant t \leqslant \beta).$$

Let S_x be the surface of revolution formed by rotating the curve C about the x-axis one round. Then the area of S_x:

$$A_x = 2\pi \overline{y} L,$$

where L is the length of the curve.

Proof Note the area element $dA_x = 2\pi y ds = 2\pi y \sqrt{(x'(t))^2 + (y'(t))^2}\, dt$. Then

$$\begin{aligned} A_x &= 2\pi \int_\alpha^\beta y \sqrt{(x'(t))^2 + (y'(t))^2}\, dt \\ &= 2\pi L \cdot \frac{1}{L} \int_\alpha^\beta y \sqrt{(x'(t))^2 + (y'(t))^2}\, dt \\ &= 2\pi L \cdot \overline{y} = 2\pi \overline{y} L. \ \square \end{aligned}$$

Using the above theorem, we can derive immediately the following

Theorem 8. 2. 4 (Theorem of Pappus for area) Let C be a plane curve with length L that lies entirely on one side of a line l in the plane. Let P be the centroid of the curve C and let r be the distance between P and l. If C is rotated about l one round, then the area A of the resulting surface is

$$A = 2\pi rL.$$

§8.2.3 Hydrostatic force and pressure

In this subsection, we will study how to use the definite integral to compute the force exerted by a liquid on a surface, for example, the force of water on a dam. The idea is to get the force from the pressure. The *pressure* in a liquid is the force per unit area exerted by the liquid. By Pascal law, we know about pressure that at any point pressure is exerted equally in all directions-up, down, sideways, and pressure increases with depth.

At a depth of h feet, the pressure p, exerted by the liquid, measured in pounds per square foot, is given by computing the total weight of a column of liquid h feet high with a base of 1 square foot. The volume of such a column of liquid is just h cubic feet. If the liquid has density ρ (mass per unit volume), then its weight per unit volume is pg, where $g \approx 9.8$ is the acceleration due to gravity. The weight of the column of liquid is ρgh, so we have: Pressure = Density • g • Depth, i. e.

$$p = \rho gh.$$

Next we need to know the relation between force and pressure. Provided the pressure is constant over a given area, we have the relation: Force=Pressure • Area, i. e.

$$F = pA = \rho ghA,$$

with the corresponding units:

$$\mathrm{N} = \mathrm{Pa} \cdot \mathrm{m}^2 = \mathrm{kg/m}^3 \cdot (9.8)\mathrm{m/s}^2 \cdot \mathrm{m} \cdot \mathrm{m}^2,$$

where N=Newton, Pa=Pascal, m=meter, kg=kilogram, s=second.

When the pressure is not constant over a surface, we divide the surface into small pieces in such a way that the pressure is nearly constant on each one. Then we can use this formula on each piece and obtain a definite integral to get the force over an entire surface. Since the pressure varies with depth, we should divide the surface into horizontal strips, each of which is at an approximately constant depth. This idea is also that of the approach of infinitesimal element, which will be illustrated in the following examples:

Example Suppose a dam has the shape of the curved trapezoid $ABCD$, which is bounded by the curve $y = f(x)$, the line $x = a$, $x = b$ $(b > a)$ and the x-axis as shown in Figure 8 - 8. Let the dam be vertically submerged in liquid with constant density ρ. Find the force on the dam due to hydrostatic pressure if the liquid level is a meters above the top of the dam.

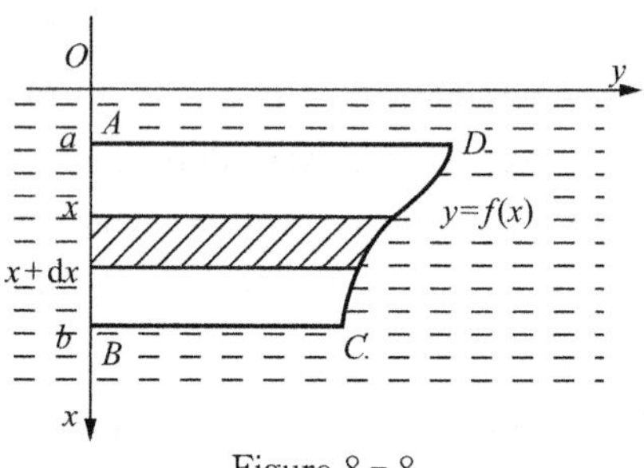

Figure 8 - 8

Solution Just as shown in the figure, we choose a vertical x-axis with origin at the surface of the liquid and the y-axis is exactly the liquid level. Let $x \in [a, b]$. Take the force element $\mathrm{d}F$ as the force on the horizontal strip between x and $x+\mathrm{d}x$. Thus

$$\mathrm{d}F = p \cdot (\mathrm{d}A) = \rho g x \cdot (\mathrm{d}A) = \rho g x \cdot (f(x)\mathrm{d}x) = \rho g x f(x)\mathrm{d}x,$$

which implies that

$$F = \int_a^b \rho g x f(x)\mathrm{d}x = \rho g \int_a^b x f(x)\mathrm{d}x.$$

Sometimes we may be given the density of the liquid as a weight per unit volume, rather than a mass per unit volume. In that case, we do not need to multiply by g because it has already be done. Since pounds are units of weight, and not mass, knowing that water weights 62.4 (lb/ft^3) tells us that for water, $\rho g = 62.4(\text{lb/ft}^3)$, and pressure at depth h is $62.4h(\text{lb/ft}^2)$.

Example A trough 14 feet long has one rectangular vertical side 14 feet by 3 feet, one rectangular inclined side 14 feet by 5 feet, and two triangular ends with sides 3 feet, 4 feet, and 5 feet (cf. Figure 8-9). If the trough is filled to the top with water, compute the force on each end and each side.

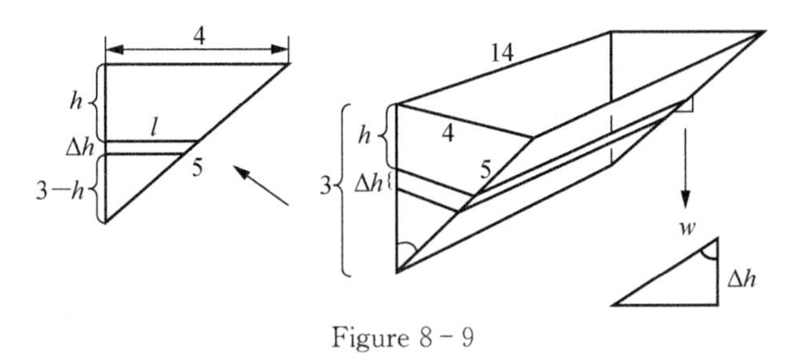

Figure 8 - 9

Solution We will do the long vertical side first. Divide the side into thin horizontal strips of width dh. Since pressure depends only on the depth h and the strips are thin and horizontal, we can assume the pressure is the same at every point in the strip. The area of the strip is $14\mathrm{d}h(\text{ft})^2$, and the force on the strip at depth h(ft) is

$$\mathrm{d}F = 62.4\ (\text{lb/ft}^2) \cdot 14\mathrm{d}h(\text{ft}^2) = 873.6h\ \mathrm{d}h(\text{lb}).$$

Thus

$$F_{\text{vertical side}} = 873.6\int_0^3 h\ \mathrm{d}h \approx 3\,931\ (\text{pounds}).$$

To compute the force on the inclined side, we use an inclined horizontal strip. Each strip has length 14, but its width is not dh. By similar triangles the width w of the strip satisfies

$$\frac{\mathrm{d}h}{w}=\frac{3}{5}, \text{ so } w=\frac{5}{3}\mathrm{d}h.$$

Therefore

$$\mathrm{d}F=(62.4h)\times 14w=(62.4h)\times 14\times\frac{5}{3}\mathrm{d}h \text{ (lb)},$$

which implies that

$$F_{\text{inclined side}}=\int_0^3 (62.4h)\times 14\times\frac{5}{3}\mathrm{d}h\approx 6\,550\text{ (pounds)}.$$

Notice that $F_{\text{inclined side}}=\frac{5}{3}\cdot F_{\text{vertical side}}$.

Finally, to compute the force on each triangular end, we again divide the end into horizontal strips. The width of the strip is $\mathrm{d}h$ as with the vertical side, but this time the length of the strip varies from 4 feet at the top of the trough to 0 feet at the bottom. By similar triangles the length l of a strip is related to the depth h by

$$\frac{l}{4}=\frac{3-h}{3}, \text{ so } l=\frac{4}{3}(3-h).$$

Thus, the area of the strip is $l\,\mathrm{d}h=\frac{4}{3}(3-h)\mathrm{d}h(\text{ft}^2)$. As before, this means the total force is

$$F_{\text{end}}=\int_0^3 (62.4h)\left[\frac{4}{3}(3-h)\right]\mathrm{d}h\approx 374\text{ (pounds)}.$$

§8.2.4 Average value of function

In practice, we often meet with the problem of finding the average value. It is easy to calculate the average value of finitely many numbers $x_1, x_2, \cdots, x_n$, $\bar{x}=\sum_{k=1}^{n} x_k/n$. But for some continuous function $f(x)$ on $[a, b]$, what should the average value of $f(x)$ on $[a, b]$ be? To answer this question, we start by introducing a partition on $[a, b]$: $P=\{x_0=a, x_1, x_2, \cdots, x_{n-1}, x_n=b\}$ such that all subintervals are of a same length $\Delta x = \Delta x_k = (b-a)/n$. Clearly, $\overline{f}_n = \frac{1}{n}\sum_{k=1}^{n} f(x_k)$ is the average value of $\{f(x_1), f(x_2), \cdots, f(x_n)\}$. Then

$$\overline{f}_n = \frac{1}{n}\sum_{k=1}^{n} f(x_k) = \frac{1}{b-a}\cdot\frac{b-a}{n}\sum_{k=1}^{n} f(x_k)$$
$$= \frac{1}{b-a}\sum_{k=1}^{n} f(x_k)\Delta x,$$

which obviously has the fashion of the Riemann sum. Note the limiting value of $\overline{f}_n$ as $n\to+\infty$ is

$$\lim_{n\to+\infty}\overline{f}_n = \frac{1}{b-a}\lim_{\Delta x\to 0}\sum_{k=1}^{n} f(x_k)\Delta x = \frac{1}{b-a}\int_a^b f(x)\mathrm{d}x,$$

So, it is reasonable for us to define the average value of $f(x)$ on $[a, b]$ as

$$\overline{f} = \frac{1}{b-a}\int_a^b f(x)\mathrm{d}x.$$

Example Find the average value of the function $f(x) = |\sin x - \cos x|$ over the interval $[0, \pi/2]$.

Solution

$$\overline{f} = \frac{1}{b-a}\int_a^b f(x)\mathrm{d}x = \frac{1}{\frac{\pi}{2}-0}\int_0^{\frac{\pi}{2}} |\sin x - \cos x|\,\mathrm{d}x$$
$$= \frac{2}{\pi}\left[\int_0^{\frac{\pi}{4}}(\cos x - \sin x)\mathrm{d}x + \int_{\frac{\pi}{4}}^{\frac{\pi}{2}}(\sin x - \cos x)\mathrm{d}x\right]$$
$$= \frac{4(\sqrt{2}-1)}{\pi}.$$

Exercises

1. Find the area of the plane region which is surrounded by the parabolas $y=x^2$ and $y = 2 - x^2$.

2. Find the area of the plane region which is surrounded by $y = |\ln x|$, $x = \frac{1}{10}$ and $x = 10$.

3. The parabola $y^2 = 2x$ separates a circle $x^2 + y^2 = 8$ into two parts. Find the ratio of the areas of these two parts.

4. Prove that the area of the ellipse $x = a \cos t$, $y = b \sin t$ $(a, b > 0)$ is $ab\pi$.

5. Find the area of the plane region which is surrounded by $x = t - t^3$, $y = 1 - t^4$.

6. Find the area of the plane region which is surrounded by $r = a \sin 3\theta$ $(a > 0)$.

7. Find the area of the plane region which is enclosed by $\sqrt{\frac{x}{a}} + \sqrt{\frac{y}{b}} = 1$ $(a, b > 0)$.

8. Find the area of the plane region which is bounded by the curve $y^2 = x^2(1 - x)$.

9. Find the area of the common part of two ellipses $\frac{x^2}{a^2} + \frac{y^2}{b^2} = 1$ and $\frac{x^2}{b^2} + \frac{y^2}{a^2} = 1$ $(a, b > 0)$.

10. Find the volume of the solid which is obtained by rotating the following plane curve one round about the given axis:

(1) $y = \sin x$ $\quad (0 \leqslant x \leqslant \pi)$, about x-axis;

(2) $x = a(t - \sin t)$, $y = a(1 - \cos t)$ $\quad (a > 0, 0 \leqslant t \leqslant 2\pi)$, about x-axis;

(3) $r = a(1 + \cos \theta)$ $\quad (a > 0)$, about polar-axis;

(4) $\frac{x^2}{a^2}+\frac{y^2}{b^2}=1$ $(a, b>0)$, about x-axis.

11. Let R be a plane region bounded by $y=\sin x$ $(0\leqslant x\leqslant\pi)$ and $y=0$. Find the volume of the solid which is obtained by rotating R one round about y-axis.

12. Find the arc lengths of the following curves:

(1) $y=x^{\frac{3}{2}}$, $0\leqslant x\leqslant 4$;

(2) $\sqrt{x}+\sqrt{y}=1$;

(3) $x=a\cos^3 t$, $y=a\sin^3 t$;

(4) $x=a(\cos t+t\sin t)$, $y=a(\sin t-t\cos t)$ $(0\leqslant x\leqslant 2\pi)$;

(5) $r=a\sin^3\frac{\theta}{3}$ $(a>0)$.

13. Find the curvatures of the following curves at the given points:

(1) $xy=4$ at point $(2, 2)$;

(2) $y=\ln x$ at point $(1,0)$;

(3) $x=a(t-\sin t)$, $y=a(1-\cos t)$ at $t=\frac{\pi}{2}$;

(4) $x=a\cos^3 t$, $y=a\sin^3 t$ at $t=\frac{\pi}{4}$.

14. Suppose a curve C is described by a polar equation $r=f(\theta)$ which is secondly derivable. Prove that the curvature of C at θ is

$$K=\frac{|r^2+2r'^2-rr''|}{(r^2+r'^2)^{3/2}},$$

and find the curvature of the circle $r=a\sin\theta$ at the point (r_1, θ_1).

15. Prove that the parabola $y=ax^2+bx+c$ attains its least curvature radius at its top point.

16. Determine the point on the curve $y=e^x$ where it attains its greatest curvature.

17. Find the areas of the surfaces of revolution obtained by rotating the following plane curve one round about the given axis:

(1) $y = \sin x$ $(0 \leqslant x \leqslant \pi)$, about x-axis;

(2) $x = a(t - \sin t)$, $y = a(1 - \cos t)$ $(a > 0,\ 0 \leqslant t \leqslant 2\pi)$, about x-axis;

(3) $r = a(1 + \cos\theta)$ $(a > 0)$, about polar-axis;

(4) $\frac{x^2}{a^2} + \frac{y^2}{b^2} = 1$ $(a,\ b > 0)$, about x-axis.

18. Find the area of the surface of revolution which is obtain by rotating the circle $x^2 + (y - R)^2 = r^2$ $(r < R)$ one round about x-axis.

19. For next problems, sketch the solid obtained by rotating each region around the indicated axis. Using the sketch, show how to approximate the volume of the solid by a Riemann sum, and hence find the volume.

(1) Region bounded by $y = x^3$, $x = 1$, $y = -1$. Axis: $y = -1$;

(2) Region bounded by $y = \sqrt{x}$, $x = 1$, $y = 0$. Axis: $x = 1$;

(3) Region bounded by the first arch of $y = \sin x$, $y = 0$. Axis: x-axis.

20. For next problems consider the region bounded by $y = e^x$, the x-axis, and the lines $x = 0$ and $x = 1$. Find the volume of the following solids.

(1) The solid obtained by rotating the region about the x-axis;

(2) The solid obtained by rotating the region about the horizontal line $y = -3$;

(3) The solid obtained by rotating the region about the horizontal line $y = 7$;

(4) The solid whose base is the given region and whose cross-sections perpendicular to the x-axis are square;

(5) The solid whose base is the given region and whose cross-sections perpendicular to the x-axis are semicircles.

21. The graph of the function $y = \cosh x = \frac{1}{2}(e^x + e^{-x})$ is called a catenary and represents the shape of a hanging cable. Find the length of this catenary between $x = -1$ and $x = 1$.

22. Find the area of a plane region which is enclosed by the curve $r = a \sin n\theta$ ($a > 0$, $n \in \mathbb{N}$).

23. For next problems, find the arc length of the given function from $x = 0$ to $x = 2$.

(1) $f(x) = \sqrt{4 - x^2}$; (2) $f(x) = \sqrt{x^3}$.

24. Find the volume of the following solids.

(1) The solid whose base is a circle with radius 2 and whose cross-sections perpendicular to the base are equilateral triangles;

(2) The solid whose base is a circle with radius 1 and whose cross-sections perpendicular to the base are squares;

(3) The solid which is obtained by rotating about the x-axis the region bounded by $y = \frac{1}{2}x^2 + 3$ and $y = 12 - \frac{1}{2}x^2$;

(4) The solid which is obtained by rotating about the x-axis the region bounded by $y = x$ and $y = x^2$.

25. Suppose a spheroid S with radius r is cut into two parts by a plane with distance h ($< r$) from the center of S. Prove that the smaller part of S has the volume

$$V = \frac{1}{3}\pi(r - h)^2(2r + h).$$

26. (1) Find the work done by a weightlifter in raising a 60-kg barbell from the floor to a height of 2 m.

(2) When a particle is at a distance x meters from the origin, a force of $\cos(\pi x/3)$ Newtons acts on it. How much work is done in moving the particle from $x=1$ to $x=2$?

(3) A spring has a natural length of 20 cm. if a 25 N force is

required to keep it stretched to a length of 30 cm. How much work is needed to stretch it from 20 cm to 25 cm?

(4) If the work required to stretch a spring 1 foot beyond its natural length is 12 foot-pound, how much work is needed to stretch it 9 inches beyond its natural length?

(5) If 6 J of work are needed to stretch a spring from 10 cm to 12 cm and another 10 J are needed to stretch it from 12 cm to 14 cm. Find the natural length of the spring.

(6) A worker on a scaffolding 75 ft above the ground needs to lift a 500 lb bucket of cement from the ground to a point 30 ft above the ground by pulling on a rope weighing 5 lb/ft. How much work is required?

27. Determine the centroid of the part of an ellipse $\frac{x^2}{a^2}+\frac{y^2}{b^2}=1$ in the first quadrant.

28. Prove: The arc length of the curve $y=\sin x\ [x\in[0,\ 2\pi]]$ is equal to the circumference of an ellipse of which the long semi-axis is $\sqrt{2}$ and the short semi-axis is 1.

29. (1) A 1 000 lb weight is being lifted to a height 10 feet off the ground. It is lifted using a rope which weighs 4 lb per foot and which is being pulled up by construction workers standing on a roof 30 feet off the ground. Find the work done to lift the weight.

(2) A rectangular water tank has length 20 ft, width 10 ft, and depth 15 ft. If the tank is full, how much work does it take to pump all the water out? (Note that 1 cubic foot of water weighs 62.4 lb.)

(3) A water tank is in the form of a right circular cylinder with height 20 ft and radius 6 ft. If the tank is half full of water, find the work required to pump all of it over the top rim. (Note that 1 cubic foot of water weighs 62.4 lb.)

(4) Suppose the tank in Problem (3) is full of water. Find the

work required to pump all of it to a point 10 ft above the top of the tank.

(5) A fuel oil tank is an upright cylinder, buried so that its circular top is 10 feet beneath ground level. The tank has a radius of 5 feet and is 15 feet high, although the current oil level is only 6 feet deep. Calculate the work required to pump all of the oil to the surface. Assume that the oil weighs 50 lb per cubic foot.

(6) A gas station stores its gasoline in a tank under the ground. The tank is a cylinder lying horizontally on its side. (In other words, the tank is not standing vertically on one of its flat ends.) If the radius of the cylinder is 4 feet, its length is 12 feet, and its top is 10 feet under the ground, find the total amount of work needed to pump the gasoline out of the tank. (Gasoline weighs 42 $\mathrm{lb/ft^3}$.)

(7) How much work is required to lift a 1 000 kg satellite to an altitude of 2×10^6 m above the surface of the earth? (The radius of the earth is 6.4×10^6 m, its mass is 6×10^{24} kg, and in these units the gravitational constant, G, is 6.67×10^{-11}.)

30. (1) Show that the escape velocity needed by an object to escape the gravitational influence of a spherical planet of density ρ and radius R is proportional to R and to $\sqrt{\rho}$.

(2) Calculate the escape velocity of an object from the moon. The acceleration due to gravity on the moon is $1.6\ \mathrm{m/sec^2}$, whereas on earth it is $9.8\ \mathrm{m/sec^2}$; the radius of the moon is approximately 1 740 km. [Hint: If g is the acceleration due to gravity, the force on an object is $F = mg = GMm/R^2$.]

31. (1) An aquarium 2 m long, 1 m wide, and 1 m deep is full of water. Find (a) the hydrostatic pressure on the bottom of the aquarium, (b) the hydrostatic force on the bottom, and (c) the hydrostatic force on one end of the aquarium.

(2) A swimming pool 5 m wide, 10 m long, and 3 m deep is

filled with seawater of density 1 030 kg/m^3 to a depth of 2. 5 m. Find (a) the hydrostatic pressure on the bottom of the pool, (b) the hydrostatic force on the bottom, and (c) the hydrostatic force on one end of the pool.

(3) A trough is filled with a liquid of density 840 kg/m^3. The ends of the trough are equilateral triangles with sides 8 m long and vertex at the bottom. Find the hydrostatic force on one end of the trough.

(4) A cube with 20 cm-long sides is sitting on the bottom of an aquarium in which the water is 1 m deep. Find the hydrostatic force on (a) the top of the cube and (b) one of the sides of the cube.

(5) Suppose that a plate is immersed vertically in a fluid with density ρ and the width of the plate is $w(x)$ at a depth of x meters beneath the surface of the fluid. If the top of the plate is at depth a and the bottom is at depth b, show that the hydrostatic force on one side of the plate is

$$F = \int_a^b \rho g x w(x) \mathrm{d}x.$$

Chapter 9 Preliminary of differential equations

An equation containing unknown $f(x)$, $f'(x)$, $f''(x)$, $\cdots$, denoted by

$$F(f(x), f'(x), f''(x), \cdots) = 0,$$

is called a *differential equation*, i. e. a differential equation is an equation in which derivatives of an unknown function $y=f(x)$ occur. Examples for such equations are

$$y' = 3x+8; \; y'' = 5y' + 7x; \; y'' + 4y' + y = 2; \; \frac{\mathrm{d}y}{\mathrm{d}x} = -ay \ln \frac{y}{b}.$$

As we shall see, many physical processes can be described by differential equations. To solve a differential equation is to find a function which satisfies the equation, in which integration plays a key role. In this chapter we explore some topics in differential equations and use our resulting knowledge to study problems from various fields.

§ 9.1 Basic concepts of differential equations

There are two kinds of differential equations: ordinary differential equations and partial differential equations. An *ordinary differential equation* involves an unknown function of a single variable and some of its derivatives, while a *partial differential equation* involves an unknown function of two or more variables and some of its partial derivatives. The *order* of a differential equation is the order of the highest derivative that appears in the equation. A differential equation of order 1 (or 2, or 3, …) is also called a *first-order* (*or second-order*, *or third-order*) differential equation. In this chapter, we only study ordinary differential equations, so when we talk about a differential equation we mean an ordinary differential equation.

For example, $f(x)=f'(x)$ is an ordinary differential equation of order 1; $y^{(3)}(x)+2y'(x)=\sin(xy)$ is an ordinary differential equation of order 3;

$$\frac{\partial^2 f(x,y)}{\partial x^2}+\frac{\partial^2 f(x,y)}{\partial y^2}=0$$

is a partial differential equation of order 2.

When a function $y=f(x)$ and its derivatives are substituted into the equation $F(f(x), f'(x), f''(x), \cdots)=0$, the equation is satisfied, then $f(x)$ is called a *solution* of the equation. Let $F(f(x), f'(x), f''(x), \cdots, f^{(n)}(x))=0$ be a differential equation of order n. A solution with n arbitrary constants is called a *general solution*, and a solution without arbitrary constants is called a *particular solution*. In many practical problems we need to find a particular solution that satisfies a condition of the form $y(x_0)=y_0$. Such a condition which determines the initial situation of the solution is called an *initial condition*, and the problem of finding a solution of the differential equation that satisfies the initial condition is called an *initial-value problem*.

Note that a solution of a differential equation can be an explicit function $y=f(x)$ or $x=g(y)$, or even an implicit function $\varphi(x,y)=0$, such as $e^{xy+\sin(xy)}+3x^2y=0$, $x=y^3+\cos(x^2y^2)$, etc.

For example, $f(x)=2e^x$ is a particular solution of the equation $f(x)=f'(x)$; $f(x)=Ce^x$, where C is an arbitrary constant, is a general solution of the equation $f(x) = f'(x)$. Finding the solution of $f(x) = f'(x)$ with an initial condition $f(0)=5$ is an initial-value problem.

§ 9.2 Differential equations of first-order

Definition An equation with the form

$$\frac{dy}{dx} = F(x, y),$$

where F is some function of the two variables x and y, is called *first-order differential equation*. Furthermore, if F can be factorized as a function of x times a function of y, then it is called a *separable equation*, i. e. a separable equation has the form

$$\frac{dy}{dx} = f(x)g(y); \text{ or } \frac{dy}{dx} = \frac{f(x)}{h(y)}; \text{ or } h(y)dy = g(x)dx.$$

Clearly, each of the above cases can be presented by $h(y)dy = g(x)dx$. To solve it we integrate both sides of the equation:

$$\int h(y)dy = \int g(x)dx.$$

This equation defines y implicitly as a function of x. If possible, we solve for y in terms of x.

Examples (1) Solve $x\,dy = y\,dx$.

Solution $$\frac{dy}{y} = \frac{dx}{x}, \int \frac{dy}{y} = \int \frac{dx}{x},$$

$$\ln|y| = \ln|x| + C_1, \left|\frac{y}{x}\right| = e^{C_1}, \frac{y}{x} = \pm e^{C_1} = C, \text{ i. e. } y = Cx.$$

(2) Solve $\dfrac{dy}{dx} + \dfrac{e^{y^2+3x}}{y} = 0$.

Solution $\frac{dy}{dx} = -\frac{e^{y^2}}{y}e^{3x}$, $\int e^{-y^2} y dy = \int -e^{3x} dx$,

$$-\frac{1}{2}e^{-y^2} = -\frac{1}{3}e^{3x} + C,$$

$$\text{or } 3e^{-y^2} - 2e^{3x} = C_1.$$

(3) Solve $(1+y^2)(e^{2x}dx - e^y dy) - (1+y)dy = 0$.

Solution $(1+y^2)e^{2x}dx - [(1+y^2)e^y + (1+y)]dy = 0$,

$$\int e^{2x} dx = \int \left(e^y + \frac{1+y}{1+y^2}\right)dy,$$

$$\frac{1}{2}e^{2x} = e^y + \arctan y + \frac{1}{2}\ln(1+y^2) + C.$$

(4) Solve $y' = y^2 \cos x$ with $y(0) = 1$.

Solution $\frac{dy}{dx} = y^2 \cos x$, $\int \frac{dy}{y^2} = \int \cos x \, dx$, $y = -\frac{1}{\sin x + C}$,

Since $y(0) = 1$, $C = -1$, and so

$$y = \frac{1}{1 - \sin x}.$$

(5) Solve $y' = p(x)y$ where $p(x)$ is a continuous function.

Solution $\int \frac{dy}{y} = \int p(x)dx$, $\ln|y| = \int p(x)dx + C$,

$$|y| = e^{\int p(x)dx + C},$$

$$y = \pm e_1^C e^{\int p(x)dx}, \text{ i.e. } y = Ce^{\int p(x)dx}.$$

(6) Suppose $x\int_0^x f(x)dx = (x+1)\int_0^x xf(x)dx \, (x > 0)$. Find $f(x)$.

Solution Taking derivative on both sides,

$$xf(x) + \int_0^x f(x)dx = (x+1)xf(x) + \int_0^x xf(x)dx,$$

$$\int_0^x f(x)dx - \int_0^x xf(x)dx = x^2 f(x),$$

$$f(x) - xf(x) = x^2 f'(x) + 2xf(x),$$

$$x^2 f'(x) = (1-3x)f(x), \int \frac{\mathrm{d}f}{f} = \int \frac{1-3x}{x^2}\mathrm{d}x,$$

$$\text{i. e. } f(x) = C\mathrm{e}^{-\frac{1}{x}} \frac{1}{x^3}.$$

(7) A tank contains 20 kg of salt dissolved in 5 000 L (liters) of water. Brine that contains 0. 03 kg of salt per liter of water enters the tank at a rate of 25 L/min. The solution is kept thoroughly mixed and drains from the tank at the same rate. How much salt remains in the tank after 30 minutes?

Solution Let $y(t)$ be the amount of salt in kg after t minutes. The problem is to find $y(30)$ given that $y(0)=20$. Let

r_i=the rate at which salt enters the tank at time t;

r_o=the rate at which salt leaves the tank at time t.

Thus

$$r_i = (0.03 \text{ kg/L}) \cdot (25 \text{ L/min}) = 0.75 \text{ kg/min};$$

$$r_o = \left(\frac{y(t)}{5\,000} \text{ kg/L}\right) \cdot (25 \text{ L/min}) = \frac{y(t)}{200} \text{ kg/min}.$$

Note that $\mathrm{d}y/\mathrm{d}t$ is the change rate of the amount of salt. Then

$$\frac{\mathrm{d}y}{\mathrm{d}t} = r_i - r_o = \frac{150 - y(t)}{200}, \text{ i. e. } \frac{\mathrm{d}y}{150 - y} = \frac{\mathrm{d}t}{200},$$

which is a separable equation. By integrating both sides, we have

$$-\ln|150 - y| = \frac{t}{200} + C.$$

Since $y(0)=20$, we have $C=-\ln 130$, therefore

$$|150 - y| = 130\mathrm{e}^{\frac{-t}{200}}.$$

Noting $y(t)$ is continuous with $y(0)=20$ and the right side is non-negative, we see that $150-y(t)>0$ and so

$$y(30) = 150 - 130e^{\frac{-t}{200}}\Big|_{t=30} \approx 38.1 \text{ kg}.$$

In some cases, the equation is not separable itself, however, by a variable transformation it can be converted into a separable equation. We set such examples as follows:

Examples (1) Solve $\frac{dy}{dx} = (x-y)^2 + 1$.

Solution Let $u = x - y$. Then

$$\frac{dy}{dx} = 1 - \frac{du}{dx},\ 1 - \frac{du}{dx} = u^2 + 1,$$

$$\int \frac{-1}{u^2} du = \int dx,\ \frac{1}{u} = x + C,\ x - y = \frac{1}{x+C}.$$

(2) Solve $(2x+3y+4)dx - (4x+6y+5)dy = 0$.

Solution Let $u = 2x + 3y$. Then

$$du = 2dx + 3dy,\ dy = \frac{1}{3}(du - 2dx),$$

$$(u+4)dx - (2u+5) \cdot \frac{1}{3}(du - 2dx) = 0,$$

$$(7u+22)dx = (2u+5)du,\ \int \left(2 - \frac{9}{7\left(u+\frac{22}{7}\right)}\right) du = \int 7dx.$$

Thus we get $9 \ln \left|2x+3y+\frac{22}{7}\right| = 7(6y - 3x + C)$.

(3) Solve $xy' + y = y(\ln x + \ln y)$ $(x,\ y > 0)$.

Solution Let $u = xy$. Then

$$\frac{du}{dx} = y + x\frac{dy}{dx},\ \left(\frac{du}{dx} - y\right) + y = y \ln(xy) = \frac{u}{x} \ln u,$$

$$\int \frac{du}{u \ln u} = \int \frac{dx}{x},\ \ln(\ln u) = \ln x + \ln C\ (C > 0),\ \ln(xy) = Cx.$$

(4) Solve $(x^2y^2 + 1)dx + 2x^2 dy = 0$.

Solution Let $u = xy$. Then

$$du = y\,dx + x\,dy,\ x\,dy = du - \frac{u}{x}dx,$$

$$(u^2+1)dx + 2x(du - \frac{u}{x}dx) = 0,$$

$$(u^2+1-2u)dx = -2x\,du,\ \int \frac{dx}{-2x} = \int \frac{du}{(u-1)^2},$$

$$\frac{1}{2}\ln|x| - \frac{1}{u-1} = C,$$

$$\text{i. e. } \frac{1}{xy-1} = \frac{1}{2}\ln|x| + C.$$

(5) Solve $(x^2 - y^2 - 2y)dx + (x^2 - y^2 + 2x)dy = 0$ ($x \neq 0$, $x \neq y$).

Solution 1 $(x^2 - y^2)d(x+y) + 2(x\,dy - y\,dx) = 0$,

$$\left[1 - \left(\frac{y}{x}\right)^2\right]d(x+y) + 2d\left(\frac{y}{x}\right) = 0.$$

Let $x + y = u$ and $y/x = v$. Then

$$(1-v^2)du + 2dv = 0,\ \int du = \int \frac{2}{v^2-1}dv,\ u = \ln\left|\frac{v-1}{v+1}\right| + C,$$

$$\text{i. e. } x + y = \ln\left|\frac{x-y}{x+y}\right| + C.$$

Solution 2 $(x^2 - y^2)d(x+y) + 2(x\,dy - y\,dx) = 0$, $d(x+y) + 2\frac{x\,dy - y\,dx}{x^2 - y^2} = 0$.

$$d(x+y) - d\left(\ln\left|\frac{x-y}{x+y}\right|\right) = 0,\ \text{i. e. } x + y - \ln\left|\frac{x-y}{x+y}\right| = C.$$

Definition A first-order differential equation $y' = f(x, y)$ is said to be *homogeneous* if $f(x, y)$ can be written as $g(y/x)$, where g is a function of a single variable. i. e.

$$\frac{dy}{dx} = g\left(\frac{y}{x}\right).$$

A homogeneous differential equation can always be transformed into a separable equation:

Let $u = y/x$. Then $y = ux$,

$$\frac{dy}{dx} = \frac{d}{dx}(ux) = u + x\frac{du}{dx}, \text{ i. e.}$$

$$u + x\frac{du}{dx} = g(u),\ x\frac{du}{dx} = g(u) - u,$$

which is clearly separable:

$$\frac{du}{g(u) - u} = \frac{dx}{x}, \text{ and so } \int \frac{du}{g(u) - u} = \int \frac{dx}{x}.$$

Examples (1) Solve the following equations

① $\frac{dy}{dx} = 2\sqrt{\frac{y}{x}} + \frac{y}{x}$;

② $\frac{dy}{dx} = 2\left(\frac{y}{x+y}\right)^2 \quad (x,\ y > 0)$.

Solution ① Let $u = y/x$. Clearly, $g(u) = 2\sqrt{u} + u$ and so

$$\int \frac{du}{(2\sqrt{u} + u) - u} = \int \frac{dx}{x},\ \sqrt{u} = \ln|x| + C,$$

$$y = xu = x(\ln|x| + C)^2;$$

② $\frac{dy}{dx} = 2\left(\frac{y/x}{1 + (y/x)}\right)^2$. Let $u = y/x$, then

$$g(u) = 2\left(\frac{u}{1+u}\right)^2,\ \int \frac{du}{2\left(\frac{u}{1+u}\right)^2 - u} = \int \frac{du}{x}.$$

Integrating both sides, we have

$$\ln(ux) + 2\arctan u = C,\ \ln y + 2\arctan\frac{y}{x} = C.$$

(2) Solve $y^2 dx + x^2 dy = xy\, dy (x,\ y > 0)$.

Solution $\frac{\mathrm{d}y}{\mathrm{d}x} = \frac{y^2}{xy - x^2} = \frac{(y/x)^2}{(y/x) - 1}$. Let $u = y/x$. Then

$$g(u) = \frac{u^2}{u-1}, \int \frac{\mathrm{d}u}{\frac{u^2}{u-1} - u} = \int \frac{\mathrm{d}x}{x}, \int \frac{u-1}{u} = \int \frac{\mathrm{d}x}{x},$$

$$u - \ln u = \ln x - \ln C, \frac{y}{x} - \ln \frac{y}{x} = \ln \frac{x}{C}, e^{\frac{y}{x}} = \frac{y}{C}, y = Ce^{\frac{y}{x}}.$$

Definition A differential equation with the form

$$\frac{\mathrm{d}y}{\mathrm{d}x} + p(x)y = q(x),$$

where $p(x)$ and $q(x)$ are continuous functions on some interval, is called a *first-order linear differential equation*. In particular, if $q(x) = 0$, it is called a *first-order homogeneous linear differential equation*; if $q(x) \neq 0$, it is called a *first-order non-homogeneous linear differential equation*.

We first solve the homogeneous equation

$$\frac{\mathrm{d}y}{\mathrm{d}x} + p(x)y = 0.$$

Clearly, it is separable and

$$\frac{\mathrm{d}y}{y} = -p(x)\mathrm{d}x, \int \frac{\mathrm{d}y}{y} = \int (-p(x))\mathrm{d}x, y = Ce^{-\int p(x)\mathrm{d}x},$$

where C is an arbitrary constant.

Now we further solve the non-homogeneous equation

$$\frac{\mathrm{d}y}{\mathrm{d}x} + p(x)y = q(x).$$

The standard method, called *constant variation method*, for solving this equation is to regard the constant C in the solution of the homogeneous equation as a function of x, i. e. $C = C(x)$, and to substitute the solution

$$y = C(x)e^{-\int p(x)dx}$$

in the non-homogeneous equation to determine $C(x)$:

$$y' = C'(x)e^{-\int p(x)dx} + C(x)e^{-\int p(x)dx}(-p(x)),$$

$$C'(x)e^{-\int p(x)dx} + C(x)e^{-\int p(x)dx}(-p(x)) + p(x)C(x)e^{-\int p(x)dx} = q(x),$$

$$C'(x)e^{-\int p(x)dx} = q(x),\ C'(x) = q(x)e^{\int p(x)dx},$$

$$C(x) = \int q(x)e^{\int p(x)dx}dx + C_1.$$

Therefore, we have the general solution of the non-homogeneous equation is

$$y = e^{-\int p(x)dx}\left(\int q(x)e^{\int p(x)dx}dx + C_1\right),$$

where C_1 is an arbitrary constant.

Examples (1) Solve the following differential equations:

$$① \ y' + \frac{y}{x} = \frac{\sin x}{x} \quad (x > 0);\ ② \ \frac{dy}{dx} = \frac{y}{2x - y^2}.$$

Solution ① $p(x) = 1/x$, $q(x) = \sin x/x$. Then

$$y = e^{-\int \frac{1}{x}dx}\left(\int \frac{\sin x}{x}e^{\int \frac{1}{x}dx}dx + C\right) = e^{-\ln x}\left(\int \frac{\sin x}{x}e^{\ln x}dx + C\right)$$

$$= \frac{1}{x}\left(\int \frac{\sin x}{x}x\ dx + C\right) = \frac{1}{x}(-\cos x + C).$$

$$② \ \frac{dx}{dy} = \frac{2x - y^2}{y} = \frac{2}{y}x - y,\ \frac{dx}{dy} + \left(-\frac{2}{y}\right)x = -y.$$

Hence, $p(x) = -2/y$, $q(x) = -y$, and so

$$x = e^{-\int p(y)dy}\left(\int q(y)e^{\int p(y)dy}dy + C\right)$$

$$= e^{-\int \frac{-2}{y} dy} \left(\int (-y) e^{\int \frac{-2}{y} dy} dy + C \right)$$

$$= e^{2 \ln|y|} \left(\int (-y) e^{-2 \ln|y|} dy + C \right)$$

$$= y^2 \left(\int (-y) \frac{1}{y^2} dy + C \right) = y^2 \left(\int \frac{-1}{y} dy + C \right)$$

$$= y^2 (-\ln|y| + C).$$

(2) Solve $y' \cos^2 x + y - \tan x = 0$.

Solution $y' + \frac{1}{\cos^2 x} y = \frac{\tan x}{\cos^2 x}$.

Thus

$$y = e^{-\int \frac{1}{\cos^2 x} dx} \left(\int \frac{\tan x}{\cos^2 x} e^{\int \frac{1}{\cos^2 x} dx} dx + C \right)$$

$$= e^{-\tan x} \left(\int \tan x \cdot e^{\tan x} d(\tan x) + C \right)$$

$$= e^{-\tan x} \left(\tan x \cdot e^{\tan x} - \int e^{\tan x} d(\tan x) + C \right)$$

$$= \tan x - 1 + C e^{-\tan x}.$$

(3) Solve $(xy^5 - x^2 y^2) dy + (x^2 - y^6) dx = 0 (x \neq 0)$.

Solution Note $xy^2(y^3 - x) dy + (x - y^3)(x + y^3) dx = 0 (x \neq 0)$. Obviously, $x = y^3$ is a particular solution. To find general solution, we may suppose $x \neq y^3$, i.e. $y^3 - x \neq 0$. Then

$$xy^2 dy - (x + y^3) dx = 0, \ \frac{1}{3} x \, d(y^3) - (x + y^3) dx = 0.$$

Let $u = y^3$,

$$x \, du - 3x \, dx - 3u \, dx = 0, \ \frac{du}{dx} + \left(-\frac{3}{x} \right) u = 3,$$

which is a first-order linear equation. So,

$$u = e^{-\int (-\frac{3}{x}) dx} \left(\int 3 e^{\int (-\frac{3}{x}) dx} dx + C \right) = e^{3 \ln|x|} \left(\int 3 e^{e^{-3 \ln|x|}} dx + C \right)$$

$$= |x|^3\left(3\int |x|^{-3}\mathrm{d}x + C\right) = x^3\left(3\int x^{-3}\mathrm{d}x + C\right)$$

$$= x^3\left[3\left(-\frac{1}{2}x^{-2}\right) + C\right] = -\frac{3}{2}x + Cx^3,$$

i. e. the general solution is $y^3 = -\frac{3}{2}x + Cx^3$.

Note A particular solution which can not be expressed by the general solution is called a *singular solution*. In the above example, $x = y^3$ is a singular solution.

Definition A differential equation with the form

$$\frac{\mathrm{d}y}{\mathrm{d}x} = p(x)y + q(x)y^n \, (n \neq 0, 1, \text{ being a constant})$$

is called a *Bernoulli Equation of order n*. A Bernoulli Equation can be converted into a first-order linear equation by a transformation.

Note

$$y^{-n}\frac{\mathrm{d}y}{\mathrm{d}x} = y^{1-n}p(x) + q(x).$$

Let $z = y^{1-n}$. Then

$$\frac{\mathrm{d}z}{\mathrm{d}x} = (1-n)y^{-n}\frac{\mathrm{d}y}{\mathrm{d}x} = (1-n)y^{-n}(p(x)y + q(x)y^n)$$

$$= (1-n)(y^{1-n}p(x) + q(x))$$

$$= (1-n)p(x)z + (1-n)q(x).$$

This is a first-order linear equation.

Examples (1) Solve

$$\frac{\mathrm{d}y}{\mathrm{d}x} = 6\,\frac{y}{x} - xy^2.$$

Solution This is a Bernoulli Equation of order $n = 2$. Let $z =$

$y^{1-n}=y^{-1}$. Then

$$\frac{\mathrm{d}z}{\mathrm{d}x}=-y^{-2}\frac{\mathrm{d}y}{\mathrm{d}x}=-y^{-2}\left(6\frac{y}{x}-xy^2\right)$$
$$=-\frac{6}{x}y^{-1}+x=-\frac{6}{x}z+x,\text{ i. e. }\frac{\mathrm{d}z}{\mathrm{d}x}+\frac{6}{x}z=x.$$

Thus, by the formula for the solution of the first-order linear equation , we have

$$y^{-1}=z=\mathrm{e}^{-\int p(x)\mathrm{d}x}\left(\int q(x)\mathrm{e}^{\int p(x)\mathrm{d}x}\mathrm{d}x+C\right)=\mathrm{e}^{-\int\frac{6}{x}\mathrm{d}x}\left(\int x\mathrm{e}^{\int\frac{6}{x}\mathrm{d}x}\mathrm{d}x+C\right)$$
$$=\frac{C}{x^6}+\frac{x^2}{8},$$

i. e. the general solution is

$$\frac{x^6}{y}-\frac{x^8}{8}=C.$$

(2) Solve

$$\frac{\mathrm{d}y}{\mathrm{d}x}=\frac{1}{xy+x^2y^3}.$$

Solution Note that

$$\frac{\mathrm{d}x}{\mathrm{d}y}=xy+x^2y^3=yx+y^3x^2,$$

which is a Bernoulli Equation of order $n=2$ with $p(y)=y$ and $q(y)=y^3$. Let $z=x^{1-n}=x^{-1}$. Then

$$\frac{\mathrm{d}z}{\mathrm{d}y}=\frac{\mathrm{d}}{\mathrm{d}y}\left(\frac{1}{x}\right)=\frac{-1}{x^2}\frac{\mathrm{d}x}{\mathrm{d}y}=\frac{-1}{x^2}(yx+y^3x^2)$$
$$=-y\left(\frac{1}{x}\right)-y^3=-yz-y^3,$$

i. e.

$$\frac{\mathrm{d}z}{\mathrm{d}y}+yz=-y^3,$$

which is a first-order linear equation. So, we have

$$\frac{1}{x}=z=\mathrm{e}^{-\int y\,\mathrm{d}y}\left(-\int y^{3}\mathrm{e}^{\int y\,\mathrm{d}y}\mathrm{d}y+C\right)$$

$$=\mathrm{e}^{-\frac{y^{2}}{2}}\left[-\mathrm{e}^{\frac{y^{2}}{2}}(y^{2}-2)+C\right]=C\mathrm{e}^{-\frac{y^{2}}{2}}-y^{2}+2.$$

(3) Solve

$$\frac{\mathrm{d}y}{\mathrm{d}x}-\frac{4}{x}y=x\sqrt{y}.$$

Solution Note

$$\frac{\mathrm{d}y}{\mathrm{d}x}=\frac{4}{x}y+x\sqrt{y},$$

which is a Bernoulli Equation of order $n=1/2$. Let $z=y^{1-(1/2)}=\sqrt{y}$. Then

$$\frac{\mathrm{d}z}{\mathrm{d}x}=\frac{1}{2\sqrt{y}}\frac{\mathrm{d}y}{\mathrm{d}x}=\frac{1}{2\sqrt{y}}\left(\frac{4}{x}y+x\sqrt{y}\right)=\frac{2}{x}\sqrt{y}+\frac{x}{2}=\frac{2}{x}z+\frac{x}{2}.$$

So,

$$\frac{\mathrm{d}z}{\mathrm{d}x}-\frac{2}{x}z=\frac{x}{2},\text{ and } z=\mathrm{e}^{-\int\left(-\frac{2}{x}\right)\mathrm{d}x}\left(\int\frac{x}{2}\mathrm{e}^{\int\left(-\frac{2}{x}\right)\mathrm{d}x}\mathrm{d}x+C\right)$$

$$=x^{2}\left(\frac{\ln|x|}{2}+C\right),$$

i. e. the general solution is

$$y=z^{2}=x^{4}\left(\frac{\ln|x|}{2}+C\right)^{2}.$$

In some cases, the differential equations can be converted into Bernoulli Equations.

(4) Solve

$$xy'\ln x\cdot(\sin y)+\cos y(1-x\cos y)=0\quad(x>0).$$

Solution $(-x)\dfrac{dy}{dx}\ln x(-\sin y)+\cos y = x\cos^2 y$,

$$-x\ln x\frac{d(\cos y)}{dx}+\cos y = x\cos^2 y.$$

Let $z=\cos y$. Then

$$-x\ln x\frac{dz}{dx}+z=xz^2,\qquad \frac{dz}{dx}=\frac{1}{x\ln x}z-\frac{1}{\ln x}z^2,$$

which is a Bernoulli Equation of order $n=2$. Let $u=z^{1-2}=1/z$. Then

$$\frac{du}{dx}=\frac{-1}{z^2}\frac{dz}{dx}=\frac{-1}{z^2}\left(\frac{1}{x\ln x}z-\frac{1}{\ln x}z^2\right)$$
$$=\frac{-1}{x\ln x}\cdot\frac{1}{z}+\frac{1}{\ln x}=\frac{-1}{x\ln x}u+\frac{1}{\ln x},$$

i. e.

$$\frac{du}{dx}+\frac{1}{x\ln x}u=\frac{1}{\ln x},$$

which is a first-order linear equation. So, we have

$$u=e^{-\int\frac{1}{\ln x}}\left(\int\frac{1}{\ln x}e^{\int\frac{1}{\ln x}dx}dx+C\right)=\frac{1}{\ln x}(x+C).$$

Hence, we have the general solution

$$\frac{1}{\cos y}=\frac{1}{\ln x}(x+C),\ \text{i. e. } (x+C)\cos y=\ln x.$$

(5) Solve

$$y'+xy^2-x^3y-2x=0.$$

Solution $2x-y'=xy^2-x^3y=-(x^5-x^3y)+(x^5+xy^2-2x^3y)=-x^3(x^2-y)+x(x^2-y)^2$.

i. e.

$$\frac{\mathrm{d}}{\mathrm{d}x}(x^2-y)=-x^3(x^2-y)+x(x^2-y)^2.$$

Let $z=x^2-y$. then we have a Bernoulli Equation of order $n=2$:

$$\frac{\mathrm{d}z}{\mathrm{d}x}=-x^3z+xz^2,$$

Let $u=z^{1-2}=1/z$.

$$\frac{\mathrm{d}u}{\mathrm{d}x}=\frac{-1}{z^2}\frac{\mathrm{d}z}{\mathrm{d}x}=\frac{-1}{z^2}(-x^3z+xz^2)=x^3z^{-1}-x=x^3u-x,$$

$$\text{i. e. } \frac{\mathrm{d}u}{\mathrm{d}x}-x^3u=-x.$$

So,

$$u=\mathrm{e}^{-\int(-x^3)\,\mathrm{d}x}\left(\int(-x)\mathrm{e}^{\int(-x^3)\mathrm{d}x}\,\mathrm{d}x+C\right)=\mathrm{e}^{\frac{x^4}{4}}\left(-\int x\mathrm{e}^{-\frac{x^4}{4}}\,\mathrm{d}x+C\right),$$

i. e. we have the general solution

$$\frac{1}{x^2-y}=\mathrm{e}^{\frac{x^4}{4}}\left(-\int x\mathrm{e}^{-\frac{x^4}{4}}\,\mathrm{d}x+C\right).$$

(6) Let

$$\frac{\mathrm{d}y}{\mathrm{d}x}=p(x)y^2+q(x)y+r(x).$$

Suppose $y=y_1(x)$ is a particular solution. Prove that the general solution is

$$y=y_1(x)+\mathrm{e}^{\int(2p(x)y_1(x)+q(x))\mathrm{d}x}\cdot\left(C-\int p(x)\mathrm{e}^{\int(2p(x)y_1(x)+q(x))\mathrm{d}x}\mathrm{d}x\right)^{-1},$$

where C is an arbitrary constant.

Proof Let $y(x)$ be a general solution and let $z=y(x)-y_1(x)$. Then

$$\frac{\mathrm{d}z}{\mathrm{d}x}+\frac{\mathrm{d}y_1}{\mathrm{d}x}=\frac{\mathrm{d}y}{\mathrm{d}x}=p(x)(z+y_1(x))^2+q(x)(z+y_1(x))+r(x).$$

Noting $y_1(x)$ is a particular solution of the equation,

$$\frac{\mathrm{d}y_1}{\mathrm{d}x} = p(x)y_1^2 + q(x)y_1 + r(x), \text{ and so}$$

$$\begin{aligned}\frac{\mathrm{d}z}{\mathrm{d}x} &= p(x)(z^2 + 2zy_1(x)) + q(x)z \\ &= (2p(x)y_1(x) + q(x))z + p(x)z^2.\end{aligned}$$

This is a Bernoulli Equation of order $n = 2$. Let $u = z^{1-2} = 1/z$.

$$\begin{aligned}\frac{\mathrm{d}u}{\mathrm{d}x} &= \frac{-1}{z^2}\frac{\mathrm{d}z}{\mathrm{d}x} = \frac{-1}{z^2}[(2p(x)y_1(x) + q(x))z + p(x)z^2] \\ &= -(2p(x)y_1(x) + q(x))\frac{1}{z} - p(x),\end{aligned}$$

i. e.

$$\frac{\mathrm{d}u}{\mathrm{d}x} + (2p(x)y_1(x) + q(x))u = -p(x).$$

So,

$$\begin{aligned}\frac{1}{y(x) - y_1(x)} &= u \\ &= \mathrm{e}^{-\int(2p(x)y_1(x)+q(x))\mathrm{d}x}\left[\int(-p(x))\mathrm{e}^{\int(2p(x)y_1(x)+q(x))\mathrm{d}x}\mathrm{d}x + C\right],\end{aligned}$$

which implies that

$$y = y_1(x) + \mathrm{e}^{\int(2p(x)y_1(x)+q(x))\mathrm{d}x} \cdot \left(C - \int p(x)\mathrm{e}^{\int(2p(x)y_1(x)+q(x))\mathrm{d}x}\mathrm{d}x\right)^{-1}. \square$$

Example 6 can be used as a formula to solve some differential equations. Next is an example.

(7) Solve $(x^2 - 1)y' + y^2 - 2xy + 1 = 0$.

Solution It is easy to see that $y_1 = x$ is a particular solution of this equation, and

$$\frac{\mathrm{d}y}{\mathrm{d}x} = -\frac{1}{x^2 - 1}y^2 + \frac{2x}{x^2 - 1}y - \frac{1}{x^2 - 1}.$$

Let

$$p(x)=-\frac{1}{x^2-1},\ q(x)=\frac{2x}{x^2-1},\ r(x)=-\frac{1}{x^2-1}.$$

Then by Example 6,

$$y=e^{\int\left[2\left(-\frac{1}{x^2-1}\right)x+\frac{2x}{x^2-1}\right]dx}\cdot\left\{C-\int\left(-\frac{1}{x^2-1}\right)e^{\int\left[2\left(-\frac{1}{x^2-1}\right)x+\frac{2x}{x^2-1}\right]dx}dx\right\}^{-1}$$

$$=x+e^{\int 0\,dx}\left[C-\int\left(\frac{-1}{x^2-1}\right)e^{\int 0\,dx}dx\right]^{-1}=x+\left(C+\int\frac{1}{x^2-1}dx\right)^{-1}$$

$$=x+\left(C+\frac{1}{2}\ln\left|\frac{x-1}{x+1}\right|\right)^{-1},\ \text{or}\sqrt{\left|\frac{x-1}{x+1}\right|}=e^{\frac{1}{y-x}-C}.$$

This general solution can also be written as

$$\sqrt{\left|\frac{x-1}{x+1}\right|}=C_1e^{\frac{1}{y-x}},$$

where C_1 is an arbitrary positive constant.

§ 9.3 Degrading method of second-order differential equations

In this section, we deal with second-order differential equations with the form $y''=f(x,y,y')$. We have no systematic way to solve such equations. However, for three special types of second-order differential equations:

$$y''=f(x);\ y''=f(x,y');\ y''=f(y,y'),$$

we do have a method called *degrading method* to solve them, the main idea of which is to degrade the second-order equation into the first-order equation.

Type 1 $y''=f(x)$:

Let $y'=p$. Then

$$p' = y'' = f(x), \text{ i. e. } \frac{dp}{dx} = f(x),$$

which is a first-order equation. So, we have

$$dp = f(x)dx, \ \frac{dy}{dx} = p = \int f(x)dx + C_1, \ dy = \left(\int f(x)dx + C_1\right)dx,$$

$$y = \int\left(\int f(x)dx\right)dx + \int C_1 dx + C_2 = \int\left(\int f(x)dx\right)dx + C_1 x + C_2.$$

Remark For the equation $y^{(n)} = f(x)$, a similar approach can be used for the solution:

$$y^{(n-1)} = \int f(x)dx + C_1, \ y^{(n-2)} = \int\left(\int f(x)dx + C_1\right)dx + C_2, \ \cdots.$$

Example Solve $y^{(4)} = \sin x + x$.

Solution $y''' = -\cos x + \frac{x^2}{2!} + C_1, \ y'' = -\sin x + \frac{x^3}{3!} + C_1 x + C_2,$

$$y' = \cos x + \frac{x^4}{4!} + C_1 \frac{x^2}{2!} + C_2 x + C_3,$$

$$y = \sin x + \frac{x^5}{5!} + C_1 \frac{x^3}{3!} + C_2 \frac{x^2}{2!} + C_3 x + C_4.$$

Noting that $C_i (i = 1, 2, 3, 4)$ represents an arbitrary constant, the general solution can be briefly written as

$$y = \sin x + \frac{x^5}{5!} + C_1 x^3 + C_2 x^2 + C_3 x + C_4.$$

Type 2 $y'' = f(x, y')$:

Let $y' = p$. Then

$$p' = y'' = f(x, y'), \text{ i. e. } \frac{dp}{dx} = f(x, y'),$$

which is a first-order equation. Suppose its solution is $p = \varphi(x, C_1)$. Thus, we have the general solution

$$y = \int \varphi(x, C_1)\mathrm{d}x + C_2.$$

Examples (1) Solve $xy'' = y'\ln y'$

Solution Let $y' = p$. Then

$$p' = y'', x\frac{\mathrm{d}p}{\mathrm{d}x} = p\ln p, \frac{\mathrm{d}p}{p\ln p} = \frac{\mathrm{d}x}{x},$$

$$\frac{\mathrm{d}(\ln p)}{\ln p} = \frac{\mathrm{d}x}{x}, \ln(\ln p) = \ln x + \ln C_1,$$

$$\ln p = C_1 x, \frac{\mathrm{d}y}{\mathrm{d}x} = p = \mathrm{e}^{C_1 x}, y = \int \mathrm{e}^{C_1 x}\mathrm{d}x + C_2 = \frac{1}{C_1}\mathrm{e}^{C_1 x} + C_2.$$

(2) Solve $2xy'y'' = (y')^2 + 1(x > 0)$.

Solution Let $y' = p$. Then

$$y'' = \frac{\mathrm{d}p}{\mathrm{d}x}, 2xp\frac{\mathrm{d}p}{\mathrm{d}x} = 1 + p^2,$$

$$\int \frac{2p}{1+p^2}\mathrm{d}p = \int \frac{\mathrm{d}x}{x}, \ln(1+p^2) = \ln x + \ln C_1,$$

$$1 + p^2 = C_1 x, \frac{\mathrm{d}y}{\mathrm{d}x} = \pm\sqrt{C_1 x - 1}, \mathrm{d}y = \pm\sqrt{C_1 x - 1}\,\mathrm{d}x,$$

$$y = \int(\pm\sqrt{C_1 x - 1})\mathrm{d}x + C_2 = \frac{2}{3C_1}(C_1 x - 1)^{3/2} + C_2.$$

Type 3 $y'' = f(y, y')$:

Let $y' = p$. Then

$$y'' = \frac{\mathrm{d}}{\mathrm{d}x}(y') = \frac{\mathrm{d}p}{\mathrm{d}x} = \frac{\mathrm{d}p}{\mathrm{d}y}\frac{\mathrm{d}y}{\mathrm{d}x} = p\frac{\mathrm{d}p}{\mathrm{d}y}, \text{ i. e. } p\frac{\mathrm{d}p}{\mathrm{d}y} = f(y, p),$$

which is a first-order equation. Suppose its solution is $p = \varphi(y, C_1)$. Thus

$$\frac{\mathrm{d}y}{\mathrm{d}x} = \varphi(y, C_1), \mathrm{d}y = \varphi(y, C_1)\mathrm{d}x,$$

and so we have the general solution

$$\int \frac{\mathrm{d}y}{\varphi(y, C_1)} = x + C_2.$$

Examples (1) Solve

$$y'' = \frac{1+(y')^2}{2y} \quad (y > 0).$$

Solution Let $y' = p$. Then

$$y'' = p\frac{\mathrm{d}p}{\mathrm{d}y},\ p\frac{\mathrm{d}p}{\mathrm{d}y} = \frac{1+p^2}{2y},\ \frac{2p\mathrm{d}p}{1+p^2} = \frac{\mathrm{d}y}{y},$$

which is separable, and so we have

$$\ln(1+p^2) = \ln y + \ln C_1,\ 1+p^2 = C_1 y,$$

$$\left(\frac{\mathrm{d}y}{\mathrm{d}x}\right)^2 = C_1 y - 1,\ \mathrm{d}y = \pm\sqrt{C_1 y - 1}\,\mathrm{d}x,$$

i. e. $$\int \frac{\mathrm{d}y}{\pm\sqrt{C_1 y - 1}} = \int \mathrm{d}x + C_2,\ \pm\frac{2}{C_1}\sqrt{C_1 y - 1} = x + C_2.$$

Finally, we have the general solution

$$\frac{4}{C_1^2}(C_1 y - 1) = (x + C_2)^2.$$

(2) Solve $yy'' - (y')^2 = y^2 y' \quad (y > 0)$.

Solution Let $y' = p$. Then

$$y'' = p\frac{\mathrm{d}p}{\mathrm{d}y},\ yp\frac{\mathrm{d}p}{\mathrm{d}y} - p^2 = y^2 p,$$

by which, we have

$$p = 0 \text{ or } \frac{\mathrm{d}p}{\mathrm{d}y} - \frac{p}{y} = y.$$

Hence,

$$p=0 \text{ or } p=e^{-\int\left(-\frac{1}{y}\right)dy}\left(\int y e^{\int\left(-\frac{1}{y}\right)dy}dy+C_1\right)$$

$$=e^{\ln y}\left(\int y e^{-\ln y}dy+C_1\right)=y\left(\int dy+C_1\right)=y(y+C_1),$$

then

$$y=C \text{ or } \frac{dy}{y(y+C_1)}=dx.$$

By integrating both sides of the second equality, we have

$$y=C \text{ or } y=\frac{C_1C_2e^{C_1x}}{1-C_2e^{C_1x}},$$

where C, C_1, C_2 are all arbitrary constants.

(3) Solve $(y''')^2-y''y^{(4)}=0$.

Solution Let $y''=p$. Then

$$y'''=\frac{dp}{dx}=p', \text{ and } y^{(4)}=p'',$$

and so we have the equation

$$(p')^2-pp''=0,$$

which is an equation of Type 3. Let $p'=z$. Then

$$p''=\frac{dz}{dx}=\frac{dz}{dp}\frac{dp}{dx}=z\frac{dz}{dp},\ z^2-pz\frac{dz}{dp}=0,$$

which implies that

$$z=p\frac{dz}{dp}, \text{ or } z=0 \quad \text{i. e. } \int\frac{1}{z}dz=\int\frac{1}{p}dp \text{ or } z=0.$$

Thus

$$\frac{dp}{dx}=z=C_1p \text{ or } \frac{dp}{dx}=z=0,$$

then

$$p = C_2 e^{C_1 x} \text{ or } p = C, \text{ i. e. } y_1'' = C_2 e^{C_1 x} \text{ or } y_2'' = C,$$

$$y_1' = \int C_2 e^{C_1 x} dx + C_3 = \frac{C_2}{C_1} e^{C_1 x} + C_3$$

or

$$y_2' = \int C\, dx + C_1 = Cx + C_1,$$

$$y_1 = \int \frac{C_2}{C_1} e^{C_1 x} dx + C_3 x + C_4 = \frac{C_2}{C_1^2} e^{C_1 x} + C_3 x + C_4$$

or
$$y_2 = \int Cx\, dx + C_1 x + C_2 = \frac{1}{2} Cx^2 + C_1 x + C_2.$$

§ 9.4 Linear differential equations of second-order

Definition A differential equation with the form

$$p(x) \frac{d^2 y}{dx^2} + q(x) \frac{dy}{dx} + r(x) y = g(x) \tag{1}$$

where $p(x)$, $q(x)$ and $r(x)$ are continuous functions, is called a *second-order linear differential equation*; moreover, if $g(x) = 0$, it is said to be *homogeneous*; if $g(x) \neq 0$, it is said to be *non-homogeneous*. The functions $p(x)$, $q(x)$ and $r(x)$ in Equation (1) are called *coeffcients* of the equation; if $p(x)$, $q(x)$ and $r(x)$ are all constants, then Equation (1) is called *constant coeffcient linear differential equation*; otherwise, it is called *variable coeffcient linear differential equation*. In Equation (1), $g(x)$ is also a known function, which is called the *free term* of the equation.

Before we discuss the structure of the solutions of Equation (1), we first make certain of the structure of the solutions of the related homogeneous equation:

$$p(x)\frac{d^2y}{dx^2}+q(x)\frac{dy}{dx}+r(x)y=0 \tag{2}$$

which is called the *complementary equation* of Equation (1).

Theorem 9.4.1 If $y_1(x)$ and $y_2(x)$ are both solutions of Equation (2), then the function

$$y(x)=C_1y_1(x)+C_2y_2(x),$$

where C_1 and C_2 are arbitrary constants, is also a solution of Equation (2).

Proof By the condition, we have

$$p(x)y_1''+q(x)y_1'+r(x)y_1=0 \text{ and } p(x)y_2''+q(x)y_2'+r(x)y_2=0,$$

which clearly imply that

$$\begin{aligned} p(x)y''+q(x)y'+r(x)y &= C_1(p(x)y_1''+q(x)y_1'+r(x)y_1) \\ &\quad +C_2(p(x)y_2''+q(x)y_2'+r(x)y_2)=0. \square \end{aligned}$$

By the precedent theorem we see that the solutions of Equation (2) possess the additivity. However, the solution with the form $C_1y_1(x)+C_2y_2(x)$ is not necessarily the general solution of Equation (2). For example, if y_1 is a solution of Equation (2), then $y_2=ky_1$ (k being a constant) is also a solution of Equation (2), but the solution $y=C_1y_1+C_2y_2$ can be rewritten as $y=Cy_1$ where $C=C_1+kC_2$, which is not the general solution of Equation (2) because it only contains one arbitrary constant. So, if two particular solutions y_1 and y_2 are not proportional, then $y=C_1y_1(x)+C_2y_2(x)$ must be the general solution of Equation (2).

Definition Let $f(x)$ and $g(x)$ be functions both defined on an interval I. If $f(x)/g(x)$ or $g(x)/f(x)$ is a constant on I, then we say f and g are *linearly dependent* on I; otherwise, they are said to be *linearly independent*. For example, x^3 and $3x^3$ are linearly dependent on any interval I; while x^3 and $x^3 \cdot \sin x$ are linearly independent on any interval I.

According to the above discussion and the notions of linear dependence and linear independence, we may derive the next theorem which gives the general solution of Equation (2), i.e. a linear combination of two linearly independent solutions of Equation (2) is its general solution.

Theorem 9.4.2 If $y_1(x)$ and $y_2(x)$ are two linearly independent solutions of Equation (2), then the general solution of Equation (2) is given by

$$y(x) = C_1 y_1(x) + C_2 y_2(x),$$

where C_1 and C_2 are two arbitrary constants.

By this theorem, we only need to get two linearly independent particular solutions of Equation (2) to find its general solution. However, it is not easy to discover particular solution to a second-order linear equation. In some simple cases, we can find them just by observation.

Examples (1) Solve $y'' - y = 0$.

Solution By observation, $y_1 = e^x$ and $y_2 = e^{-x}$ are two particular solutions. Since $y_1/y_2 \neq$ constant and also $y_2/y_1 \neq$ constant, y_1 and y_2 are linearly independent. Therefore, the general solution to the equation is

$$y = C_1 e^x + C_2 e^{-x},$$

where C_1 and C_2 are two arbitrary constants.

(2) Solve $(x-1)y'' - xy' + y = 0$.

Solution By observation, $y_1 = x$ and $y^2 = e^x$ are two particular solutions. Since $y_1/y_2 \neq$ constant and also $y_2/y_1 \neq$ constant, y_1 and y_2 are linearly independent. Therefore, the general solution to the equation is

$$y = C_1 x + C_2 e^x,$$

where C_1 and C_2 are two arbitrary constants.

In some cases, if we can discover one particular solution $y_1 = y_1(x)$. To find another particular solution y_2, let $y_2 = C(x) y_1(x)$ and substitute it for y in the original equation to determine $C(x)$. This method is illustrated in the following examples:

(3) Solve $(\cos x) y''(\sin x) y' + (\sec x) y = 0$.

Solution By observation, $y_1 = \cos x$ is a particular solution. To determine another particular solution which is linearly independent of y_1, let $y_2 = C(x) y_1 = C(x)\cos x$ and substitute it for y in the original equation. We have

$$(\cos x) C''(x) - (\sin x) C'(x) = 0.$$

Let $C' = p$. Then $(\cos x) p'(x) - (\sin x) p(x) = 0$, which is a separable equation. Thus, we have

$$p = C_1 \sec x \text{ (for convenience, taking } C_1 = 1),$$

$$C'(x) = \sec x,\ C(x) = \int \sec x dx = \ln | \sec x + \tan x | + C_2,$$

which implies that

$$y_2 = \cos x \ln | \sec x + \tan x |$$

is another particular solution. Thus the general solution to the equation is

$$y = C_1 \cos x + C_2 \cos x \ln | \sec x + \tan x |.$$

(4) Let $y_1 = e^{2x}$ be a particular solution of the equation $(x^2 + 1) y'' - 2xy' - (ax^2 + bx + c) y = 0$. First determine the values of a, b and c and then find the general solution of the equation.

Solution Substituting $y_1 = e^{2x}$, $y_1' = 2e^{2x}$ and $y_1'' = 4e^{2x}$ for the y

of the equation, we have

$$e^{2x}(4x^2-4x+4)=e^{2x}(ax^2+bx+c).$$

Comparing the coefficients of this equality, we get $a=4$, $b=-4$ and $c=4$, and so the equation becomes $(x^2+1)y''-2xy'-(4x^2-4x+4)y=0$.

Let $y_2=C(x)e^{2x}$ be the another particular solution. So

$$(x^2+1)C''(x)+2(2x^2-x+2)C'(x)=0.$$

Let $p(x)=C'(x)$. Then

$$(x^2+1)\frac{dp}{dx}+2(2x^2-x+2)p=0,\ p=(x^2+1)e^{-4x},$$

i. e. $$C(x)=\int(x^2+1)e^{-4x}dx=-\frac{1}{4}(x^2+\frac{x}{2}+\frac{9}{8})e^{-4x}.$$

Therefore,

$$y_2=C(x)e^{2x}=-\frac{1}{4}(x^2+\frac{x}{2}+\frac{9}{8})e^{-2x},$$

and so the general solution required is

$$y=C_1e^{2x}+C_2(x^2+\frac{x}{2}+\frac{9}{8})e^{-2x}.$$

Now we turn to the structure of the solutions of Equation (1). First, we have the following simple property.

Theorem 9.4.3 (Additivity rule) Let $y_i(x)(i=1, 2)$ be a solution of the equation $p(x)y''+q(x)y'+r(x)y=g_i(x)(i=1, 2)$. Then y_1+y_2 is a solution of the equation

$$p(x)y''+q(x)y'+r(x)y=g_1(x)+g_2(x).$$

Proof Just notice that

$$p(x)y_i''+q(x)y_i'+r(x)y_i=g_i(x)\quad(i=1, 2)$$

implies

$$p(x)(y_1+y_2)''+q(x)(y_1+y_2)'+r(x)(y_1+y_2)=g_1(x)+g_2(x).\ \square$$

Theorem 9.4.4 Suppose $y_c(x)$ is the general solution of Equation (2) and $y_p(x)$ is a particular solution of Equation (1). Then

$$y=y_c(x)+y_p(x)$$

is the general solution of Equation (1).

Proof By Additivity rule, we see immediately that $y_c(x)+y_p(x)$ is a solution of Equation (1).

Since y_c has two arbitrary constants and y_p has no arbitrary constants, then y_c+y_p must have two arbitrary constants. Therefore

$$y=y_c(x)+y_p(x)$$

is the general solution of Equation (1). $\square$

Example Find the general solution of the equation $y''+y=x^2$.

Solution By observation, $y_1=\cos x$ and $y_2=\sin x$ are two linearly independent particular solution of the equation $y''+y=0$. Thus $y_c=C_1\cos x+C_2\sin x$ is the general solution of the equation $y''+y=0$. It is easy to check that $y_p=x^2-2$ is a particular solution of the equation $y''+y=x^2$. By the above theorem, $y=C_1\cos x+C_2\sin x+x^2-2$ is the general solution of the equation $y''+y=x^2$.

Generally speaking, it is hard to find a particular solution of Equation (1). However, if we have found the general solution of Equation (2), we can use an efficient method, called *constant variation method*, to find a particular solution of Equation (1). We now introduce this method as follows:

Let $y_c(x)=C_1y_1(x)+C_2y_2(x)$ be the general solution of Equation (2), i. e. $y_1(x)$ and $y_2(x)$ are two linearly independent particular solutions of (2). Convert the constants C_1 and C_2 into functions $C_1(x)$ and $C_2(x)$ and substitute

$$y_p(x) = C_1(x)y_1(x) + C_2(x)y_2(x)$$

for the y in Equation (1) to determine $C_1(x)$ and $C_2(x)$:

$$y'(x) = C_1(x)y_1'(x) + C_2(x)y_2'(x) + C_1'(x)y_1(x) + C_2'(x)y_2(x).$$

In order to avoid higher order derivative of C_1 and C_2 and to simplify the calculation, we may impose

(i) $C_1'(x)y_1(x)+C_2'(x)y_2(x) = 0$ as a constraint, and then we have

$$y'(x) = C_1(x)y_1'(x) + C_2(x)y_2'(x) \text{ and}$$
$$y''(x) = C_1(x)y_1''(x) + C_2(x)y_2''(x) + C_1'(x)y_1'(x) + C_2'(x)y_2'(x).$$

Substitute y, y' and y'' for y in Equation (1). Noting y_1 and y_2 are two particular solutions of Equation (2), we have

(ii) $$p(x)(C_1'(x)y_1'(x) + C_2'(x)y_2'(x)) = g(x).$$

Solve the equations (i) and (ii) for $C_1(x)$ and $C_2(x)$ as follows:

$$\begin{cases} C_1'(x)y_1(x) + C_2'(x)y_2(x) = 0, \\ C_1'(x)y_1'(x) + C_2'(x)y_2'(x) = \dfrac{g(x)}{p(x)}. \end{cases}$$

Since $y_1(x)$ and $y_2(x)$ are two linearly independent particular solutions of (2), the determinant

$$\begin{vmatrix} y_1 & y_2 \\ y_1' & y_2' \end{vmatrix} = y_1y_2' - y_2y_1' \neq 0,$$

which is called Wronski determinant of y_1 and y_2, and denoted by

$$W(y_1, y_2) = \begin{vmatrix} y_1 & y_2 \\ y_1' & y_2' \end{vmatrix}.$$

Thus, by Cramer rule we have

$$\frac{dC_1(x)}{dx}=\frac{-\frac{g(x)}{p(x)}y_2(x)}{W(y_1,y_2)};\ \frac{dC_2(x)}{dx}=\frac{\frac{g(x)}{p(x)}y_1(x)}{W(y_1,y_2)},$$

and furthermore,

$$C_1(x)=\int\frac{-\frac{g(x)}{p(x)}y_2(x)}{W(y_1,y_2)}dx+C_3;\ C_2(x)=\int\frac{\frac{g(x)}{p(x)}y_1(x)}{W(y_1,y_2)}dx+C_4,$$

where C_3 and C_4 are two arbitrary constants. Since we just want to find a particular solution, so we may take $C_3=C_4=0$. Therefore, we obtain a particular solution of Equation (1):

$$y_p=y_1(x)\int\frac{-\frac{g(x)}{p(x)}y_2(x)}{W(y_1,y_2)}dx+y_2(x)\int\frac{\frac{g(x)}{p(x)}y_1(x)}{W(y_1,y_2)}dx.$$

Example Find the general solution of $(x-5)y''+(4-x)y'+y=(x-5)^2e^x$.

Solution By observation, we see the general solution of $(x-5)y''+(4-x)y'+y=0$ is $y_c=C_1e^x+C_2(4-x)$. Let

$$y_p=C_1(x)e^x+C_2(x)(4-x).$$

Then by constant variation method, noting $y_1(x)=e^x$, $y_2(x)=4-x$, $p(x)=x-5$, $g(x)=(x-5)^2e^x$ and $W=y_1y'_2-y_2y'_1=e^x(-1)-(4-x)e^x=e^x(x-5)$, we have

$$\begin{aligned}y_p&=y_1(x)\int\frac{-\frac{g(x)}{p(x)}y_2(x)}{W(y_1,y_2)}dx+y_2(x)\int\frac{\frac{g(x)}{p(x)}y_1(x)}{W(y_1,y_2)}dx\\&=e^x\int\frac{-\frac{(x-5)^2e^x}{x-5}(4-x)}{e^x(x-5)}dx+(4-x)\int\frac{\frac{(x-5)^2e^x}{x-5}e^x}{e^x(x-5)}dx\\&=e^x\int(x-4)dx+(4-x)\int e^x dx=e^x\left(\frac{x^2}{2}-4x\right)+(4-x)e^x\end{aligned}$$

$$= \left(\frac{x^2}{2} - 5x + 4\right)e^x.$$

So, the general solution of the original equation is

$$y = C_1 e^x + C_2(4 - x) + \left(\frac{x^2}{2} - 5x + 4\right)e^x.$$

§ 9.5 Second-order linear equations with constant coefficients

As we see in the above section, it is usually difficult to find solutions of a second-order linear differential equation. But it becomes much easier to solve the equation if the coefficient functions $p(x)$, $q(x)$ and $r(x)$ are constant functions, i. e. the equation has the form

$$a\frac{d^2y}{dx^2} + b\frac{dy}{dx} + cy = g(x) \qquad (1)$$

where a, b and c are constants.

We first consider the solution of the related homogeneous equation

$$a\frac{d^2y}{dx^2} + b\frac{dy}{dx} + cy = 0 \qquad (2)$$

which is called the *complementary equation* of the original non-homogeneous equation (1) and plays an important role in the solution of Equation (1).

It is not hard to think of some likely candidates for particular solutions of Equation (2) if we state the equation verbally. We are looking for a function y such that a constant times y'' plus another constant times y' plus a third constant times y is equal to 0. Note that for the exponential function $y = e^{rx}$ (where r is a constant) we have $y' = re^{rx}$ and $y'' = r^2 e^{rx}$. If we substitute these expressions into

Equation (2), we see that $y = e^{rx}$ is a solution if

$$ar^2 e^{rx} + bre^{rx} + ce^{rx} = (ar^2 + br + c)e^{rx} = 0.$$

But $e^{rx} \neq 0$. So, $y = e^{rx}$ is a solution of Equation (2) if

$$ar^2 + br + c = 0. \tag{3}$$

Hence, if we select r as a root of Equation (3), then $y = e^{rx}$ must be a solution of equation (2). The algebraic equation (3) is called the *characteristic equation* (or the *auxiliary equation*) of Equation (2), denoted briefly as CE. Thus, the problem of finding the solutions of Equation (2) is induced into that of finding the solutions of its characteristic equation (3).

The roots of the characteristic equation (3) can be found by using the quadratic formula:

$$r_1 = \frac{-b + \sqrt{b^2 - 4ac}}{2a}; \; r_2 = \frac{-b - \sqrt{b^2 - 4ac}}{2a}.$$

We distinguish three cases according to the sign of the discriminant $\Delta = b^2 - 4ac$.

Case 1. $b^2 - 4ac > 0$

In this case the characteristic equation (3) has two distinct real roots r_1 and r_2, and so $y_1 = e^{r_1 x}$ and $y_2 = e^{r_2 x}$ are two linearly independent particular solutions of Equation (2). Thus, the general solution of Equation (2) is:

$$y = C_1 e^{r_1 x} + C_2 e^{r_2 x}.$$

Case 2. $b^2 - 4ac = 0$

In this case the characteristic equation (3) has two equal real roots $r_1 = r_2 = -b/2a$, denoted by $r = -b/2a$, the common value of r_1 and r_2.

We know that $y_1 = e^{rx}$ is one solution of equation (2). We now verify that $y_2 = xe^{rx}$ is also a solution:

$$\begin{aligned} ay''_2 + by'_2 + cy_2 &= (2ar + b)e^{rx} + (ar^2 + br + c)xe^{rx} \\ &= 0 \cdot e^{r_2 x} + 0 \cdot xe^{rx} = 0. \end{aligned}$$

Since $y_1 = e^{rx}$ and $y_2 = xe^{rx}$ are linear independent, the general solution of Equation (2) is:

$$y = C_1 e^{rx} + C_2 xe^{rx}.$$

Case 3. $b^2 - 4ac < 0$

In this case the characteristic equation (3) has a pair of conjugate complex roots

$$r_1 = \alpha + i\beta;\ r_2 = \alpha - i\beta,$$

where α and β are real numbers with

$$\alpha = \frac{-b}{2a};\ \beta = \frac{\sqrt{4ac - b^2}}{2a}.$$

So,

$$y_1 = e^{\alpha + i\beta},\ y_2 = e^{\alpha - i\beta}$$

are two linear independent solutions of Equation (2), and the general solution of Equation (2) is:

$$y = C_1 e^{(\alpha + i\beta)x} + C_2 e^{(\alpha - i\beta)x} = e^{\alpha x}(C_1 e^{i\beta x} + C_2 e^{-i\beta x}).$$

Using Euler formula

$$e^{i\theta} = \cos\theta + i\sin\theta,$$

we may rewrite the general solution of Equation (2) as

$$\begin{aligned} y &= C_1 e^{\alpha x}(\cos\beta x + i\sin\beta x) + C_2 e^{\alpha x}(\cos\beta x - i\sin\beta x) \\ &= e^{\alpha x}[(C_1 + C_2)\cos\beta x + i(C_1 - C_2)\sin\beta x] \\ &= e^{\alpha x}(C_3\cos\beta x + C_4\sin\beta x), \end{aligned}$$

where $C_3 = C_1 + C_2$, $C_4 = \mathrm{i}(C_1 - C_2)$.

In fact, it is easy to check that $e^{\alpha x} \cos \beta x$ and $e^{\alpha x} \sin \beta x$ are two linearly independent particular solutions of Equation (2).

Now, we summarize the discussion for all three cases in the following theorem:

Theorem 9.5.1 (1) If Equation (3) has two distinct real roots r_1 and r_2, then Equation (2) has the general solution

$$y = C_1 e^{r_1 x} + C_2 e^{r_2 x};$$

(2) If Equation (3) has two equal real roots $r_1 = r_2 = r$, then Equation (2) has the general solution

$$y = C_1 e^{rx} + C_2 x e^{rx};$$

(3) If Equation (3) has a pair of conjugate complex roots

$$r_1 = \alpha + \mathrm{i}\beta;\ r_2 = \alpha - \mathrm{i}\beta,$$

then Equation (2) has the general solution

$$y = e^{\alpha x}(C_1 \cos \beta x + C_2 \sin \beta x).$$

Example Find the general solution of the following equations:

(1) $y'' + y' - 2y = 0$;
(2) $4y'' - 20y' + 25y = 0$;
(3) $y'' + 6y' + 13y = 0$.

Solution (1) CE: $r^2 + r - 2 = 0$, roots: $r_1 = 1$; $r_2 = -2$,
general solution: $y = C_1 e^{x} + C_2 e^{-2x}$.
(2) CE: $4r^2 - 20r + 25 = 0$, roots: $r_1 = r_2 = 5/2$,
general solution: $y = (C_1 + C_2 x) e^{(5/2)x}$.
(3) CE: $r^2 + 6r + 13 = 0$, roots: $r_{1,2} = -3 \pm 2\mathrm{i}$,
general solution: $y = e^{-3x}(C_1 \cos 2x + C_2 \sin 2x)$.
Remark The above consequence can also be generalized to find

the general solution of a higher order linear differential equation with constant coefficients. We illustrate this as follows:

Let

$$y^{(n)}+a_1y^{(n-1)}+\cdots+a_iy^{(n-i)}+\cdots+a_ny=0$$

be a homogeneous linear differential equation with order n. Then we define its *characteristic equation* (CE) as

$$r^n+a_1r^{n-1}+\cdots+a_ir^{n-i}+\cdots+a_n=0.$$

(1) If r_0 is a single real root of the CE, then the corresponding particular solution is e^{r_0x};

(2) If r_0 is a j-multiple real root of the CE ($j>1$), then the corresponding particular solutions are e^{r_0x}, xe^{r_0x}, $x^2e^{r_0x}$, $\cdots$, $x^{j-1}e^{r_0x}$;

(3) If $\alpha\pm i\beta$ is a single pair of complex conjugate roots of the CE, then the corresponding particular solutions are $e^{\alpha x}\cos\beta x$, $e^{\alpha x}\sin\beta x$;

(4) If $\alpha\pm i\beta$ is a j-multiple pair of complex conjugate roots of the CE ($j>1$), then the corresponding particular solutions are $e^{\alpha x}\cos\beta x$, $e^{\alpha x}\sin\beta x$, $xe^{\alpha x}\cos\beta x$, $xe^{\alpha x}\sin\beta x$, ..., $x^{j-1}e^{\alpha x}\cos\beta x$, $x^{j-1}e^{\alpha x}\sin\beta x$.

All these particular solutions are linearly independent, and the general solution is the linear combination of the particular solutions.

Example Find the general solutions of the following equations:

(1) $y^{(5)}-y^{(4)}+y^{(3)}-y''=0$;

(2) $y^{(7)}-4y^{(6)}+11y^{(5)}-6y^{(4)}-25y'''+88y''-115y'+50=0$.

Solution (1) CE: $r^5-r^4+r^3-r^2=0$, i. e. $r^2(r-1)(r^2+1)=0$; roots: $r_1=0$, $r_2=0$, $r_3=1$, $r_4=i$, $r_5=-i$. So, the general solution is

$$\begin{aligned}y&=C_1e^{0\cdot x}+C_2xe^{0\cdot x}+C_3e^{1\cdot x}+C_4e^{0\cdot x}\cos x+C_5e^{0\cdot x}\sin x\\&=C_1+C_2x+C_3e^x+C_4\cos x+C_5\sin x.\end{aligned}$$

(2) CE: $r^7-4r^6+11r^5-6r^4-25r^3+88r^2-115r+50=0$; roots:

$r=-2$, 1, 1, $1\pm 2\mathrm{i}$, $1\pm 2\mathrm{i}$. So, the corresponding particular solutions are: e^{-2x}, e^x, xe^x, $e^x\cos 2x$, $xe^x\cos 2x$, $e^x\sin 2x$, $xe^x\sin 2x$, and the general solution is

$$y=C_1e^{-2x}+C_2e^x+C_3xe^x+C_4e^x\cos 2x$$
$$+C_5xe^x\cos 2x+C_6e^x\sin 2x+C_7xe^x\sin 2x.$$

Now we are going to find the general solution of Equation (1). We know from Section 9.4 that the general solution of Equation (1) is $y=y_c(x)+y_p(x)$, where $y_c(x)$ is the general solution of Equation (2) and $y_p(x)$ is a particular solution of Equation (1). So, we only need to find y_p here. There are two methods for finding a particular solution y_p. One is the constant variation method which was introduced in Section 9.4. We now illustrate *the method of undetermined coefficients*. The constant variation method works for every function $g(x)$ but is usually more difficult to apply in practice, while the method of undetermined coefficients is straightforward but works only for a restricted class of functions $g(x)$:

$$g(x)=P_1(x)e^{\alpha x}\cos\beta x+P_2(x)e^{\alpha x}\sin\beta x\ (\neq 0),$$

where $P_1(x)$ and $P_2(x)$ are polynomials of x, α and β are constants. For convenience, we suppose $p(x)\equiv 1$, i.e. we just consider the equation: $y''+by'+cy=g(x)$ (note that its complementary equation is $y''+by'+cy=0$ with the characteristic equation: $r^2+br+c=0$), and we discuss the type of $g(x)$ in four cases.

Case 1. $\alpha=\beta=0$, i.e. $g(x)=P_1(x)$ is a polynomial with degree n.

ⓘ if $c\neq 0$, i.e. the characteristic equation has no zero root, then set

$$y_p=\sum_{k=0}^{n}a_kx^k;$$

ⓘⓘ if $c=0$, $b\neq 0$, i. e. the characteristic equation has exactly one zero root, then set

$$y_p = x\sum_{k=0}^{n} a_k x^k;$$

ⓘⓘⓘ if $b=c=0$, i. e. the characteristic equation has two zero roots, then set

$$y_p = x^2\sum_{k=0}^{n} a_k x^k.$$

By substituting the above expressions of y_p for y in the equation $y''+by'+cy=g(x)$ and comparing the coefficients on both sides, we can determine all a_k, and also the particular solution y_p.

Example Find particular solutions of the following equations:

(1) $y''-2y'-3y=3x+1$;

(2) $y''-4y'=5$;

(3) $y''=1$.

Solution (1) Note $c=-3\neq 0$ and $g(x)=3x+1$. Set $y_p=A+Bx$, where the constants A and B are to be determined. Then by substitution we have

$$0-2B-3(A+Bx)=3x+1,$$

which implies $A=1/3$ and $B=-1$, i. e. a particular solution

$$y_p=\frac{1}{3}-x.$$

(2) Note $c=0$, $b=-4\neq 0$, and $g(x)=5$. Set $y_p=x\cdot A=Ax$. Then

$$0-4A=5, \text{ i. e. } A=-\frac{5}{4}.$$

So, a particular solution

$$y_p = -\frac{5}{4}x.$$

(3) Note $b = c = 0$ and $g(x) = 1$. Set $y_p = x^2 \cdot A = Ax^2$. Then

$$2A = 1, \text{ i. e. } A = \frac{1}{2}.$$

So, a particular solution

$$y_p = \frac{1}{2}x^2.$$

Case 2. $\alpha \neq 0, \beta = 0$, i. e. $g(x) = P_1(x)e^{\alpha x}$, where $P_1(x)$ is a polynomial with degree n.

ⓘ if α is not a root of the characteristic equation, then set

$$y_p = e^{\alpha x} \sum_{k=0}^{n} a_k x^k;$$

ⓘⓘ if α is a single root of the characteristic equation, then set

$$y_p = x e^{\alpha x} \sum_{k=0}^{n} a_k x^k;$$

ⓘⓘⓘ if α is a multiple root of the characteristic equation, then set

$$y_p = x^2 e^{\alpha x} \sum_{k=0}^{n} a_k x^k.$$

In this case, also by the substitution of y_p for y in the equation $y'' + by' + cy = g(x)$, we can determine all a_k.

Example Find particular solutions of the following equations:

(1) $y'' + 5y' + 6y = e^{-x}$;

(2) $y'' + 3y' + 2y = 3xe^{-x}$;

(3) $y'' - 6y' + 9y = (x+1)e^{3x}$.

Solution (1) Note $\alpha = -1$ is not a root of the characteristic equation $r^2 + 5r + 6 = 0$ and $P(x) = 1$. Set $y_p = Ae^{-x}$. By substituting

y_p for y in the original equation, we have

$$Ae^{-x} + 5(-Ae^{-x}) + 6Ae^{-x} = e^{-x},$$

which implies $A = 1/2$, i.e. a particular solution

$$y_p = \frac{1}{2}e^{-x}.$$

(2) Note $\alpha = -1$ is a single root of the characteristic equation $r^2 + 3r + 2 = 0$ and $P(x) = 3x$. Set $y_p = x(Ax + B)e^{-x}$. By substitution, we have

$$(2Ax + B + 2A)e^{-x} = 3xe^{-x},$$

which implies $A = 3/2$, $B = -3$, i.e. a particular solution

$$y_p = \left(\frac{3}{2}x^2 - 3x\right)e^{-x}.$$

(3) Since $\alpha = 3$ is a multiple root of the characteristic equation $r^2 - 6r + 9 = 0$ and $P(x) = x + 1$, we may set $y_p = x^2(Ax + B)e^{3x}$. Substituting y_p for y in the equation, we have

$$6Ax + 2B = x + 1,$$

which implies $A = 1/6$, $B = 1/2$, i.e. a particular solution

$$y_p = \frac{1}{2}x^2\left(\frac{1}{3}x + 1\right)e^{3x}.$$

Case 3. $\alpha = 0$, $\beta \neq 0$, i.e. $g(x) = P_1(x)\cos\beta x + P_2(x)\sin\beta x$, where $P_1(x)$ and $P_2(x)$ are polynomials with degree l and m respectively.

① if $i\beta$ is not a root of the characteristic equation, then set

$$y_p = R_1(x)\cos\beta x + R_2(x)\sin\beta x,$$

where $R_1(x)$ and $R_2(x)$ are both polynomials with a same degree $n = \max\{l, m\}$;

ⅱ if $i\beta$ is a root of the characteristic equation, then set

$$y_p = x(R_1(x)\cos\beta x + R_2(x)\sin\beta x),$$

where $R_1(x)$ and $R_2(x)$ are both polynomials with a same degree $n = \max\{l, m\}$.

Example Find particular solutions of the following equations:

(1) $y'' + y = \sin x$;

(2) $y'' + y = 2x \sin x \cos x$.

Solution (1) Note $g(x) = \sin x$, i. e. $P_1(x) = 0$, $P_2(x) = 1$, $\beta = 1$. Since $i\beta = i$ is a root of $r^2 + 1 = 0$ and $n = \{l, m\} = \{0,0\} = 0$, we may set $y_p = x(A \cos x + B \sin x)$. Then $y'_p = A \cos x - Ax \sin x + B \sin x - Bx \cos x$ and $y''_p = -2A \sin x - Ax \cos x + 2B \cos x - Bx \sin x$. Substituting y_p for y in the equation, we have

$$y''_p + y_p = -2A \sin x + 2B \cos x = \sin x.$$

Comparing coefficients on two sides, we get $A = -1/2$, $B = 0$. Thus we have a particular solution

$$y_p = -\frac{1}{2}x \cos x.$$

(2) Note $g(x) = 2x \sin x \cos x = x \sin 2x$, i. e. $P_1(x) = x$, $P_2(x) = 0$, $\beta = 2$. Since $i\beta = 2i$ is not a root of the characteristic equation $r^2 + 1 = 0$ and $n = \{l, m\} = \{0, 0\} = 0$, we may set $y_p = (Ax + B)\cos 2x + (Cx + D)\sin 2x$. By substitution, we can determine: $A = 0$, $B = -1/3$, $C = 0$, $D = 4/9$. Thus a particular solution

$$y_p = -\frac{1}{3}\cos 2x + \frac{4}{9}\sin 2x.$$

Case 4. $\alpha \neq 0$, $\beta \neq 0$, i. e. $g(x) = e^{\alpha x}(P_1(x)\cos\beta x + P_2(x)\sin\beta x)$, where $P_1(x)$ and $P_2(x)$ are polynomials with degree l and m respectively.

ⓘ if $\alpha+i\beta$ is not a root of the characteristic equation, then set

$$y_p = e^{\alpha x}[R_1(x)\cos\beta x + R_2(x)\sin\beta x],$$

where $R_1(x)$ and $R_2(x)$ are both polynomials with a same degree $n = \max\{l, m\}$;

ⓘⓘ if $\alpha+i\beta$ is a root of the characteristic equation, then set

$$y_p = xe^{\alpha x}[R_1(x)\cos\beta x + R_2(x)\sin\beta x],$$

where $R_1(x)$ and $R_2(x)$ are both polynomials with a same degree $n = \max\{l, m\}$.

In this case, by substituting y_p for y in the equation $y'' + by' + cy = g(x)$ and comparing two sides of the equation, we can determine all the coefficients of the polynomials R_1 and R_2, and further the particular solution y_p.

Example Find a particular solution of the equation $y'' - 2y' + 2y = e^x \cos x$.

Solution Note $g(x) = e^x \cos x$, i. e. $P_1(x) = 0$, $P_2(x) = 1$, $\alpha = 1$, $\beta = 1$. Since $\alpha+i\beta = 1+i$ is a root of $r^2 - 2r + 2 = 0$ and $n = \{l, m\} = \{0,0\} = 0$, we may set $y_p = xe^x(A\cos x + B\sin x)$. Then $y'_p = e^x(1+x)(A\sin x + B\cos x) + xe^x(A\cos x - B\sin x)$ and $y''_p = 2e^x(A\sin x + B\cos x) + 2(x+1)e^x(A\cos x - B\sin x)$. By substitution, we have

$$2e^x(A\cos x - B\sin x) = e^x\cos x,$$

which implies that $A = 1/2$, $B = 0$. So, a particular solution

$$y_p = \frac{1}{2}e^x\sin x.$$

Note If the free term $g(x)$ in Equation (1) is a sum of different types mentioned above, we may use the additivity rule to find a particular solution, and further the general solution.

Example Find the general solution of the equation $y''+y=x^2+x\cos x$.

Solution CE: $r^2+1=0$, roots: $r=\pm i$. Thus, the general solution of the equation $y''+y=0$ is

$$y_c=C_1\cos x+C_2\sin x.$$

Let y_{p1} be a particular solution of $y''+y=x^2$. Then y_{p1} has a form $y_{p1}=Ax^2+Bx+C$. By substitution, we may determine that $A=1$, $B=0$, $C=-2$, and so $y_{p1}=x^2-2$.

Let y_{p2} be a particular solution of $y''+y=x\cos x$. Then, y_{p2} has a form

$$y_{p2}=x[(Ax+B)\cos x+(Cx+D)\sin x].$$

By substituting y_{p2} in the equation, we will have $A=0$, $B=C=1/4$, $D=0$, i. e.

$$y_{p2}=\frac{1}{4}x(\cos x+x\sin x).$$

Hence, we obtain the general solution of the original equation

$$\begin{aligned} y &= y_c+(y_{p1}+y_{p2}) \\ &= C_1\cos x+C_2\sin x+\left[x^2-2+\frac{1}{4}x(\cos x+x\sin x)\right]. \end{aligned}$$

§ 9.6 Euler Equation

We know that it is usually difficult to find a solution of a linear differential equation which has function coefficients. But for some special kind of equations with function coefficients, we can use transformation of variable to convert them into equations with constant coefficients. Euler equation is such a kind of equation which can be solved in this way.

Definition An equation with a form

$$x^2 y'' + axy' + by = f(x) \tag{1}$$

is called an *Euler equation* of order 2, where a and b are constants. Generally, an Euler Equation of order n can be expressed as

$$x^n y^{(n)} + a_1 x^{n-1} y^{(n-1)} + \cdots + a_{n-1} xy' + a_n y = f(x) \tag{2}$$

where $a_i (i = 1, 2, \cdots, n)$ are constants. We now show the way to solve Euler equations:

Let $x = e^t$. Then $t = \ln x$. Regarding y as a function of t, since

$$y' = \frac{dy}{dx} = \frac{dy}{dt} \cdot \frac{dt}{dx} = \frac{1}{x}\frac{dy}{dt},\ xy' = \frac{dy}{dt},$$

$$y'' = \frac{d^2 y}{dx^2} = \frac{d}{dx}\left(\frac{1}{x} \cdot \frac{dy}{dt}\right) = -\frac{1}{x^2}\frac{dy}{dt} + \frac{1}{x}\frac{d^2 y}{dt^2} \cdot \frac{1}{x} = \frac{1}{x^2}\left(\frac{d^2 y}{dt^2} - \frac{dy}{dt}\right),$$

$$x^2 y'' = \frac{d^2 y}{dt^2} - \frac{dy}{dt}.$$

So, Equation (1) can be changed into

$$\frac{d^2 y}{dt^2} + (a-1)\frac{dy}{dt} + by = f(e^t) \tag{3}$$

which is a second-order linear equation with constant coefficients. So, we find the relationship between y and t, and then find the relationship between y and x by substitution $t = \ln x$.

Example Solve the equation $x^2 y'' - 2xy' + 2y = x^3$.

Solution Let $x = e^t$. Then $t = \ln x$, and so by the above expression (3), we have

$$\frac{d^2 y}{dt^2} - 3\frac{dy}{dt} + 2y = e^{3t}.$$

CE: $r^2 - 3r + 2 = 0$, roots: $r_1 = 1$, $r_2 = 2$. So, the general solution of the complementary equation $x^2 y'' - 2xy' + 2y = 0$ is: $y_c = C_1 e^t +$

$C_2 e^{2t}$. Set $y_p = Ae^{3t}$, a particular solution of the original equation. By substitution, we have $A = 1/2$ and so $y_p = (1/2)e^{3t}$. Hence, the general solution required is

$$y = C_1 e^t + C_2 e^{2t} + \frac{1}{2}x^3.$$

In the process of converting Euler equation into constant coefficient equation by the variable transformation $x = e^t$, for convenience we may use the symbol D to express the derivative operation with respect to t:

Denote

$$Dy := \frac{d}{dt}y = \frac{dy}{dt}.$$

Then

$$D^2 y = D(Dy) = \frac{d}{dt}(Dy) = \frac{d}{dt}\left(\frac{dy}{dt}\right) = \frac{d^2 y}{dt^2}, \ \cdots, \ D^n y = \frac{d^n y}{dt^n},$$

and furthermore we have

$$xy' = x\frac{dy}{dx} = x\frac{dy}{dt}\frac{dt}{dx} = xDy \cdot \frac{1}{x} = Dy,$$

$$\begin{aligned}
x^2 y'' &= x^2 \frac{d^2 y}{dx^2} = x^2 \frac{d}{dx}\left(\frac{dy}{dx}\right) = x^2 \frac{d}{dx}\left(\frac{dy}{dt}\frac{dt}{dx}\right) \\
&= x^2\left[\frac{d}{dx}\left(\frac{dy}{dt}\right) \cdot \frac{dt}{dx} + \frac{dy}{dt} \cdot \frac{d}{dx}\left(\frac{dt}{dx}\right)\right] \\
&= x^2\left[\frac{d}{dt}\left(\frac{dy}{dt}\right) \cdot \left(\frac{dt}{dx}\right)^2 + \frac{dy}{dt} \cdot \frac{d}{dx}\left(\frac{dt}{dx}\right)\right] \\
&= x^2\left[D^2 y \cdot \left(\frac{1}{x}\right)^2 + Dy\left(\frac{-1}{x^2}\right)\right] \\
&= D^2 y - Dy = D(D-1)y,
\end{aligned}$$

in general,

$$x^k y^{(k)} = D(D-1)(D-2)\cdots(D-k+1)y.$$

Thus, the Euler equation (2) is converted into:

$$D(D-1)\cdots(D-n+1)y + a_1 D(D-1)\cdots(D-n+2)y + \cdots + a_{n-1}Dy + a_n y = f(e^t) \tag{4}$$

which is a constant coefficient linear equation. The characteristic equation of the complementary equation is

$$r(r-1)\cdots(r-n+1) + a_1 r(r-1)\cdots(r-n+2) + \cdots + a_{n-1}r + a_n = 0. \tag{5}$$

Example Fin the general solution of the equation $x^3 y''' + x^2 y'' - 4xy' = 3x^3$.

Solution Let $x = e^t$. Then the equation can be converted into

$$D(D-1)(D-2)y + D(D-1)y - 4Dy = 3e^{2t},$$

i. e.

$$(D^3 - 2D^2 - 3D)y = 3e^{2t}.$$

CE: $r^3 - 2r^2 - 3r = 0$, roots: $r = 0, -1, 3$. So, the general solution of the complementary equation:

$$y_c = C_1 + C_2 e^{-t} + C_3 e^{3t} = C_1 + \frac{C_2}{x} + C_3 x^3.$$

Setting the particular solution $y_p = Ae^{2t}$, and substituting it for y in the equation:

$$\frac{d^3 y}{dt^3} - 2\frac{d^2 y}{dt^2} - 3\frac{dy}{dt} = 3e^{2t},$$

we can determine $A = -1/2$, i. e. $y_p = -(1/2)e^{2t} = -(1/2)x^2$. Therefore, the general solution is

$$y = C_1 + \frac{C_2}{x} + C_3 x^3 - \frac{1}{2}x^2.$$

Exercises

1. Solve the following equations:

(1) $y' = \dfrac{x\sqrt{x^2+1}}{ye^y}$;　　(2) $y' = \dfrac{\ln x}{xy + xy^3}$;

(3) $x^2 y' + y = 0$;　　(4) $(x^2+1)y' = xy$.

2. Solve the following differential equations with initial values:

(1) $(\ln y)^2 y' = x^2 y$, $y(2) = 1$;

(2) $y' = x\sin x$, $y(\pi/2) = 0$;

(3) $(1+e^x)yy' = e^x$, $y|_{x=1} = 1$;

(4) $y' = e^{2x-y}$, $y|_{x=0} = 0$;

(5) $y' = y^2 + 1$, $y(1) = 0$;

(6) $y' = e^{x-y}$, $y(0) = 1$;

(7) $\dfrac{du}{dt} = \dfrac{2t+1}{2(u-1)}$, $u(0) = -1$;

(8) $\dfrac{dy}{dt} = \dfrac{ty+3t}{t^2+1}$, $y(2) = 2$.

3. Which of the following equations are homogeneous?

(1) $x(y')^2 - 2yy' + x = 0$;

(2) $(x^2 - y^2)dx + (x^2 + y^2)dy = 0$;

(3) $(7x - 6y)dx + (x+y)dy = 0$;

(4) $x^2 + 1 + 2xyy' = 0$;

(5) $(x^2 + y^2)y' = 2xy + x^2 y$;

(6) $y' = \ln y - \ln x$;

(7) $\sqrt{x^2+y^2}\,dx + ydy = 0$　$(x > 0)$.

4. Solve the following homogeneous equations:

(1) $y' = \dfrac{x-y}{x}$;

(2) $y' = \dfrac{x+y}{x-y}$;

(3) $xy' = y + xe^{y/x}$;

(4) $xy'\sin\dfrac{y}{x} = y\sin\dfrac{y}{x} - x$.

5. Which of the following equations are linear?

(1) $y' + x^2 y = y^2$;

(2) $x^2 y' - y + x = 0$;

(3) $xy' = x - y$;

(4) $yy' = \sin x$.

6. Solve the following linear differential equations:

(1) $y' - 3y = e^x$;

(2) $y' + 4y = x$;

(3) $y' - 2xy = x$;

(4) $xy' + 2y = e^{x^2}$;

(5) $y'\cos x = y\sin x + \sin 2x$, $x \in (-\pi/2, \pi/2)$;

(6) $e^x + xy = xy'$;

(7) $y' + 2xy = x^2$;

(8) $y' = x\sin 2x + y\tan x$, $x \in (-\pi/2, \pi/2)$;

(9) $y' - y\tan x = 1$, $x \in (-\pi/2, \pi/2)$;

(10) $xy' + xy + y = e^{-x}$ $(x > 0)$.

7. Find the general solutions of the following equations:

(1) $y'' = \sin 2x$;

(2) $y''' = xe^x$;

(3) $y'' = 2y'$;

(4) $yy'' - 2(y')^2 = 0$;

(5) $y'' - a^2 y = 0$;

(6) $4xy'' = 4y' + y''$;

(7) $y'' = x + \sin x$;

(8) $y'' = \dfrac{1}{1+x^2}$;

(9) $4xy'' = 4y' + y''$;

(10) $xy'' = y' \ln \frac{y'}{x}$;

(11) $yy'' - (y')^2 = y^2 y'$;

(12) $2xy'y'' = 1 + (y')^2$;

(13) $y''' = y''$.

8. Solve the following initial-value problems:

(1) $y^2 y'' + 1 = 0$, $y|_{x=1} = 1$, $y'|_{x=1} = 0$;

(2) $y'' - a(y')^2 = 0$, $y|_{x=0} = 0$, $y'|_{x=0} = -1$;

(3) $y'' - y'^2 = 0$, $y|_{x=0} = 0$, $y'|_{x=0} = -1$;

(4) $y^3 y'' + 1 = 0$, $y|_{x=1} = 1$, $y'|_{x=1} = 0$;

(5) $yy'' = y'^2 - y'^3$, $y|_{x=1} = 1$, $y'|_{x=1} = -1$;

(6) $xy'' + xy'^2 - y' = 0$, $y|_{x=2} = 2$, $y'|_{x=2} = 1$.

9. Find the general solutions of the following equations:

(1) $y' = \frac{y}{x} + \ln \frac{y}{x}$;

(2) $xy' - x \sin \frac{y}{x} - y = 0$;

(3) $(x+y)y' + (x-y) = 0$;

(4) $y' = \frac{y}{y-x}$;

(5) $\frac{dy}{dx} = \frac{y^2 + 2xy}{x^2}$;

(6) $\frac{dy}{dx} = \frac{x+y}{x}$;

(7) $\frac{dy}{dx} - \tan x \cdot y = e^{\sin x} \quad \left(|x| < \frac{\pi}{2}\right)$;

(8) $x \frac{dy}{dx} + y = x \sin x^2 + 5x$.

10. Solve the following initial-value problems:

(1) $y' + 5y = -4e^{-3x}$, $y(0) = -4$;

(2) $xy' + y = y^2 x^2 \ln x$, $y(1) = \frac{1}{2} \quad (0, +\infty)$;

(3) $y' - 4y = 2e^x \sqrt{y}$, $y(0) = 2$;

(4) $\cos x y' + y = 1$, $y\left(\frac{\pi}{4}\right) = 2 \quad \left(0 < x < \frac{\pi}{2}\right)$;

(5) $y' + y = x + e^x$, $y(0) = 0$;

(6) $xy' - 3y = x^2$, $x > 0$, $y(1) = 0$;

(7) $x^2y' + 2xy = \cos x$, $y(\pi) = 0$;

(8) $xy' - \frac{y}{x+1} = x$, $x > 0$, $y(1) = 0$.

11. Solve the following Bernoulli Equations:

(1) $xy' + y = -xy^2$;

(2) $y' + y = xy^3$;

(3) $y' + \frac{2y}{x} = \frac{y^3}{x^2}$;

(4) $y' - y = \frac{x^2}{y}$.

12. Find general solutions of the following equations:

(1) $(x+1)\frac{\mathrm{d}y}{\mathrm{d}x} - ny = \mathrm{e}^x(x+1)^{n+1}$;

(2) $(x^2+1)\frac{\mathrm{d}y}{\mathrm{d}x} + 2xy = 4x^2$;

(3) $\frac{\mathrm{d}y}{\mathrm{d}x} - 3xy - xy^2 = 0$;

(4) $(y^2 - 6x)y' + 2y = 0$;

(5) $(x - 2xy - y^2)y' + y^2 = 0$;

13. Let $\int_0^1 \varphi(tx)\mathrm{d}x = (\varphi(t))^2$. Find $\varphi(x)$.

14. Check that $y_1 = \mathrm{e}^{x^2}$ and $y_2 = x\mathrm{e}^{x^2}$ are both solutions of the equation $y'' - 4xy' + (4x^2 - 2)y = 0$; give the general solution of this equation.

15. Check that $y_1 = \mathrm{e}^x$ is a solution of the equation $(1 + 2x - x^2)y'' + (x^2 - 3)y' + 2(1 - x)y = 0$, determine the coefficients of $y_2 = ax^2 + bx + c$ such that y_2 is another solution of this equation, and then give the general solution of this equation.

16. (1) Check that $y = C_1\mathrm{e}^x + C_2\mathrm{e}^{2x} + \frac{1}{12}\mathrm{e}^{5x}$ (C_1 and C_2 being arbitrary constants) is the general solution of the equation $y'' - 3y' + 2y = e^{5x}$;

(2) Check that $y = C_1 \cos 3x + C_2 \sin 3x + \frac{1}{32}(4x \cos x + \sin x)$ (C_1 and C_2 being arbitrary constants) is the general solution of the equation $y'' + 9y = x \cos x$;

(3) Check that $y = C_1 x^5 + \frac{C_2}{x} - \frac{x^2}{9} \ln x$ (C_1 and C_2 being arbitrary constants) is the general solution of the equation $x^2 y'' - 3xy' - 5y = x^2 \ln x$.

17. Given one solution of the following homogeneous equations, find another solution of the equation which is linearly independent of the given solution, and then find the general solution:

(1) $(x^2 + 1)y'' - 2xy' + 2y = 0$, $y_1 = x$;

(2) $(2x - 1)y'' - (2x + 1)y' + 2y = 0$, $y_1 = e^x$;

(3) $(x^3 - x^2)y'' - (x^3 + 2x^2 - 2x)y' + (2x^2 + 2x - 2)y = 0$, $y_1 = x^2$.

18. Find by observation a particular solution of the following equations, and then find the general solution:

(1) $x^2(\ln x - 1)y'' - xy' + y = 0$;

(2) $(3x + 2x^2)y'' - 6(1 + x)y' + 6y = 0$;

(3) $(\sin x - \cos x)y'' - 2(\sin x)y' + (\sin x + \cos x)y = 0$.

19. Find a particular solution of each of the following equations by constant variation method, and then find the general solution:

(1) $y'' + y = \tan x$;

(2) $(x^2 + 1)y'' - 2xy' + 2y = 6(x^2 + 1)^2$.

20. Find the general solutions of the following differential equations:

(1) $y'' - 3y' + 2y = 0$;

(2) $3y'' - 8y' - 3y = 0$;

(3) $y''+2y'+10y=0$;

(4) $y''+25y=0$;

(5) $y''-4y'=0$;

(6) $2y''+y'=0$;

(7) $y''+2y'-y=0$;

(8) $y''-y'+2y=0$;

(9) $y'''-y''+y'-y=0$;

(10) $y^{(4)}+8y''+16=0$.

21. Solve the following initial value problems:

(1) $y''+3y'-4y=0$, $y(0)=2$, $y'(0)=-3$;

(2) $y''-2y'+2y=0$, $y(0)=1$, $y'(0)=2$;

(3) $y''-2y'-3y=0$, $y(1)=3$, $y'(1)=1$;

(4) $y''+9y=0$, $y\left(\frac{\pi}{3}\right)=0$, $y'\left(\frac{\pi}{3}\right)=1$;

(5) $y''-4y'+29y=0$, $y(0)=0$, $y'(0)=15$;

(6) $y''-4y'+13y=0$, $y(0)=0$, $y'(0)=3$.

22. Find the general solutions of the following differential equations:

(1) $y''+8y'=8x$;

(2) $y''-y'-6y=\cos 3x$;

(3) $y''-4y'+4y=\mathrm{e}^{-x}$;

(4) $y''+36y=2x^2-x$;

(5) $y''+3y'=3x\mathrm{e}^{-3x}$;

(6) $y''-2y'+5y=\mathrm{e}^{x}\sin 2x$;

(7) $y''-4y'+4y=\sin 2x$;

(8) $y''+4y'=x\cos x$;

(9) $y''-2y'+y=\mathrm{e}^{x}+x^2+\sin x$;

(10) $y''-y=\sin^2 x$.

23. Find the particular solutions of the following differential equations with initial conditions:

(1) $y''-2y'+5y=x+\sin 3x$, $y(0)=1$, $y'(0)=2$;

(2) $y''-y=xe^{3x}$, $y(0)=0$, $y'(0)=1$;

(3) $y''-3y'+2y=5$, $y(0)=1$, $y'(0)=2$;

(4) $y''+y=2\cos x$, $y(0)=1$, $y'(0)=0$.

24. Find the general solutions of the following Euler equations:

(1) $x^2y''-3xy'+3y=0$;

(2) $x^2y''+xy'+y=0$;

(3) $9x^2y''+3xy'+y=0$;

(4) $x^3y'''+2x^2y''-xy'+y=0$;

(5) $x^3y'''-x^2y''-6xy'+18y=0$;

(6) $x^2y''-5xy'+8y=2x^3$;

(7) $x^2y''+xy'+4y=2x\ln x$;

(8) $(x+1)^2y''-2(x+1)y'+2y=0$;

(9) $(2x-3)^2y''-6(2x-3)y'+12y=0$.

25. Suppose a, b and c are all positive constants and $y(x)$ is a solution of the equation $ay''+by'+cy=0$. Prove that $\lim\limits_{x\to+\infty}y(x)=0$.

26. Suppose $\varphi(x)$ is a continuous function such that

$$\varphi(x)=e^x+\int_0^x\varphi(t)dt-x\int_0^x\varphi(t)dt,$$

find $\varphi(x)$.